ADVANCES IN ELECTROCHEMISTRY AND ELECTROCHEMICAL ENGINEERING

VOLUME 9

ADVANCES IN ELECTROCHEMISTRY AND ELECTROCHEMICAL ENGINEERING

•

ADVANCES IN ELECTROCHEMISTRY AND ELECTROCHEMICAL ENGINEERING

•

VOLUME 9

Optical Techniques in Electrochemistry

Edited by ROLF H. MULLER

Inorganic Materials Research Division
Lawrence Berkeley Laboratory
and Department of Chemical Engineering;
University of California, Berkeley

A Wiley-Interscience Publication

JOHN WILEY & SONS, New York · London · Sydney · Toronto

Library of Congress Catalog Card Number: 61–15021

ISBN 0–471–20585–0

Printed in the United States of America

10 9 8 7 6 5 4 3 2 1

Introduction to the Series

This series was planned to make available authoritative reviews in the area of electrochemical phenomena, and to bridge the gap between electrochemistry as a part of physical chemistry and electrochemical engineering. Increased interest in electrochemistry and its applications renders this publication timely.

Chapters will vary considerably in their method of approach and length. Some chapters will be definitive while others, dealing with unsettled problems in rapidly evolving areas, will have to be somewhat tentative. Coverage over a number of volumes should be quite comprehensive. Emphasis will be placed in successive volumes either on the physicochemical or the engineering approach, although some volumes may cover both aspects. Specific problems in electroanalytical chemistry will not be included, except for certain theoretical questions not usually considered by analytical chemists.

Paul Delahay
Charles W. Tobias

Preface

The application of optical techniques has recently opened new avenues in electrochemical research. As a result of the present resurgence in optical physics some classical optical techniques have been substantially modified; others are entirely new.

Electrochemists are particularly interested in these techniques because optical observations are nondestructive, provide data that are independent of other electrochemical techniques, and can often be gathered simultaneously with the latter.

This book provides reviews on selected techniques of current interest. It grew out of part of a Gordon conference organized in 1968 by Paul Delahay. The first two chapters deal with reflection spectroscopy, which has been introduced rather recently and is now developing at a rapid pace. Ellipsometry is treated in the next two chapters. Although this technique has been in existence for a long time, its utility is just being realized because of new advances. Interferometry, a classical optical technique that has seen a large instrumental development, is discussed in Chapters 5 and 6. New theoretical refinements may provide broadened utility for this technique, particularly in its novel version of holographic interferometry. The concluding chapter on microscopy demonstrates that refinement of an established technique can substantially enlarge its usefulness.

I am grateful to the authors for their contributions and their patience.

Rolf H. Muller

Berkeley, California
November 1972

Contributors to Volume 9

WILFORD N. HANSEN
Department of Physics, Utah State University, Logan, Utah

J. KRUGER
Institute for Materials Research, National Bureau of Standards, Washington, D.C.

J. D. E. MCINTYRE
Bell Laboratories, Murray Hill, New Jersey

ROLF H. MULLER
Inorganic Materials Research Division, Lawrence Berkeley Laboratory and Department of Chemical Engineering; University of California, Berkeley, California

A. C. SIMON
Electrochemistry Branch, Naval Research Laboratory, Washington, D.C.

V. S. SRINIVASAN
Department of Chemistry, Bowling Green State University, Bowling Green, Ohio

Contents

ADVANCES IN ELECTROCHEMISTRY AND
ELECTROCHEMICAL ENGINEERING

VOLUME 9

Internal Reflection Spectroscopy in Electrochemistry

WILFORD N. HANSEN
Department of Physics, Utah State University

Contents

I. Introduction

This chapter is concerned with the optical and spectroscopic observation of the region within about one wavelength of the electrochemical interface, without any interference whatever from the rest of the electrolyte. This unique capability is made possible by internal reflection techniques which bring the radiation through the electrode to form an "evanescent" wave in the electrolyte at the interface. This intense wave does not propagate into the electrolyte, but its intensity decays rapidly as the point of consideration passes from the electrode into the electrolyte. All material within the depth of penetration of this wave will interact with it, the greatest interaction being where the electromagnetic electric fields are highest, i.e., in the region of the compact double layer itself.

This region close to the electrode is most important to electrochemistry. The idea is to observe electroactive species within this region as they come and go during electrochemical manipulation. By "observe" we mean to detect, identify, and/or quantitatively measure concentration of said species. The use of light is usually nondestructive, and the interaction time is shorter than that required for a chemical reaction. Since detection of light can also be very fast, the method is ideally suited for kinetic studies. It therefore seems that the use of internal reflection techniques can aid significantly in bringing powerful optical methods such as absorption spectroscopy and ellipsometry to the study of the electrochemical interface.

The most obvious and simple observation is that of an absorbing species in the electrolyte diffusing to or from the electrode. Optical theory presented later shows how quantitative spectra of such species can be obtained. Since a typical depth of 1000 Å is sampled by the evanescent wave, it is the surface concentration of a typical diffusion profile that is measured. The theory relating to this type of measurement has been confirmed beautifully by experiment, as will be discussed later.

But the most exciting application of the method is the study of surface reactions, i.e., reactions within the compact double layer itself, which is the very heart of electrochemistry. Internal reflection permits the optical and spectroscopic observation of this region, but care must be taken that the signal is not swamped by the other 99% of the electrolyte layer sampled. A number of ways of accomplishing this come to mind, but almost no experimental results have yet been reported in the literature. An exception to this is the study of changes in the metal electrode itself, i.e., electronic changes occurring within the compact double layer within 1 Å of the interface.

Much of the material that follows is relevant optical theory. Part of it is very old, part of it is new, and part of it occurs here for the first time in print. All of it has been selected and arranged to form the theoretical basis of the internal reflection method applied to electrochemistry. It is absolutely essential to the understanding and development of the method. As theory is considerably ahead of experiment at the present time the greater part of the chapter will be devoted to theory. After the optical theory is presented, a number of experimental results which appear in the literature will be discussed, not as a comprehensive survey of the work done to date, but as a systematic correlation of theory and observed reality. So far, almost no anomalies have been found and the correspondence between theoretical prediction, and experimental observation is nearly perfect. While extensive experimental effort will no doubt produce anomalies, it will surely confirm the prediction of the theory that optical methods can be used to investigate even the compact double layer in significant detail.

II. Basic Optical Equations

1. The Fresnel Coefficients

Starting with Maxwell's equations for a linear isotropic substance, equations for Fresnel's coefficients of reflection and transmission at a phase boundary can easily be derived by using the continuity requirements of the electric and magnetic field vectors (2,19). These equations are basic to the present subject. For nonmagnetic materials they are given for polarization perpendicular, $\perp$, and parallel, $\|$, to the plane of incidence:

$$r_{\perp jk} = \frac{\xi_j - \xi_k}{\xi_j + \xi_k}, \qquad t_{\perp jk} = \frac{2\xi_j}{\xi_j + \xi_k} \tag{1}$$

$$r_{\| jk} = \frac{\hat{n}_k{}^2\xi_j - \hat{n}_j{}^2\xi_k}{\hat{n}_k{}^2\xi_j + \hat{n}_j{}^2\xi_k}, \qquad t_{\| jk} = \frac{2\hat{n}_j\hat{n}_k\xi_j}{\hat{n}_k{}^2\xi_j + \hat{n}_j{}^2\xi_k} \tag{2}$$

The coefficient r is the ratio of the complex electric field amplitude of the reflected wave to the incident wave at the boundary between phases j and k, j being the incident phase. The coefficient t is the corresponding ratio for the transmitted wave at the boundary.

These equations apply at any boundary in a stack of parallel plane bounded phases (called a stratified medium).

The index of refraction for phase i is indicated by n, and is generally complex being given by $\hat{n}_i = n_i - ik_i$ where n is the ordinary real index of refraction and k is the extinction coefficient. (An alternate definition is $\hat{n}_i = n_i(1 - i\kappa_i)$ where κ is called the attenuation index.) We define for phase i, $\xi \equiv \hat{n} \cos \theta$ where θ is the angle between the normal to the interface and the direction of wave propagation. In general θ_i is complex and this simple interpretation is true only in a formal sense. With the equations cast in this form it is a simple matter to keep track of complex roots. Snell's law in general form is

$$\hat{n}_i \sin \theta_i = n_1 \sin \theta_1 \qquad (3)$$

where i is any phase in the stratified medium and 1 is the incident phase. Therefore, ξ_i for any phase is given by $\xi_i = (\hat{n}_i^2 - n_1^2 \sin^2 \theta_1)^{1/2}$. The incident phase is assumed to be transparent. As ξ_i is in general complex there is no way of knowing in which quadrant of the complex plane it will lie (i.e., which root to take) except by reference to the physics involved. It can be shown that for $\hat{n} = n - ik$, it is always true that

$$\operatorname{Re} \xi_i \geq 0, \qquad \operatorname{Im} \xi_i \leq 0 \qquad (4)$$

For a discussion of this point see Appendix B of Ref. 10. By using this simple unambiguous rule, calculations are greatly facilitated. Using this scheme, phase jumps on reflection, i.e., arg r, will be positive and be between 0 and 2π, as displayed in most optics books.

The Fresnel formulas for a three-phase system are given by 10

$$r_\perp = \frac{r_{\perp 12} + r_{\perp 23} e^{-2i\beta}}{1 + r_{\perp 12} r_{\perp 23} e^{-2i\beta}}, \qquad t_\perp = \frac{t_{\perp 12} t_{\perp 23} e^{-i\beta}}{1 + r_{\perp 12} r_{\perp 23} e^{-2i\beta}} \qquad (5)$$

$$r_\parallel = \frac{r_{\parallel 12} + r_{\parallel 23} e^{-2i\beta}}{1 + r_{\parallel 12} r_{\parallel 23} e^{-2i\beta}}, \qquad t_\parallel = \frac{t_{\parallel 12} t_{\parallel 23} e^{-i\beta}}{1 + r_{\parallel 12} r_{\parallel 23} e^{-2i\beta}} \qquad (6)$$

where $\beta \equiv 2\pi(h/\lambda)\xi_2$, h/λ being the thickness of the second phase in wavelengths. Fresnel equations for any number of phases are given in Ref. 10, page 387, and for brevity will not be repeated here. (Note, however, that in Ref. 10, $\hat{n}_i$ and ξ_i are conjugates of those defined here.) They are rather complicated, but can be programmed easily for a computer.

2. *Reflectance and Transmittance*

From the Fresnel coefficients, it is a simple matter to obtain reflectance, R, transmittance, T, and corresponding phase changes. Transmittance is the fraction of energy incident in the first phase that crosses the final interface into the semiinfinite final phase, f. Its formula is complicated somewhat by the change in beam direction. Reflectance is measured in the first (incident) phase. The formulas for R and T for any number of phases are given by

$$R = |r|^2, \qquad T_\perp = \frac{\operatorname{Re} \xi_f}{\xi_1} |t_\perp|^2, \qquad T_\parallel = \frac{\operatorname{Re} (\xi_f/\hat{n}_f^2)}{\xi_1} |\hat{n}_f t_\parallel|^2 \quad (7)$$

the proper r and t being inserted for each polarization. Also, for reflection and transmission involving any number of phases, the angular changes in phase are given by

$$\delta^r = \arg r, \qquad \delta^t = \arg t \qquad (8)$$

where $\arg z$ is the angle that is the argument of the complex number z. It is found that δ ranges all the way from zero to 2π rad and is best calculated using the arctan function and the real and imaginary parts of r and t. Note that $\arg z = \tan^{-1} (\operatorname{Im} z/\operatorname{Re} z)$ for $\operatorname{Re} z > 0$ and $\arg z = \tan^{-1} (\operatorname{Im} z/\operatorname{Re} z) + \pi$ for $\operatorname{Re} z < 0$. For $\operatorname{Re} z = 0$, $\arg z = \pi/2$ where $\operatorname{Im} z > 0$ and $\arg z = -\pi/2$ where $\operatorname{Im} z < 0$. For $\operatorname{Re} z = 0 = \operatorname{Im} z$, $\arg z$ is undefined. The explicit formulas for phase changes in terms of real quantities are very complicated, offering little physical insight, and it is just as well to leave them in their simpler complex form. For computer calculations (generally the only practical kind), the complex form is easily handled. For R and T, however, some physical insight can be gained from the real arithmetic in certain cases. They are given here for the general three-phase case for both polarizations.

$$R = \frac{R_{12} + R_{23} e^{4 \operatorname{Im} \beta} + R_{12}^{1/2} R_{23}^{1/2} e^{2 \operatorname{Im} \beta} 2 \cos (\delta_{23}^r - \delta_{12}^r - 2 \operatorname{Re} \beta)}{1 + R_{12} R_{23} e^{4 \operatorname{Im} \beta} + R_{12}^{1/2} R_{23}^{1/2} e^{2 \operatorname{Im} \beta} 2 \cos (\delta_{12}^r + \delta_{23}^r - 2 \operatorname{Re} \beta)} \qquad (9)$$

$$T = Q \frac{|t_{12}|^2 |t_{23}|^2 e^{2 \operatorname{Im} \beta}}{1 + R_{12} R_{23} e^{4 \operatorname{Im} \beta} + R_{12}^{1/2} R_{23}^{1/2} e^{2 \operatorname{Im} \beta} 2 \cos (\delta_{23} + \delta_{12} - 2 \operatorname{Re} \beta)} \qquad (10)$$

where Q is $(\operatorname{Re} \xi_3 | \xi_1)$ for perpendicular polarization and $[|\hat{n}_3/n_1|^2 \operatorname{Re} (\xi_3/\hat{n}_3^2)/(\xi_1/n_1^2)]$ for parallel polarization.

The optical equations above are not exactly like those in Ref. 10 because of the definition of $\hat{n} \equiv n - ik$ in the present paper instead of $\hat{n} \equiv n(1 + i\kappa)$. This choice has been made to meet the latest recommendations in nomenclature. Using a minus instead of a plus sign changes the ξ's, r's, t's, etc., to their complex conjugates, makes Im β always negative, and changes the signs of all δ's.

3. *Ellipsometric Equations for Internal Reflection*

The basic ellipsometric measurement is to analyze the ellipticity of the internally reflected beam after it has left the sample region. Comparisons with the incident beam, of intensity or phase, are not made. The two quantities $\Delta = \delta_{\parallel}{}^{r} - \delta_{\perp}{}^{r}$ and $\psi = \tan^{-1}|r_{\parallel}/r_{\perp}|$ are determined.

In other words, the quantity measured is the ratio of the Fresnel reflection coefficients

$$\rho = \frac{r_{\parallel}}{r_{\perp}} = \tan \psi e^{i\Delta} \tag{11}$$

Thus changes in the optical properties at the electrochemical interface will cause changes in ψ and Δ. In the present case the purpose is to see small changes in the interfacial region during electrochemical manipulation. It is actually the change in Δ and the change in ψ that are observed. The changing species need not be optically absorbing. Ellipsometry is especially sensitive to small changes at the interface—being able to detect the equivalent of a 0.1 Å thick film or less.

It should be noted that if measurements are made at angles of incidence greater than the critical angle [$\theta_c \equiv \sin^{-1}(n_f/n_1)$ where f indicates the final phase] and there are no absorbing phases, $|\rho|$ will always be unity giving $\psi = 45°$. In this case the only useful data are Δ and $\Delta\Delta$ where $\Delta\Delta$ is the change in Δ due to changes at the interface. This is not the case, however, when making internal reflection measurements in electrochemistry because in this case there is always an optically absorbing electrode. The amount of energy the electrode absorbs may be a sensitive function, for example, of the refractive index of the double layer region. In fact, it has been shown (14) that internal reflection ellipsometry can be an order of magnitude more sensitive than external reflection ellipsometry to changes in the double layer region.

III. Optical Absorption and the Mean Square Field Strength

1. Poynting's Vector and Energy Loss

It is extremely useful to interpret the Poynting vector defined by

$$\mathbf{S} = \mathbf{E} \times \mathbf{H} \ (\mathrm{W/m^2}) \tag{12}$$

as the intensity of electromagnetic energy flow at a point, i.e., the energy passing per second per unit area in the direction of $\mathbf{E} \times \mathbf{H}$. Here $\mathbf{E}$ and $\mathbf{H}$ are the *real* electric and magnetic fields. For present purposes we are interested in time averages since radiation frequencies are very high. It can easily be shown (19) that the time average Poynting vector is given by

$$\mathbf{S}_{\text{ave}} = \tfrac{1}{2}\mathrm{Re}\,(\hat{\mathbf{E}} \times \hat{\mathbf{H}}^*) \tag{13}$$

where the field vectors now are assumed to be in their usual complex form, and the asterisk signifies complex conjugate. This real vector, which is most conveniently represented in this form for a harmonic electromagnetic field, represents the time average flux of electromagnetic energy at any given point.

The rate of absorption of radiation (i.e., conversion of electromagnetic energy into heat) at any point in a medium is given by $-\nabla\cdot\mathbf{S}_{\text{ave}}$, where we are again taking the time average. We also know that $-\nabla\cdot\mathbf{S}_{\text{ave}} = \tfrac{1}{2}\sigma\hat{\mathbf{E}}\cdot\hat{\mathbf{E}}^*$, where σ is the effective electrical conductivity at optical frequencies, and that the time average of the square of the electric field, $\langle E^2\rangle$, is given by $\tfrac{1}{2}\hat{\mathbf{E}}\cdot\hat{\mathbf{E}}^*$. We can therefore write

$$\text{rate of energy dissipation} = -\nabla\cdot\mathbf{S}_{\text{ave}} = \sigma\langle E^2\rangle \ \mathrm{W/m^3} \tag{14}$$

Light intensity is also proportional to $\langle E^2\rangle$, so, as expected, the rate of absorption is proportional to light intensity. The main point to be made here is that if we know the magnitude of $\langle E^2\rangle$ at various points in a medium, we know where the energy will be absorbed. We can thus gain physical insight into the mechanism and in turn design systems to optimize sensitivity, etc.

2. Exact Equations for the Mean Square Fields

To study this subject in general it is necessary to derive equations for $\langle E^2\rangle$ at any point in a stratified medium. This has been done in Ref. 10, is rather complicated, and will not be repeated here. Some less general cases will be discussed, and some surprising results given, which illustrate the insight afforded by this powerful method.

A. Two-Phase System. First consider the simple plane bounded two-phase system. Such a system is represented in Fig. 1. The vectors $\mathbf{S}_i$, $\mathbf{S}_r$, and $\mathbf{S}_t$ represent the average Poynting vectors of the incident, reflected, and transmitted radiation. The critical angle ($\theta_c = \sin^{-1}(n_2/n_1)$), in a strict sense, is defined only when both phases are transparent. However, so long as $k \ll 1$, the idea of a critical angle remains a useful concept even though we no longer have total reflection when $\theta_1 \geq \theta_c$.

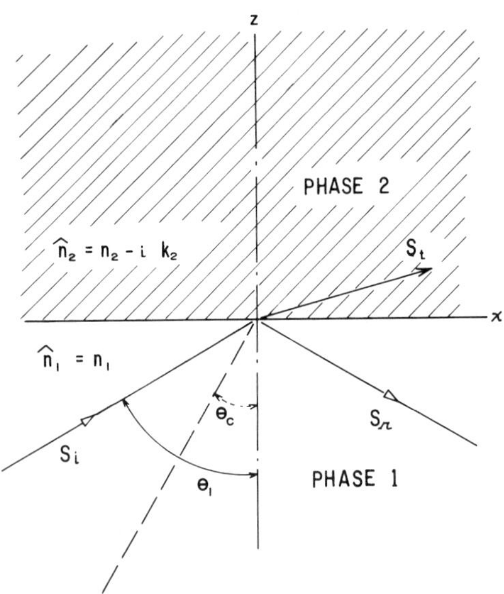

Fig. 1. Two-phase system.

At angles of incidence larger than θ_c there is no refracted beam if the second phase is transparent. This does not mean that $\mathbf{S}_t$ is zero, for it is still large close to the boundary. It is parallel to the boundary for $k_2 = 0$ and acquires a z component if $k_2 > 0$. The angle $\mathbf{S}_t$ makes with the z axis is *not* given by Snell's law. Recall that in the case of total reflection there is an intense "evanescent" wave whose intensity falls rapidly with z, typically to $1/e$ in one-tenth of a wavelength. This physical picture is essentially correct so long as $k \ll 1$.

The formulas for $\langle E^2 \rangle$ in a two-phase system are now given. They are like those of Ref. 10 with appropriate changes to fit the present nomenclature. The symbols $\perp$ and $\parallel$ refer to polarizations with **E** perpendicular to or parallel to the plane of incidence. Subscript 1 or 2 refers to phase 1 or phase 2, and subscripts x and z refer to the coordinates in Fig. 1. The mean square field given is the ratio of the one indicated to that of the incident beam, making the equations dimensionless and simpler.

$$\langle E_{\perp 1}{}^2 \rangle = [(1 + R_\perp) + 2R_\perp{}^{1/2} \cos (\delta_\perp{}^r + 4\pi(z/\lambda)\xi_1)] \quad (15)$$

$$\langle E_{\perp 2}{}^2 \rangle = |t_\perp|^2 \exp (4\pi(z/\lambda) \operatorname{Im} \xi_2) \quad (16)$$

$$\langle E^2_{\parallel 1x} \rangle = \cos^2 \theta[(1 + R_\parallel) - 2R_\parallel{}^{1/2} \cos (\delta_\parallel{}^r + 4\pi(z/\lambda)\xi_1)] \quad (17)$$

$$\langle E^2_{\parallel 1z} \rangle = \sin^2 \theta[(1 + R_\parallel) + 2R_\parallel{}^{1/2} \cos (\delta_\parallel{}^r + 4\pi(z/\lambda)\xi_1)] \quad (18)$$

where

$$r_\perp = R_\perp{}^{1/2} \exp (i\delta_\perp{}^r), \qquad r_\parallel = R_\parallel{}^{1/2} \exp (i\delta_\parallel{}^r) \quad (19)$$

$$\langle E^2_{\parallel 2x} \rangle = \left| \frac{\xi_2}{\hat{n}_2} t_\parallel \right|^2 \exp \left(4\pi \frac{z}{\lambda} \operatorname{Im} \xi_2\right) \quad (20)$$

$$\langle E^2_{\parallel 2z} \rangle = \left| \frac{n_1 \sin \theta_1}{\hat{n}_2} t_\parallel \right|^2 \exp \left(4\pi \frac{z}{\lambda} \operatorname{Im} \xi_2\right) \quad (21)$$

The z direction is normal to the interface and away from the incident phase. The x direction lies in the plane of incidence and the plane of the interface and is oriented so that the x component of Poynting's vector is positive. The y direction is perpendicular to the x, z plane and oriented so that x, y, z form a right-handed triad. Perpendicular polarization has only a y component whereas parallel polarization has both x and z components. The total $\langle E^2 \rangle$ for parallel polarization is thus the sum of x and z components, giving

$$\langle E_{\parallel 2}{}^2 \rangle = (|\sin \theta_2|^2 + |\cos \theta_2|^2)|t_\parallel|^2 \exp (4\pi(z/\lambda) \operatorname{Im} \xi_2) \quad (22)$$

All of these equations are necessary to the understanding of optical absorption at an interface. For the particular case of two transparent phases, the fields at the interface are given by

$$\left.\begin{aligned} \langle E_{\perp 1}{}^2 \rangle = \langle E_{\perp 2}{}^2 \rangle &= \frac{4\xi_1{}^2}{(\xi_1 + \xi_2)^2} \qquad \text{for} \quad \theta_1 \le \theta_c \\ &= \frac{4\xi_1{}^2}{n_1{}^2 - n_2{}^2} \qquad \text{for} \quad \theta_1 \ge \theta_c \end{aligned}\right\} \quad (23)$$

where θ_c is the critical angle. Of course, there is no critical angle if $n_2 \geq n_1$. In this case only the first equation applies. The corresponding equations for parallel polarization are

$$\left.\begin{aligned} \langle E_{\|2}{}^2 \rangle &= 4\left(\frac{\xi_1}{n_2 \cos\theta_1 + n_1 \cos\theta_2}\right)^2 \quad \text{for} \quad \theta_1 \leq \theta_c \\ \langle E_{\|2}{}^2 \rangle &= 4(\sin^2\theta_2 - \cos^2\theta_2)\left(\frac{\xi_1{}^2}{(n_2 \cos\theta_1)^2 - (n_1 \cos\theta_2)^2}\right) \\ &\qquad \text{for} \quad \theta_1 \geq \theta_c \end{aligned}\right\} \tag{24}$$

The time average of the x component is $\cos\theta_2 \cos^*\theta_2$ times the far right factors in Eq. 24. For the z component it is $\sin\theta_2 \sin^*\theta_2$ times the same factors. Notice that for $\theta_1 \leq \theta_c$, θ_2 is real. For $\theta_2 > \theta_c$, θ_2 is complex, $\sin\theta_2$ is real, and $\cos\theta_2$ is purely imaginary. It is useful to note several special cases. From Eqs. 23 and 24 we see that at the critical angle,

$$\left.\begin{aligned} \langle E_\perp{}^2 \rangle &= 4 \\ \langle E_{\|x}{}^2 \rangle &= 0 \\ \langle E_{\|z}{}^2 \rangle &= 4(n_1/n_2)^2 \end{aligned}\right\} \quad \theta_1 = \theta_c \tag{25}$$

At grazing incidence all fields go to zero. At normal incidence,

$$\left.\begin{aligned} \langle E_\perp{}^2 \rangle &= \langle E_\|{}^2 \rangle = 4\left(\frac{n_1}{n_1 + n_2}\right)^2 \\ &= \langle E_x{}^2 \rangle = \langle E_y{}^2 \rangle \\ \langle E_{\|z}{}^2 \rangle &= 0 \end{aligned}\right\} \quad \theta_1 = 0 \tag{26}$$

Some surprising characteristics are immediately evident from these equations. Fields are usually larger at the interface than in the original incident beam, and they are sometimes orders of magnitude larger. This means that sensitivity to interesting molecular species is also enhanced over what it is in ordinary transmission methods. The orientation of the electric field vector with respect to the interface is determined by choice of polarization and angle of incidence. For perpendicular polarization both x and z components exist, but for $k_2 \ll 1$ and near $\theta_1 = \theta_c$ the x component is almost zero, so the resultant field is normal to the interface. For $\theta_1 > \theta_c$, $\langle E^2 \rangle$ decays rapidly with z, typically as $\exp(-10z/\lambda)$. Interaction with the electromagnetic wave is only through these electric fields; hence this interaction becomes negligible within a wavelength or so of the interface. Specific examples of fields at the interface and further details are presented in a later section.

B. Three-Phase System. While the incident phase itself, i.e., the internal reflection element, may be an electrical conductor such as Si or Ge, and may be used directly as the electrode, it is more usual to use a thin electrode on a transparent substrate. The reason is that good conductors are also good absorbers of light, making the use of thin layer electrodes necessary for internal reflection. Therefore, we must usually deal with at least three phases. Such a system is represented in Fig. 2.

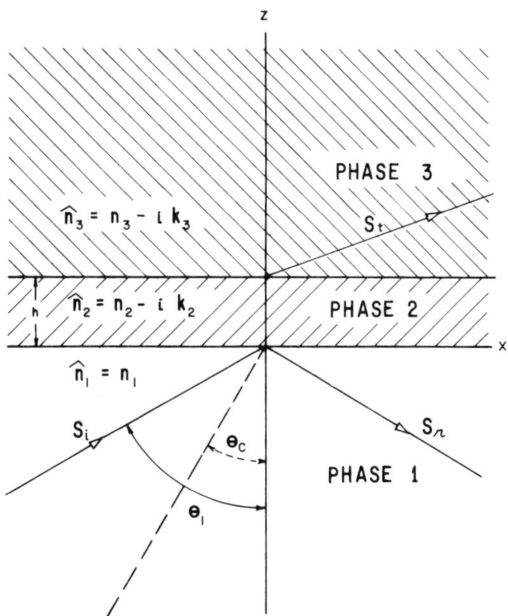

Fig. 2. Three-phase system.

The electromagnetic waves in the incident and final phases are related in about the same way as in the two-phase case. The magnitude of $\mathbf{S}_t$ depends on phase 2, but its direction does not. The critical angle is well defined regardless of the optical constants of phase 2 (It may even be a metal!), so long as $k_3 \ll 1$. At $\theta_1 > \theta_c$ an evanescent wave again exists near the boundary in the third phase. Its general nature, except for amplitude and relative phase, is independent of the presence of phase 2.

It is especially important to know the mean square electric fields in the third phase at the boundary, this being the double layer region. For perpendicular polarization

$$\langle E_{\perp 3}{}^2 \rangle = |t_\perp|^2 \exp\left[(4\pi/\lambda) \text{ Im } \xi_3(z - h)\right] \tag{27}$$

where z and h are as indicated in Fig. 2. For parallel polarization

$$\langle E_{\parallel 3}{}^2 \rangle = (|\sin\theta_3|^2 + |\cos\theta_3|^2)|t_\parallel|^2 \exp\left[(4\pi/\lambda) \text{ Im } \xi_3(z - h)\right] \tag{28}$$

and the x and z components are given by

$$\langle E^2_{\parallel 3x} \rangle = \left|\frac{\xi_3}{\hat{n}_3} t_\parallel\right|^2 \exp\left[(4\pi/\lambda) \text{ Im } \xi_3 \, (z - h)\right] \tag{29}$$

$$\langle E^2_{\parallel 3z} \rangle = \left|\frac{n_1 \sin\theta_1}{\hat{n}_3} t_\parallel\right|^2 \exp\left[(4\pi/\lambda) \text{ Im } \xi_3(z - h)\right] \tag{30}$$

The t's are given by Eqs. 5 and 6.

Much of the nature of the electromagnetic wave in the third phase becomes apparent upon examining Eqs. 27–30. It will be noticed that these equations are identical to Eqs. 16 and 20–22 for the two-phase case, if things are taken as a function of distance into the final phase and it is remembered that t is taken at the final boundary. In fact, these equations apply to the final phase no matter how many intermediate phases there are, i.e., they apply to a general stratified medium. This means that the angle of refraction, for example, in the final phase is independent of the presence of intermediate phases. The value of t, however, is very much dependent on the intermediate phases. In fact, it is possible for $|t^2|$ (which given $\langle E^2 \rangle$ at the boundary) to be orders of magnitude larger than in the incident wave. The multiphase can significantly enhance the spectral sensitivity of the system. Examples are shown later.

Note that for a transparent final phase, at less than critical angle, $\text{Im } \xi_f = 0$ giving $\langle E^2 \rangle = |t^2|$ throughout the final phase, and there is no attenuation of refracted beam intensity, just as expected. At $\theta_1 > \theta_c$, however, $\text{Im } \xi_f$ is large and attenuation is very rapid in any case. This will be discussed in detail in the section on penetration depth. Note, also, that the ratio of $\langle E^2_{\parallel 3x} \rangle$ to $\langle E^2_{\parallel 3z} \rangle$ is independent of the presence of the second phase. In particular, at the critical angle, $\langle E^2_{\parallel 3x} \rangle = 0$ and the field is entirely normal to the interface regardless of how absorbing phase 2 may be. This is seen immediately from Eq. 29 because $\xi_3 = 0$ at $\theta_1 = \theta_c$.

In a three-phase system, if the second phase had about the same optical properties as phase 1 or were thin enough, the three-phase case would reduce approximately to the two-phase case. Neither situation obtains, however, for real electrode systems, and the electrode has a profound effect upon the fields at the interface. That such a complicated optical system would ever yield simple manageable spectra could hardly be expected. It will be shown, however, that a law analogous to Beer's law can be derived for absorption in the electrolyte phase. It will also be shown that simple, transmission-like spectra can be obtained using thin metal electrodes.

C. $\langle E^2 \rangle$ Changes Across an Interface. At a boundary between two phases, when at least one phase is conducting, the normal component of the electric displacement, **D**, may be no longer continuous across the boundary. There is no simple general relationship between the electric fields in the two phases. For strictly harmonic fields, however, there *is* a simple relationship between $\langle E^2 \rangle$ in the two phases at the boundary. This relationship can be derived as follows.

At the boundary between two phases, one or both of which may be conducting, we know from the integral form of Maxwell's first equation that

$$\hat{z} \cdot (\mathbf{D}_2 - \mathbf{D}_1) = \rho \qquad (\text{real } D)$$

where **D** is the electric displacement in real form, $\hat{z}$ is a unit vector normal to the boundary, and ρ is the surface charge density. For linear isotropic media of real dielectric constant ε, this can be written in terms of real E as,

$$\varepsilon_2 E_{2z} - \varepsilon_1 E_{1z} = \rho$$

The flow of charge at the boundary must satisfy the equation of continuity giving

$$\hat{z} \cdot (\mathbf{J}_2 - \mathbf{J}_1) = -\frac{\partial \rho}{\partial t}$$

or from Ohm's law, $\mathbf{J} = \sigma \mathbf{E}$, we have

$$\sigma_2 E_{2z} - \sigma_1 E_{1z} = -\frac{\partial \rho}{\partial t}$$

All of the above equations are real. If the fields are harmonic, as is true for electromagnetic radiation, $\partial\rho/\partial t$ will also be harmonic and so will ρ. We can therefore rewrite the equations in complex exponential notation, keeping in mind that the real fields are the real part of the complex fields, etc. In particular, differentiation of the $e^{i\omega t}$ factor is the same as multiplication by $i\omega$. Thus we can write

$$\varepsilon_2\hat{E}_{2z} - \varepsilon_1\hat{E}_{1z} = \hat{\rho}$$

and

$$\sigma_2\hat{E}_{2z} - \sigma_1\hat{E}_{1z} = -i\omega\hat{\rho}$$

or

$$i\frac{\sigma_2}{\omega}\hat{E}_{2z} - i\frac{\sigma_1}{\omega}\hat{E}_{1z} = \hat{\rho}$$

where the $\hat{E}$'s and $\hat{\rho}$ are complex. This gives

$$\left(\varepsilon_2 - i\frac{\sigma_2}{\omega}\right)\hat{E}_{2z} = \left(\varepsilon_1 - i\frac{\sigma_1}{\omega}\right)\hat{E}_{1z}$$

Using the concept of a complex dielectric constant (2), $\hat{\varepsilon} = \varepsilon - i(\sigma/\omega)$, this becomes

$$\hat{\varepsilon}_2\hat{E}_{2z} = \hat{\varepsilon}_1\hat{E}_{1z}$$

Note that $D_z = \varepsilon E_z$ is *not* continuous across the boundary in general if at least one $\sigma \neq 0$. Recall that

$$\langle E^2\rangle = \tfrac{1}{2}(\hat{\mathbf{E}}\cdot\hat{\mathbf{E}}^*) \tag{31}$$

For the normal component at the interface we therefore have

$$\hat{\varepsilon}_1\hat{\varepsilon}_1^*\langle E_1^{\,2}\rangle_z = \hat{\varepsilon}_2\hat{\varepsilon}_2^*\langle E_2^{\,2}\rangle_z \tag{32}$$

both sides of which are real. In terms of the complex index of refraction, since $\hat{\varepsilon} = \hat{n}^2$ and $\hat{n}\hat{n}^* = n^2 + k^2$,

$$(n_1^{\,2} + k_1^{\,2})^2\langle E_1^{\,2}\rangle_z = (n_2^{\,2} + k_2^{\,2})^2\langle E_2^{\,2}\rangle_z \tag{33}$$

Therefore $\langle E^2\rangle_z$ jumps by a constant factor as the boundary is crossed. This simple relationship is extremely useful since optical effects of interest, such as absorption, are proportional to $\langle E^2\rangle$, making $\langle E^2\rangle$ the quantity of interest, rather than E, H, or $\langle H^2\rangle$.

Equation 33 can lead to some interesting results. To illustrate, choose the air-gold boundary in the infrared. For air $n = 1$, $k = 0$. Representative of gold at about 5 μ in the infrared are $n = 10$, $k = 30$. From Eq. 33

$$\langle E_g^{\,2}\rangle_z = \frac{n_{\text{air}}^{\,4}}{(n_g^{\,2} + k_g^{\,2})^2}\langle E_{\text{air}}^{\,2}\rangle_z = 10^{-6}\langle E_{\text{air}}^{\,2}\rangle_z \tag{34}$$

The mean square field decreases by a factor of one million as we pass from air to gold. This makes possible some interesting and useful phenomena such as the passing of electromagnetic fields through an apparently opaque gold film.

Both the x and the y components of **E**, being tangential, are continuous across the boundary, and therefore $\langle E^2 \rangle_x$ and $\langle E^2 \rangle_y$ are also continuous for absorbing phases as well as transparent. Therefore

$$\left.\begin{aligned} \langle E_1{}^2 \rangle_x &= \langle E_2{}^2 \rangle_x \\ \langle E_1{}^2 \rangle_y &= \langle E_2{}^2 \rangle_y \end{aligned}\right\} \tag{35}$$

Another interesting case is that of a high index transparent material like germanium in the infrared. Consider a thin film of germanium of $n = 4$, $k = 0$ surrounded by a field $\langle E^2 \rangle$ in air.

$$\langle E_{\mathrm{Ge}}{}^2 \rangle_z = 1/256 \langle E_{\mathrm{air}}{}^2 \rangle_z \tag{36}$$

whereas

$$\left.\begin{aligned} \langle E_{\mathrm{Ge}}{}^2 \rangle_x &= \langle E_{\mathrm{air}}{}^2 \rangle_x \\ \langle E_{\mathrm{Ge}}{}^2 \rangle_\perp &= \langle E_{\mathrm{air}}{}^2 \rangle_\perp \end{aligned}\right\} \tag{37}$$

In order to get a high field inside such a material it is wise to arrange for large tangential fields at the interface.

3. Examples of Fields

A. Two-Phase Systems. An examination of $\langle E^2 \rangle$ under various circumstances leads to some surprising and useful results. The simple case of internal reflection from a glass-air interface is illustrated in Fig. 3. For both perpendicular polarization, $\langle E_y{}^2 \rangle$, and parallel polarization, $\langle E_x{}^2 \rangle + \langle E_z{}^2 \rangle$, the values of $\langle E^2 \rangle$ at the surface of the glass in the air phase are always larger than in the incident beam except at grazing incidence beyond $\theta = 70°$. In such a system the largest value always occurs at the critical angle. For perpendicular polarization the maximum value is always 4, but for parallel polarization it is about 9 in this case, in accordance with Eq. 25. At $\theta = \theta_c$, parallel polarization gives only fields perpendicular to the interface, and perpendicular polarization gives fields only parallel to the interface. This can be useful in studying the orientation of absorbing molecules adsorbed at the interface. Figure 4 gives a similar plot for a silicon-air interface. The outstanding feature is the large value of $\langle E_z{}^2 \rangle$ at θ_c. This makes for intense interaction with species at the interface. Any absorbing species dispersed in the

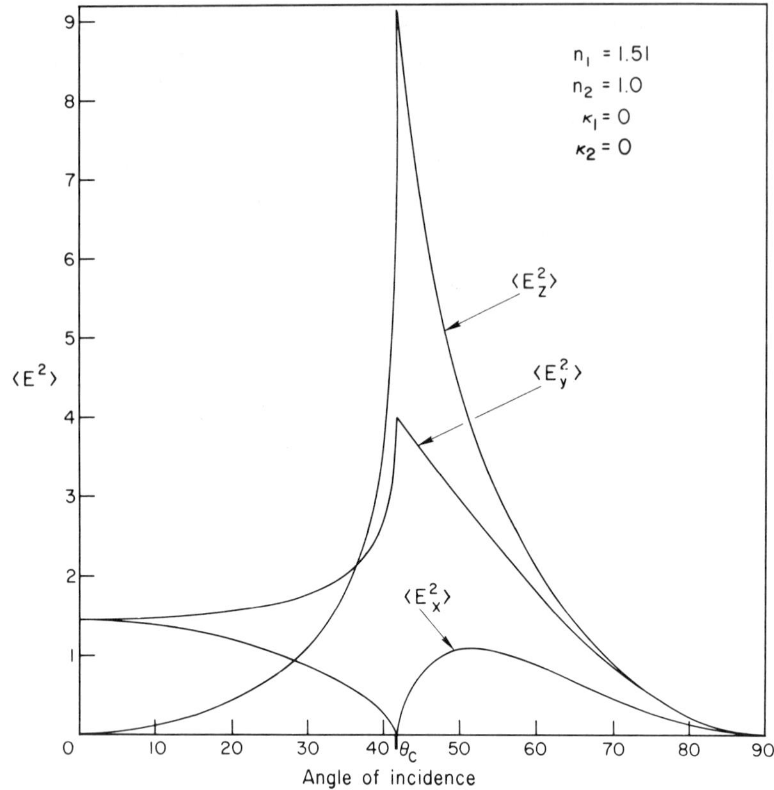

Fig. 3. Mean square fields at the surface of glass during internal reflection. Magnitude is relative to a value of unity for the incident beam. $\langle E_y^2 \rangle$ is for $\perp$ polarization, $\langle E_x^2 \rangle + \langle E_z^2 \rangle$ is total for $\|$ polarization. $n_1 = 1.51$, $n_2 = 1.0$, $\kappa_1 = \kappa_2 = 0$.

final phase or concentrated at the interface will experience the field there and will extract energy at a rate proportional to $\langle E^2 \rangle$.

For comparison, fields produced by external reflection from gold in the infrared are included in Fig. 5. The fields are taken at the interface in air which is the incident phase in this case. They are the result of interference of incident and reflected waves. The theoretical maximum of $\langle E_z^2 \rangle$ is four times that in the incident beam. The tangential components are small because the interference at the boundary is such that the fields of incident and reflected waves nearly cancel at the surface.

In the case of internal reflection, for $\theta > \theta_c$ where there is no

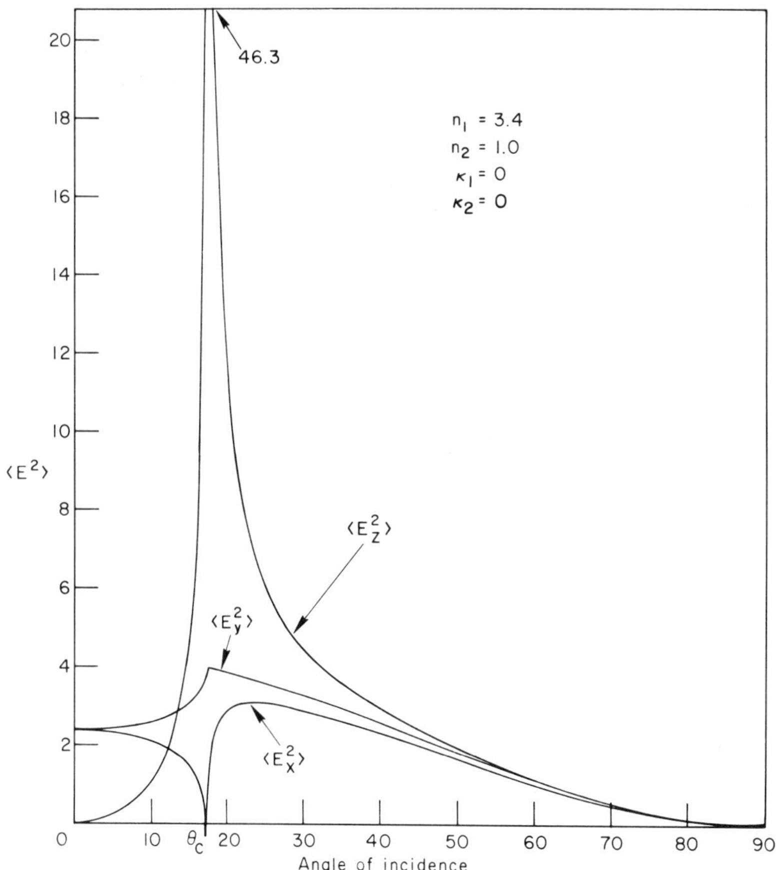

Fig. 4. Mean square fields at the surface of silicon during internal reflection; $n_1 = 3.4$, $n_2 = 1.0$, $\kappa_1 = \kappa_2 = 0$.

refracted beam, absorption in the final phase will have little effect on $\langle E_f^2 \rangle$ so long as $k_f \ll 1$. For example, Fig. 3 still applies even if the final phase is an absorbing dye solution.

B. Three-Phase Systems. As a first example consider all three phases transparent with $n_2 > n_1 > n_3$. In Fig. 6 is given $\langle E_y^2 \rangle$ at the boundary in the third phase as a function of optical thickness of phase 2 and the absorption of phase 3. This particular calculation is for $n_1 = 1.527$, $n_2 = 1.880$, $n_3 = 1.341$, $\theta_1 = 72.59°$, $\kappa_2 = 0$, $\kappa_3 = 0$–0.019, and $hn_2/\lambda = 0$–3.0. The upper curve is for

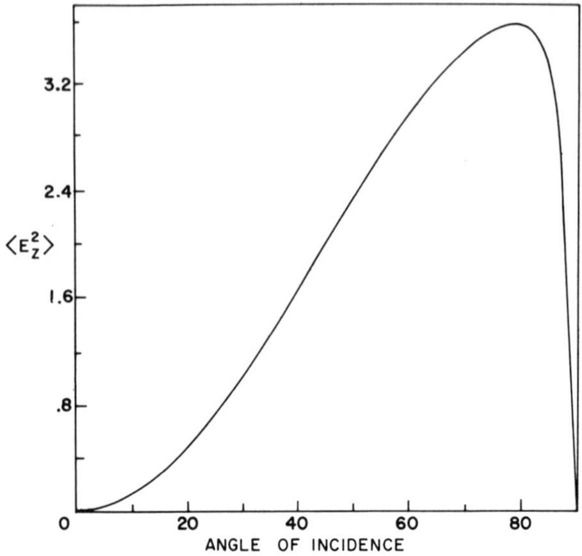

Fig. 5. Mean square fields normal to interface at the surface of gold during external reflection; $n_1 = 1$, $n_2 = 3$, $\kappa_1 = 0$, $\kappa_2 = 10$.

$\kappa_3 = 0$. (We use κ instead of $k \equiv n\kappa$ because our computer programs are set up in terms of κ.) Large periodic variation with film thickness is observed. A corresponding large variation in spectral sensitivity would be expected. The values at far left correspond to ordinary two-phase attenuated total reflection. Note that a change of 0.001 in κ_3, a large change, causes little change in $\langle E_y{}^2 \rangle$.

Figure 7 shows the reflectance of the same system. The periodic nature of the sensitivity is evident. Since the abscissa also represents the effect of changing λ with fixed thickness, each curve is the reflection spectrum that would be obtained if the optical constants at the third phase were constant with wavelength. By contrast a transmission spectrum would be a straight line. It is evident that to make sense of a spectrum with bandwidth greater than a fraction of one period would require a knowledge of the optical parameters of phase 2 and some calculations.

A real electrode system of interest is tin oxide on glass. This case corresponds to the above case except that the tin oxide layer is itself absorbing with $\kappa_2 \cong 0.005$ in the visible region. Figure 8 shows the corresponding $\langle E_y{}^2 \rangle$ for this case, again as a function of

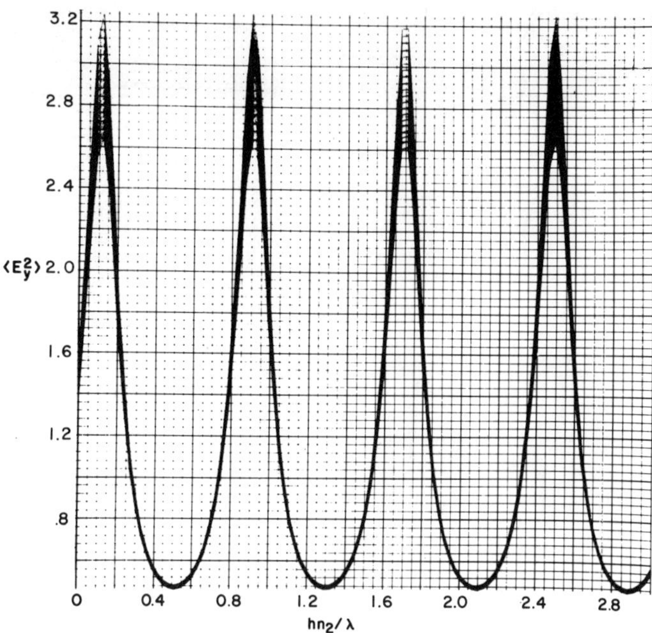

Fig. 6. Mean square fields at the surface of nonabsorbing tin oxide on glass with aqueous third phase, perpendicular polarization, $n_1 = 1.527$, $n_2 = 1.88$, $n_3 = 1.231$, κ_3 varies from 0 for the uppermost curve to 19×10^{-3} for the lowest, in 20 equal steps of 1×10^{-3}; $\theta_1 = 72.59°$.

electrode thickness or inverse wavelength. In addition to the simple periodicity, an attenuation factor has been added. This is seen from Eqs. 27 and 6 since β is no longer real. The corresponding reflectance of this system is shown in Fig. 9. The uppermost curve is not 1.0 as it was in Fig. 7. Except at small hn_2/λ there is more energy absorbed by the electrode than by phase 3. The convenient way to record the spectrum, however, is to experimentally record the difference between the $\kappa_3 = 0$ case and the case of interest. For given electrode thickness and wavelength, which is the practical situation when monitoring the concentration of a species of interest, it can be seen from the figure that the change in reflectance is nearly linear with κ_3. Hence, it is only necessary to make sure that the sensitivity is high enough. It will be shown later that the negative logarithm of R is more nearly linear with κ_3 than is R itself.

For parallel polarization the curves are similar to those for perpendicular polarization discussed above. They are, however, less

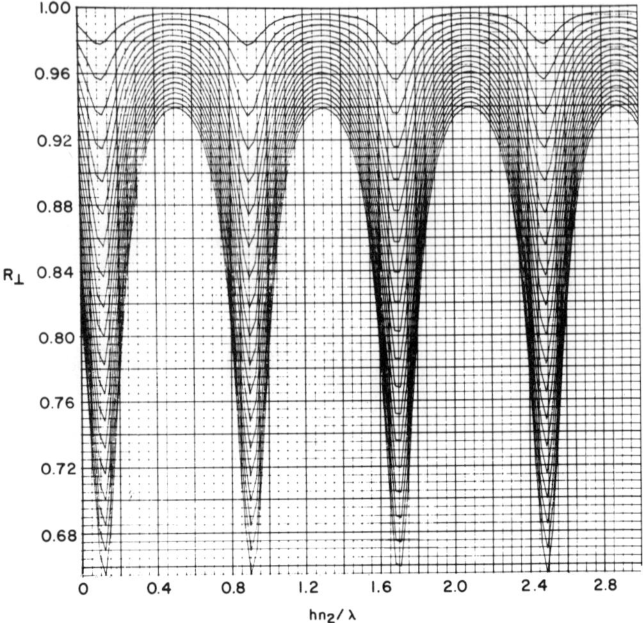

Fig. 7. Internal reflection spectra of a phase 3 with κ_3 constant with wavelength. Here κ_3 varies as in Fig. 6. κ_2 is 0 for this exact calculation, but the spectra are independent of κ_2 for $(\kappa_2\, hn_2/\lambda) \ll 1$. Other parameters are as in Fig. 6.

steep. This can be seen in Fig. 10 where $R_{\parallel}$ is given for the absorbing tin oxide electrode case. The sensitivity to change in κ_3 is somewhat larger than for $R_{\perp}$ in regions of highest sensitivity, but is especially so in low sensitivity regions.

The electric fields and the corresponding spectral sensitivities are also a function of angle of incidence. In fact, the fields at large angles of incidence may be much larger with the tin oxide film present than they are in its absence, even if the film is absorbing. Large angles are desirable to maximize the interface area illuminated by a given beam. As seen in Figs. 3 and 4, $\langle E^2 \rangle$ drops off rapidly in a two-phase system. If $\langle E^2 \rangle$ could be kept high while taking advantage of the geometric factor of large angles, the sensitivity could be enhanced, and the convenience of a single reflection system could be used more often. Such is the case for tin oxide on glass, as shown in Fig. 11. Here h/λ varies from 0.9 to 1.1, the bottom curve at the extreme left corresponding to 0.9.

As another example of a three-phase system and a useful internal

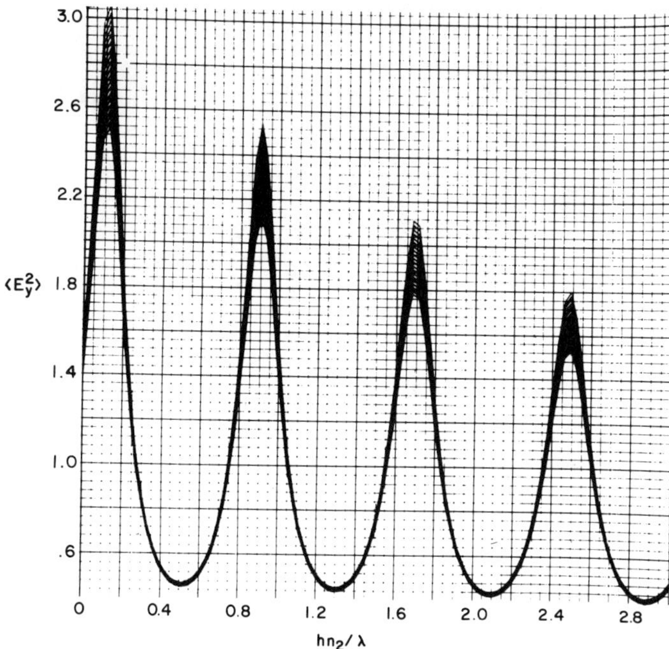

Fig. 8. Mean square fields for perpendicular polarization at the surface of an absorbing tin oxide electrode on glass, $\kappa_2 = 0.005$. Other parameters are as in Fig. 6.

reflection electrode, a thin metal film will be considered as phase 2. The optical constants chosen are typical of copper at 5 μ wavelength, $n_2 = 3$, $\kappa_2 = 10$. Similar results are obtained for thin films of gold, silver, platinum, etc. For metals in the visible region, results are somewhat similar but not so dramatic. Phase 1 is a transparent dielectric of index 2.4. Phase 3 has an index typical of a nonaqueous electrolyte, $n_3 = 1.6$, and is transparent. Figure 12 plots the $\langle E_z^2 \rangle$ values at the interface in the third phase for three film thicknesses, $h/\lambda = 0$, 8×10^{-4}, and 16×10^{-4}. The no-film case has the lowest maximum and the thickest film has the highest maximum. The 8×10^{-4} film corresponds to $h = 40$ Å, which is about the thinnest film that can be made at present with good electrical conductivity.

The surprising and dominant feature of Fig. 12 is the large field enhancement. The factor of 40 over that in the incident beam makes possible an enormous increase in sensitivity. Even though the films

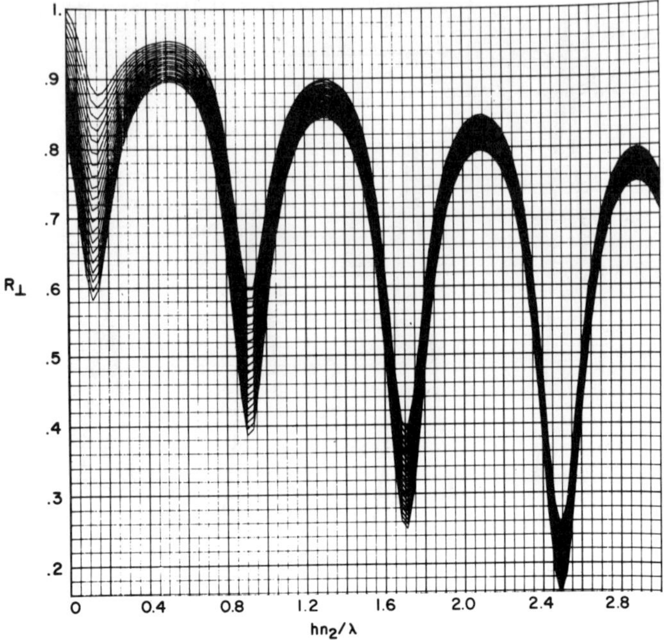

Fig. 9. Internal reflection spectra of a phase 3 with $\kappa_2 = 0.005$. Other parameters as in Fig. 6. Top curve is for no absorption in phase 3.

transmit only a few percent at normal incidence, by choosing angle and polarization properly a huge evanescent field can be established on the far side of the electrode. By comparison $\langle E_y{}^2 \rangle$ is relatively constant with angle, at about 0.1. The component of $\langle E_{\parallel}{}^2 \rangle$ parallel to the film is also small. Recall that the ratio of $\langle E_z{}^2 \rangle$ to $\langle E_x{}^2 \rangle$ is not affected by the metal film. This means that $\langle E_x{}^2 \rangle$ must also be enhanced. However, at the critical angle $\langle E_x{}^2 \rangle = 0$, and just beyond it where enhancement is greatest, $\langle E_x{}^2 \rangle$ is still small and remains so even with enhancement. Its maximum value for the thickest film is 0.9. Only at angles just larger than critical and with parallel polarization does the electromagnetic energy pass through the metal film to give a large field which is nearly normal to the interface.

IV. Derivation of Tractable Absorption Equations

The general equations given above, such as Eq. 9, give exactly the amount of energy absorbed under various conditions. They are

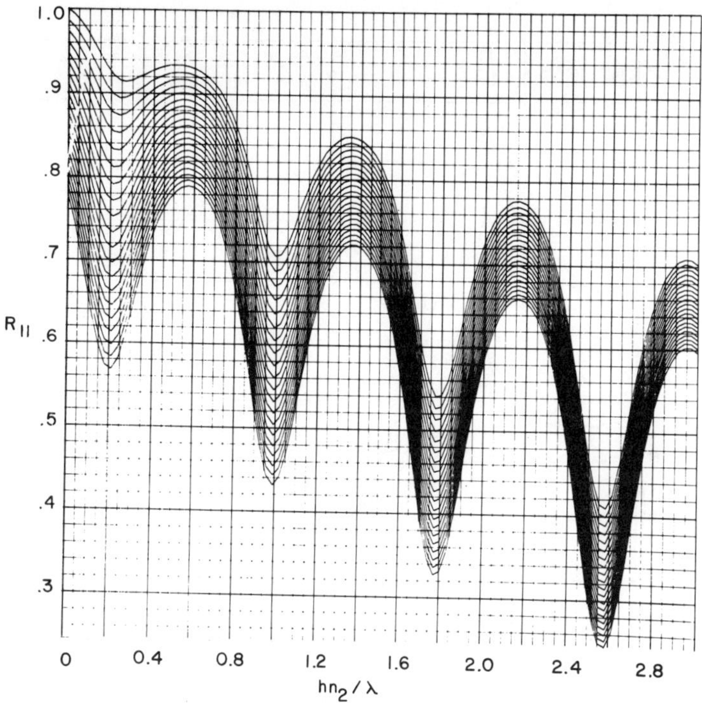

Fig. 10. Internal reflection spectra of a phase 3. As in Fig. 9, but for $\parallel$ polarization.

so cumbersome, however, that it is hard to gain physical insight by examining them analytically. Much can be gained by examining various cases numerically using a computer, and this is a standard procedure. A series of calculations can be made, for example, and the results plotted to show the functional dependence of R on one of the optical parameters. It is also possible to use sophisticated numerical techniques to solve for the optical properties of the system if reflection data are known. For example, three different reflectance measurements can be made at various angles and/or polarizations, and the data used to calculate three unknown optical properties of the system, such as n, k, and the thickness of a thin absorbing layer. This brute force method is extremely useful, but it is cumbersome and offers little physical insight. Where simpler approximate equations and procedures are adequate they are to be preferred. In the following sections some simple approximate

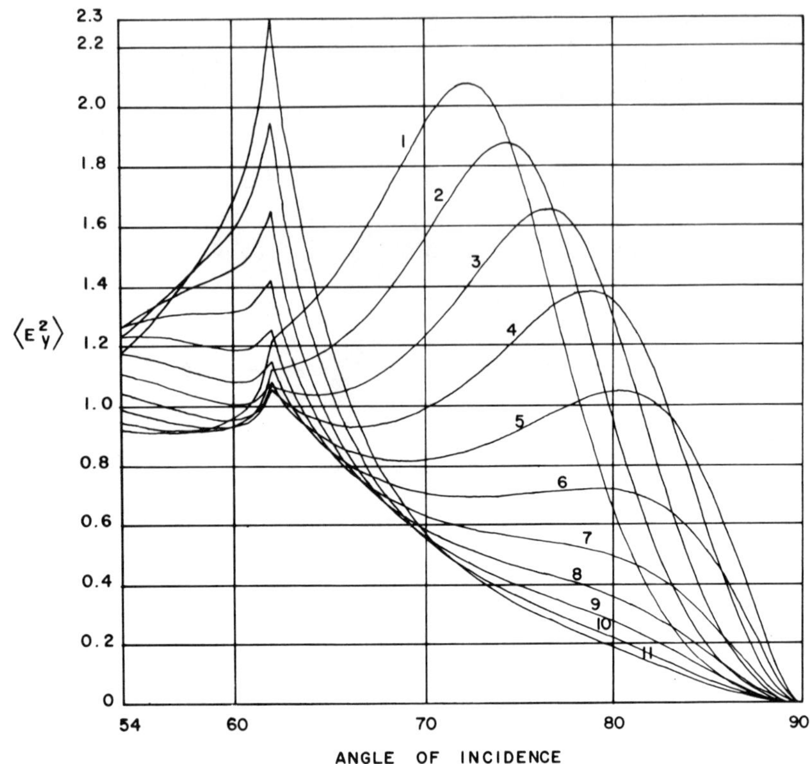

Fig. 11. $\langle E_y{}^2 \rangle$ at the surface of an absorbing tin oxide on glass electrode, as a function of angle of incidence. Curves 1–11 progress in uniform steps from thickness $h/\lambda = 0.9$ to 1.1. Other parameters are $n_1 = 1.51$, $n_2 = 1.88$, $\kappa_2 = .00532$, $n_3 = 1.34$, $\kappa_3 = 0$.

absorption equations are derived which give physical insight into a number of useful cases including those encountered in the application of reflection techniques to electrochemistry.

1. Penetration Depth

From Eqs. 16, 22, 27, and 28 we see that the mean square field intensities in the final phase of a stratified system fall to $1/e$ of their boundary value at a distance $\lambda/(-4\pi \operatorname{Im} \xi_f)$ into the phase. We define this as the penetration depth, d, of radiation into the final phase, f, and

$$d = \lambda/(-4\pi \operatorname{Im} \xi_f) = \lambda \operatorname{Re} \xi_f / n_f k_f \tag{38}$$

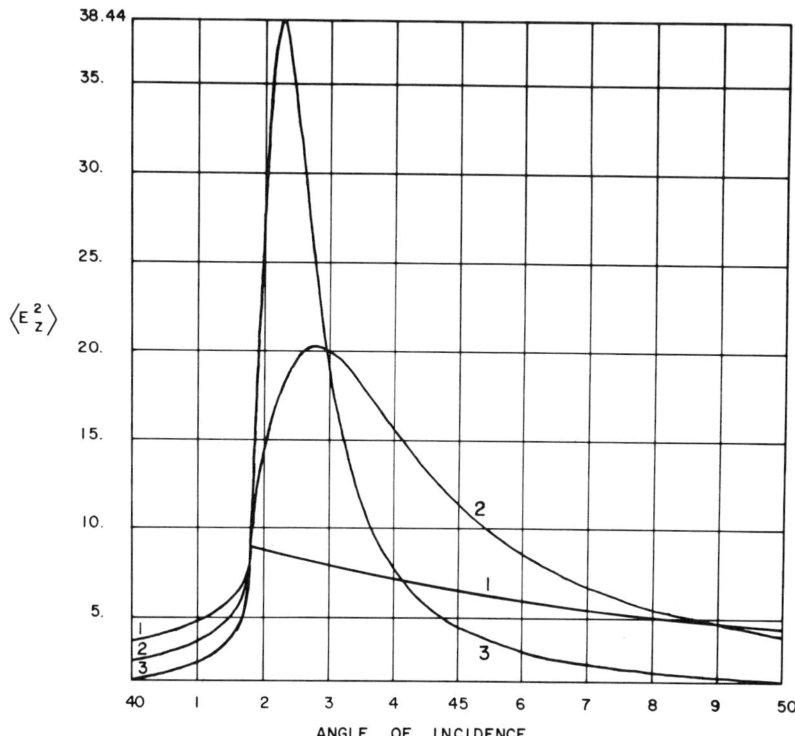

Fig. 12. $\langle E_z^2 \rangle$ at the surface of a thin gold electrode at λ about 5 μ. Curve 1 is for no gold; curve 2 is for gold $h/\lambda = 8 \times 10^{-4}$; curve 3 is for $h/\lambda = 16 \times 10^{-4}$. Other parameters are: $n_1 = 2.4$, $n_2 = 3.0$, $\kappa_2 = 10$, $n_3 = 1.6$, and $\kappa_3 = 0$.

(Recall that Im ξ is always negative.) This quantity is just one-half the quantity known as "skin depth" in electromagnetic theory. Note that the second form of the equation is indeterminate when the final phase is transparent, i.e., when $k_f = 0$. Also note that in this case, for $\theta_1 > \theta_c$

$$d = \lambda/[4\pi(n_1^2 \sin^2 \theta_1 - n_f^2)^{1/2}] \qquad k_f = 0, \theta_1 > \theta_c \tag{39}$$

From these simple expressions it is evident that penetration depth is independent of polarization and is infinite when $\theta_1 \leq \theta_c$ and $k_f = 0$; large when $\theta_1 \leq \theta_c$ and k_f is small; of the order of 0.1λ for k_f small, and $\theta_1 > \theta_c$ and θ_1 not close to θ_c; of the order of λ for k_f small, and $\theta_1 > \theta_c$ and θ_1 close enough to θ_c so that $(n_1^2 \sin^2 \theta_1 - n_f^2)^{1/2} \cong 0.1$; and is small (typically $\lambda/100$) when Im ξ_f is large, as is typical of metals at any angle of incidence.

The magnitude of $\langle E^2 \rangle$ always falls off exponentially in the final phase. Therefore, d tells us the distribution of $\langle E^2 \rangle$, including the thickness of the layer involved.

2. *Effective Thickness and an Absorption Law for Internal Reflection*

A. Absorbing Final Phase. In internal reflection techniques for obtaining absorption-like spectra of phase f, the sensitivity of the method depends on the penetration depth and the field strength at the phase boundary. We define the effective thickness, b_e, as the thickness of phase f that would have to be passed through by the incident light beam to give the same loss of energy by transmission as the loss experienced during reflection. Since the rate of energy absorption at any point is proportional to $\langle E^2 \rangle$, the total absorption is a volume integral, and for unit cross section of the beam

$$\begin{aligned} \langle E_T{}^2 \rangle b_e &= \frac{1}{\cos\theta_1} \int_0^\infty \langle E_f{}^2 \rangle \exp\left(-\frac{1}{d} z\right) dz \\ &= \frac{1}{\cos\theta_1} \langle E_f{}^2 \rangle d \end{aligned} \tag{40}$$

where z is the normal distance from the final boundary into the final phase, $\langle E_T{}^2 \rangle$ is the mean square field in the sample (phase f) in transmission and $\langle E_f{}^2 \rangle$ is the mean square field in the sample at the boundary in the case of reflection. If we neglect light intensity loss due to reflection (antireflection coatings can be imagined) for a given light intensity in two different phases j and k, $n_j \langle E_j{}^2 \rangle = n_k \langle E_k{}^2 \rangle$, implying that $\langle E_T{}^2 \rangle = (n_1/n_f) \langle E_1{}^2 \rangle$.

Substituting in Eq. 40 gives

$$b_e = \frac{n_f/n_1}{\cos\theta_1} \frac{\langle E_f{}^2 \rangle}{\langle E_1{}^2 \rangle} d \qquad \text{(bulk phase)} \tag{41}$$

If we allow $\langle E_1{}^2 \rangle$ to be unity as before we have

$$b_e = \frac{n_f}{n_1 \cos\theta_1} \langle E_f{}^2 \rangle d \tag{42}$$

The fields can readily be calculated from equations given earlier and d from Eq. 38. From this equation for b_e it is possible to derive approximate equations for absorption in the final phase when $k_f \ll 1$, which is usually the case.

The effective thickness is a measure of the sensitivity of the internal reflection method for absorption measurements. In fact, from our definition of b_e, and Beer's law we have

$$A = ab_e c = \frac{\alpha_f n_f d}{\ln 10 n_1 \cos \theta_1} \langle E_f{}^2 \rangle \tag{43}$$

where the absorptivity, a, and concentration, c, have their usual meaning, and A is reflection absorbance ($A \equiv -\log_{10} R$), a measure of the energy lost on reflection. The usefulness of this equation depends on the constancy of b_e during a measurement, and to some extent on the ease of calculating b_e for a given system. If $k_f \ll 1$ throughout the experiment, which is usually the case, the presence of the absorbing species in the final phase has little influence on n_f, $\langle E_f{}^2 \rangle$, or d. So b_e is constant and can be calculated once and for all, if desired, by simply ignoring k_f. Therefore, we have just derived an analogue of Beer's law, for internal reflection spectroscopy. Any number of intermediate phases might be involved, and they may be arbitrarily inhomogeneous in the direction normal to the interface. This equation was first derived for a two-phase system by a different method in Ref. 12, pp. 819 ff. It will be derived again later in a fashion more rigorous than the present one, as a limiting case of Eq. 84.

For very strong absorption bands, such as a concentrated solution of certain water-soluble dyes in the visible, n_f will vary appreciably with wavelength. A good approximate rule (12) is that $[\Delta(n_f{}^2)]_{\max} = n_f{}^2 \kappa_{\max}$. To get a feeling for what might be expected, consider the case of electrochemical experiments with *o*-tolidene (9). It gives rise to a very absorbing species, having a molar absorptivity of ca. 45,000 liter/(mole-cm). Yet in the 5 mM concentration typical of electrochemical experiments $\kappa_{\max}$ was equal to ca. 0.001. This gives a maximum change in an aqueous solution, with nominal $n_f = 1.34$, of $(\Delta(n_f{}^2))_{\max} = 0.00134$ or $\Delta n_f \cong 0.0007$. This change in n_f will usually cause a fractional change in reflection absorbance of less than 0.007, and that much only at the place in the absorption band where Δn_f is maximum. At the center of a classic band another approximate rule (12) gives $\Delta(n_f{}^2) = n_f{}^2 \kappa_{\max}^2$. For the present example this gives $\Delta(n_f{}^2) = 1.34 \times 10^{-6}$ or $\Delta n_f \cong 0.7 \times 10^{-6}$, which is completely negligible. For the exceptional case where Δn_f is so large that it cannot be ignored, there are still two wavelengths in the absorption band for which n_f (and therefore b_e) is constant with concentration. The points ordinarily lie to the short wavelength side of band center and with one close to the wavelength $\kappa_{\max}$. At these points we can expect Eq. 43 to remain linear in concentration to a $\kappa_{\max}$ of greater than 0.1, provided the chemical nature of the absorbing species doesn't change, i.e., provided absorptivity, a, remains constant. Similar restrictions are necessary for the Beer's law to hold in ordinary transmission spectroscopy.

Equation 43 is a new absorption law. It holds the same importance for internal reflection spectroscopy that Beer's law does for transmission spectroscopy. As indicated above, it holds over a wide range. It will not hold, however, when $\kappa_{\max}$ approaches unity or higher. (Of course, neither will Beer's law.) It is also not evident that it will hold for a general stratified medium where one or more intermediate phases are strongly absorbing. The problem here is not with n_f, or d, which can be treated as above, but with the change in the absorption of the intermediate phases with κ of the final phase. This in turn can also change $\langle E_f^2 \rangle$ which changes b_e. Under many conditions, however, Eq. 43 still holds, but with a new formula for b_e to replace Eq. 41. These cases are considered in detail later, where it will be shown that Eq. 43 holds when there is a constant intermediate layer such as a thin film electrode.

B. Thin Films. Effective thickness can also be defined for an absorbing thin film. Whereas in Eq. 40 it was necessary to integrate from 0 to ∞, in the case of a thin layer the volume integral is for a volume of the film of unit cross section. If the film is thin enough $\langle E^2 \rangle$ can be considered constant throughout and analogous to Eq. 43,

$$\langle E_T^2 \rangle b_e = \frac{1}{\cos \theta_1} \langle E_F^2 \rangle h \tag{44}$$

where h is the actual thickness of the film and $\langle E_F^2 \rangle$ is the field in the film. This equation holds for any film if $\langle E_F^2 \rangle$ is the time and space average field in the film. Thus in place of Eq. 42 we can write

$$b_e = \frac{n_F}{n_1 \cos \theta_1} \langle E_F^2 \rangle h \qquad \text{(thin film)} \tag{45}$$

and we again have Eq. 43, which is best expressed in this case as

$$A = \frac{\alpha}{\ln 10} (b_e) \tag{46}$$

where α is the absorption coefficient of the film, related to its extinction coefficient by $\alpha = 4\pi k/\lambda$. Note that in the case of a bulk phase as in Eq. 42, b_e is proportional to λ through d. However, in the case of a thin film as in Eq. 45, there is no such dependence on wavelength, and A is proportional to α. In this case, therefore, the spectrum of A will resemble a transmission spectrum, which is essentially a plot of α versus λ. Again, if $k_F \ll 1$, the fields can be calculated ignoring the film absorption, and b_e will be constant with k_F as well as with wavelength, neglecting small changes in n_F. Even when k_F is not small it will be found possible to derive tractable absorption equations for very thin films.

3. Absorption by a Thin Film or Multiple Thin Films

Absorption by a thin film at the interface, when the angle of incidence is greater than critical is given by Eq. 45 and 46. A more rigorous derivation of this absorption law for thin films follows.

The rate of absorption of radiant energy at any point in the film is given by Eq. 14 as

$$-\nabla \cdot \mathbf{S}_{\text{ave}} = \sigma\langle E^2\rangle = 2n_F k_F \omega\langle E^2\rangle \text{ (mks units)} \tag{47}$$

where ω is the angular frequency of the radiation. For a section of the film the (incident radiant power) $-$ (reflected radiant power) $=$ (rate of energy absorption by film) $= \int_\tau \nabla \cdot \mathbf{S}_{\text{ave}}\, d\tau$ where τ is the volume. This leads immediately to

$$I_0 - I = \frac{2\omega}{\text{area}} \int_\tau n_F k_F \langle E^2\rangle \, d\tau \tag{48}$$

where I_0 is the incident light intensity and area is the cross-sectional area of the beam. If we now define $\langle E^2\rangle_F$ as the time-space average of $\langle E^2\rangle$ in the film, i.e.,

$$\langle E^2\rangle_F \equiv \frac{1}{\tau}\int_\tau \langle E^2\rangle \, d\tau \tag{49}$$

and if the film is homogeneous, we can write

$$I_0 - I = 2\omega n_F k_F \langle E^2\rangle_F \frac{h}{\cos\theta_1} \tag{50}$$

where h is the thickness of the film. Or

$$I_0 - I = C n_F \alpha_F \langle E^2\rangle_F \frac{h}{\cos\theta_1} \tag{51}$$

where α_F is the absorption coefficient of the film and C is a constant.

If we now divide both sides of this equation by the intensity of the incident beam in the initial phase we get

$$\frac{I_0 - I}{I_0} = \frac{n_F \alpha_F}{n_1} \frac{\langle E^2 \rangle_F}{\langle E^2 \rangle_1} \frac{h}{\cos \theta_1} \tag{52}$$

or

$$1 - R = \frac{h n_F \alpha_F}{n_1 \cos \theta_1} \langle E^2 \rangle_F \tag{53}$$

where $R = I/I_0$ is the reflectivity of the system with the absorbing film present and $\langle E^2 \rangle_1$ is taken as unity. So far all of these equations are exact, and they apply to any film regardless of thickness or optical constants. From Eq. 53 we have clear insight into the physics of the situation, because the influence of each factor involved in the absorption is clearly shown.

The term that complicates things is $\langle E^2 \rangle_F$. In general it is a very difficult quantity to predict. For very thin films, however, we can closely predict $\langle E^2 \rangle_F$ without knowing the optical constants of the film and without elaborate multiphase calculations.

The first approximation is the reasonable assumption that

$$\langle E^2 \rangle_F \cong \frac{1 + R}{2} \langle E^2 \rangle_F^\circ \tag{54}$$

for thin films, where $\langle E^2 \rangle_F^\circ$ is the value for a nonabsorbing film of the same refractive index, n_F. The quantity $\langle E^2 \rangle_F^\circ$ can in turn be simply related to the field at the interface with no film present. This latter quantity is easily calculated. From Eqs. 53 and 54 we have

$$2 \frac{1 - R}{1 + R} = \frac{n_F}{n_1} \alpha_F \langle E^2 \rangle_F^\circ \frac{h}{\cos \theta_1} \tag{55}$$

and since the left-hand side closely approximates ln $(1/R)$ we can write

$$A \equiv \log_{10} \frac{1}{R} = \frac{n_F \alpha_F h}{\ln 10 n_1 \cos \theta_1} \langle E^2 \rangle_F^\circ \tag{56}$$

where A is the reflection absorbance. In terms of the fields at the boundary in the first phase, which are easily determined, we have

$$A_\perp = \frac{n_F \alpha_F h}{\ln 10 n_1 \cos \theta_1} \langle E^2 \rangle_{y1}^\circ \tag{57}$$

$$A_\parallel = \frac{n_F \alpha_F h}{\ln 10 n_1 \cos \theta_1} \left(\langle E^2 \rangle_{x1}^\circ + \frac{n_1^4}{(n_F^2 + k_F^2)^2} \langle E^2 \rangle_{z1}^\circ \right) \tag{58}$$

And for the special case of θ equal to the critical angle,

$$\left.\begin{aligned} A_{\perp} &= \frac{4}{\ln 10} \frac{h}{n_1 \cos \theta_c} n_F \alpha_F \\ A_{\parallel} &= \frac{4}{\ln 10} \frac{h}{n_1 \cos \theta_c} n_F \alpha_F \frac{n_1^2 n_3^2}{(n_F^2 + k_F^2)^2} \end{aligned}\right\} \theta_1 = \theta_c \qquad (59)$$

The validity of Eq. 56 can be tested by comparing numerical results obtained from it with those obtained using exact equations for a three-phase system. This will now be done. The three-phase system comprises a transparent high index initial phase, a thin absorbing layer, and a transparent low index final phase. Films of

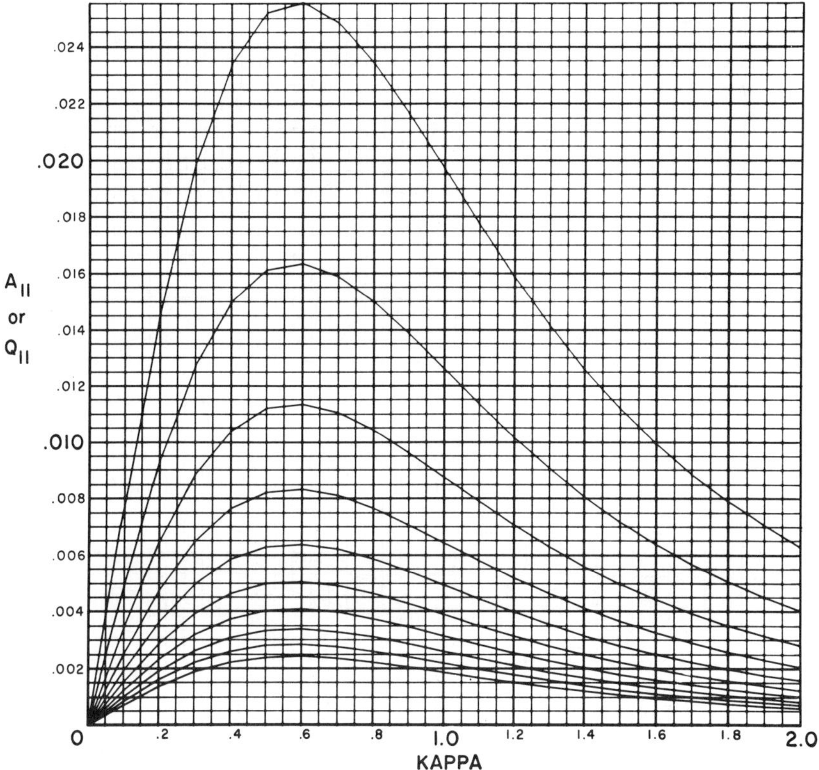

Fig. 13. Absorbance of a thin film in a three-phase system. $A_{\parallel}$ is from exact equations, $Q_{\parallel}$ is from Eq. 56, 58, or 59, and n_2 varies in steps of 0.2 from 0.8 for the top curve to 2.6 for the bottom curve. Other parameters include $h/\lambda = 0.001$, $\theta = \theta_c = 30°$, $n_1 = 2$, $n_3 = 1$, and $\kappa_3 = 0$, κ_2 used as abscissa.

thickness 0.001 and 0.01 wavelengths are used. The thickest is 500 Å at 5 μ wavelength. The thinner the film the more closely the approximate equations will be obeyed. The incident phase has refractive index 2.0, and the final phase has index 1.0. Results do not change if all phases are multiplied by a common factor, making it possible to relate the curves given to the case of an electrolyte as third phase. This conversion from one set of refractive indices to another is facilitated if the attenuation index, κ, is used as the absorption parameter of the film, because, unlike the extinction coefficient, k, κ is invariant from set to set. Recall that $n\kappa = k$.

In Figs. 13 through 18 the exact absorbance A, and approximate absorbance, Q, are plotted against κ as the abscissa and n_F as

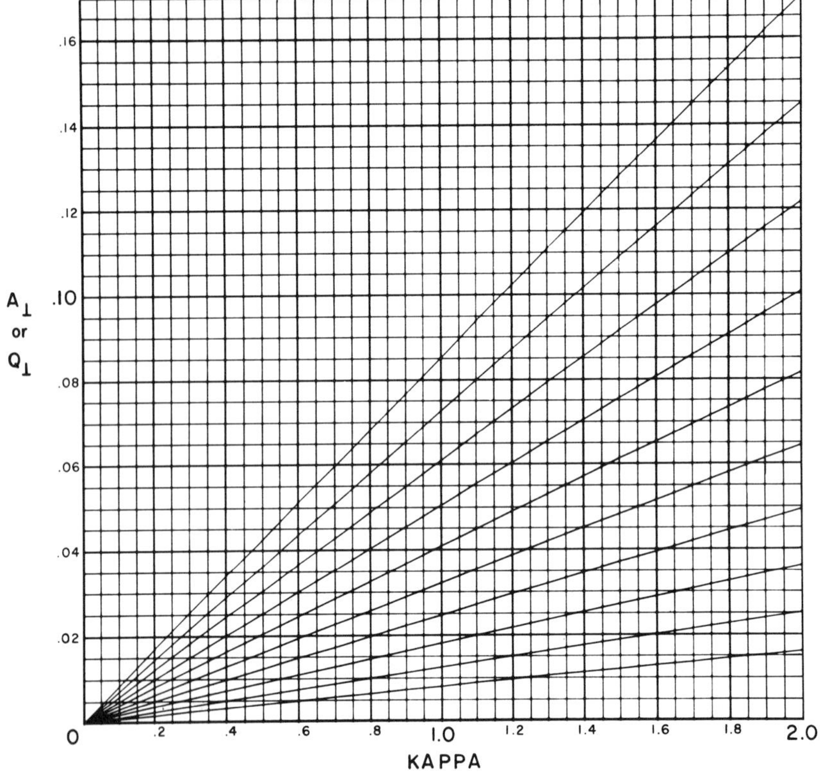

Fig. 14. Absorbance of a thin film in a three-phase system. $A_\perp$ and $Q_\perp$ are again equal. Parameters are as in Fig. 13. Here the top curve is for $n_2 = 2.6$, the bottom for $n_2 = 0.8$.

family parameter. In Fig. 13 are shown results for a thickness of 0.001, parallel polarization, and critical angle of incidence. Agreement of Q with the exact A is so close for all n_F and κ_F involved that they superimpose on the graph. The tabular numerical data show typical errors of less than one part per thousand. Figure 14 shows similar results for perpendicular polarization. Again the approximate equations are essentially exact for the range of film parameters covered.

These families of curves are valuable beyond just testing the validity of the approximate equations. They show in detail absorbance behavior over a wide range of optical properties. For example, they show the sensitivity of internal reflection to an absorbing adsorbed

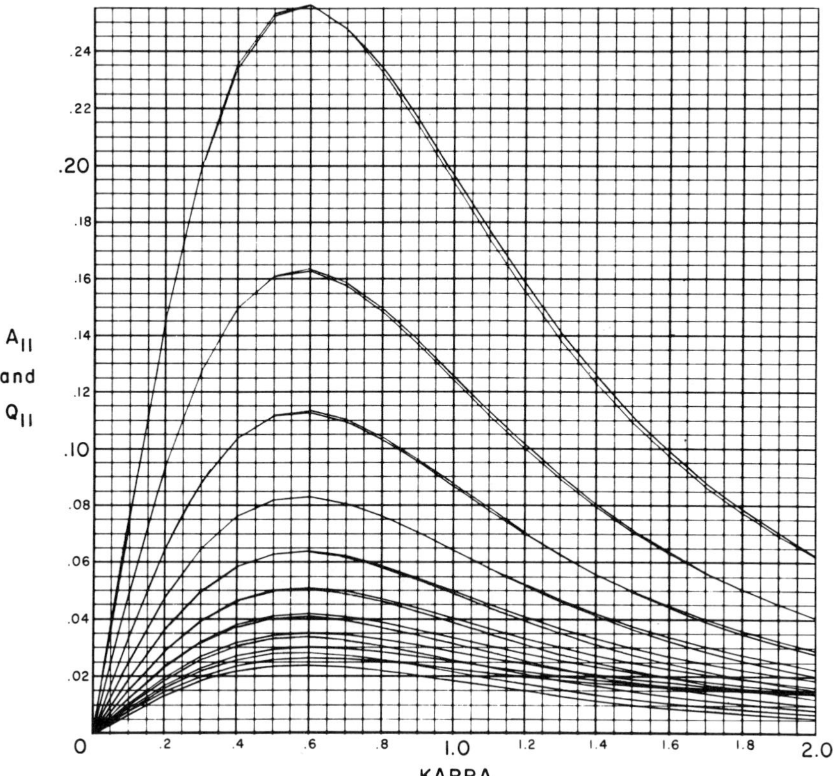

Fig. 15. Absorbance of a thin film. Same as Fig. 13 except $h/\lambda = 0.01$. $Q_{||}$ deviates above $A_{||}$ somewhat in the upper curves, and below $A_{||}$ in the lower curves.

monomolecular layer of known molar absorptivity. They also show which polarization to use, whether it be maximum sensitivity or linearity with respect to kappa that is most desired. Note that $A_{\parallel}$ is a maximum for low film index, whereas $A_{\perp}$ is a maximum for high.

Similar graphs, but with the thickness increased by a factor of 10, are shown in Figs. 15 and 16. Limits in the optical parameters beyond which large deviations occur are clearly indicated. But the approximate equations are shown to be useful over an enormous range. Strongly absorbing organics in the infrared typically have a maximum κ less than 0.5. Figures 15 and 16 indicate that the approximate equations can even be used for some metal films in the visible. They even apply for a 50 Å gold film at 0.5 μ wavelength.

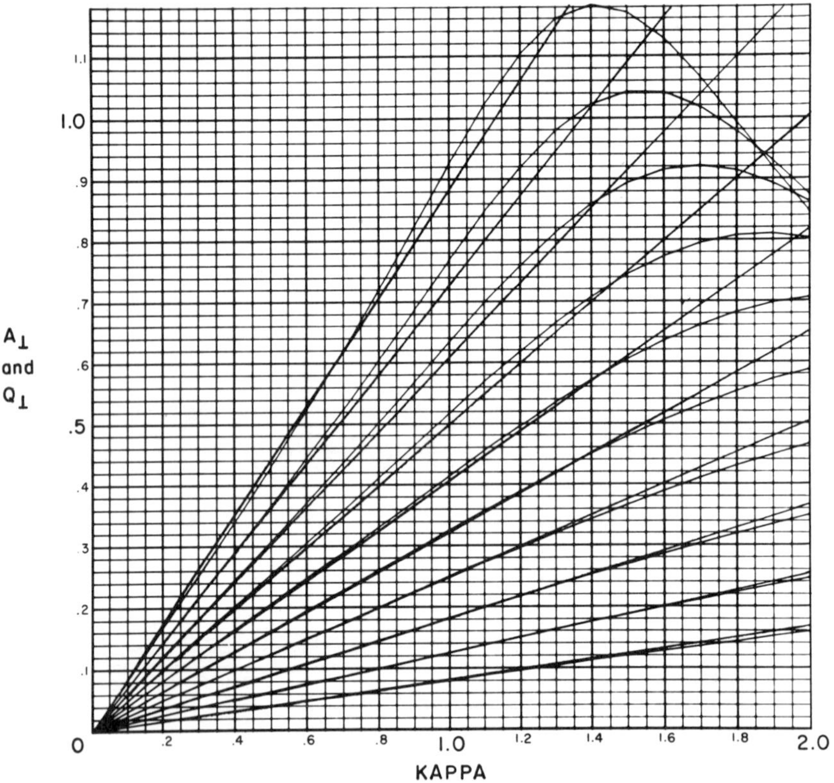

Fig. 16. Absorbance of a thin film. Same as Fig. 14 except $h/\lambda = 0.01$. The straight lines are $Q_{\perp}$.

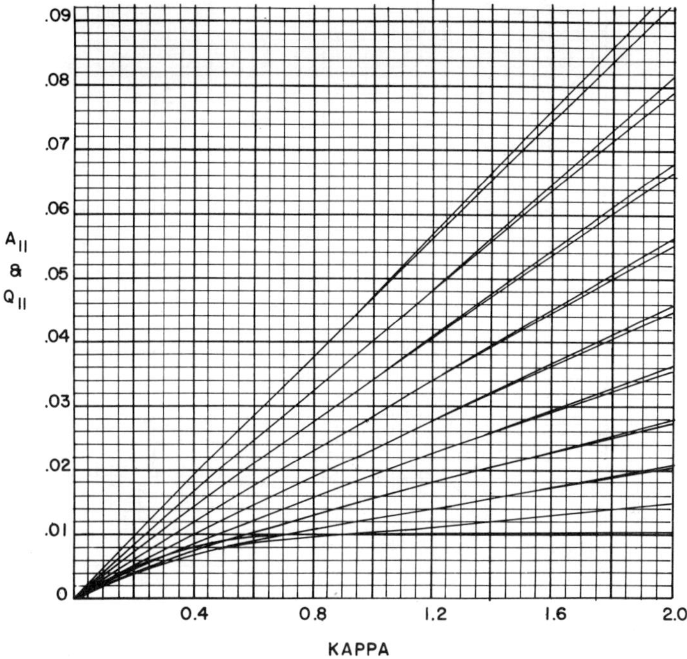

Fig. 17. Absorbance of thin film at angle far from θ_c. $A_{\|}$ is exact; $Q_{\|}$ is from Eq. 56. $\theta = 45°$; other parameters as in Fig. 13. Here $n_2 = 2.6$ is top pair of curves. $Q_{\|}$ is below $A_{\|}$ in the pairs.

They cannot be used, however, for gold films in the infrared, or for metal films with $\kappa \gg 1$.

Figures 17 and 18 show similar data, but at an angle considerably larger than critical. In this case the approximate equations are accurate to within about 1% for perpendicular polarization to $\kappa \cong 1.0$ and for parallel polarization to $\kappa \cong 2.0$. At $\theta_1 = 45°$ and $h/\lambda = 0.01$, deviation is several percent by the time κ has reached 0.2. At greater κ and/or thickness we must use the exact equations.

Let us now consider the case of two or more absorbing thin films at the interface. An example of this case is a colored layer electro-adsorbed on a thin layer electrode.

The derivation of absorption equations can proceed as above. Equation 51 can be written

$$I_0 - I = \frac{C}{\cos \theta_1} \sum_i n_i \alpha_i \langle E^2 \rangle_i h_i \tag{60}$$

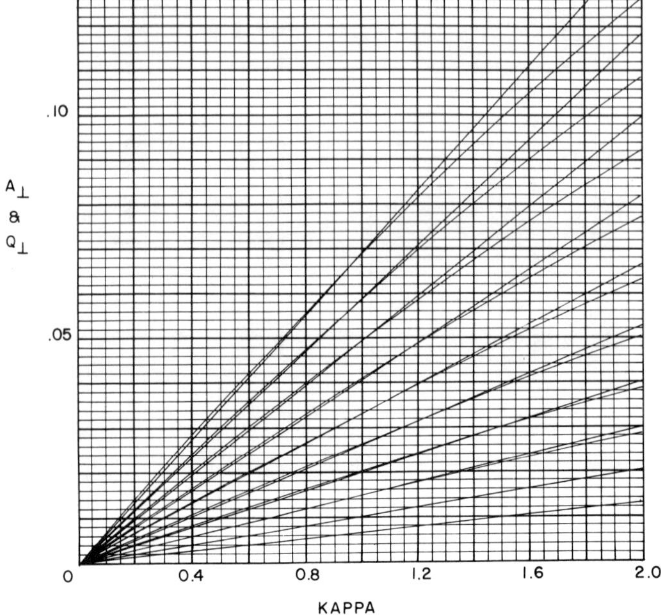

Fig. 18. Absorbance of thin film at angle far from θ_c. As in Fig. 17, except for $\perp$ polarization. The straight lines are $Q_\perp$; $n_2 = 2.6$ is top pair.

and Eq. 53 becomes

$$1 - R = \frac{1}{n_1 \cos \theta_1} \sum_i n_i \alpha_i \langle E^2 \rangle_i^\circ h_i \tag{61}$$

As before, we now introduce the approximation

$$\langle E^2 \rangle_i \cong \frac{1 + R}{2} \langle E^2 \rangle_i^\circ \tag{62}$$

where $\langle E^2 \rangle_i^\circ$ is the field when all phases are transparent, and Eq. 56 becomes, as expected

$$A \equiv \log_{10} \frac{1}{R} = \frac{1}{\ln 10 n_1 \cos \theta_1} \sum_i n_i \alpha_i \langle E^2 \rangle_i^\circ h_i \tag{63}$$

We see that the absorbancies are additive. The corresponding generalization of Eqs. 57, 58, and 59 are therefore obvious. To a first approximation we can expect the above equations for the multiple film case to hold as long as the combined thickness, average index, and combined absorbance are within the range of validity for a single film.

Equation 63 is valid for any number of films, which means that the optical properties can vary arbitrarily in a direction perpendicular to the interface. It is only necessary to have some average values for the optical parameters under the summation sign.

The practical way to record the spectrum of some species of interest, say an absorbing molecular species which is being generated or destroyed at the electrochemical interface, is to measure the change in absorbance due to the presence of the species, i.e.,

$$A_j = \frac{n_j \alpha_j h_j}{\ln 10 n_1 \cos \theta_1} \langle E^2 \rangle_j^{\circ} \tag{64}$$

To derive a somewhat more accurate equation than Eq. 64 go back to the exact equation, Eq. 60. Take as reference intensity the intensity with all films absorbing except film j, write

$$I_{\kappa_j=0} - I = \frac{C}{\cos \theta_1} n_j \alpha_j \langle E^2 \rangle_j h_j \tag{65}$$

and

$$R_0 - R = R_0(1 - R_j) = \frac{1}{n_1 \cos \theta_1} n_j \alpha_j \langle E^2 \rangle_j h_j \tag{66}$$

where

$$R_0 \equiv \frac{I_{\kappa_j=0}}{I_0}, \qquad R_j = \frac{R}{R_0}$$

Now make the approximation

$$\langle E^2 \rangle_j \simeq \frac{1 + R_j}{2} \langle E^2 \rangle_j' \tag{67}$$

where $\langle E^2 \rangle_j'$ now has reference to the relative field in phase j when all phases are present and all phases are absorbing except phase j. Equation 67 will often be valid even when Eq. 62 is made invalid by the presence of a strongly absorbing phase other than j. Therefore, when Eq. 67 is valid, we can write

$$A_j = \frac{2}{\ln 10} \cdot \frac{1 - R_j}{1 + R_j} = \frac{n_j \alpha_j h_j}{\ln 10 R_0 n_1 \cos \theta_1} \langle E^2 \rangle_j' \tag{68}$$

Equation 68 takes into account the absorption of other phases, but it assumes that this absorption does not depend on the absorpton in phase j.

There is also a problem of consistency between Eqs. 64 and 68. Using Eq. 62, $\langle E^2 \rangle_j \cong [(1 + R_0)/2]\langle E^2 \rangle_j^\circ$ and substitution in Eq. 68 does not give exactly Eq. 64, but substitution of $\langle E^2 \rangle_j' \cong R_0 \langle E^2 \rangle_j$ would. This discrepancy arises from the approximation of Eq. 62; hence for R_0 appreciably less than unity, Eq. 68 is preferred over Eq. 64.

A serious problem in trying to derive an equation for the absorption of one layer in the presence of other strongly absorbing layers is that a change in the layer being studied changes the absorption in the other layers. To account for this change in a general way, one must use the exact equations and computer calculations. It is possible, however, to derive approximate equations similar to those above, which take into account the presence of even a strongly absorbing film and which are valid for many cases of interest. This derivation follows.

Suppose we have a highly absorbing film, phase a, on a transparent incident phase. On the fixed phase a there is a variable film, phase x, followed by the final phase which is transparent. The observed light intensity after reflection at $\theta_1 \geq \theta_c$ is given by

$$I_{\text{obs}} = I_0 - I_a - I_x = I_0 - I_x - (I_a^\circ - \Delta I_a) \tag{69}$$

where I_a and I_x are the radiant power per unit interface area being absorbed by phase a and phase x, I_a° is the value of I_a when phase x is absent, and ΔI_a is the change in I_a due to the presence of film x. In the absence of phase x

$$I_{\text{obs}} \rightarrow I_{\text{obs}}^\circ = I_0 - I_a^\circ \tag{70}$$

and

$$\frac{I_{\text{obs}}^\circ - I_{\text{obs}}}{I_0} = \frac{I_x - \Delta I_a}{I_0} \tag{71}$$

From the above discussions, we know that

$$\frac{I_x}{I_0} = \frac{n_x \alpha_x h_x}{n_1 \cos\theta_1} \langle E^2 \rangle_x = \frac{n_x \alpha_x h_x}{n_1 \cos\theta_1} \frac{1 + R_x}{2} \langle E^2 \rangle_x^\circ \tag{72}$$

where $R_x \equiv I_{\text{obs}}/I_{\text{obs}}^\circ$ and $\langle E^2 \rangle_x$ is the space-time average field in phase x (recall $\langle E^2 \rangle$ of incident wave is unity) at time of measuring I_{obs}, and $\langle E^2 \rangle_x^\circ$ is the field in phase x in the limit as x disappears (i.e., at time of measuring I_{obs}°). Likewise

$$\frac{I_a}{I_0} = \frac{n_a \alpha_a h_a}{n_1 \cos\theta_1} \langle E^2 \rangle_a = \frac{n_a \alpha_a h_a}{n_1 \cos\theta_1} \frac{1 + R_x}{2} \langle E^2 \rangle_a^\circ \tag{73}$$

where $\langle E^2\rangle_a^\circ$ is the space-time average field in a as phase x vanishes. Let

$$P' \equiv \frac{n_a\alpha_a h_a}{n_1 \cos\theta_1}$$

and write

$$\frac{I_a}{I_0} = P'\langle E^2\rangle_a^\circ - P'\frac{1-R_x}{2}\langle E^2\rangle_a^\circ \tag{74}$$

$$= \frac{I_a^\circ}{I_0} - \frac{\Delta I_a}{I_0} \tag{75}$$

$$\frac{I_x}{I_0} - \frac{\Delta I_a}{I_0} = \frac{n_x\alpha_x h_x}{n_1\cos\theta_1}\frac{1+R_x}{2}\langle E^2\rangle_x^\circ - P'\frac{1-R_x}{2}\langle E^2\rangle_a^\circ \tag{76}$$

or since

$$\frac{(I_x/I_0) - (\Delta I_a/I_0)}{I_{\text{obs}}^\circ/I_0} = \frac{(I_{\text{obs}}^\circ/I_0) - (I_{\text{obs}}/I_0)}{I_{\text{obs}}^\circ/I_0} = 1 - R_x \tag{77}$$

where $I_{\text{obs}}^\circ/I_0 \equiv R_0$, then

$$1 - R_x = \frac{n_x\alpha_x h_x}{R_0 n_1\cos\theta_1}\frac{1+R_x}{2}\langle E^2\rangle_x^\circ - \frac{P'}{R_0}\frac{1-R_x}{2}\langle E^2\rangle_a^\circ \tag{78}$$

or

$$A_x \cong \frac{2}{\ln 10}\frac{1-R_x}{1+R_x}$$

$$= \frac{n_x\alpha_x h_x}{\ln 10 R_0 n_1\cos\theta_1}\langle E^2\rangle_x^\circ - \frac{P'}{2R_0}\frac{2}{\ln 10}\frac{1-R_x}{1+R_x}\langle E^2\rangle_a^\circ \tag{79}$$

Now let

$$P = P'\frac{\langle E^2\rangle_a^\circ}{2R_0} = \frac{2\pi n_a k_a h_a}{\lambda n_1\cos\theta_1 R_0}\langle E^2\rangle_a^\circ \tag{80}$$

and write

$$A_x = \frac{n_x\alpha_x h_x}{(1+P)R_0 \ln 10 n_1\cos\theta_1}\langle E^2\rangle_x^\circ \tag{81}$$

The quantity P, which does not depend on phase x, corrects for the change in absorption by phase a due to absorption by phase x.

Equation 81 should be compared to Eqs. 56 and 68. It reduces to Eq. 68 if $P \ll 1$, which will be the case as phase a becomes very thin and/or not very absorbing. It reduces to Eq. 56 as phase a

disappears, making $P \to 0$ and $R_0 \to 1$. The factor $\langle E^2 \rangle_a^\circ$ in P should really be the space-time average in phase a. Since P is only a correction term, it is usually sufficient to calculate a value anywhere in phase a. However, if phase a is several quarter wavelengths thick, as may be the case when it is not very absorbing, standing waves will be established in this layer such that the value of $\langle E^2 \rangle_a^\circ$ varies greatly from point to point. We have found by calculation and experiment that even in this case, it is safe to use the value in phase a at the a, x boundary.

The meaning of $\langle E^2 \rangle_x^\circ$ can also be relaxed. Strictly speaking, it is the field in phase x under the standard conditions under which I_{obs}° is measured. Under these conditions $I_x = 0$, which means that either α_x or h_x approaches zero. Either case may be used in calculating $\langle E^2 \rangle_x^\circ$. The calculated values of R_0 and P will also depend somewhat on the choice, but the net result in Eq. 81 will be small. In the case of a phase x with κ less than unity it is probably more accurate to choose $\alpha_x = 0$ as the standard state. In this derivation it is intended that phase x be so thin that the fields do not vary appreciably from one side to the other, so $\langle E^2 \rangle_x^\circ$ can be taken anywhere within phase x.

From Eq. 81, equations for both polarizations can be written down at once for a variable phase x.

$$A_\perp = \frac{n_x \alpha_x h_x}{(1 + P) R_0 \ln 10 n_1 \cos \theta_1} \langle E^2 \rangle_\perp^\circ \tag{82}$$

$$A_\parallel = \frac{n_x \alpha_x h_x}{(1 + P) R_0 \ln 10 n_1 \cos \theta_1} \langle E^2 \rangle_\parallel^\circ \tag{83}$$

The fields for the standard state for both polarizations refer to either vanishingly thin phase x or nonabsorbing phase x, whichever is most convenient, as discussed above.

The presence of the intermediate phase a makes it impossible to write down simple equations like Eq. 59 in the present case. It is still true, however, that near the critical angle, the electric field vector in the final phase is normal to the interface for parallel polarization. If phase x is not too thick, the same is true in it. For perpendicular polarization, the electric field vector is parallel to the interface, always.

Equation 81 and those derived from it are most useful from an electrochemical point of view. They are sufficiently powerful to deal with actual internal reflection electrode systems such as those involving doped tin oxide and gold films. Their range of validity can be checked by comparison with the exact equations of Ref. 10. By this means and by direct experimental observation it has been shown that they are valid for the electrodes mentioned in the optical range where glass is transparent (near-UV, visible, and near-IR). Equation 83 is valid for gold and other thin metal electrodes in the vibrational infrared provided the angle used is close to critical.

The most striking feature of Eq. 81 is that the absorbance is proportional to the amount of absorbing material in phase x. The amount of absorber is of course directly related to h_x or to α_x via concentration. This feature is by no means obvious for such a complicated system. Nor can it be seen from the exact equations. For those who are tempted to say they suspected it all along, let it be added that it is usually not true. Large deviations from Eq. 81 also involved large deviations from linearity.

A second striking feature of Eq. 81 becomes clear when numbers are supplied for the fields, as for example those given in Fig. 12. The sensitivity of A_x to absorbing material at the interface may be many times that given by Beer's law for transmission.

A third feature is that the spectrum of A_x as a function of wavelength will vary with $\langle E^2 \rangle_x^\circ$ which may fluctuate over a wide range, and to a lesser extent with P and R_0. If the optical parameters of all phases except x do not vary much, $A_\perp$ and $A_\parallel$ as a function of wavelength will appear as they did in Eqs. 57 and 58, except for a constant factor. $A_\perp$ will appear like a transmission spectrum multiplied by n_x, while near the critical angle $A_\parallel$ will have the additional factor $n_1{}^2 n_3{}^2/(n_x{}^2 + k_x{}^2)^2$, which shifts the spectrum maximum to shorter wavelengths.

A partial monolayer of crystal violet was adsorbed on glass and on a gold film electrode, and examined by internal reflection. All features examined were found in quantitative agreement with Eqs. 82 and 83. That included expected sensitivity, relative sensitivity for the two polarizations, and relative spectral shapes. The adsorbed material had a different spectrum than crystal violet in solution, as expected, and the peak for $A_\parallel$ was shifted to shorter wavelengths as discussed above. It is interesting that for a partial monomolecular layer, an absorbance change of 0.15 was observed for a single reflection. This proves that the sensitivity of the method is easily adequate for the study of adsorption, including electroadsorption.

4. Absorption by the Final Phase

For absorption in the semi-infinite final phase in the presence of a strongly absorbing fixed intermediate phase, a treatment analogous to the derivation of Eq. 81 can be made. The results can be written down by inspection as

$$A_f = \frac{n_f \alpha_f d}{(1 + P)R_0 \ln 10 n_1 \cos \theta_1} \langle E^2 \rangle_f^\circ \tag{84}$$

which includes the cases

$$A_\perp = \frac{n_f \alpha_f d}{(1 + P)R_0 \ln 10 n_1 \cos \theta_1} \langle E^2 \rangle_\perp^\circ \tag{85}$$

and

$$A_\parallel = \frac{n_f \alpha_f d}{(1 + P)R_0 \ln 10 n_1 \cos \theta_1} \langle E_2 \rangle_\parallel^\circ \tag{86}$$

The $\langle E^2 \rangle^\circ$ values are taken at the boundary in the last phase and refer to $\alpha_f = 0$, but with n_f the same as in the equations. R_0, P, and penetrating depth, d, refer to these same standard conditions.

An example of absorbance spectra taken through a thin gold electrode is given in Fig. 19. It will be seen that the basic attenuated total reflection (ATR) shape is maintained, a remarkable feature in light of the rapidly changing properties of the gold film, and the large absorption by the gold film. Note also the increased sensitivity in the case of parallel polarization and the decreased sensitivity of perpendicular polarization relative to ATR. These features are predicted in quantitative detail by Eqs. 85 and 86 when the optical constants of gold, etc., are inserted.

The range of validity of Eqs. 85 and 86 has been examined by numerical comparison with the exact equations of Ref. 10. This is a complex problem, but some results are available, and certain generalities can be made.

First consider the case of classical attenuated total reflection, ATR, in which the intermediate phase is absent. In this case $P \to 0$ and $R_0 \to 1$. (Note that Eq. 84 then corresponds to Eq. 43 except for a difference in field which is slight so long as $\kappa_f \ll 1$. In any case the field in Eq. 84 is the easiest to use as well as the most accurate.) Data for the typical case of a glass reflection element in the ATR region are given in Table I. The absorbance values listed are for a single reflection. For N reflections the single value is multiplied by N. It is seen that Eqs. 85 and 86 are essentially exact for $\kappa_f \leq 0.001$,

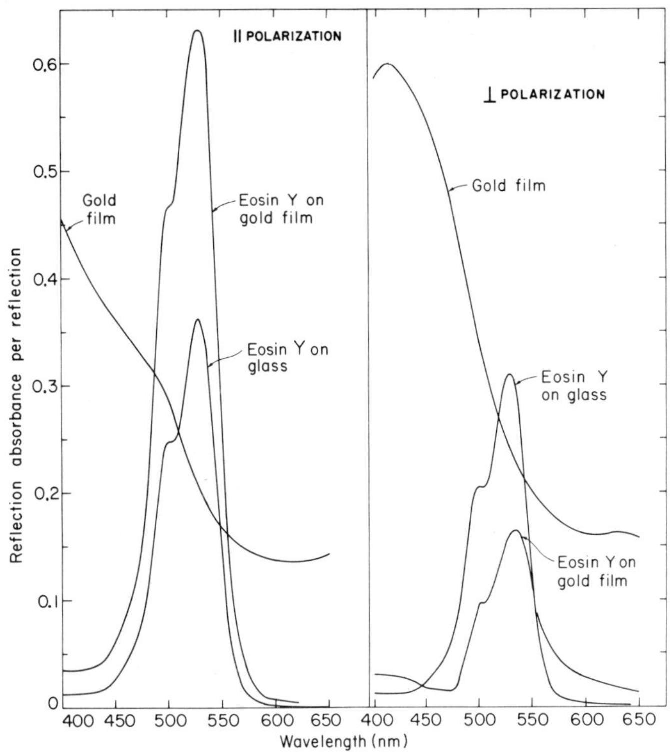

Fig. 19. The internal reflection spectrum of eosin y dye solution taken through a gold film electrode and compared to an ATR spectrum of the same solution. The reflection spectrum of the gold film itself is also given.

TABLE I
Validity of Eq. 84, Classical ATR Case

$A_{\parallel}$ (exact)	$A_{\parallel}$ (Eq. 86)	$A_{\perp}$ (exact)	$A_{\perp}$ (Eq. 85)	κ_f	Other parameters
0.0068595	0.0068598	0.0054194	0.0054197	0.001	$\theta_1 = 70°$
0.06828	0.068598	0.053914	0.054197	0.01	$n_1 = 1.536$
0.1347	0.1372	0.1062	0.1084	0.02	$n_f = 1.34$
0.496	0.686	0.373	0.542	0.1	$(\theta_c = 60.74°)$
0.0100412	0.0100419	0.0041252	0.0041256	0.001	$\theta_1 = 38.88°$
0.0997	0.100419	0.0410	0.041256	0.01	$n_1 = 2.5$
					$n_f = 1.5$
					$(\theta_c = 36.87°)$

TABLE II
Validity of Eq. 84, Tin Oxide Case

$A_{\parallel}$ (exact)	$A_{\parallel}$ (Eq. 86)	$A_{\perp}$ (exact)	$A_{\perp}$ (Eq. 85)	h/λ	κ_f	Other parameters
1.085×10^{-m}	1.087×10^{-m}	0.629×10^{-m}	0.633×10^{-m}	0.1	0.1×10^{-m}, $m \geq 2$	
1.083×10^{-1}	1.087×10^{-1}	0.629×10^{-1}	0.633×10^{-1}	0.1	0.01	
0.94	1.087	0.60	0.633	0.1	0.1	
1.081×10^{-m}	1.070×10^{-m}	0.767×10^{-m}	0.692×10^{-m}	0.5	0.1×10^{-m}, $m \geq 2$	$\theta_1 = 70°$ $(\theta_c = 60.74°)$ $n_1 = 1.536$ $n_f = 1.34$ $n_2 = 1.88$ $\kappa_2 = 0.005$ $\lambda =$ ca. 500 nm
1.080×10^{-1}	1.070×10^{-1}	0.769×10^{-1}	0.692×10^{-1}	0.5	0.01	
0.7814×10^{-m}	0.7680×10^{-m}	0.2614×10^{-m}	0.2453×10^{-m}	1.0	0.1×10^{-m}, $m \geq 2$	
0.7820×10^{-1}	0.7680×10^{-1}	0.2615×10^{-1}	0.2453×10^{-1}	1.0	0.01	
0.738	0.7680	0.234	0.2453	1.0	0.1	
0.4736×10^{-m}	0.4614×10^{-m}	0.2216×10^{-m}	0.1946×10^{-m}	2.0	0.1×10^{-m}, $m \geq 2$	
0.4698×10^{-1}	0.4614×10^{-1}	0.2179×10^{-1}	0.1946×10^{-1}	2.0	0.01	
0.35	0.4614	0.15	0.1946	2.0	0.1	
1.2156×10^{-4}	0.88×10^{-4}	0.6040×10^{-4}	0.3531×10^{-4}	3.0	0.00001	
1.2155×10^{-3}	0.88×10^{-3}	0.6044×10^{-3}	0.3531×10^{-3}	3.0	0.0001	
1.2174×10^{-2}	0.88×10^{-2}	0.6051×10^{-2}	0.3531×10^{-2}	3.0	0.001	
1.2350×10^{-1}	0.88×10^{-1}	0.6112×10^{-1}	0.3531×10^{-1}	3.0	0.01	
2.50×10^{-1}	1.76×10^{-1}	1.23×10^{-1}	0.706×10^{-1}	3.0	0.02	

accurate to a fraction of a percent at $\kappa_f = 0.01$, and deviate by 30% or so at $\kappa_f = 0.1$. The deviation from linearity behaves similarly.

Table II presents data for an absorbing tin-oxide-on-glass electrode in the visible. Absorbance values predicted by Eqs. 85 and 86 are compared with exact calculations for various κ_f and for various thicknesses of electrodes.

Various important generalities are evident. Equation 84 predicts linearity of κ_f with respect to α_f when n_f is constant, a good assumption for $\kappa_f \ll 1$. An absorbance change as small as about 10^{-6} can be measured, so Table II shows that linearity holds over a wide useful range, corresponding to a variation in concentration of a factor of 10^{-6}. This linearity holds even for a thickness of $h/\lambda = 3$ where four-fifths of the incident energy is absorbed by the electrode itself. This linearity means that a Beer-type law holds even for this complicated system.

Linearity holds better than absolute accuracy, but Eq. 84 gives the correct sensitivity $(dA/d\kappa_f)$ within a few percent except for the thickest film considered. Where the sensitivity is off, it is off by a factor independent of κ_f. It will also be noted that $A_{\parallel}$ is usually more accurate than $A_{\perp}$, although not necessarily more linear.

By comparison of Table II and Table I it is seen that the sensitivity is usually enhanced by the presence of the tin oxide layer, in keeping with previous discussion. Of course the angle of incidence could be chosen to give enhancement for $h/\lambda = 2.0$ just as well as for the other thicknesses. The angle of 70° was chosen simply because it is a convenient angle to work with for a glass-electrode-aqueous electrolyte system.

Table III gives data for another useful electrode system, a gold film in the visible. The thickness of 0.01 corresponds to 50 Å at 500-nm wavelength. Zero thickness is included here for easy comparison. It corresponds to ATR. Absolute accuracy is always off by a few percent for $h/\lambda = 0.01$, by 10% or so for $A_{\parallel}$ at $h/\lambda = 0.02$, and by even more for $A_{\perp}$ at $h/\lambda = 0.02$. Linearity is better and is always essentially perfect for $\kappa_f \leq 0.001$. It is this linearity that makes a spectrum look like a plot of κ_f versus λ. Note that the sensitivity is enhanced by the presence of the gold film, and the enhancement is greater the greater the gold thickness. This enhancement only occurs for parallel polarization and only at angles slightly greater than critical. Perpendicular polarization loses sensitivity rapidly with increased film thickness.

TABLE III
Validity of Eq. 84, Gold Film in Visible

$A_{\parallel}$ (exact)	$A_{\parallel}$ (Eq. 86)	$A_{\perp}$ (exact)	$A_{\perp}$ (Eq. 85)	h/λ	κ_f	Other parameters
1.2248×10^{-m}	1.2248×10^{-m}	0.9522×10^{-m}	0.9522×10^{-m}	0	0.1×10^{-m}, $m \geq 2$	
1.211×10^{-1}	1.2248×10^{-1}	0.940×10^{-1}	0.9522×10^{-1}	0	0.01	$\theta_1 = 65°$
0.723	1.2248	0.530	0.9522	0	0.1	$(\theta_c = 60.74°)$
						$n_1 = 1.536$
1.676×10^{-m}	1.6087×10^{-m}	0.41×10^{-m}	0.449×10^{-m}	0.01	0.1×10^{-m}, $m \geq 2$	$n_2 = 1.130$
1.669×10^{-1}	1.6087×10^{-1}	0.38×10^{-1}	0.449×10^{-1}	0.01	0.01	$\kappa_2 = 2.150$
1.14	1.6087	0.12	0.449	0.01	0.1	$n_f = 1.340$
2.098×10^{-m}	1.786×10^{-m}	0.143×10^{-m}	0.217×10^{-m}	0.02	0.1×10^{-m}, $m \geq 2$	
2.105×10^{-1}	1.786×10^{-1}	0.123×10^{-1}	0.217×10^{-1}	0.02	0.01	
2.5	1.786	0.017	0.217	0.02	0.1	

TABLE IV
Validity of Eq. 84, Gold Film in Infrared

$A_{\parallel}$ (exact)	$A_{\parallel}$ (Eq. 86)	$A_{\perp}$ (exact)	$A_{\perp}$ (Eq. 85)	h/λ	κ_f	Other parameters
6.80×10^{-m}	6.82×10^{-m}	0.099×10^{-m}	0.100×10^{-m}	6×10^{-4}	$0.1 \times 10^{-m}, m \geq 3$	$\theta_1 = 37.47°$
6.79×10^{-2}	6.82×10^{-2}	0.098×10^{-2}	0.100×10^{-2}	6×10^{-4}	0.001	$= \theta_c + 0.6°$
6.75×10^{-1}	6.82×10^{-1}	0.090×10^{-1}	0.100×10^{-1}	6×10^{-4}	0.01	$n_1 = 2.5$
0.7	6.82	0.037	0.100	6×10^{-4}	0.1	$n_2 = 2.0$
						$\kappa_2 = 18.0$
2.80×10^{-4}	3.03×10^{-4}	1.01×10^{-6}	1.03×10^{-6}	0.002	0.00001	$n_3 = 1.5$
2.79×10^{-3}	3.03×10^{-3}	1.00×10^{-5}	1.03×10^{-5}	0.002	0.0001	$\lambda =$ ca. 5μ
2.63×10^{-2}	3.03×10^{-2}	1.00×10^{-4}	1.03×10^{-4}	0.002	0.001	
0.84×10^{-1}	3.03×10^{-1}	0.91×10^{-3}	1.03×10^{-3}	0.002	0.01	
−0.05	3.03	0.40×10^{-2}	1.03×10^{-2}	0.002	0.1	

Table IV shows results for a gold film in the infrared. In particular, the optical constants of gold at 5 μ (1) were used. The thicknesses correspond to 30 Å for the thinnest and 100 Å for the other. A gold film of 30 Å is about the thinnest that can be made continuous with high electrical conductivity. The 100 Å gold film is easily made. For work at this wavelength, glass substrates must be replaced by infrared transmitting materials such as Irtran II. For this reason n_1 was taken as 2.5. The angle of incidence must be chosen carefully in the present case. On the one hand high sensitivity occurs only for a small angular range greater than θ_c, and on the other hand sensitivity can change rapidly with θ in this region. The angle chosen for Table IV is roughly optimum. The resulting sensitivity is much higher than the ATR case for parallel polarization. For perpendicular polarization the sensitivity is orders of magnitude less. All of these factors are accurately predicted by the approximate equations.

While only a limited number of cases have been studied in detail, they included phase combinations very different in their optical properties. Since the approximate equations work so well in the cases tested, it is expected that they will work for any thin film electrode that isn't too thick, and with the angle of incidence chosen with some care.

V. Application to Electrochemistry

1. *Techniques and Apparatus*

Applications of internal reflection methods to electrochemistry are new, and only a few experiments have been tried. Some of the results obtained have been surprising, but experiment and theory are in beautiful agreement.

The optical aspects of the techniques and apparatus are based on aspects of internal reflection spectroscopy (IRS) developed in recent years. A review of IRS through 1966 along with a good bibliography is given in the book by Harrick (15). A review of the brief history of spectroscopic observations in electrochemistry per IRS up to September 1967 is given by Hansen (13). The present writing includes work through 1969.

A simple IRS-electrochemical cell for fixed angle observation is shown in Fig. 20. This particular cell can easily be placed in the (large) sample compartment of a spectrophotometer, or lined up with a laser beam. It permits multiple reflections at the electrode

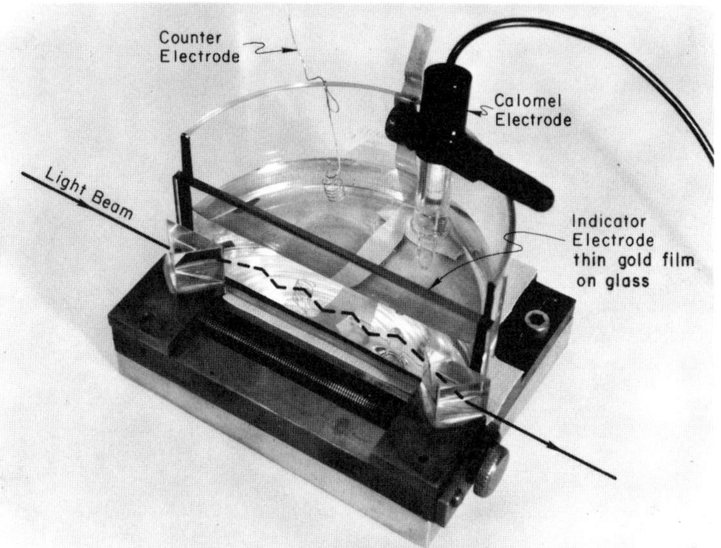

Fig. 20. IRS-electrochemical cell.

solution interface, the number of reflections and angle of incidence depending on geometry. The electrochemical cell can easily be made more sophisticated by adding, for example, tubes with fritted disk bottoms to isolate the electrodes. A sophisticated version of this type of cell, which has a number of advantages including low electrical time constant is described by Kuwana and co-workers (22). With the angle of incidence say ten degrees greater than critical, the penetration depth is typically a tenth of a wavelength. So the evanescent wave will sample the interface and diffusion region to about that depth. The idea is to observe changes in spectra during electrochemical manipulation. Of course the IRS plate could be infrared-transmitting for work there, but keep in mind that if the solvent is absorbing at the wavelength of interest it might dominate so that small effects are unobservable. On the other hand, the electrolyte can be opaque due to an electroactive solute, and changes still be observed near the electrode as well as in the case of a transparent electrolyte.

For a single reflection at variable angle of incidence a cell of the type shown in Fig. 21 was used by Prostak and Hansen in conjunction with a variable angle IRS device (18). This design permitted the use of a small electrode area (by IRS standards) of a few square millimeters, with a resultant low electrical time constant.

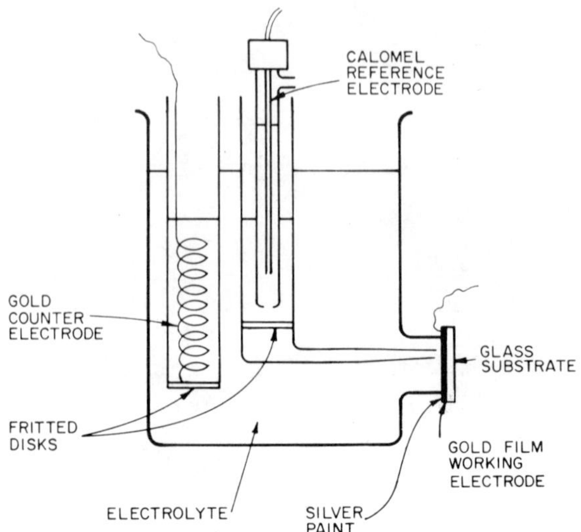

Fig. 21. Small electrode area IRS-electrochemical cell.

It is handy for investigating phenomena as a function of angle of incidence and at kilohertz frequencies.

It is evident from the optical theory presented in this paper that many different types of cells could be used, and in particular that the above basic designs could have many different types of electrodes and still give understandable spectroscopic data. Besides gold and tin oxide films, which have been used most extensively, platinum, silver, and copper films have been used. It is possible for the IRS element itself (e.g., the plate in Fig. 20) to be an electrical conductor. A germanium plate has been used this way in the infrared (16). Unfortunately, such an electrode is active electrochemically. It also becomes opaque when its electrical conductivity is made high.

There are special design problems for an electrochemical cell coupled with IRS. The main one is getting an electrode with high electrical conductivity and low light absorption. To a certain extent, these two properties are incompatible and limitations are imposed by theory. Tricks can be used, however, such as using the right angle of incidence and polarization to give a low field inside a gold film, as was described earlier. Another problem is the size of the electrode required to accommodate the beam. Again, where low time constants are necessary, special focusing and masking are used to reduce the electrode size. Some characteristics of a low power CW laser beam are ideal for IRS studies, particularly collimation and high intensity. Such a beam can be focused on a very small electrode area. For studies at a fixed wavelength, this is the method of choice.

2. *Observation of the Diffusion Region*

As indicated above, the evanescent wave of internal reflection samples the electrolyte to a distance from the interface of approximately one-tenth wavelength. In the visible then this depth is typically 50 nm. Species with a diffusion coefficient of 10^5 cm^2/sec, formed at the electrode at a fixed concentration for a time period of only 1 sec, will form a concentration profile which decreases by about 1% in 400 nm. In the same distance $\langle E^2 \rangle$ will have decayed to less than one-thousandth of its surface value. So we see that for changes with time constants approximating a second or longer, the concentration seen by IRS is the *surface* concentration.

The work of Hansen, Kuwana, and Osteryoung (7) illustrates the quantitative monitoring of the surface concentration of an electroactive species during extensive electrochemical manipulation. Potential sweep, potential step, chronopotentiometric, and potentiometric techniques were used to test the utility and accuracy of the method. The results were a vivid confirmation of theory, with surprising clarity and sensitivity. All results were shown to agree quantitatively with optical and electrochemical theory, except for some second order changes in absorbance which are now known to have been due to changes in the electrode itself.

In these studies a cell similar to that shown in Fig. 20 was used. It was placed in the sample compartment of an accurate spectrophotometer and coupled to electrochemical controlling and recording equipment. Some of their results (7) using chronopotentiometry are discussed as an example. The electrolyte was an aqueous solution to 0.1 F in HCl and 5 m*M* in *o*-tolidine. In acid media *o*-tolidine is oxidized to a yellow compound whose maximum absorptivity is at 4380 Å. The yellow compound is monitored by setting the spectrophotometer at 4380 Å and recording the changes in absorbance. The chronopotentiometric potential-time and absorbance-time plots with anodic current equal to 1 ma for a time period, τ, of 90 sec followed by a cathodic current of 1 mA, are given in Fig. 22. Theory (5) predicts a transition time of 30 sec after reversing to cathodic current, in agreement with the observed

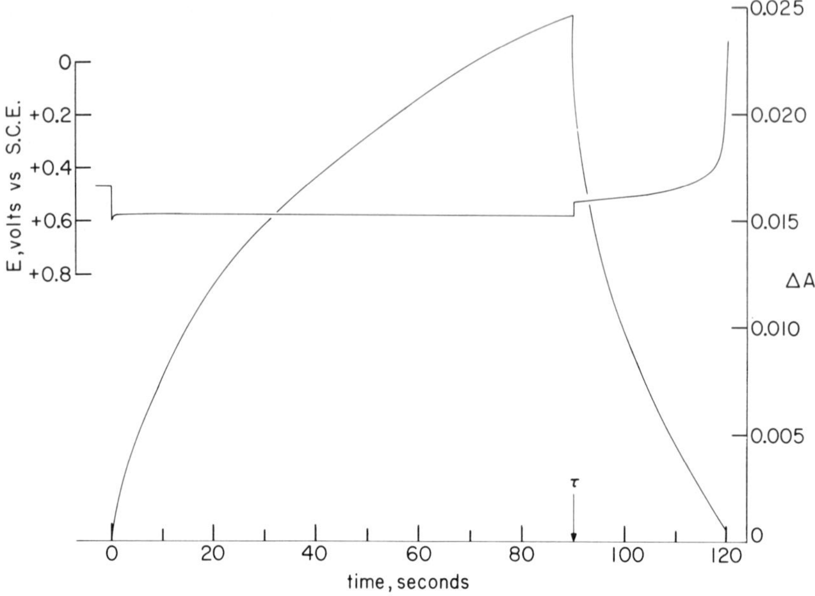

Fig. 22. Change in reflection absorbance, ΔA, and potential, E, vs. time during anodic and cathodic chronopotentiometry with solution 2 mM in o-tolidine.

value. Figure 23 is a plot of ΔA versus $t^{1/2}$ and $[t^{1/2} - 2(t - \tau)^{1/2}]$ for data obtained from Fig. 22. The optical theory discussed in earlier sections predicts that ΔA will be strictly proportional to surface concentration of the absorbing species (provided its absorptivity is constant with concentration, which is reasonable). Electrochemical diffusion theory (6) predicts a linear relationship between concentration and the time functions given above, and that the slopes of ΔA versus time function will be the same for both cases. These aspects are confirmed by the experimental data of Fig. 23.

Although it was not done in Ref. 7, an absolute calculation of the expected absorbance can be made using the optical parameters of the experimental setup, Eq. 86, and the absorptivity of oxidized o-tolidine.

Conditions used in Ref. 7 were: phase 1 was glass with $n_1 = 1.53$, phase 2 was the thin-film tin oxide electrode, and phase 3 was dilute aqueous solution with $n_3 = 1.34$ and $\kappa_3 \simeq 0$. The light beam was reflected five times in series from the interface. Also $\theta_1 = 72.6°$ $> \theta_c$ and $\lambda = 4380$ Å. Under these conditions Eq. 38 shows the

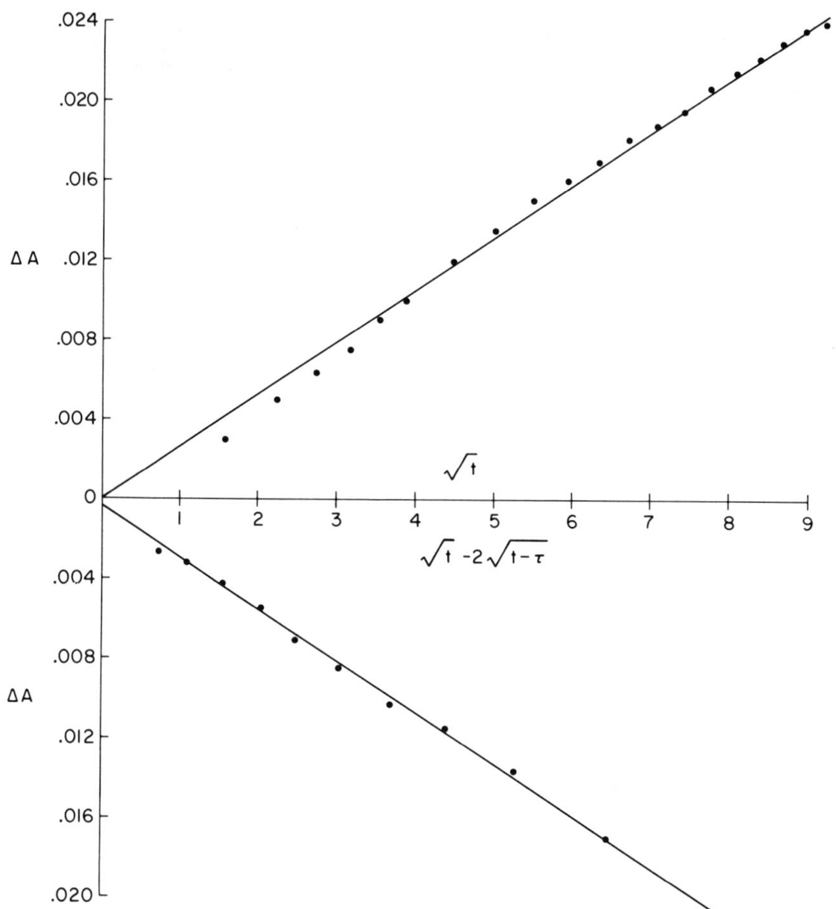

Fig. 23. ΔA vs. $t^{1/2}$ and ΔA vs. $[t^{1/2} - 2\sqrt{t - \tau}]$ for data of Fig. 22. τ is time of reversal.

penetration depth $d = \lambda/7.3 = 600$ Å. From electrochemical diffusion equations we know that at this distance from the electrode the concentration of absorbing species is essentially the same as at the surface. Therefore, Eq. 43 holds, with c simply the surface concentration $C_{\text{II}}{}^s$ of Reference 7. From Ref. 7

$$\Delta A = kC_{\text{II}}{}^s \tag{87}$$

and from Eq. 43

$$\Delta A = ab_e c = ab_e C_{\text{II}}{}^s \tag{88}$$

The observed ΔA of Eq. 87 will agree with the ΔA calculated by means of Eq. 88 if $k = ab_e$. From Eq. 42 we calculate b_e using the d calculated above, and the value of $\langle E_f^2 \rangle$ calculated for the particular electrode used. This value was 3.2 for the parallel polarization used. The molar absorptivity of the colored species was determined to be about 45,000 l/(mole cm) in a separate experiment. Therefore

$$ab_e = a \times 5d \times \langle E_f^2 \rangle \frac{n_f}{n_1 \cos \theta_1}$$

$$= 45{,}000 \times \frac{5 \times 4380 \times 10^{-8}}{7.3} \times 3.2 \frac{1.34}{1.53 \times 0.299}$$

$$= 13 \text{ liter/mole}$$

This should equal the value of k determined by measuring ΔA in the electrochemical experiments. Potential step measurements gave $k = 14$, and chronopotentiometric runs gave $k = 12$. The agreement is excellent.

In the work of Prostak, Mark, and Hansen (17) a gold-film-on-glass electrode was used instead of tin oxide. It was found that the same types of measurements could be made with either electrode, but the usable potential ranges differed somewhat, and the strong optical interference present with tin oxide was absent with gold. An interesting feature of this work was that in the case of tris (4, 7-dimethyl-1, 10-phenanthroline) ironII sulphate, the surface concentration of reactant was monitored even though the solution was opaque. Only the layer within about 500 Å of the electrode was seen by the light, and when the surface concentration was low, the absorbance was low. Monitoring this change with an opaque bulk solution is impossible with any transmission technique, including thin layer transmission.

3. Studies of Adsorbed Species

The optical theory strongly suggests that species adsorbed on the electrode surface can readily be studied by means of IRS. Some experimental data indirectly confirm this, but no detailed study has yet appeared in the literature. The very intense spectrum of a monolayer of methylene blue adsorbed on glass is reported (8), and from the optical study given above there is no doubt that the spectrum would be even more intense on a tin oxide or gold electrode. Figure 3 of Ref. 17 shows adsorption on a gold film electrode. The solid +0.94V curve is due to adsorbed DM ferroin. Unpublished work in our laboratory indicates that a monomolecular layer with an absorptivity equal to or greater than about 10 liter/mole cm can be seen. This means that one-thousandth of a monolayer of a species with absorptivity of 10,000 can be seen. Many substances have this or higher absorptivities.

The optical effect of an adsorbed layer is accurately described by Eq. 81. To get the sensitivity, i.e., the change in absorbance with respect to some quantity at the interface which changes, it is only necessary to take the partial derivative of absorbance with respect to that quantity. A change in the concentration of light-absorbing species at the interface is common, and it may occur with little change in refractive index. As an example, consider the gold film of Table III. To use the data of this case in which the final phase is absorbing, for the case of an absorbing film adsorbed on the gold, we have only to replace d in Eq. 84 with the h_x of Eq. 81 (and make some adjustments in $\langle E^2 \rangle$, which we ignore here). Thus for a film whose thickness, h_x, is equal to d, the sensitivities of Table III apply. If the minimum detectable absorbance change is $\Delta A = 10^{-6}$, the minimum κ detectable using parallel polarization, in a film of thickness d is $\kappa = 5 \times 10^{-6}$. A highly absorbing film such as adsorbed crystal violet has a κ of the order of unity. Thus a very small fraction of a monolayer can be detected. If we set the lower limit of detection at $\Delta A = 10^{-4}$, which is realistic when no special signal-enhancing techniques are employed, the minimum κ is 5×10^{-4}. This may still be orders of magnitude less than a monolayer. In an actual case of a monolayer of crystal violet adsorbed on a gold film, we observe an absorbance change of 0.1, confirming the calculations just given.

It is also possible to adjust optical parameters so that the absorbance is sensitive to refractive index changes in the vicinity of the interface. Thus it appears possible to study certain adsorption phenomena even though no absorbing species are involved. The change in absorbance comes from the change in energy absorbed by the thin film electrode itself. Various cases can easily be examined if a computer program is written incorporating the equations for a stratified medium discussed above. Examples of this type of calculation can be seen in Fig. 3 of Ref. 11 and Fig. 4 of Ref. 22. A little numerical exploration will reveal conditions for high sensitivity.

4. Kinetic Studies

The principles of internal reflection spectroscopy have been used to develop techniques to study the kinetics of fast homogeneous reactions involving electrogenerated species by Winograd and Kuwana (21). In their method, one of the active species is identified and quantitatively monitored near the electrode within the penetration depth of the radiation by means of internal reflection and the use of an optically thin electrode. Conditions are chosen to make the spectrophotometric method very sensitive to small changes in concentration of the compound of interest. By making the electrochemical manipulation repetitive they are able to use signal-averaging techniques to greatly increase the signal to noise ratio.

One reaction studied by Kuwana and Winograd was the reduction of methyl viologen. In a chronoamperometric experiment, starting with a 2.5×10^{-4} *M* solution of the ion MV^{+2} in 0.5 *M* tetraethylammonium perchlorate in acetonitrile electrolyte, they first stepped the potential to a point where the monocation radical would be formed by a one-electron reduction. Data were obtained by applying a 2-m sec potential pulse with repetition rate of 50 Hz for 60 sec for signal enhancement. The radical was monitored at 605 nm through an optically thin electrode. With the arrangement used, the radical was observable in the "cell" determined by the radiation penetration depth (of the order of 0.1λ) within a few μsec after the potential step. At the end of the pulse the absorbing radical diffused quickly out of the thin cell so that conditions were restored for another pulse. It was found that observed absorbance versus time behavior agreed with that calculated using optical and diffusion theory. In a second chronoamperometric experiment, the potential was stepped to a point where MV° is produced at the electrode at a diffusion-controlled rate by a two-electron reduction of MV^{+2}. In this case, a homogeneous reaction occurs to again form the monocation radical which is again monitored, i.e.,

$$MV^{+2} + MV^{\circ} \underset{k_b}{\overset{k_f}{\rightleftharpoons}} 2MV^{\dagger}$$

where k_f and k_b are the rate constants of the forward and reverse reactions. Within the thin optical cell that is being examined spectroscopically, the radical is being formed by the above reaction; and it is being depleted by the reverse reaction as well as by diffusion to the bulk and to the electrode. Its concentration will depend on k_f and be less than in the first case where all of the MV^{+2} that diffused to the electrode was converted to the radical. We see that if k_f is large enough so that an appreciable concentration of MV^+ builds up in the cell, and if it is not too fast, it will be a function of the observed absorbance. The value of k_f was determined by a curve-fitting technique as $k_f \geq 3 \times 10^9/M$ sec. They represent this as the upper limit attainable with their present electrochemical cell and instrumentation.

This and future kinetic studies using internal reflection must involve a knowledge of the basic optics to be successful. To illustrate, consider the penetration depth, d, of the evanescent wave, given by Eq. 38 or 39. Recall that it was independent of the nature of intermediate layers such as optically thin electrodes. It is even completely unaffected by fickle changes such as unknown electroadsorption or thickness changes in the electrode itself. Also consider the linear dependence of absorbance on the amount of absorbing material in the "cell" bounded by the electrode surface and the penetration depth. Not even a qualitative understanding of spectral observations can be had without considering these points—and they are by no means obvious. For a quantitative understanding it is necessary to calculate the optical sensitivity. This requires the use of either the exact equations of Section II or the approximate equations such as Eq. 56 or 81. These in turn require a knowledge of the optical constants of all phases. All these steps are straightforward once the basics are understood, and a few computer programs are available for the optical calculations. But they must not be ignored.

5. *Optical Changes in the Electrode*

Even if chemically inert, an electrode may be more than a passive source of electrons. Recently it has been found (14) that the optical properties of an electrode change with electrochemical potential. This is so because the electronic structure near the surface is a function of field, which causes the Fermi level to change. This is undoubtedly a universal effect, but the change in the optical constants depends on details of the electronic structure of the particular electrode involved, and is sometimes strongly dependent on wavelength.

On the one hand this effect is an unwelcome intruder. It can complicate measurements, especially high sensitivity measurements. It has caused researchers in the past to make wrong interpretations, such as those made by Walker (20). (Walker thought he was observing the hydrated electron through changes in his laser light intensity. What he actually saw was the electroreflectance change in his silver electrode.) Such misinterpretation is an annoyance in ellipsometry. Since it is here to stay, the only effective way to handle it is to understand it. On the other hand it can be used to advantage. It represents a new probe of the electronic structure of metals, and can be used to study the electrode itself.

In the case of gold, the effect has been studied in some detail (11). It was found to be maximum at a photon energy approximately 2.5 eV (wavelength approximately 0.5 μ) where with a 50 Å thick gold electrode the change in internal reflectance was as much as 7% with double layer charging, no electrons crossing the interface. This was with perpendicular polarization. Under the conditions used, there was much less effect on parallel polarization, and near the critical angle it went to zero. The way to avoid the effect in this case is simply to work with parallel polarization near the critical angle. In the case of tin oxide electrodes the effect has also been observed (22). Here it was also avoided when desired by choosing optical parameters properly.

Acknowledgment

Support of this work by the National Science Foundation, Grant No. GP13767, is gratefully acknowledged.

List of Symbols

A	Reflectance absorbance $A \equiv \log_{10} 1/R$
ΔA	Change in A as standard reflector is replaced by sample
ATR	Attenuated total reflection
b_e	Effective thickness
d	Penetration depth of radiation
$\mathbf{D}$	Electric displacement vector
$\mathbf{E}$, E, $\hat{\mathbf{E}}$	Electric field vector, electric field magnitude, and complex electric field vector respectively
$\hat{\mathbf{E}}^*$	Complex conjugate of the complex electric field vector
$\langle E^2 \rangle$	Mean square electric field strength
$\langle E_{\parallel 1}{}^2 \rangle_x$	Mean square electric field strength of the x component of parallel polarized light in phase 1
$\mathbf{H}$	Magnetic field vector
I	Radiant power
$\mathbf{S}$	Poynting's vector

$\mathbf{S}_i$, $\mathbf{S}_r$, $\mathbf{S}_t$	Poynting's vector in the incident, reflected, and transmitted beams respectively
IRS	Internal reflection spectroscopy
i	$\sqrt{-1}$
i, j, k	Subscripts denoting phase
Re, Im	Real and imaginary parts of a complex quantity
x, y, z	Right handed coordinate system; x is then in the plane of incidence and the first interface, z is perpendicular to the interface, and y is perpendicular to the plane of incidence
n	Refractive index
$\hat{n}$	Complex refractive index
k	Extinction coefficient
κ	Attenuation index
α	Absorption coefficient
ε, $\hat{\varepsilon}$	Dielectric constant, real and complex
σ	Electrical conductivity at optical frequencies
ρ, $\hat{\rho}$	Charge density, real and complex
θ, θ_c	Angle of incidence; critical angle of incidence
λ, ω	Wavelength in vacuum; angular frequency in vacuum
h_i	Thickness of phase i
ξ_i	The complex index of refraction times the cosine of the complex angle of refraction, all for phase i
β	The quantity $2\pi(h/\lambda)\xi$
$\perp$	Symbol for perpendicular polarization
$\parallel$	Symbol for parallel polarization
r	Amplitude reflection coefficient
R	Reflectance
t	Amplitude transmission coefficient, also time
T	Transmittance
δ^r, δ^t	Phase changes on reflection and transmission
Δ, $\Delta\Delta$	Relative phase change, $\delta_{\parallel}-\delta_{\perp}$, and change in Δ
$\tan \psi$	Relative amplitude attenuation

References

1. Bennett, H. E., and J. M. Bennett, in *Optical Properties and Electronic Structure of Metals and Alloys*, F. Abelès, Ed., Wiley, New York, 1966.
2. Born, M., and E. Wolf, *Principles of Optics*, Pergamon Press, New York, 1959.
3. Born, M., and E. Wolf, *ibid.*, pp. 611 ff.
4. Corson, D. R., and P. Lorrain, *Introduction to Electromagnetic Fields and Waves*, Freeman, San Francisco, 1962, p. 334.
5. Delahay, P., *New Instrumental Methods in Electrochemistry*, Interscience, New York, 1954, Ch. 8.
6. *Ibid.*, pp. 193 ff.
7. Hansen, W. N., T. Kuwana, and R. A. Osteryoung, *Anal. Chem.*, **38**, 1810 (1966).
8. Hansen, W. N., and J. A. Horton, *Anal. Chem.*, **36**, 783 (1964).

9. Hansen, W. N., R. A. Osteryoung, and T. Kuwana, *J. Am. Chem. Soc.*, **88**, 1062 (1966).
10. Hansen, W. N., *J. Opt. Soc. Am.*, **58**, 380 (1968).
11. Hansen, W. N., and Arnold Prostak, *Phys. Rev.*, **174**, 500 (1968).
12. Hansen, W. N., *Spectrochimica Acta*, **21**, 815 (1965).
13. Hansen, W. N., "Spectroscopic Observation of an Electrochemical Reaction by Means of Internal Reflection," in *Modern Aspects of Reflectance Spectroscopy*, W. W. Wendlandt and H. G. Hecht, Eds., Plenum Press, New York, 1966.
14. Hansen, W. N., *Surf. Sci.*, **16**, 206 (1969).
15. Harrick, N. J., *Internal Reflection Spectroscopy*, Interscience, New York, 1967.
16. Mark, H. B., and B. S. Pons, *Anal. Chem.*, **38**, 119 (1966).
17. Prostak, Arnold, H. B. Mark, and W. N. Hansen, *J. Phys. Chem.*, **72**, 2576 (1968).
18. Prostak, Arnold, "The Thin Film Gold Electrode," Ph.D. Thesis, University of Michigan, 1969.
19. Stratton, J. A., *Electromagnetic Theory*, McGraw-Hill, New York, 1941.
20. Walker, D. C., *Anal. Chem.*, **39**, 896 (1967).
21. Winograd, N., and T. Kuwana, *J. Am. Chem. Soc.*, **92**, 224 (1970).
22. Winograd, N., and T. Kuwana, *J. Electroanal. Chem.*, **23**, 333 (1969).

Specular Reflection Spectroscopy of the Electrode-Solution Interphase

J. D. E. McINTYRE
Bell Laboratories
Murray Hill, New Jersey

Contents

I. Introduction and Historical Survey

As the fundamental mechanisms of electrochemical reactions and electrocatalysis are investigated ever more closely, the need increases for new physicochemical techniques that can resolve the details of processes occurring at the electrode-electrolyte interface with *molecular* specificity. Many heterogeneous reactions of the "electrocatalytic" type (115,222,265), e.g., the hydrogen and oxygen evolution reactions and electro-organic oxidations, are of technological importance in electrochemical energy conversion, metallic corrosion and deposition, and chemical synthesis. These electrocatalytic reactions proceed through transition states exhibiting strong interaction with the electrode substrate and are characterized by multiple-step reaction mechanisms involving one or more adsorbed intermediate species. A complete kinetic description of such processes requires a knowledge of the chemical identity of these species together with their adsorption isotherms and the kinetic and mass transport parameters which characterize the individual reaction steps. Correlation of the catalytic properties of the electrode substrate material with the rate and course of heterogeneous reactions on its surface requires, in addition, a knowledge of the structure and electronic properties of the surface atomic layers. While voltammetric measurements can provide information concerning surface coverage and reaction rate, these methods are limited to the extent that the electrical quantity measured (potential, current or charge) corresponds to the *cumulative* action of the individual steps. Direct identification of individual intermediate species is only possible in the simplest cases, e.g., the H-adatom in the hydrogen evolution reaction. Information concerning the catalytic influence of the solid substrate is generally limited to a correlation of measured exchange-current densities with the *d*-band character of the metal, and the dependence of work functions and heats of adsorption on the crystal planes exposed (35). A definite need exists for complementary *in situ* physicochemical methods that can provide further detailed information concerning: (i) the chemical identity and kinetic behavior of adsorbed species; (ii) their interaction with and coverage of the substrate surface; (iii) the solid-state properties of the substrate material in its surface region; and (iv) the structure and constitution of the electrical double layer.

Modern physical methods of surface characterization such as low-energy electron diffraction (86,87,100,101,205), field-electron and field-ion emission microscopy (206,212,213), Auger-electron spectroscopy (135,292), and ion-neutralization spectroscopy (118,119, 120) have enabled the details of molecular processes on solid surfaces (in gaseous ambients at very low pressure) to be studied with microscopic and even atomic resolution. Such methods are not, of course, applicable to studies of the surface of a metal or semiconductor electrode immersed in a liquid electrolyte. Further, results from gas-solid adsorption measurements must be extended with caution to analogous liquid-solid interfaces owing to the possible competitive adsorption of other solution constituents.

For *in situ* characterization of an electrode surface, optical spectroscopy is one of the most versatile of currently available physicochemical methods. Although individual details of molecular processes on the surface cannot be resolved on a scale much smaller than the wavelength of the incident radiation, optical techniques can, nevertheless, be employed to monitor changes in surface coverage at the submonolayer level—a result of the extreme sensitivity of the reflection coefficients to small changes in phase angle. In addition, the wavelength selectivity and rapid response of these methods facilitate kinetic and mechanistic studies of surface processes. In the following sections, a brief historical review is given of the development and application of modern spectroscopic techniques for surface studies.

1. Optical Absorption Spectroscopy

Optical absorption spectroscopy has played a fundamental role in studies of the mechanisms of heterogeneous catalysis at solid-gas interfaces because it affords a direct means of observing the interactions and perturbations occurring as a result of adsorption at the surface, defining the nature of active substrate sites, and determining the chemical identity and structure of adsorbed species. Infrared spectroscopy has proved particularly useful in this respect, because the experimental spectra contain absorption bands characteristic of the quantized vibrational modes of structural groups of atoms, whose frequencies are relatively independent of the constitution of the molecule as a whole. Shifts of these frequencies are employed to investigate intermolecular interactions with the environment, overtone bands being particularly sensitive to external influences. Electronic transitions of a number of chromophoric groups are observed in the visible-ultraviolet wavelength region. Examples include organic chromophores such as the π-electrons in unsaturated olefins and aromatic compounds, carbonyl and azo groups, carbonium ions, carbanions, and free radicals (156,184). Electronic absorption spectra have also been employed extensively to obtain information about the distribution of electron energy levels and chemical bonding in complex metal ions (154), particularly those of the transition elements. Many of these species whose spectra are known are of importance in electrochemistry.

In optical studies of gas adsorption on metals, the adsorbent material is usually dispersed in the form of small particles (50–100 Å in diameter) on inert, nonporous supporting spheres of silica or alumina (~200 Å in diameter), in order to obtain a sufficient quantity of adsorbate material for analysis by infrared transmission spectroscopy. The vast literature on infrared spectra of adsorbed species has been extensively reviewed (61,77,78,121,184,187,275). In the visible-ultraviolet region, intense scattering precludes use of finely divided metal adsorbents, but a number of transmission studies have been made using thin, porous metal films evaporated onto quartz windows (160,161,162,163). Only a few studies have been made by reflection spectroscopy of molecules adsorbed on unsupported metals, owing to the difficulty of attaining adequate detection sensitivity, a problem which is particularly severe in the infrared region (112,113). To overcome this difficulty, multiple specular reflection cells were developed in 1959 by Francis and Ellison (89), who studied infrared spectra of metal stearate films on metal mirrors, and by Pickering and Eckstrom (226), who recorded infrared spectra of hydrogen chemisorbed on rhodium mirrors. Recent developments and applications of these methods have been discussed by Poling (230).

2. *Ellipsometry*

Ellipsometry is a technique that measures the polarization state of light reflected from optically smooth surfaces. Using polarized obliquely incident radiation, the optical constants of sample materials in the near-IR, visible, and near-UV wavelength regions can be determined with great precision. Since the pioneering investigations of Tronstad (280,281,282), ellipsometry has also been well-recognized for its ability to detect very thin surface films on metals immersed in liquid electrolytes. The historical development of this method has been described by Winterbottom (300). In recent years, it has been widely employed for *in situ* kinetic studies of film formation and dissolution (6,33,34,138,139,179,217,241,242,262,288). Particular emphasis has been placed on the determination of the mechanism of passivation of corroding metals, an aspect recently reviewed by Hayfield (140).

Although ellipsometry has provided much information about the kinetics of surface film growth, it has provided disappointingly little concerning the *spectroscopic* properties of these films; ellipsometric measurements have almost invariably been made at one or two isolated wavelengths in the visible region. The feasibility of employing this polarimetric method to study electronic absorption spectra of surface films of monomolecular thickness was first demonstrated by Bartell and Churchill in 1961 (18), but subsequent applications (19,57,67,193) have been sparse indeed. This is due, in large part, to the tedium of measuring the characteristic ellipsometric angles Ψ and Δ when manual ellipsometer-nulling techniques are employed, each measurement requiring several minutes. This problem will be resolved as fast, scanning, and automatically balancing, ellipsometric spectrometers become more readily available.

3. *Specular Reflection Spectroscopy*

Specular reflection spectroscopy is a technique extensively employed to determine the optical properties of solids over a wide photon energy range (0–25 eV). To avoid the need for polarizing prisms, which are unavailable for wavelengths shorter than ca. 1200 Å (25,110), reflected intensity measurements are made with light *normally* incident on the test surface. Characteristic optical constants are deduced from the experimental reflection spectra by applying the Kramers-Kronig transform relationships (173,177). With modern photodetection systems, it is now possible to measure minute intensity changes both rapidly and accurately with relative ease. This development has recently led to the adoption of specular reflection spectroscopy, in both normal and oblique incidence modes, as a *differential* technique for studying the optical properties of the electrode-solution interphase over a broad spectral range.

Historically, however, the application of reflection spectroscopy to electrochemical studies had an inauspicious beginning. In the early 1900s, the inability of Müller and Koenigsberger (168,215) to distinguish between the reflecting powers of active and passive iron was frequently cited (see, e.g., Byers, 48) as evidence in opposition to Faraday's oxide-film theory of passivity (84). By the mid-1930s, the objections had been largely removed (cf. Hedges, 144) by the reflection studies of Freundlich, Patscheke, and Zocher (91) on oxide-free iron films, and the ellipsometric studies of Tronstad and co-workers (280,281,282,283,284). With the advent of high-quality, recording double-beam spectrophotometers in the 1950s, reflection spectroscopy was developed as a method for measuring the thickness of anodically formed, transparent oxide films on valve metals (cf. Young, 302).

Currently, specular reflection spectroscopy is being actively applied (17,31,171,201,227,272,274) for *in situ* studies of the formation and optical behavior of monolayer films on surfaces, and for detecting intermediates and products of heterogeneous chemical and electrochemical reactions. The development of this method for *in situ* electrochemical investigations stems largely from the preliminary reflectance studies of Koch *et al.* on platinum electrodes (166, 167) in the mid-1960s, and the extensive use of electric-field modulation spectroscopy (54) to reveal structure in the optical properties of solids, dating from the original work of Seraphin *et al.* (255,256) on the electroreflectance of germanium.

The purpose of this chapter is to provide a background of the theory and techniques required for the application of differential specular reflection spectroscopy to the study of the optical properties of the electrode-solution interphase, and to survey the recent experimental results in this field. The intent is also to provide a comprehensive (but not exhaustive) compilation of relevant references from the fields of optical, solid-state and surface physics, surface chemistry, and electrochemistry. Emphasis is placed on an interfacial region of ca. 10 Å thickness—the zone of greatest importance in surface chemistry and physics. The discussion is primarily concerned with the measurement of interphase spectra *via external* reflection, where the light beam from the monochromator is transmitted through the aqueous or organic electrolyte solution enroute to and from the mirror electrode surface. The applications of this technique are thus restricted to wavelength regions where the ambient phase is effectively transparent. In the infrared region (2.5–25 μ), the universal presence of strong solvent-absorption bands precludes use of an electrolyte layer greater than a few microns in thickness (13), a value unsuitable for most electrochemical investigations. Although infrared spectra of adsorbed molecules are of fundamental importance in surface studies, this area may best be approached by internal reflection spectroscopy, a topic discussed elsewhere in this volume (131). The present chapter is thus mainly concerned with the absorption spectra of monolayer surface films in the ultraviolet, visible, and near infrared regions—the wavelength range 0.2 μ to 1.0 μ.

This rapidly developing field is passing through a transition stage. Most of the work covered in this review comprises work published or in press by the end of 1970. Early work was devoted primarily to establishing new experimental techniques and requisite optical theory for application of this method. The state of the art at that time is well-conveyed by the papers presented at the Faraday Society Symposium on Optical Studies of Adsorbed Layers at Interfaces, held in London in December 1970. Interpretation of the physical significance of *in situ* reflection spectra in electrochemical studies is still at a very early stage by comparison with the sophisticated treatments that have been made in optical studies of solid-state band structure and heterogeneous catalysis. Many of the early interpretations of experimental reflection spectra may therefore be subject to extension, modification, or revision as further understanding is developed. Computational methods for determining the actual absorption spectra of monolayers of chemisorbed species have recently been developed (171,201), and rapid progress should now be made in elucidating the detailed features of the electronic interactions of these species with the substrate surface.

The outline of the chapter is as follows. In Section II, we present a brief summary of the principles of physical optics which determine the reflectivity of a multiphase stratified system. All mathematical relations presented are consistent with the optical conventions proposed (214) at the Symposium on Recent Developments in Ellipsometry, held in Lincoln, Nebraska in 1968. Emphasis is placed on the influence of the electric-field strength in the standing wave formed at the reflecting interface, an understanding of which is essential for correct interpretation of reflection spectra. Section III deals with experimental apparatus and techniques employed in spectro-electrochemical studies, and includes a review of methods for determining the characteristic optical constants of the substrate and surface film materials. Section IV discusses the origins of the electroreflectance effect in metals. An understanding of this effect is basic to the interpretation of optical effects due to adsorption, and for deducing the electronic properties of the surface atom layers of metal electrocatalysts. Section V reviews the results of recent specular reflection studies of electrode surface films in the monolayer thickness range. Finally, Section VI discusses possible future applications of this technique.

II. Physical Optics

1. *Optical Constants*

When an electromagnetic plane wave is propagated in a nonabsorbing medium, the instantaneous value of the electric field vector is given by the time-dependent solution of Maxwell's equations, as

$$\mathbf{E} = \mathbf{E}^{\circ} \exp\left[i\left(\omega t - \frac{2\pi n}{\lambda}\,\mathbf{s}\cdot\mathbf{r}\right)\right] \tag{1}$$

where $\mathbf{E}^{\circ}$ is the wave amplitude, $\omega = 2\pi\nu$ is the angular frequency, $\mathbf{r}$ is a position vector referred to an arbitrary coordinate system, $\mathbf{s}$ is a unit vector in the propagation direction, and λ is the *vacuum* wavelength. The index of refraction of the medium, n, is defined by

$$n = \frac{c}{v} = (\varepsilon\mu)^{1/2} \tag{2}$$

where c is the velocity of light *in vacuo*, v is the phase velocity of the wave in the medium, and ε and μ are, respectively, the optical frequency dielectric constant and magnetic permeability of the material.

In an absorbing phase, the wave is attenuated as it travels through the medium, such that

$$\mathbf{E} = \mathbf{E}^\circ \exp\left[i\left(\omega t - \frac{2\pi n}{\lambda}\,\mathbf{s}\cdot\mathbf{r}\right)\right] \exp\left(-\frac{2\pi k}{\lambda}\,\mathbf{s}\cdot\mathbf{r}\right)\Bigg] \tag{3}$$

where the *extinction coefficient*, $k \geq 0$. By analogy to Eq. 1, the complex refractive index is defined as†

$$\hat{n} = n - ik \tag{4}$$

The light intensity, I, which is proportional to the mean-square electric field strength, $\langle E^2 \rangle$, is attenuated according to the Lambert law

$$I = I^\circ(1 - R)e^{-\alpha z} \tag{5}$$

where I° is the intensity of light incident on the surface of the absorbing medium, R is the reflectivity of the interface, α is the *absorption coefficient*, and z is the distance of penetration into the medium measured in a direction normal to the surface plane ($z = 0$). Thus

$$\alpha = \frac{4\pi k}{\lambda} = \frac{4\pi n\kappa}{\lambda} \tag{6}$$

The *penetration depth* of the radiation into the medium is defined as the distance $1/\alpha$.

In an absorbing phase, the dielectric constant is also complex and can be written as

$$\hat{\varepsilon} = \varepsilon' - i\varepsilon'' \tag{7}$$

where ε' and ε'' are the real and imaginary parts of $\hat{\varepsilon}$. Then

$$\hat{n} = (\mu\hat{\varepsilon})^{1/2} \tag{8}$$

† In some optical literature, the complex refractive index is expressed in terms of the *attenuation index*, κ, as $\hat{n} = n(1 - i\kappa)$.

and since the magnetic permeability is in general real,

$$\varepsilon' = (n^2 - k^2)/\mu \tag{9a}$$

$$\varepsilon'' = 2nk/\mu \tag{9b}$$

The magnetic permeability can usually be taken as unity in the optical frequency range.

The optical properties of each phase are completely specified by the set of optical constants: μ, ε', and ε'' (or by the alternate set: μ, n, and k). The constants ε' and ε'' are related to the frequency-dependent polarizability, α (not to be confused with the absorption coefficient), and conductivity, σ, viz. (23)

$$\varepsilon' = 1 + 4\pi[\alpha_{\text{intraband}} + \alpha_{\text{interband}} + \alpha_{\text{lattice absorption}} + \alpha_{\text{localized states}} + \alpha_{\text{core potential}} + \cdots] \tag{10a}$$

$$\varepsilon'' = \frac{4\pi}{\omega}[\sigma_{\text{intraband}} + \sigma_{\text{interband}} + \cdots] \tag{10b}$$

Now σ, the real part of the complex conductivity, is directly proportional to the energy absorbed by the material and the components of σ are nonzero only in the frequency ranges where absorption *via* a particular mechanism occurs. The components of the polarizability α, however, may be nonzero in the energy ranges outside those where absorption takes place. The value of the refractive index n and its frequency dispersion are, thus, partially determined by absorption processes occurring in remote energy regions. The primary tasks of solid-state spectroscopy are the determination of the frequency dependence of the optical constants of the experimental system, the chemical identification of the absorbing species, and the correlation of this information with the electronic energy band structure of the system.

2. *Reflectivity of a Multilayer System*

In the simplest approximation, the electrode-electrolyte interfacial region can be described in terms of a two-phase system with a plane boundary, consisting of a transparent solution phase and an absorbing, film-free substrate, both of which are assumed to be homogeneous, isotropic, and semi-infinite in extent. A three-phase stratified system with plane parallel boundaries is formed when a homogeneous, isotropic thin film of uniform thickness is generated on the substrate surface. The film may consist of specifically adsorbed

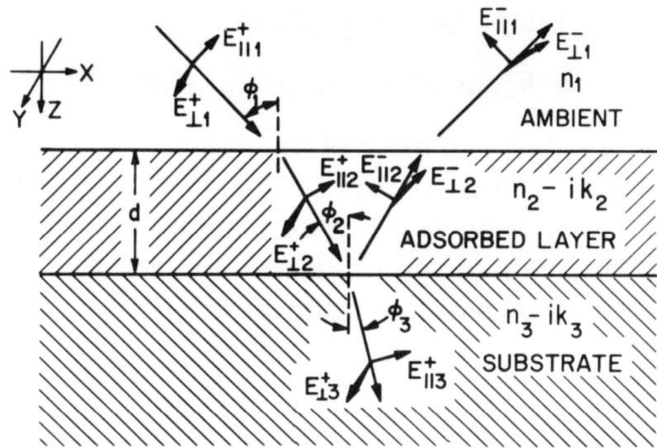

Fig. 1. Electric-field vectors in a three-phase system.

ions; adsorbed chemical reactants, intermediates or products; a surface film such as an oxide; or a very thin surface layer of the substrate material itself, whose optical properties have been perturbed by the application of an external stress or field or by interaction with an adsorbed species. Such a system is illustrated schematically in Fig. 1.

In Fig. 1, the electric field strength of the incident plane wave propagating in the positive z direction through the transparent ambient phase, (1), is denoted by $\mathbf{E}_1{}^+$. Within the thin film phase, (2), $\mathbf{E}_2{}^+$ represents the instantaneous local resultant sum of the electric field vectors of all the waves propagating in the positive z direction; $\mathbf{E}_2{}^-$ is the corresponding resultant of the multiply-reflected negative-going waves (cf. Heavens (143)). In the semiinfinite substrate phase (3), there are no reflected waves; the transmitted waves have the local resultant field $\mathbf{E}_3{}^+$. The amplitudes of the resultants in any arbitrary phase (j) are denoted by $E_j{}^{\circ +}$ and $E_j{}^{\circ -}$.

A. Fresnel Coefficients and Reflectivity Relations. The Fresnel reflection coefficient of the interface of two contiguous phases is defined as the ratio of the complex amplitudes of the electric-field vectors of the reflected and incident waves. Its value is a function of the polarization of the incident beam and the angle of incidence. For obliquely incident radiation it is useful to define† for any phase

† The quantities $\hat{n}$, $\hat{\varepsilon}$, and ξ employed in this article are complex conjugates of the analogous quantities defined for an $e^{-i\omega t}$ field time-dependence. Since the field is attenuated in an absorbing medium, it can readily be shown that ξ must lie in the fourth quadrant, i.e., Re $\xi \geq 0$ and Im $\xi_j \leq 0$.

j, the angular-dependent quantity, ξ_j, as

$$\xi_j = \hat{n}_j \cos \phi_j = (\hat{n}_j^2 - n_1^2 \sin^2 \phi_1)^{1/2} \tag{11}$$

where ϕ_1 and n_1 are the *real* angle of incidence and index of refraction of the transparent ambient phase, (1), while $\hat{n}_j$ is the *complex* refractive index of phase j. If phase j is absorbing, the angle ϕ_j is complex. The Fresnel coefficients of the interface between two phases j and k for perpendicular and parallel polarized radiation are then given by (269)

$$r_{\perp jk} = \frac{E_{\perp j}^{\circ -}}{E_{\perp j}^{\circ +}} = \frac{\mu_k \xi_j - \mu_j \xi_k}{\mu_k \xi_j + \mu_j \xi_k} \tag{12a}$$

$$r_{\parallel jk} = \frac{E_{\parallel j}^{\circ -}}{E_{\parallel j}^{\circ +}} = \frac{\hat{\varepsilon}_k \xi_j - \hat{\varepsilon}_j \xi_k}{\hat{\varepsilon}_k \xi_j + \hat{\varepsilon}_j \xi_k} \tag{12b}$$

At normal incidence ($\phi_1 = 0°$), $r_{\perp jk} = -r_{\parallel jk}$.

The Fresnel coefficients can be expressed in polar form as

$$r_{\perp jk} = |r_{\perp jk}| e^{i\delta^r_{\perp jk}} \tag{13a}$$

$$r_{\parallel jk} = |r_{\parallel jk}| e^{i\delta^r_{\parallel jk}} \tag{13b}$$

where $\delta^r_{\perp jk}$, $\delta^r_{\parallel jk}$ represent the absolute phase changes on reflection for perpendicular- and parallel-polarized radiation, respectively. If the time-dependence of the field is taken as $e^{+i\omega t}$ as in Eq. 1, the phase changes on reflection are always *positive*, such that $0 \leq \delta^r_{jk} \leq 2\pi$. These quantities are related to the real and imaginary parts of the Fresnel coefficients by

$$\delta^r_{jk} = \arg (r_{jk}) = \tan^{-1} \left[\frac{\mathrm{Im}\,(r_{jk})}{\mathrm{Re}\,(r_{jk})} \right] \tag{14}$$

and are best evaluated with an electronic computer, taking care to specify the magnitude and sign of *both* arguments. The subscripts denoting the beam polarization are not indicated in Eq. 14 because the forms of the relations are identical for both modes. In applying this and subsequent relations where the polarization mode is not specified, the use of Fresnel coefficients, reflectivities, and phase changes corresponding to the appropriate plane of polarization is implied.

The reflectivity of the interface jk is given by

$$R_{jk} = |r_{jk}|^2 = r_{jk} \cdot r^*_{jk} \tag{15}$$

where r^*_{jk} is the complex conjugate of r_{jk}. Figure 2 illustrates the family of normal incidence isoreflectance circles for the interface

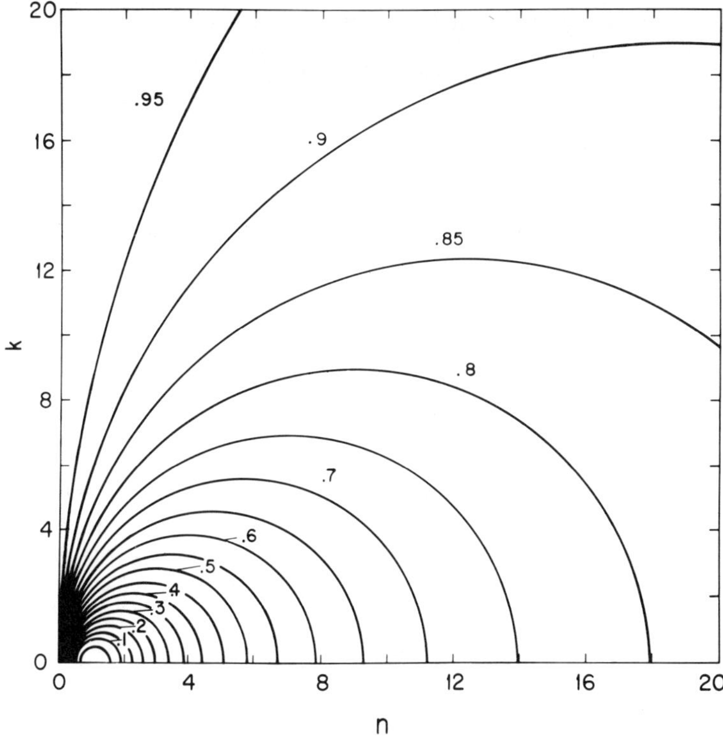

Fig. 2. Relation between normal-incidence reflectivity and optical constants.

between air and an absorbing material (210). For each curve, there is an infinite number of pairs of the optical constants, n and k, which yield the same value of R.

The equivalent Fresnel reflection coefficient of a multiphase system is defined by analogy to the two-phase system as the ratio of the amplitudes of the incident and reflected waves in the transparent first phase. For a three-phase system: ambient (1)/thin film (2)/substrate (3), the equivalent Fresnel coefficient is

$$r_{123} = \frac{E_1^{\circ -}}{E_1^{\circ +}} = \frac{r_{12} + r_{23}e^{-2i\beta}}{1 + r_{12}r_{23}e^{-2i\beta}} \tag{16}$$

where β, a complex quantity representing the change in phase and amplitude of the beam during one traversal of the thin film phase of thickness d, is given by

$$\beta = \frac{2\pi\hat{n}_2 d \cos \phi_2}{\lambda} \tag{17}$$

The reflectivity of the three-phase system is given by

$$R_{123} = |r_{123}|^2 = \frac{\{R_{12} + R_{23}e^{4\,\mathrm{Im}\,(\beta)} + 2R_{12}^{1/2}R_{23}^{1/2}e^{2\,\mathrm{Im}\,(\beta)} \cos[\delta_{12}^r - \delta_{23}^r + 2\,\mathrm{Re}\,(\beta)]\}}{\{1 + R_{12}R_{23}e^{4\,\mathrm{Im}\,(\beta)} + 2R_{12}^{1/2}R_{23}^{1/2}e^{2\,\mathrm{Im}\,(\beta)} \cos[\delta_{12}^r + \delta_{23}^r - 2\,\mathrm{Re}\,(\beta)]\}} \tag{18}$$

Since the reflectivity of a film-free two-phase system, R_{13}, is identical to that of a three-phase system with $d = 0$, we can write

$$R_{13} = R(0) \tag{19a}$$

$$R_{123} = R(d) \tag{19b}$$

Although spectrophotometric measurement of the absolute reflectivities, $R(0)$ and $R(d)$, is very difficult when the reflecting substrate surface is enclosed in an experimental cell, it is possible to make accurate measurements of the ratio $R(d)/R(0)$ since various optical system errors occur as common factors and cancel. However, even for our idealized model of the electrode-solution interface, the exact theoretical equations for $R(d)$ and $R(0)$ given above are too complicated to let us discern precisely how the dielectric constant and film thickness of an intermediate phase affect the reflectivity. In many surface studies, however, the thickness of the thin film phase is much less than the wavelength of the incident radiation. It is then possible to make first-order approximations, analogous to those employed by Drude (71,72) for the ellipsometry of "transition layers," which lead to simplified general relations for the reflectivity ratio, $R(d)/R(0)$.

B. Linear Approximation Theory. When $d \ll \lambda$, the ratios of the Fresnel coefficients for three-phase and two-phase systems are given, to first order in d/λ, by (203)

$$\frac{r_{\perp 123}}{r_{\perp 13}} = 1 - \frac{4\pi i\, dn_1 \cos\phi_1}{\lambda}\left(\frac{\mu_3\hat{\varepsilon}_2 \cos^2\phi_2 - \mu_2\hat{\varepsilon}_3 \cos^2\phi_3}{\mu_3\varepsilon_1 \cos^2\phi_1 - \mu_1\hat{\varepsilon}_3 \cos^2\phi_3}\right) \tag{20a}$$

$$\frac{r_{\parallel 123}}{r_{\parallel 13}} = 1 - \frac{4\pi i\, dn_1 \cos\phi_1}{\lambda}\left(\frac{\mu_2\hat{\varepsilon}_3 \cos^2\phi_2 - \mu_3\hat{\varepsilon}_2 \cos^2\phi_3}{\mu_1\hat{\varepsilon}_3 \cos^2\phi_1 - \mu_3\varepsilon_1 \cos^2\phi_3}\right) \tag{20b}$$

The corresponding reflectivity ratios are

$$\frac{R_\perp(d)}{R_\perp(0)} = 1 + \frac{8\pi\, dn_1 \cos\phi_1}{\lambda} \operatorname{Im}\left(\frac{\mu_3\hat{\varepsilon}_2 \cos^2\phi_2 - \mu_2\hat{\varepsilon}_3 \cos^2\phi_3}{\mu_3\varepsilon_1 \cos^2\phi_1 - \mu_1\hat{\varepsilon}_3 \cos^2\phi_3}\right) \tag{21a}$$

and

$$\frac{R_\parallel(d)}{R_\parallel(0)} = 1 + \frac{8\pi\, dn_1 \cos\phi_1}{\lambda} \operatorname{Im}\left(\frac{\mu_2\hat{\varepsilon}_3 \cos^2\phi_2 - \mu_3\hat{\varepsilon}_2 \cos^2\phi_3}{\mu_1\hat{\varepsilon}_3 \cos^2\phi_1 - \mu_3\varepsilon_1 \cos^2\phi_3}\right) \tag{21b}$$

It is convenient to express these results in terms of the normalized reflectivity change defined as

$$\frac{\Delta R}{R} = \frac{R(d) - R(0)}{R(0)} = \frac{R(d)}{R(0)} - 1 \tag{22}$$

Assuming $\mu_1 = \mu_2 = \mu_3 = 1$ in the optical frequency range, these expressions simplify to

$$\left(\frac{\Delta R}{R}\right)_\perp = \frac{8\pi\, dn_1 \cos\phi_1}{\lambda} \operatorname{Im}\left(\frac{\hat{\varepsilon}_2 - \hat{\varepsilon}_3}{\varepsilon_1 - \hat{\varepsilon}_3}\right) \tag{23a}$$

and

$$\left(\frac{\Delta R}{R}\right)_\parallel = \frac{8\pi\, dn_1 \cos\phi_1}{\lambda} \operatorname{Im}\left\{\left(\frac{\hat{\varepsilon}_2 - \hat{\varepsilon}_3}{\varepsilon_1 - \hat{\varepsilon}_3}\right)\left[\frac{1 - (\varepsilon_1/\hat{\varepsilon}_2\hat{\varepsilon}_3)(\hat{\varepsilon}_2 + \hat{\varepsilon}_3)\sin^2\phi_1}{1 - (1/\hat{\varepsilon}_3)(\varepsilon_1 + \hat{\varepsilon}_3)\sin^2\phi_1}\right]\right\} \tag{23b}$$

At normal incidence these further reduce to

$$\frac{\Delta R}{R} = \frac{8\pi\, dn_1}{\lambda} \operatorname{Im}\left(\frac{\hat{\varepsilon}_2 - \hat{\varepsilon}_3}{\varepsilon_1 - \hat{\varepsilon}_3}\right) \equiv \frac{8\pi\, dn_1}{\lambda} \operatorname{Im}\left(\frac{\hat{\varepsilon}_2 - \varepsilon_1}{\varepsilon_1 - \hat{\varepsilon}_3}\right) \tag{24}$$

This linear approximation theory is valid for both internal and external specular reflection modes and for *all* angles of incidence. The normalized reflectivity changes are expressed in terms of the fundamental parameters, μ_j and $\hat{\varepsilon}_j$, which characterize the response of the system to elementary excitations by electromagnetic radiation, and the simple form of the relationships enables considerable physical insight to be gained into the optical properties of very thin film systems. The applications of this theory to a number of systems of interest in surface chemistry and physics have been described in detail (203).

C. Specular Reflection. The dependence of the specular (external) reflectivity of a film-free absorbing material (in air) on its optical constants (n, k) and the angle of incidence and polarization of an incident monochromatic light beam is illustrated in Fig. 3

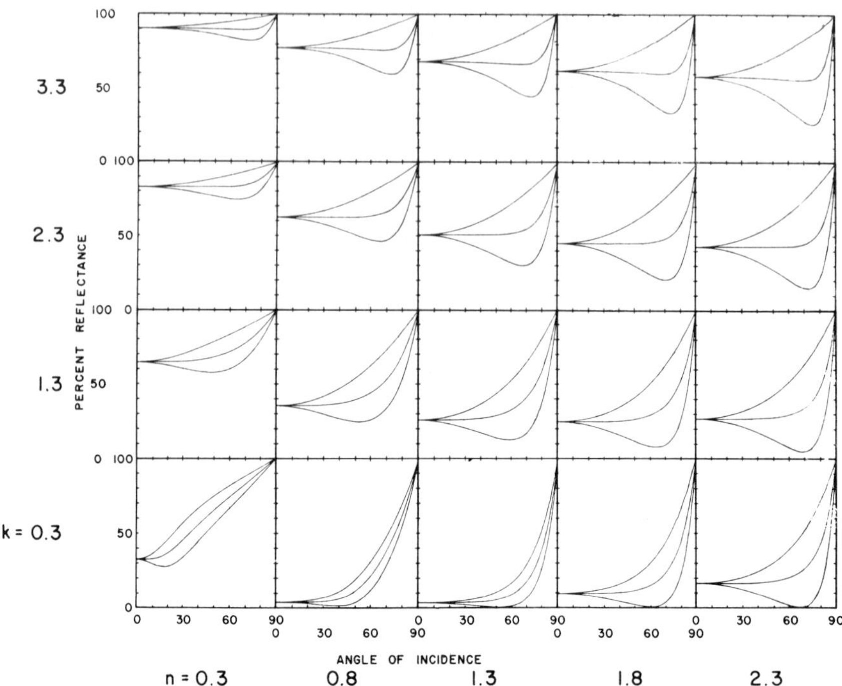

Fig. 3. Reflectivity vs. angle of incidence curves calculated using the optical constants shown by the large numbers. In each square, the upper curve is $R_{\perp}$, the lower curve is $R_{\parallel}$ and the center curve is the arithmetic average for natural, unpolarized radiation: $R_N = (R_{\perp} + R_{\parallel})/2$. After Hunter (149).

(149). For a film-free surface, $R_{\perp}$ increases monotonically from its minimum value at $\phi_1 = 0°$ to unity when $\phi_1 = 90°$, whereas $R_{\parallel}$ exhibits a minimum at the pseudo-Brewster angle. The arithmetic average reflectivity for natural, unpolarized radiation, $R^N = (R_{\perp} + R_{\parallel})/2$, remains nearly constant to large values of ϕ_1.

Figure 4*a* illustrates the angular dependence of the reflectivity of a moderately reflecting metal covered with a strongly absorbing surface film approximately one monomolecular layer in thickness ($d/\lambda = 1.0 \times 10^{-3}$). The optical constants of this system are typical of a metal such as platinum, covered by a monolayer of a semiconducting oxide film and immersed in an aqueous solution, at wavelengths in the visible region. The variation of $\Delta R/R$ as a function of ϕ_1 and polarization is shown in Fig. 4*b* for the same interface. With oblique incidence and perpendicular polarization, the film detection

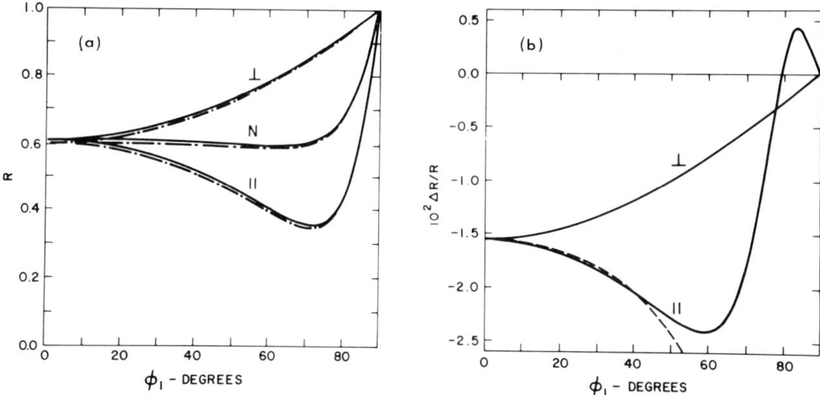

Fig. 4. (*a*) Specular reflectivity of a bare metal substrate (——) and of a substrate covered by ca. one monolayer of an absorbing film (–·–·) as a function of the angle of incidence and polarization of the incident beam. The optical parameters are: $n_1 = 1.333$; $n_2 = 3.0$, $k_2 = 1.5$; $n_3 = 2.0$, $k_3 = 4.0$; $d/\lambda = 1.0 \times 10^{-3}$. (*b*) The normalized change in reflectivity $\Delta R/R$ from *a*, produced by ca. one monolayer of an absorbing film ($d/\lambda = 1.0 \times 10^{-3}$). For perpendicular polarization, $\Delta R/R$ varies as $\cos \phi_1$. For parallel polarization, the change can be approximated by a $1/\cos \phi_1$ variation for small ϕ_1 (the dashed curve). After McIntyre and Aspnes (203).

sensitivity decreases monotonically with $\cos \phi_1$ from its maximum value at normal incidence to zero at grazing incidence (cf. Eq. 23a). With parallel-polarized radiation, the variation of $\Delta R/R$ is more complicated. In the above example, to a good approximation (cf. Fig. 4*b*), $|\Delta R/R|_{\parallel}$ increases with ϕ_1 as $(1/\cos \phi_1)$ in the range $0° \leq \phi_1 \leq 45°$. At high values of ϕ_1, $(\Delta R/R)_{\parallel}$ exhibits a minimum, a maximum, and a sign change. Such anomalous behavior does not occur for all sets of optical constants, but measurements at angles of incidence greater than the pseudo-Brewster angle are subject to large experimental errors and should be interpreted with discretion.

The range of validity of the linear-approximation theory is illustrated in Fig. 5 for the same interface, together with that for an expansion to second order in d/λ. It is evident from the inset that the linear approximation is very good indeed for surface films in the monolayer thickness range. For parallel polarization, the linear and quadratic plots are virtually coincident.

It is apparent, from the form of Eqs. 23a and 23b, that the differential reflection spectra of surface films on absorbing substrates are not transmission-like but are strongly influenced by the optical

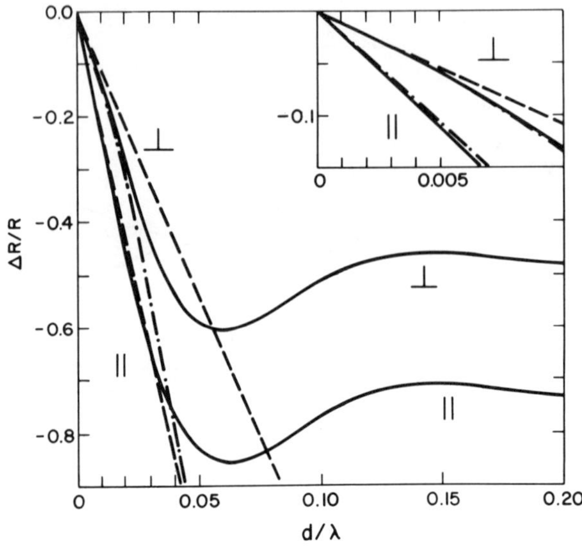

Fig. 5. Comparison of linear (– – –) and quadratic (–·–·) approximations of $\Delta R/R$ to the exact (——) expression for an absorbing film on a metal substrate, for parallel and perpendicular polarizations and an angle of incidence $\phi_1 = 45°$. The optical constants are the same as for Fig. 4. The portion of the figure near the origin is expanded in the inset. After McIntyre and Aspnes (203).

properties of the substrate and the d/λ scaling factor. The film absorption bands themselves are further distorted by anomalous dispersion of the real part of the film refractive index, as is the case for internal reflection spectroscopy (133).

For a weakly absorbing film on a strongly absorbing substrate, $\varepsilon_2'' \ll \varepsilon_3''$, and thus

$$\operatorname{Im} \frac{\hat{\varepsilon}_2 - \hat{\varepsilon}_3}{\varepsilon_1 - \hat{\varepsilon}_3} \approx -\frac{\varepsilon_3''({n_2}^2 - {k_2}^2 - {n_1}^2)}{(\varepsilon_1 - \varepsilon_3')^2 + \varepsilon_3''^2} \tag{25}$$

If $k_2 \ll n_2$, it is evident that $(\Delta R/R)_\perp$ is very small and insensitive to a variation of k_2 at all angles of incidence. The reflectivity change as a function of wavelength is determined largely by the optical properties of the substrate and little information can be gained concerning the absorptive properties of the film. For parallel polarization, however, the term $[1 - (1/\hat{\varepsilon}_3)(\varepsilon_1 + \hat{\varepsilon}_3) \sin^2 \phi_1]$, which appears in the denominator of Eq. 23b, approaches zero at high angles of incidence, with the result that $(\Delta R/R)_\parallel \gg (\Delta R/R)_\perp$ for such a system. Since $\cos \phi_1 = 0$ at $\phi_1 = 90°$, it is evident that

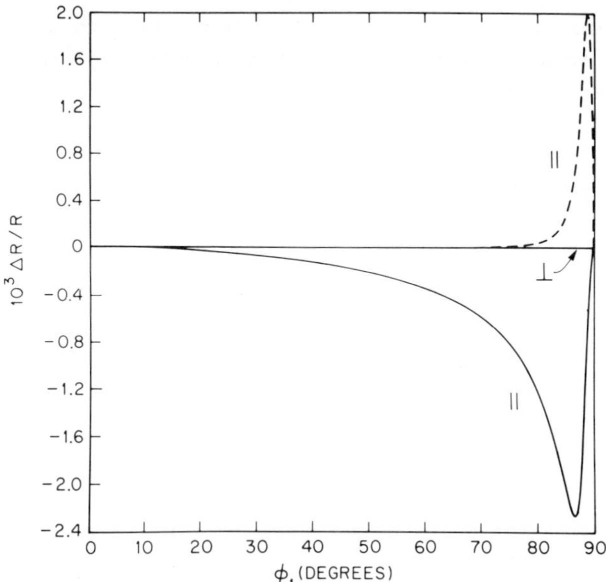

Fig. 6. Normalized infrared reflectivity change produced by deposition of transparent (– – –) and absorbing (——) films, ca. one monolayer in thickness, on the surface of a highly reflecting substrate, as a function of the angle of incidence and polarization of the incident beam. The optical parameters are: $n_1 = 1.0$; $n_2 = 1.3$, $k_2 = 0.0$ (– – –), $k_2 = 0.1$ (——); $n_3 = 3.0$, $k_3 = 30.0$; $d/\lambda = 1.0 \times 10^{-4}$.

$(\Delta R/R)_{\parallel}$ must exhibit a maximum or a minimum at very high angles of incidence. This is indeed the case, as illustrated in Fig. 6, for an adsorbed layer of monolayer thickness in spectral regions where the film is transparent ($k_2 = 0$) and absorbing ($k_2 > 0$), respectively. The optical constants in this example are typical of those encountered in the infrared spectroscopy of adsorbed layers on metallic substrates. This phenomenon is due, in part, to the relative magnitudes of the electric field strengths at the mirror surface, and is discussed further in Section II.3.

Features of the internal and external reflection spectra of very thin films on transparent substrates are discussed in reference (203).

D. Variation of the Reflectivity with the Optical Constants of Film and Substrate. Owing to the complex form of the reflectivity relations, it is often difficult to predict intuitively what changes will be produced by an alteration of the optical constants of the film or substrate. It is helpful, therefore, to examine the effects that result from a variation of the constants individually in a typical electrochemical system.

Figures 7*a* and 7*b* illustrate the computed variation of $\Delta R/R$ for a three-phase system, with $d/\lambda = 1.0 \times 10^{-3}$ and $\phi_1 = 45°$, as a function of the thin film optical constants, n_2 and k_2. The substrate optical constants are typical of a metal such as platinum in the visible wavelength region. Both positive *and* negative values of $\Delta R/R$ are observed in these plots. In Fig. 7*a*, a sign reversal of $\Delta R/R$ occurs at low values of n_2. While values of $n_2 < 1$ are not common, they do occur in regions of strong anomalous dispersion and are often observed for metallic films (136). In Fig. 7*b*, $\Delta R/R$ is not a monotonic function of k_2 when $k_2 \leq 5.0$. It is also evident that $\Delta R/R$ is relatively insensitive to the value of k_2. This implies that weak absorption bands of a monomolecular film will be difficult to resolve by specular reflection spectroscopy. The conditions for sign reversal and intersection of the plots of $(\Delta R/R)_\perp$ and $(\Delta R/R)_\parallel$ can be readily deduced from the linear approximation relations, Eqs. 23a and 23b.

Figures 8*a* and 8*b* illustrate the corresponding variation of $\Delta R/R$

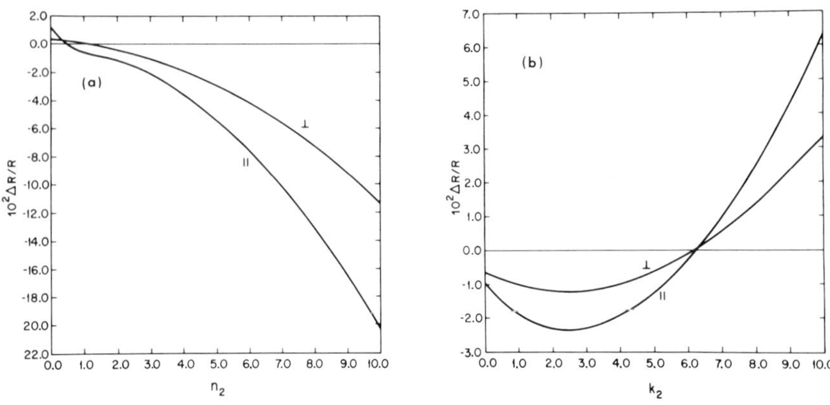

Fig. 7. Variation of the normalized UV-visible reflectivity change of a moderately reflecting substrate covered by ca. one monolayer of an absorbing film, as a function of the index of refraction and extinction coefficient of the film. (*a*) Optical parameters: $n_1 = 1.333$; $k_2 = 1.5$; $n_3 = 2.0$, $k_3 = 4.0$; $d/\lambda = 1.0 \times 10^{-3}$; $\phi_1 = 45.0°$. (*b*) Optical parameters: $n_1 = 1.333$; $n_2 = 3.0$; $n_3 = 2.0$, $k_3 = 4.0$; $d/\lambda = 1.0 \times 10^{-3}$; $\phi_1 = 45.0°$.

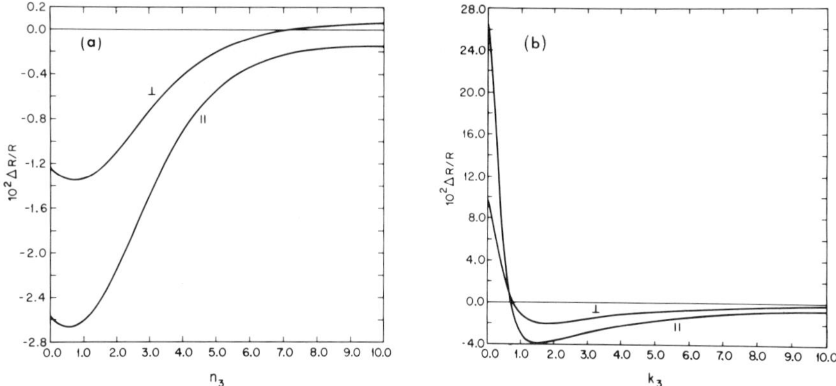

Fig. 8. Variation of the normalized UV-visible reflectivity change of an absorbing substrate, covered by ca. one monolayer of an absorbing film, as a function of the index of refraction and extinction coefficient of the substrate. (*a*) Optical parameters: $n_1 = 1.333$; $n_2 = 3.0$, $k_2 = 1.5$; $k_3 = 4.0$; $d/\lambda = 1.0 \times 10^{-3}$; $\phi_1 = 45.0°$. (*b*) Optical parameters: $n_1 = 1.333$; $n_2 = 3.0$, $k_2 = 1.5$; $n_3 = 2.0$; $d/\lambda = 1.0 \times 10^{-3}$; $\phi_1 = 45.0°$.

as a function of the substrate optical constants, n_3 and k_3. The optical constants n_2 and k_2 are typical of a semiconducting thin film. In Fig. 8*a*, there are two values of n_3 which yield the same value of $\Delta R/R$ in the region $0 \leq n_3 \leq 1.5$. $(\Delta R/R)_\perp$ exhibits a zero-crossing when

$$\frac{\varepsilon_2'' - \varepsilon_3''}{\varepsilon_2' - \varepsilon_3'} = -\frac{\varepsilon_3''}{\varepsilon_1 - \varepsilon_3'} \tag{26}$$

At high values of n_3, the computed value of $\Delta R/R$ becomes insensitive to the variation of n_3. In Fig. 8*b*, it is apparent that deposition of a strongly absorbing thin film on a dielectric substrate ($k_3 = 0$) produces a large reflectivity increase. As k_3 increases, $\Delta R/R$ falls and becomes negative. Again, there are two values of k_3 which give the same value of $\Delta R/R$ in the region of $0.8 \leq k_3 \leq 10$. At high values of k_3, the film detection sensitivity is low and insensitive to variations of k_3.

It should be emphasized that the results shown in Figs. 7 and 8 are representative only of the particular sets of optical constants employed and an angle of incidence of 45°. Strikingly different results may be observed for different values of ϕ_1 and/or other sets of constants. Equations 23a and 23b are sometimes useful for estimating the values of $\Delta R/R$ without use of a computer.

E. Reference Reflectivities for Differential Reflection Spectroscopy. In infrared reflection studies of adsorbed molecules on metal surfaces, the characteristic absorption bands are narrow compared to those in the UV-visible region. It is, thus, often useful to compare the reflectivity of the film-covered substrate in the region of a film absorption band, to the background reflectivity of the three-phase system in a nearby wavelength region where the thin film is transparent. It can usually be assumed that n_2 varies only slowly with λ outside the band. Such a procedure has been employed by Greenler (112,113) to analyze the characteristic features of infrared reflections spectra. His "absorption factor" is defined, according to the notation used in the present chapter, by

$$A_G = \frac{[R(d)]_{k_2=0} - [R(d)]_{k_2>0}}{[R(d)]_{k_2=0}}$$

$$= \frac{R(0)}{[R(d)]_{k_2=0}} \left[\left(\frac{\Delta R}{R} \right)_{k_2=0} - \left(\frac{\Delta R}{R} \right)_{k_2>0} \right] \tag{27}$$

In general, the ratio $R(0)/[R(d)]_{k_2=0}$ is close to unity. Thus in the example illustrated in Fig. 6, the depth of the film absorption band in a spectrum of $R_{\parallel}(d)$ versus $\hbar\omega$ can be estimated closely from the distance between the dashed and solid curves.

In the UV-visible wavelength region, the above procedure is not generally applicable, since the reflectivity change produced by deposition of a transparent surface film is comparable to that for an absorbing film for *all* angles of incidence, as illustrated in Fig. 9. Near normal incidence, the change in energy absorption by the substrate *per se* is greater than the energy loss in the film for the example illustrated. This behavior differs significantly from that in the infrared region (cf. Fig. 5), where the change in absorption by the substrate is negligibly small except at very high angles of incidence. Further, film absorption bands tend to be broad in the UV-visible range. The background reflectivity of the film-free substrate thus affords the most useful reference for this wavelength region. The latter convention is employed exclusively in this chapter.

It is important to note these differences in choice of reference reflectivity when comparing computations of the normalized reflectivity change, $\Delta R/R$, since the notations used are often confusingly similar.

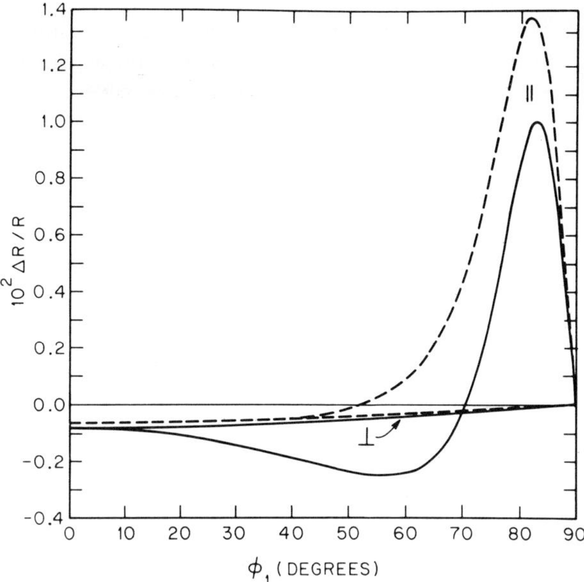

Fig. 9. Normalized UV-visible reflectivity change produced by deposition of transparent (– – –) and absorbing (——) films, ca. one monolayer in thickness, on the surface of a moderately reflecting substrate, as a function of the angle of incidence and polarization of the incident beam. The optical parameters are: $n_1 = 1.0$; $n_2 = 1.3$, $k_2 = 0.0$ (– – –), $k_2 = 0.1$ (——); $n_3 = 2.0$, $k_3 = 4.0$; $d/\lambda = 1.0 \times 10^{-3}$.

3. *Energy Absorption in Thin Film Systems*

When electromagnetic radiation is incident on a multiphase-stratified system, conservation of energy requires that the sum of the energy fluxes in the reflected and transmitted beams, plus the sum of the rates of energy dissipation in any absorbing phases, be equal to the energy flux in the incident beam. For a beam of unit intensity incident on a bare substrate

$$1 = R + A_s + T \tag{28a}$$

where R and T are the reflected and transmitted beam intensities in the transparent ambient phase, and A_s is the rate of energy absorption by the substrate phase. For a film-covered substrate

$$1 = R' + A_f + A_s' + T' \tag{28b}$$

where the primed symbols denote terms analogous to those in Eq.

28a and A_f is the rate of energy loss in the surface film. For a semi-infinite absorbing substrate, $T = T' = 0$, and we have from Eqs. 28a and 28b

$$\Delta R + A_f + \Delta A_s = 0 \tag{29}$$

The form of the differential reflection spectrum, ΔR versus $\hbar\omega$, is thus governed by two factors—absorption in the thin film *per se*, and the change in absorption in the bulk substrate resulting from deposition of the film on its surface. The magnitudes of these factors can be estimated from linear approximation theory. From Eq. 23, we find for normal incidence

$$\frac{\Delta R}{R} = \frac{8\pi d\, n_1}{\lambda}\left[\frac{(\varepsilon_1 - \varepsilon_2')\varepsilon_3'' - (\varepsilon_1 - \varepsilon_3')\varepsilon_2''}{(\varepsilon_1 - \varepsilon_3')^2 + \varepsilon_3''^2}\right] \tag{30}$$

Since the energy absorption in phase j is proportional to ε_j'' (cf., Eq. 31 below), the relative contributions of ΔA_s and A_f to the fractional reflectivity change are clearly discernible.†

Two limiting cases can be distinguished. (i) If the surface film is transparent or weakly absorbing, $A_f \approx 0$ and $\Delta R \approx -\Delta A_s$. In this case, the shape of the differential reflection spectrum is determined primarily by the absorptive properties of the substrate, as previously mentioned in Section II.2c. This is illustrated in Fig. 7*b* by the sections of the curves near the $k_2 = 0$ axis. (ii) If the substrate reflectivity is very high (i.e., $R(0) \approx 1$), very little energy is dissipated in this phase and hence $\Delta A_s \approx 0$. The spectrum of ΔR versus $\hbar\omega$ then resembles a distorted transmission spectrum of the thin film material, as is evident from Fig. 8*b* for high values of k_3.

In intermediate cases where the film is absorbing but the substrate is only a moderately good reflector ($R(0) \approx 0.5$), the two effects are not readily separable. The situation is further complicated by the inclusion of a factor due to the variable background reflectivity of the substrate in experimental spectra of $\Delta R/R$ versus $\hbar\omega$. In order to ascertain the true absorption spectrum of the thin film material *per se*, it is first necessary to compute the characteristic optical constants of the film. Methods for determining these constants are discussed in Section III.

A. Electric Field Strengths at a Reflecting Interface. Further insight into the physics of the absorption processes in the

† Comparison of Eqs. 27 and 30 reveals that the energy loss in the film is not given rigorously by Eq. 27 unless $k_2^2 \ll (n_1^2 - n_2^2)$.

thin film and substrate phases can be gained by consideration of the local field strengths in the reflected and transmitted waves. When electromagnetic radiation traverses an absorbing phase j, the rate of energy loss from the wave per unit volume of the medium is, according to Poynting's theorem (152,270)

$$-\dot{W}_j = \sigma_j(\nu)\langle E_j^2\rangle = \frac{c}{4\pi} n_j\alpha_j\langle E_j^2\rangle = \frac{\omega}{4\pi} \varepsilon_j''\langle E_j^2\rangle \tag{31}$$

where $\sigma_j(\nu)$ is the electrical conductivity of the phase at the optical frequency ν and $\langle E_j^2\rangle$ is the local mean square electric field strength, defined by

$$\langle \mathbf{E}_j^2\rangle = \langle[\mathrm{Re}\ (\mathbf{E}_j)]^2\rangle = \tfrac{1}{2}|\mathbf{E}_j^\circ|^2 \tag{32}$$

Since the energy dissipation in a very thin absorbing film is small, there is little attenuation of the incident field strength within the film. It is instructive, therefore, to consider the properties of the electric field near a reflecting film-free surface.

When light is specularly reflected from the interface of a two-phase system, the electric fields of the coherent incident and reflected plane waves sum vectorially to produce a standing-wave electric field in the incident phase. The boundary conditions of electromagnetic theory require that the components of the electric field which are tangential to the surface be continuous across the interface. The normal components of the electric displacement, $\mathbf{D} = \hat{\varepsilon}\mathbf{E}$, must also be continuous. These conditions define the magnitudes of the field strengths at the reflecting interface.

Relations for the mean-square electric field strengths in a two-phase system have been given by Hansen (128) for $e^{-i\omega t}$ field time-dependence.† In the Nebraska convention (214), the ratio of the mean square electric field strength in the standing wave in phase 1 to that in the incident plane wave becomes, for perpendicular polarization,

$$\frac{\langle E_{\perp 1}^2\rangle}{\langle E_{\perp 1}^{+2}\rangle} = (1 + R_\perp) + 2R_\perp^{1/2} \cos\left[\delta_1^r + 4\pi\left(\frac{z}{\lambda}\right)\xi_1\right] \tag{33}$$

where $r_\perp = R_\perp^{1/2} \exp(i\delta_\perp^r)$ and z is negative in phase 1. For parallel polarization, the field-strength ratios for the components tangential

† Exact relations for the electric-field strengths in three-phase and multi-phase systems have also been derived by Hansen (128) but are too complicated in form to present here.

and normal to the surface plane ($z = 0$) are, respectively,

$$\frac{\langle E^2_{\parallel 1x}\rangle}{\langle E^{+2}_{\parallel 1}\rangle} = \cos^2 \phi_1 \left\{ (1 + R_\parallel) - 2R_\parallel^{1/2} \cos \left[\delta^r_\parallel + 4\pi\left(\frac{z}{\lambda}\right)\xi_1 \right] \right\} \tag{34a}$$

$$\frac{\langle E^2_{\parallel 1z}\rangle}{\langle E^{+2}_{\parallel 1}\rangle} = \sin^2 \phi_1 \left\{ (1 + R_\parallel) + 2R_\parallel^{1/2} \cos \left[\delta^r_\parallel + 4\pi\left(\frac{z}{\lambda}\right)\xi_1 \right] \right\} \tag{34b}$$

where $r_\parallel = R_\parallel^{1/2} \exp (i\delta_\parallel{}^r)$. When $\phi_1 = 45°$, the mean square value of the total parallel field in phase 1, $\langle E^2_{\parallel 1}\rangle = \langle E^2_{\parallel 1x}\rangle + \langle E^2_{\parallel 1z}\rangle$, is constant, independent of z/λ.

At the surface of a highly reflecting metal (e.g., in the infrared) the field-strength ratios in phase 1 can be approximated by (89,130)

$$\frac{\langle E^2_{\perp 1}\rangle}{\langle E^{+2}_{\perp 1}\rangle} \approx \frac{4 \cos^2 \phi_1}{k_3{}^2} \tag{35a}$$

$$\frac{\langle E^2_{\parallel 1x}\rangle}{\langle E^{+2}_{\parallel 1}\rangle} \approx \frac{4}{k_3{}^2} \tag{35b}$$

$$\frac{\langle E^2_{\parallel 1z}\rangle}{\langle E^{+2}_{\parallel 1}\rangle} \approx 4 \sin^2 \phi_1 \tag{35c}$$

These relations are useful for estimating the magnitude of ΔR_a. It should be noted, however, that Eqs. 35 are only valid when k_3 is very large and $R \approx 1$. They yield inaccurate results for metals of lower reflectivity (e.g., in the UV-visible region) and the exact relations, Eqs. 33 and 31, should then be employed to calculate the field strengths.

Using Eqs. 30 and 31 (or Eqs. 35 when applicable), we can estimate the field strengths in the surface film and determine the contribution by absorption in this phase to the total reflectivity change produced by its deposition on the substrate. The decrease in energy flux in the reflected light beam due to absorption in the surface film is

$$\Delta I_a = -\frac{c}{4\pi} \frac{d}{\cos \phi_1} \alpha_2 n_2 \langle E^2_2\rangle \tag{36}$$

where $(d/\cos \phi_1)$ is the volume of the film sampled by a light beam of unit cross-sectional area. The energy flux in the incident beam is

$$I^\circ = \frac{c}{4\pi} n_1 \langle E_1^{+2} \rangle \tag{37}$$

In the thin film phase, since $d \ll \lambda$, both the energy loss and phase change are small; as a result the field is only slightly attenuated. Hence to a good approximation, in the region: $-d \leq z \leq 0$,

$$\langle E_{\perp 1}^2 \rangle \approx \langle E_{\perp 2}^2 \rangle \approx \langle E_{\perp 3}^2 \rangle \tag{38a}$$

$$\langle E_{\parallel 1x}^2 \rangle \approx \langle E_{\parallel 2x}^2 \rangle \approx \langle E_{\parallel 3x}^2 \rangle \tag{38b}$$

$$\varepsilon_1{}^2 \langle E_{\parallel 1z}^2 \rangle \approx |\hat{\varepsilon}_2|^2 \langle E_{\parallel 2z}^2 \rangle \approx |\hat{\varepsilon}_3|^2 \langle E_{\parallel 3z}^2 \rangle \tag{38c}$$

For perpendicular polarization, the component of the reflectivity change due to absorption in the film is

$$\begin{aligned} \Delta R_{\perp a} = \frac{\Delta I_{\perp a}}{I_\perp{}^\circ} &\approx -\frac{\alpha_2 n_2}{n_1} \left(\frac{d}{\cos \phi_1} \right) \frac{\langle E_{\perp 1}^2 \rangle}{\langle E_{\perp 1}^{+2} \rangle} \\ &\approx -\frac{2\pi \varepsilon_2''}{n_1 \cos \phi_1} \left(\frac{d}{\lambda} \right) \frac{\langle E_{\perp 1}^2 \rangle}{\langle E_{\perp 1}^{+2} \rangle} \end{aligned} \tag{39a}$$

Similarly, for parallel polarization

$$\Delta R_{\parallel a} \approx -\frac{\alpha_2 n_2}{n_1} \left(\frac{d}{\cos \phi_1} \right) \left(\frac{\langle E_{\parallel 1x}^2 \rangle}{\langle E_{\parallel 1}^{+2} \rangle} + \frac{\varepsilon_1{}^2}{|\hat{\varepsilon}_2|^2} \frac{\langle E_{\parallel 1z}^2 \rangle}{\langle E_{\parallel 1z}^{+2} \rangle} \right) \tag{39b}$$

The relative sensitivities of specular reflection and transmission spectroscopy for detecting film absorption can be compared as follows. Consider an ultrathin surface film hypothetically isolated from the substrate and suspended *in vacuo*. Since the reflectivity of such a film is virtually zero (cf. Eq. 16), $\langle E_1{}^2 \rangle \approx \langle E_1^{+2} \rangle$. Hence the transmissivity changes due to absorption in the film are, approximately,

$$\Delta T_{\perp a} \approx -\frac{\alpha_2 n_2}{n_1} \left(\frac{d}{\cos \phi_1} \right) \tag{40a}$$

$$\Delta T_{\parallel a} \approx -\frac{\alpha_2 n_2}{n_1} \left(\frac{d}{\cos \phi_1} \right) \left(\cos^2 \phi_1 + \frac{\varepsilon_1}{|\hat{\varepsilon}_2|^2} \sin^2 \phi_1 \right) \tag{40b}$$

The sensitivity ratio for equal angles of incidence and perpendicular polarization is determined solely by the field-strength ratio at the substrate surface:

$$\frac{\Delta R_{\perp a}}{\Delta T_{\perp a}} \approx \left. \frac{\langle E_{\perp 1}^2 \rangle}{\langle E_{\perp 1}^2 \rangle} \right|_{z=0} \tag{41a}$$

For external reflection, the theoretical maximum value of $\langle E_{\perp 1}^2 \rangle / \langle E_{\perp 1}^{+2} \rangle|_{z=0}$ is 1.0, but this value is approached only with transparent, weakly reflecting substrates. When UV-visible radiation is externally reflected at 45° from an aqueous-electrolyte/gold-electrode interface, $\langle E_{\perp 1}{}^2 \rangle / \langle E_{\perp 1}^{+2} \rangle|_{z=0} \approx 0.4$ (171), whereas the field-strength ratio for a highly reflecting air/metal interface in the infrared is virtually *nil* for perpendicular polarization, since $\delta_\perp{}^r \approx 180°$ (cf. Eq. 33).

For parallel polarization, the corresponding sensitivity ratio is

$$\frac{\Delta R_{\parallel a}}{\Delta T_{\parallel a}} \approx \left[\frac{\langle E_{\parallel 1x}^2 \rangle}{\langle E_{\parallel 1}^{+2} \rangle} + \frac{\varepsilon_1{}^2}{|\hat{\varepsilon}_2|^2} \frac{\langle E_{\parallel 1z}^2 \rangle}{\langle E_{\parallel 1}^{+2} \rangle}\right] \Big/ \left[\cos^2 \phi_1 + \frac{\varepsilon_1{}^2}{|\hat{\varepsilon}_2|^2} \sin^2 \phi_1\right] \tag{41b}$$

The theoretical maximum value of $\langle E_{\parallel 1x}^2 \rangle / \langle E_{\parallel 1}^{+2} \rangle|_{z=0}$ for external reflection is 1.0, whereas that of $\langle E_{\parallel 1z}^2 \rangle / \langle E_{\parallel 1}^{+2} \rangle|_{z=0}$ is $4 \sin^2 \phi_1$. Thus under optimum conditions the reflection mode has the greater sensitivity for measuring surface film absorption.

The magnitudes of the electric-field intensities at a reflecting interface and the form of differential reflection spectra depend strongly on the optical-frequency conductivity and reflectivity of the substrate material. Their characteristic features are often remarkably different in the infrared and UV-visible spectral regions and it is therefore convenient to discuss these regions separately.

B. Infrared Reflection Spectra. At low optical frequencies, the electrical conductivity of a metal approaches its dc value (σ_0(esu) $\approx$

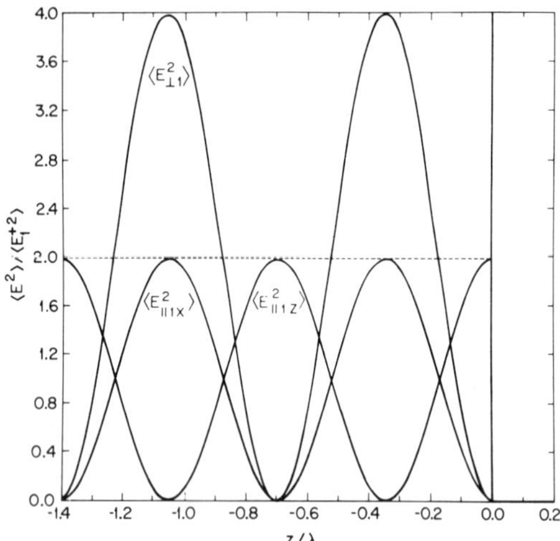

Fig. 10. Mean-square electric field strengths in the standing wave at a highly reflecting metal surface. The optical parameters are: $n_1 = 1.0$; $n_3 = 3.0$, $k_3 = 30.0$; $\phi_1 = 45.0°$.

10^{17} sec^{-1}) and its reflectivity is very close to unity. A metal with a high conductivity cannot support an electric field and, as a result of the boundary conditions, the tangential field components, $\langle E_{\perp 1}^2 \rangle$ and $\langle E_{\parallel 1x}^2 \rangle$ are near zero at the surface. Figure 10 illustrates the form of the standing-wave patterns produced for light incident on the surface of a highly reflecting metal at 45°. $\langle E_{\perp 1}^2 \rangle$ and $\langle E_{\parallel 1x}^2 \rangle$ exhibit nodes at the surface, while $\langle E_{\parallel 1z}^2 \rangle$ has an antinode. All field components are vanishingly small inside the metallic phase. Such behavior is typical of specular reflection at metals in the infrared spectral region, where in general $k_3 > 10$. From these considerations, it is evident that the sensitivity for spectroscopic detection of a surface film on a highly reflecting metal substrate is virtually *nil* at normal incidence, and at *all* angles of incidence if the radiation is perpendicularly polarized. Significantly, however, for parallel-polarized radiation the ratio $\langle E_{\parallel 1z}^2 \rangle / \langle E_{\parallel 1}^{+2} \rangle |_{z=0}$ rises to nearly 2.0 at $\phi_1 = 45°$, approaches the theoretical maximum value of 4.0 at high angles of incidence, regresses to 2.0 at the principal angle (where $\delta_{\parallel}{}^r \approx 3\pi/2$), and finally becomes zero again at grazing incidence. These features are illustrated in Fig. 11. The infrared spectrum of an adsorbed layer

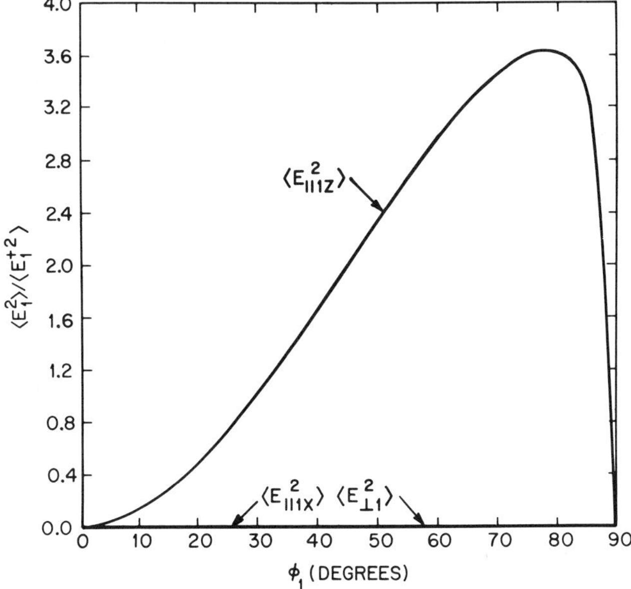

Fig. 11. Variation of the standing-wave fields at a highly reflecting surface as a function of the angle of incidence and polarization of the incident beam. The optical parameters are: $z = 0$; $n_1 = 1.0$; $n_3 = 3.0$, $k_3 = 30.0$.

whose dipoles are oriented either randomly or normal to the surface can thus be markedly enhanced by use of parallel-polarized radiation at oblique incidence. Owing to the $(1/\cos \phi_1)$ factor in Eq. 39b, the maximum loss occurs at a higher angle of incidence than does the field-strength maximum (cf., Figs. 6 and 11).

To gain a feeling for the relative magnitudes of the minute intensity changes involved in infrared reflection spectroscopy, consider the adsorption of a monomolecular film of an organic compound with a moderately strong infrared absorption band ($n_2 = 1.3$, $k_2 = 0.1$, $d/\lambda = 1.0 \times 10^{-4}$) on the surface of a highly reflecting metal substrate ($n_3 = 3.0$, $k_3 = 30.0$). At $\phi_1 = 70°$, the value of $\Delta R_{\|a}$ is calculated to be -5.7×10^{-4}, whereas at normal incidence this term is only -7.1×10^{-7}! The total reflectivity change, ΔR, includes, in addition to the film absorption component, ΔR_a, terms due to the reflection from the front surface of the film, the altered reflectivity of the metal and its different rate of energy absorption, and phase change effects. The net reflectivity changes for the examples above are: $\Delta R_{\|}(70°) = -5.5 \times 10^{-4}$ and $\Delta R(0°) = -1.1 \times 10^{-6}$, illustrating well the validity of our previous assumptions.

Such low intensity changes necessitate the use of multiple-reflection systems and expansion of the spectrometer transmission scale to achieve adequate detection sensitivity. Numerous examples of the use of such techniques have been reported in infrared spectroscopic studies of molecules adsorbed on metal surfaces (cf., Poling, 230).

C. Ultraviolet-Visible Reflection Spectra. In the UV-visible spectral region ($\lambda \approx 0.2$–$0.8\ \mu$), the film detection sensitivity is markedly increased for three reasons. For very thin films, the normalized reflectivity change produced by film formation is linearly proportional to d/λ and at UV-visible wavelengths this ratio is typically an order of magnitude greater than in the infrared region (2.5–25 μ). Secondly, the optical frequency conductivity of metals falls rapidly as the frequency is increased with the result that the conduction electrons can no longer screen the electromagnetic field effectively. This is illustrated clearly in Fig. 12, which shows the standing-wave fields formed near the surface of a metal with optical constants similar to those of platinum in the visible region. The variation of the field components with angle of incidence is shown in Fig. 13. In contrast to their infrared behavior, the tangential field components, $\langle E_{\perp 1}^2 \rangle$ and $\langle E_{\| 1x}^2 \rangle$ do not exhibit nodes at the surface but are now of sufficient magnitude to interact strongly with adsorbed species. Even at normal incidence, the mean-square field strength

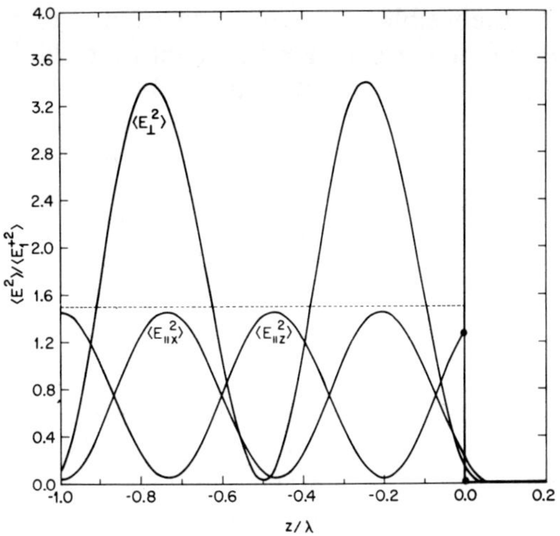

Fig. 12. Standing-wave fields at a moderately reflecting surface. The optical parameters are: $n_1 = 1.333$; $n_3 = 2.0$, $k_3 = 4.0$; $\phi_1 = 45.0°$.

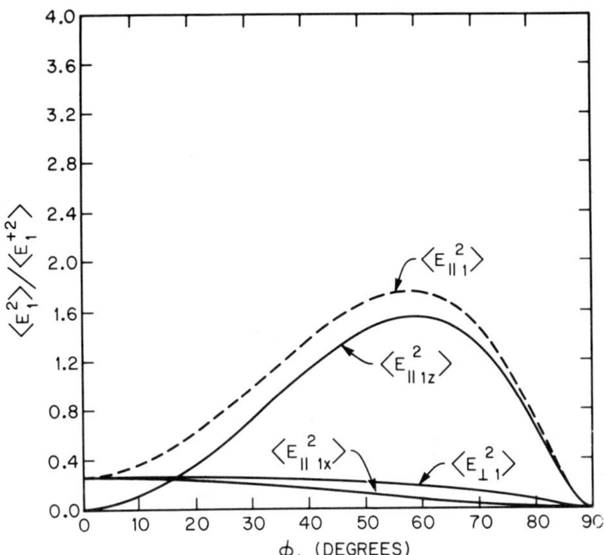

Fig. 13. Variation of the standing-wave fields at a moderately reflecting surface as a function of the angle of incidence and polarization of the incident beam. The optical parameters are: $z = 0$; $n_1 = 1.333$; $n_3 = 2.0$, $k_3 = 4.0$.

is 26% of that in the incident wave. Since the absorption coefficients of materials in the visible-UV range are usually of the same order of magnitude or much larger than in the infrared (20), the sensitivity of film detection is principally limited by the small volume of material sampled by the light beam, rather than by the lack of field intensity at the substrate surface. This fact is of considerable importance since it permits optical studies of surfaces to be made using well-characterized electrode substrate materials (e.g., bulk metals) whose properties are reproducible and not subject to time variations or sensitive to experimental conditions.

For very strongly absorbing films ($k_2 > 1$) in the UV-visible region, $|\hat{\varepsilon}_2| \gg \varepsilon_1$, and the absorption loss in the film is primarily due to interaction with the field components tangential to the surface plane, $\mathbf{E}_{\perp 2}$ and $\mathbf{E}_{\parallel 2x}$ (171). Appreciable absorption occurs even at normal incidence in this case, in marked contrast to infrared behavior. Figure 14 illustrates the attenuation of the various field components

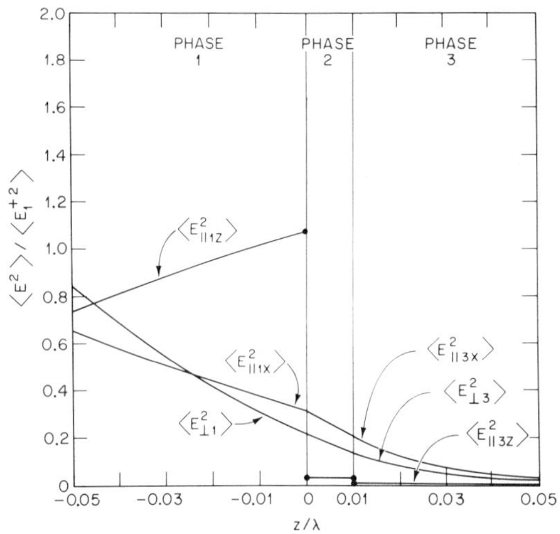

Fig. 14. Variation of the electric field intensities with penetration depth for external reflection in a thin-film system. The optical parameters are: $n_1 = 1.333$; $n_2 = 3.0$, $k_2 = 1.5$; $n_3 = 2.0$, $k_3 = 4.0$; $d/\lambda = 1.0 \times 10^{-2}$; $\phi_1 = 45.0°$.

in a strongly absorbing surface film ca. ten monolayers thick ($d/\lambda = 1.0 \times 10^{-2}$). The Z field within the film is very small, owing to the field-matching conditions. As noted previously, the field attenuation across the film is almost negligible for films only one monolayer thick.

The third reason for the increased film detection sensitivity in this spectral region is due to the appreciable energy absorption by the substrate. Provided there is sufficient mismatch between the film refractive index and that of the ambient, a significant reflectivity change will result from deposition of a transparent film ($k_2 = 0$) on the surface of a moderately reflecting substrate (cf. Fig. 9). The marked influence of substrate absorption on the form of UV-visible differential reflection spectra has seldom been taken into account in their interpretation.

4. Reflectometry in the Submonolayer Region

In the preceding phenomenological descriptions, the thin surface film has been treated as a homogeneous isotropic phase with an index of refraction and dielectric constant characteristic of those of a bulk phase.

For films of monomolecular thickness, we may expect the effective index of refraction and dielectric constant of the surface layer to differ appreciably from those for bulk material, owing to the reduction of the polarizability of the adsorbed species by the strong static field of the substrate surface dipole. Further, for surface films which only partially cover the substrate, the validity of the concept of a macroscopic index of refraction seems questionable, since the layer is not a continuum. It will be recalled, however, that details of molecular behavior on the surface cannot be resolved on a scale much smaller than the wavelength of the incident radiation—a distance which is much greater than the mean intermolecular separation at the smallest detectable surface coverage (ca. 1%). Experimental observations of the optical behavior of monolayer films thus correspond to a statistical average of the effects due to occupation of available adsorption sites over a macroscopic area. Since the optical-frequency electromagnetic field which polarizes the individual adatoms in the submonolayer film is externally applied (*via* the incident light beam), it is quite legitimate to define an *effective* index of refraction and dielectric constant for this layer (24). These quantities may vary with surface coverage. Since chemisorbed species may themselves change their state as a function of coverage (e.g., from adions to adatoms), interpretation of ellipsometric and reflectometric studies of submonolayer films is often a difficult task.

We now review briefly the features of macroscopic and microscopic models for the optical behavior of monolayer films.

A. Macroscopic Models. When the surface is fractionally covered, the mean film thickness is defined as

$$\bar{d} = \frac{1}{A} \sum_{i}^{n} d_i \, \delta A_i \tag{42}$$

where d_i is the local thickness and δA_i is the area of an island of film on the substrate of area A. If the film forms uniformly in the submonolayer range, d_i is constant and equal to the molecular diameter d_o. The surface coverage is defined as

$$\theta = \frac{\sum \delta A_i}{A} \tag{43}$$

and the effective thickness is therefore

$$\bar{d} = \theta d_o \tag{44}$$

From a macroscopic viewpoint, the optical behavior of the monolayer can be considered in terms of two extreme models.

(i) Under optical excitation, the film responds as a *continuous* phase, whose thickness $\bar{d}$ varies with surface coverage, but whose dielectric constant, $\hat{\varepsilon}_2$, and density are constant, independent of θ. Then, for normal incidence

$$\frac{\Delta R}{R} = \theta \frac{8\pi n_1 d_o}{\lambda} \operatorname{Im} \frac{\hat{\varepsilon}_2 - \hat{\varepsilon}_3}{\varepsilon_1 - \hat{\varepsilon}_3} \tag{45}$$

i.e., the fractional reflectivity change varies linearly with θ. If, however, the heat of adsorption of a light-absorbing species varies with surface coverage in a different manner for the ground and excited energy states of a particular transition, the absorption band will exhibit a frequency shift (39) and $\hat{\varepsilon}_2$ will vary with θ. $\Delta R/R$ will no longer exhibit a linear dependence on surface coverage, and the continuum model will not be valid.

(ii) The film is taken as the region between the plane of the substrate surface and a plane one molecular diameter above it. The film thickness is then constant, equal to d_o, but its density and refractive index vary continuously with surface coverage. The functional dependence of $\Delta R/R$ on surface coverage is more complex in this case. If the Clausius-Mossotti relation (44)

$$\frac{\hat{\varepsilon} - 1}{\hat{\varepsilon} + 2} = \frac{4\pi}{3} \sum N_i \alpha_i \tag{46}$$

is assumed to be valid for monolayer films (where N_i is the number of ions, atoms, or molecules of type i per unit volume and α_i is the complex polarizability), the variation of the film dielectric constant with surface coverage is given by

$$\hat{\varepsilon}_2(\theta) = \frac{1 + 2\gamma\theta}{1 - \gamma\theta} = 1 + 3\gamma\theta + 3\gamma^2\theta^2 + \cdots \tag{47}$$

where

$$\gamma = \left[\frac{\hat{\varepsilon}_2 - 1}{\hat{\varepsilon}_2 + 2}\right]_{\theta=1} \tag{48}$$

At low coverages, $\hat{\varepsilon}_2$ varies linearly with θ but at high coverages, higher order terms in the expansion must be retained. Extension of the Clausius-Mossotti relation (applicable to bulk phases) to a phase of monomolecular thickness adsorbed on a foreign substrate is, however, a procedure of very dubious validity. Nevertheless, Smith (264) has demonstrated excellent agreement between calculated values of the refractive indices of transparent monolayers of carboxylic acids and hydrocarbons physically adsorbed on mercury and the corresponding indices of the bulk materials.

B. Microscopic Models. Microscopic models for the optical behavior of monolayer films have been proposed by Strachan (268) and Sivukhin (263). The layer is treated as a two-dimensional assembly of scattering centers which act as radiating Hertz dipoles. Both treatments lead to relations of the same phenomenological form as the Drude relations for the macroscopic three-dimensional continuum film. The applicability of these models has recently been discussed by Bootsma and Meyer (41).

Archer (9) has pointed out that the scattering indices define the oscillator strengths per unit area and it is therefore reasonable to assume that they are proportional to the surface coverage. Since the first-order relations of Strachan and Sivukhin for the Fresnel coefficients are of the same form as the linear approximation relations, Eqs. 20a and 20b, it follows that the fractional reflectivity change $\Delta R/R$ should also be directly proportional to surface coverage provided there are no interactions among the adsorbed entities.

MacDonald and Barlow (188,189) have derived an expression

for the effective microscopic dielectric constant of a two-dimensional hexagonal array of polarizable entities of the form†

$$\varepsilon_1 = 1 + J\left[\left(\frac{\sigma}{8}\right)\xi^3 - \frac{1}{4} + \left(\frac{\pi}{2\sqrt{3}}\right)\frac{\{p(\xi)\}^2\{1 + F(\xi)\}}{[1 + \{p(\xi)/\xi\}^2]^{3/2}}\right] \tag{49}$$

where $\xi \equiv z/r_1$ is the ratio of the normal outward distance from the dipole array, z, to the nearest neighbor distance, r_1; σ is a lattice sum parameter (278), equal to 11.034 for a hexagonal array; $J \equiv \alpha/\beta^3$ where β is the distance between the electrical centers of adsorbed dipoles and the conducting imaging surface. The function $p(\xi)$ is defined by

$$\frac{\Psi_I(\xi)}{\Psi_\infty} = \left[1 + \left(\frac{p(\xi)}{\xi}\right)^2\right]^{-1/2} \tag{50}$$

where $\Psi_I(\xi)$ is the potential arising from an ideal-dipole cut-off model (107) at a position z along the perpendicular line through the center of a circular vacancy corresponding to a removed array element, and Ψ_∞, the potential for a uniform, smeared ideal dipole layer of strength $N\mu$ per unit area, is given (for $z > 0$) by

$$\Psi_\infty = 2\pi N\mu \tag{51}$$

where μ is the dipole moment of a discrete dipole. Finally, $F(\xi)$ is defined as

$$F(\xi) = -\frac{d \ln p(\xi)}{d \ln (\xi)} \tag{52}$$

For details of the calculation of the $p(\xi)$ function, the original paper (189) should be consulted.

For a hexagonal array of spherical adions or adatoms, $\theta = \xi^2$, and to a good approximation (264), Eq. 49 reduces to

$$\varepsilon_1 \approx 1 + J\{1.37\theta^{3/2} + 0.32/[1 + 0.36/\theta]^{3/2}\} \tag{53}$$

The quantity J may be evaluated by assuming $\varepsilon_1 = n_b^2$ at full coverage ($\theta = 1$), where n_b is the bulk refractive index of the film material. Using this method for an ellipsometric study of the physical adsorption of pentane on mercury, Smith (264) obtained close agreement between values of the dielectric constant ε_1, calculated for the above microscopic two-dimensional model, and n_θ^2, calculated for the three-dimensional continuum model, as shown in Fig. 15.

† MacDonald and Barlow's original notation is retained here and should not be confused with similar symbols previously employed in this chapter.

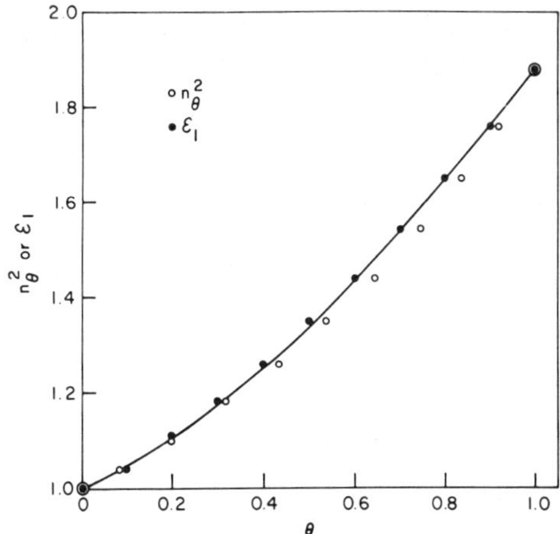

Fig. 15. Comparison of the dependence on surface coverage (θ) of the effective microscopic dielectric constant, ε_1, of a monomolecular film of pentane on mercury, and the value of $n_\theta{}^2$ calculated for a three-dimensional continuum model. After Smith (264).

Microscopic models for the optical behavior of chemisorbed layers on electrode surfaces have yet to be applied. Interpretation is much more difficult than for physical adsorption systems because of the frequency shifts of absorption bands, observed, e.g., for oxygen adsorption on platinum (201), or the possible valence variation of adsorbing metal atoms on foreign metal substrates. It is just these features, however, that can lead to a detailed picture of the adsorbed layer.

5. *Comparison of External and Internal Reflection Modes*

Internal reflection spectroscopy has recently received considerable attention as a technique for investigating the electrode-solution interphase (124,125,126,190,191,238,299), partly because it affords the possibility of enhancing the sensitivity of detection of electrode reaction products, either on the surface or within a very thin solution layer, ca. $\lambda/3$ in thickness, adjacent to it. Semitransparent electrodes suitable for such investigations have been fabricated (126,191,238) by vacuum deposition of thin metal films, 50–100 Å in thickness, on the surface of a transparent glass or quartz internal reflection element (IRE).

In the three-phase system: IRE(1)/metal film(2)/electrolyte(3), the critical angle for internal reflection, ϕ_c, is determined solely by the refractive indices of the initial and final phases, because Snell's law is operative at all interfaces. Thus

$$n_1 \sin \phi_1 = \hat{n}_2 \sin \phi_2 = n_3 \sin \phi_3 \tag{54}$$

and

$$\phi_c = \sin^{-1} (n_3/n_1) \tag{55}$$

For angles of incidence $\phi_1 \geq \phi_c$ and very thin metal films ($d_2 \leq$ 200 Å), the mean-square field strengths $\langle E^2_{\parallel 3x} \rangle$ and $\langle E^2_{\parallel 3z} \rangle$ at the rear (metal-electrolyte) interface are actually *greater* than their counterpart fields, $\langle E^2_{\parallel 1x} \rangle$ and $\langle E^2_{\parallel 1z} \rangle$, at the front surface of the metal film. This enhancement is a result of the field-matching boundary conditions of electromagnetic theory (cf. Section II.3A) and the surprising result that for sufficiently thin metal films, $\langle E^2_{\parallel 2x} \rangle$ and $\langle E^2_{\parallel 2z} \rangle$ *increase* with increasing penetration depth of the evanescent standing wave in the absorbing metal film. This phenomenon is illustrated in Fig. 16, a computer-generated plot based on Hansen's exact equations for the electric field strengths in a stratified three-phase system (128).

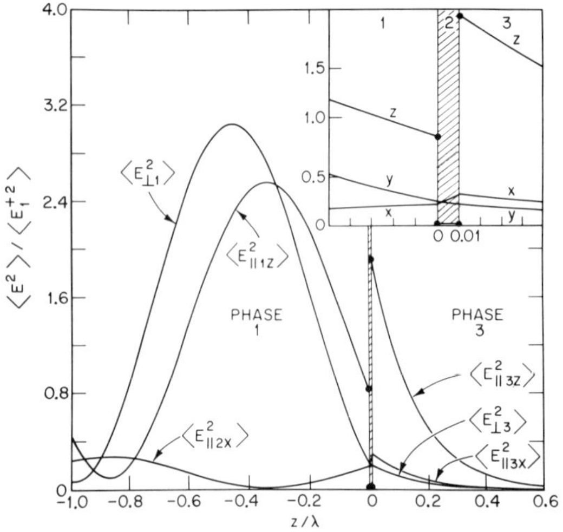

Fig. 16. Variation of the electric-field intensities for internal reflection at a thin metal-film electrode as a function of distance from the interface. No energy is transmitted into the final electrolyte phase. The inset illustrates the details of the field variation within the metal film. The optical parameters are: $n_1 = 1.52$; $n_2 = 2.0$, $k_2 = 4.0$; $n_3 = 1.333$, $k_3 = 0.0$; $d/\lambda = 1.0 \times 10^{-2}$; $\phi_1 = 72.0°$, $\phi_c = 61.28°$.

When $d_2 \ll \lambda$, the normal (Z-field) component remains nearly constant across the film and Eq. 38c provides a good approximation for the ratio of the Z-field strengths in the initial and final phases. In the internal reflection mode, $n_1 > n_3$ and the relative enhancement for a transparent or weakly absorbing electrolyte is given by

$$\frac{\langle E_{\parallel 3z}^2 \rangle |_{z=d_2}}{\langle E_{\parallel 1z}^2 \rangle |_{z=0}} \approx \frac{n_1^4}{n_3^4} \tag{56}$$

The higher the refractive index of the IRE, the greater is the enhancement of the Z field. In the UV-visible region, optical materials suitable for IRE's have indices in the range: $1.4 \leq n_1 \leq 1.8$ (134). Since the growth of the X field in very thin films is much greater than that of the Z field, Eq. 38b does not provide a good approximation for the ratio $\langle E_{\parallel 3x}^2 \rangle / \langle E_{\parallel 1x}^2 \rangle$. Near the critical angle of the system illustrated in Fig. 16 ($\phi_c = 61.28°$), the enhancement of the X field is actually greater than 10, but the absolute value of $\langle E_{\parallel 3x}^2 \rangle |_{z=d}$ is very small and is, in fact, less than the corresponding X field for external reflection at a bulk metal electrode.

It is of interest to compare the characteristic features of differential internal reflection spectra of monolayer surface films with those for the external mode examined in previous sections. The equivalent Fresnel reflection coefficient for a four-phase system: IRE(1)/metal film(2)/surface film(3)/electrolyte(4), is given by (143)

$$r_{1234} = \frac{r_{12} + r_{23}e^{-2i\beta_2} + r_{34}e^{-2i(\beta_2+\beta_3)} + r_{12}r_{23}r_{34}e^{-2i\beta_3}}{1 + r_{12}r_{23}e^{-2i\beta_2} + r_{12}r_{34}e^{-2i(\beta_2+\beta_3)} + r_{23}r_{34}e^{-2i\beta_3}} \tag{57}$$

where the coefficients r_{jk} are given by Eqs. 12 and β_2 and β_3 are defined by analogy to Eq. 17. The form of the Fresnel coefficient, r_{124}, for the film-free system, is given by Eq. 16. For the internal mode, therefore

$$\frac{\Delta R}{R} = \left| \frac{r_{1234}}{r_{124}} \right|^2 - 1 \tag{58}$$

The angular variation of the normalized internal reflectivity changes, corresponding to the deposition of transparent and strongly absorbing films of monolayer thickness on the outer surface of a semitransparent thin metal-film electrode, is illustrated in Fig. 17. Corresponding values of $\Delta R/R$ for the external mode and a thick bulk-metal electrode are shown in Fig. 18.

For the four-phase system, the critical angle for attenuated total reflection is again determined solely by the refractive indices of the initial and final phases, namely, $\phi_c = \sin^{-1}(n_4/n_1)$. When $\phi_1 > \phi_c$ there is no propagating wave in the final phase, $T = T' = 0$ (cf. Section II.3), and Eq. 29 is again applicable. In this case, the value of $(\Delta R/R)_\perp$ is primarily determined by ΔA_s, the change in absorption by the metallic substrate film. This reflects the fact that energy absorption in a metal is mainly a result of interactions with the field components tangential to the surface, $\langle E_{\perp 2}^2 \rangle$ and $\langle E_{\parallel 2x}^2 \rangle$, the normal component, $\langle E_{\parallel 2z}^2 \rangle$, being very small within the metal (cf. Fig. 16). In contrast, for parallel polarization and $\phi_1 \geq 70°$, the value of $(\Delta R/R)_\parallel$ is primarily determined by A_f, the absorption in the surface film. At lower angles of incidence, there is a significant contribution of both ΔA_s and A_f to $\Delta R_\parallel$.

Comparison of Figs. 17 and 18 reveals that for the example illustrated, the film detection sensitivity of the parallel-polarized internal reflection spectrum is greatest, owing to the enhancement of $\langle E_{\parallel 4}^2 \rangle$ at the metal-electrolyte interface. Here it is the X component of the parallel field which is responsible for the increased sensitivity, the Z field within the film being lower by a factor

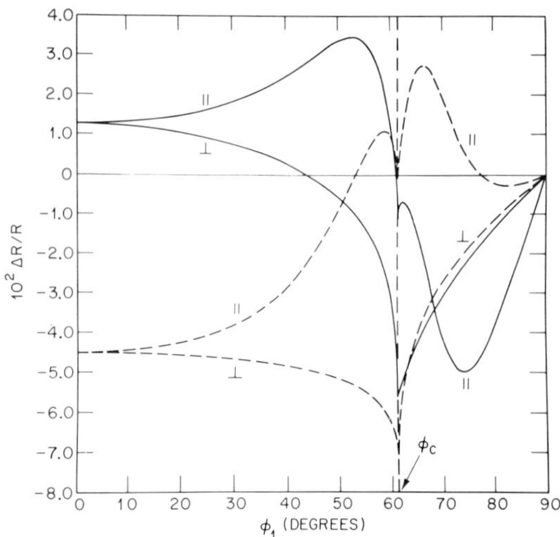

Fig. 17. Angular variation of the internal reflectivity changes produced by adsorption of ca. one monolayer of a transparent (– – –) or absorbing (——) species (phase 3) onto the outer surface of a semitransparent metal-film substrate (phase 2), which is deposited on a transparent internal reflection element (1). The ambient medium is phase (4). The optical parameters are: $n_1 = 1.52$; $n_2 = 2.0$, $k_2 = 4.0$, $d_2/\lambda = 1.0 \times 10^{-2}$; $n_3 = 3.0$, $k_3 = 0.0$(– – –), $k_3 = 1.5$ (——), $d_3/\lambda = 1.0 \times 10^{-3}$; $n_4 = 1.333$, $k_3 = 0.0$.

$(\varepsilon_4/|\hat{\varepsilon}_3|)^2 \approx 0.025$ than that in the solution. The relative contributions of the two components to the energy absorption by the film vary as the inverse square of the dielectric constant of this phase. For weakly absorbing films of lower refractive index, the Z field will have the greater influence.

In the above example, the maximum values of $|\Delta R/R|_{\parallel}$ for the internal and external modes differ by a factor of 2.1. From a practical viewpoint, this sensitivity ratio only marginally favors the internal reflection mode.† The principal advantage of this mode is that it yields *transmission-like* absorption spectra of surface films when high angles of incidence and parallel polarization are employed.

Against these apparent advantages of internal reflection spectroscopy must be weighed the considerations that vacuum-deposited metal films, 100 Å or less in thickness, have densities and optical and electrical properties that may differ significantly from those of

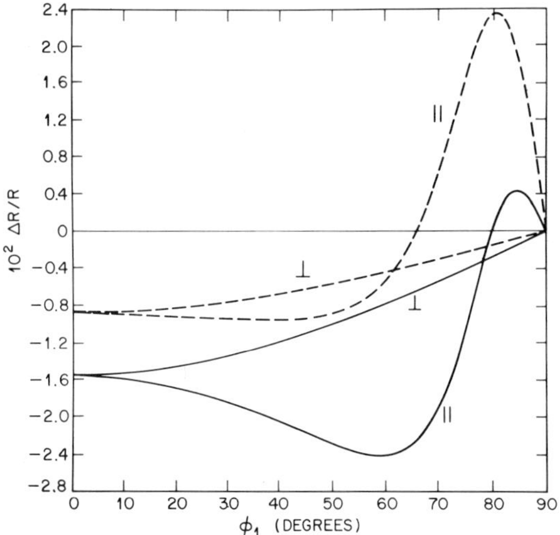

Fig. 18. Angular variation of the external reflectivity changes produced by ca. one monolayer of a transparent (– – –) or absorbing (——) film (phase 2) on the surface of a bulk metal substrate (phase 3). The ambient medium is phase (1). The optical parameters are: $n_1 = 1.333$; $n_2 = 3.0$, $k_2 = 0.0$ (– – –), $k_2 = 1.5$ (——); $d/\lambda = 1.0 \times 10^{-3}$; $n_3 = 2.0$, $k_3 = 4.0$.

† Greater detection sensitivity is obtained for a metal-film thickness $d_2 \approx 2 \times 10^{-2}\lambda$, but the contrast is reduced owing to increased absorption by the metal.

the bulk metal. When the thickness is very small, the film approximates more nearly to an array of islands dispersed within a dielectric medium (141,240). In investigations of the physical state of an adsorbed layer, a knowledge of the real and imaginary parts of the film dielectric constant, $\hat{\varepsilon}_2$, and the energy loss function, Im $\hat{\varepsilon}_2^{-1}$, is required to permit interpretation of the structural features of experimental reflection spectra in terms of the various possible electronic excitation processes (76). Although computational procedures for determining thin film optical constants have been developed for the external reflection mode (171,201), no such methods are available as yet for internal reflection spectroscopy with semitransparent metal-film electrodes. A further difficulty associated with the latter method is the need for an accurate measurement of the metal-film thickness.

Clearly, both internal and external reflection methods have unique advantages. Internal reflection spectroscopy is favored: (i) for kinetic studies of coupled homogeneous chemical reactions near the electrode surface (299); (ii) when solutions are highly colored or opaque (e.g., in the infrared); and perhaps (iii) for chemical identification of adsorbed species. Specular (external) reflection spectroscopy seems best suited for detailed studies of the optical properties of adsorbed layers, particularly those involving reaction of the metal substrate (e.g., oxide formation); (iv) kinetic studies of adsorption/desorption processes, and (v) electroreflectance studies.

III. Experimental Apparatus and Techniques

1. *Instrumentation for Spectro-Electrochemical Studies*

Differential and modulated reflectance spectroscopy involve the measurement of minute changes in reflectance. Typically, values of $\Delta R/R$ lie in the range 10^{-6} to 10^{-1}. Accurate measurement of $\Delta R/R$ demands high stability and linear response of photodetectors and amplifiers and, frequently, the ability to discriminate against high system noise levels. In the present section, the requirements of optical cells, electrodes, spectrometers, and associated electronic apparatus designed for spectro-electrochemical investigations are discussed.

A. Optical Cells. The fundamental requirements of a cell for *quantitative* specular reflection spectroscopy of electrode surfaces are essentially the same as for ellipsometry: (i) a uniform current density must be maintained on the test electrode surface, which should be optically smooth and flat; (ii) the incident plane-polarized, collimated monochromatic light beam should strike the electrode surface at an oblique angle (45° to 75°), and precise optical alignment of the test electrode with reference to the plane of polarization and direction of propagation of the beam must be possible; (iii) cell windows should be transparent over the wavelength range of interest, free from strain birefringence, and positioned so as to avoid spurious effects from stray reflections and minimize absorption by the electrolyte; (iv) the cell should be gas-tight for operation under a controlled or inert atmosphere and should have provision for stirring and/or purging of the electrolyte by gas bubbling, without interfering with the cell optics; (v) the cell should be constructed from chemically inert materials to maintain electrolyte purity and should have provision for isolation of counter and reference electrodes.

The decision of whether to employ a single- or multiple-reflection cell can be made from the following considerations. In a cell with N reflections from the test electrode surface(s), the total reflectance change caused by film deposition is

$$\Delta R_N = R^N(d) - R^N(0) = R^N(0)\left[\left(1 + \frac{\Delta R}{R}\right)^N - 1\right] \quad (59)$$

If the detection system is shot-noise limited and $\Delta R/R \ll 1$, the signal-to-noise ratio (S/N) is optimum for measurement of ΔR_N and $\Delta R_N/R^N(0)$ when $R^N(0) = 1/e^2 = 0.135$, i.e., the maximum sensitivity is obtained when the background energy is reduced after N reflections to $1/e^2$ of its incident value. Conversely, if the detection system has a constant noise level which is independent of light intensity (e.g., thermal detector noise, amplifier noise, mechanical noise in the recorder pen mechanism, etc.), ΔR_N and $\Delta R_N/R^N(0)$ have maximum values when $R^N(0) = 1/e = 0.368$. Under actual operating conditions, the optimum value of $R^N(0)$ may lie between these two extremes. For electrode substrate materials such as Pt, Au, Cu, or Ni, $0.2 \leq R \leq 0.7$ over most of the visible-ultraviolet wavelength range and the optimum value of S/N can be approached by employing a cell with only one or two reflections. Satisfaction of conditions (i), (ii), and (iv) above, as well as relative ease of construction, indicates that use of a single-reflection cell is often most advantageous.

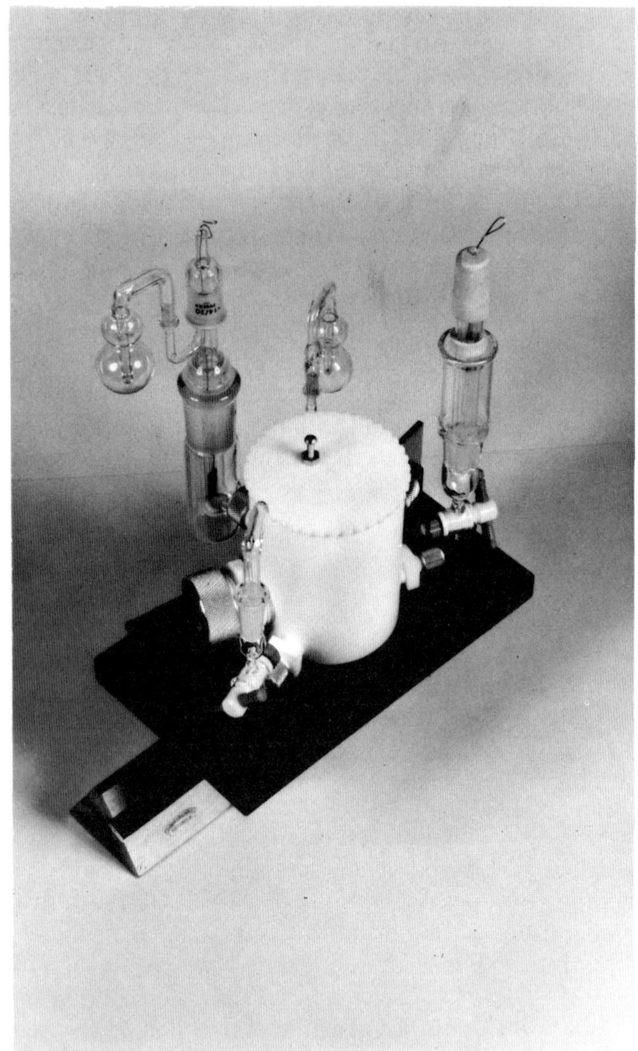

Fig. 19. Optical cell for specular reflection spectroscopy of metal-electrode surfaces. A single reflection is employed, with an angle of incidence of 70°. The front surface of the mirror electrode lies in a diametric plane of the cell.

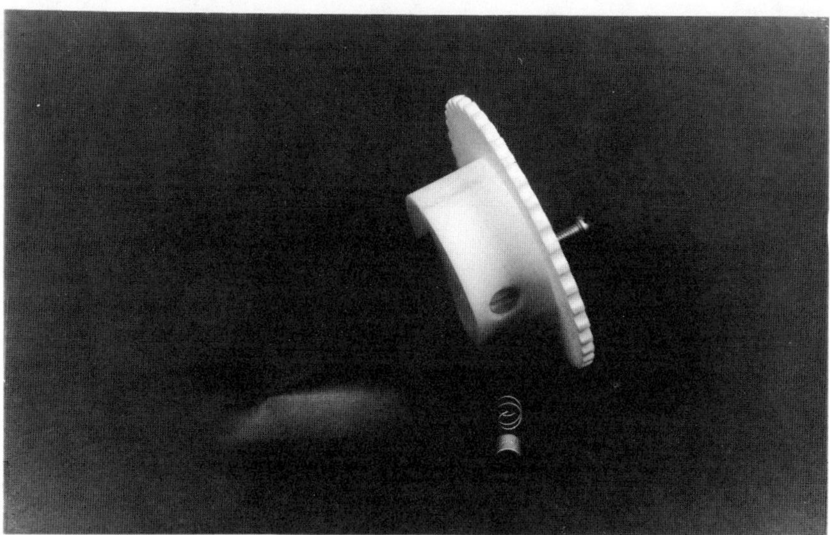

Fig. 20. Teflon cell top and evaporated-metal-film electrode from the optical cell of Fig. 19. Electrical contact is made through the gold-plated screw and spring assembly.

Figure 19 illustrates the design, construction and mounting of an optical cell used in the author's laboratory for quantitative specular reflection spectroscopy of electrode surfaces employing a 70° angle of incidence and a single reflection. The complete assembly can be fitted into the extended cell compartment of a Cary 14 spectrophotometer, which is equipped with a polarizing prism. The method of electrode mounting is shown in Fig. 20. Similar cells have been employed for a 45° incident angle (201). Such cells are readily adaptable for normal incidence use by simply rotating the cell top. A single-reflection cell (51), which facilitates variation of the angle of incidence, is illustrated in Fig. 21. The cylindrical cell optics serve to collimate the incident beam.

A multiple-reflection cell of the type shown in Fig. 22*a*, employing as many as 20 reflections, was used by Takamura *et al.* (274) to obtain enhanced resolution of small reflectivity changes. A rectangular multiple-reflection cell with parallel entrant and exit beams and a variable angle of incidence has been described by Barrett and Parsons (17). Rather than using multiple reflections, it is preferable to enhance resolution, whenever feasible, by electronic expansion of the spectrometer transmission scale. The latter procedure

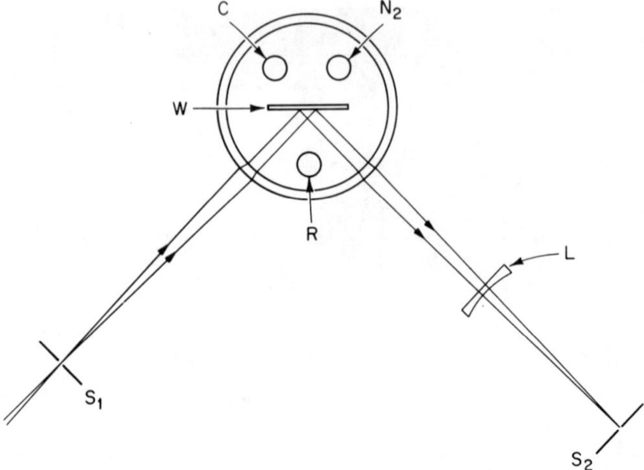

Fig. 21. Optical cell with cylindrical optics and a variable angle of incidence: W, working electrode; R, reference electrode; C, Pd-tubing counter electrode; S_1, source slit; L, negative cylindrical lens; and S_2, photodetector slit. After Cahan (51).

does not alter the S/N ratio since both quantities are amplified to the same extent. Use of multiple-reflection cells in the UV region has severe drawbacks owing to the low energy throughput ($\propto R^N I^\circ$), the degraded S/N value ($\propto R^{N/2} I^{\circ\ 1/2}$), and strong spectral distortion due to stray radiation from the monochromator and scattering from the electrode surfaces.

A novel multiple-reflection cell designed for spectroscopic detection of electrochemically generated solvated electrons has been described by Walker (291) and is illustrated in Fig. 22*b*. The inside surface of a 7-cm diameter cylindrical glass tube was coated with a vacuum-evaporated film of metallic silver, which acted as a mirror electrode. A concentric platinum wire served as counter electrode. The narrow, intense, parallel light beam from a continuous gas laser was directed in a grazing spiral path around the mirror electrode surface so as to make innumerable specular reflections. This technique produces a very long transmission path length through the solution layer adjacent to the electrode surface and optimizes the sensitivity of detection of transient intermediate species in the diffusion layer. Owing to the curvature of the electrode surface, however, a 1-mm diameter laser beam can have a range of angles of incidence as great as 14° in this optical configuration. The plane of

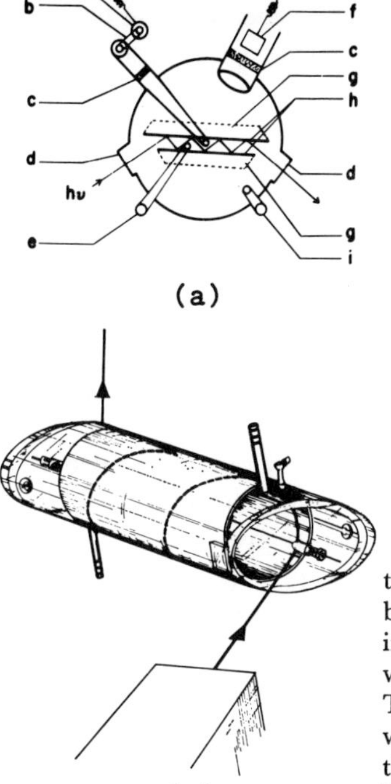

Fig. 22. (*a*) Multiple specular reflection electrochemical cell: *a*, reference electrode; *b*, salt bridge; *c*, fritted glass; *d*, optical window; *e*, gas inlet; *f*, counter electrode; *g*, Teflon holder; *h*, working electrode; and *i*, gas outlet. After Takamura et al. (274). (*b*) Electrolysis cell with cylindrical mirror electrode, showing continuous specular reflection of a laser beam. After Walker (291).

incidence of the linearly polarized incident beam can, likewise, vary by as much as 10° relative to its plane of polarization, so that, after a single reflection, the beam is *elliptically* polarized. After many reflections, severe scattering from minute surface irregularities produces a depolarized beam with a poorly defined angle of incidence. Caution must therefore be observed in interpreting experimental results from such a cell. As shown in Section II.2C, anomalous changes can arise at very high angles of incidence for the parallel-polarized component of the beam (cf. Fig. 5). Walker attributed the observed light intensity changes in his experiments at 6328 Å to absorption by solvated electrons which, he proposed (289,290,191), were generated as intermediates in the hydrogen evolution reaction. An alternative explanation of this optical effect is offered in Section IV.

Optical cells designed for "thin-layer" spectro-electrochemical studies have been described by Kissinger and Reilley (164). These cells are suitable for spectroscopic studies of soluble reaction products by double-pass *transmission* spectroscopy, since the metal electrode surface acts as a mirror, and offer an alternative to the use of optically transparent electrodes. They are not as well suited for investigation of the optical properties of adsorbed layers by *reflection* spectroscopy, however, owing to the considerations mentioned at the beginning of this section.

B. Mirror Electrodes. The preparation of large optically smooth, flat samples of bulk metals is difficult, time-consuming, and often expensive. Since the depth of penetration of light in a metal is small (ca. 100 Å), the measured optical constants are extremely sensitive to the method of surface preparation. Mechanical polishing leaves a damaged surface region. In some cases, it is possible to remove this layer by annealing, chemical etching, or electropolishing. However, the latter procedures are difficult (if not impossible) to carry out for just those materials which are of greatest interest as catalytic electrode substrates—the transition metals. In addition to lattice disorder in the surface region, surface roughness and surface films must also be avoided if the measured optical parameters are to represent accurately the optical properties of the bulk material. Various procedures suitable for mounting, insulating, and optically aligning bulk specimens in experimental cells have been described (9,56,109,179).

Vacuum-evaporated or sputtered metal films offer an attractive alternative to the use of bulk samples. Such films can be conveniently deposited on optically polished glass or quartz substrates, and the desired electrode shape can be obtained by suitable masking during the deposition process. Substrate surfaces should be extremely smooth to avoid errors in the measured reflectance caused by surface roughness. RMS surface roughnesses as low as 2.7 Å have been reported for fused quartz samples polished using a bowl-feed method (21,64). A high degree of smoothness of the film surface is also essential for detecting preferential orientation of dipoles in an adsorbed monomolecular film. If the surface roughness is large compared to the film thickness, anisotropy of the optical properties of the adsorbed layer may be obscured since the film tends to follow the contours of the surface (132).

Metal film thicknesses of ca. 0.5–1.0 μ have been found satisfactory for use as mirror electrodes (201). Such films typically have resistances less than 0.2 ohm per square and optical properties characteristic of the bulk metal. Optimum results are obtained by employing very high deposition rates and/or ultrahigh vacuum (10^{-9} torr) conditions. Reference works on vapor deposition processes (see, e.g., Ref. 236) should be consulted for procedural details of thin film preparation. Metal films deposited on substrates at room temperature are often under residual stress and annealing at high temperatures may be required to relieve this stress and impart bulk optical and electrical properties to the thin film. Recrystallization of the surface layers of metal film deposits probably occurs to some extent when surface oxide layers are electrochemically formed and reduced.

Owing to the lateral and normally directed forces which act on a metal electrode surface maintained at a potential removed from the point of zero charge, and the volume changes resulting from dissolution of evolved hydrogen, vapor-deposited metal films may peel away from the substrate surface when they are employed as electrodes. This may be prevented by use of a thin metal or metal-oxide undercoating. Vapor-deposited undercoatings of titanium, ca. 200 Å in thickness, have been employed successfully for fabricating adherent noblemetal (Au, Ag, Ni, Pt, Pd) mirror electrodes for specular reflection spectroscopy (201).† Transparent undercoatings of Bi_2O_3 and PbO (several hundred angstroms in thickness) have been employed for the manufacture of optically transparent electrodes (238).

C. Spectrophotometers. Since absorption bands measured by reflection spectroscopy tend to be broad, high resolution is not required for most measurements; a resolving power of 1–10 Å over the UV-visible range is generally adequate. Owing to their freedom from drift problems, high accuracy, and general convenience, commercial double-beam instruments are often suitable for measurements on a long time scale and where values of $\Delta R/R$ in the range 1–20% are to be measured. Choice of an instrument with a suitably large sample compartment is obviously essential. The optical system should produce a beam of low divergence which can be collimated by masking at a focal point without severe reduction of beam intensity. The S/N level of the sample beam can be improved by using an optical attenuator in the reference beam, thereby forcing the slits to open wider and increasing the energy level in the sample beam. Scattered light with wavelengths remote from the nominal measurement wavelength can produce serious errors in recorded reflection

† A sputtered niobium undercoating has recently been reported (51) to be more inert to electrolytic etching than is titanium.

spectra, particularly in the UV region where calcite prism polarizers have low transmission. Instruments with *double* monochromators afford the highest spectral purity. Application of filters and other procedures for reducing the amount of scattered light and blocking second or higher-order diffracted radiation from grating instruments have recently been reviewed by Bennett and Bennett (22).

When a reflectance change is to be monitored at constant wavelength as a function of a variable such as potential, charge, or time, provision must be made for recording the optical signal externally. This can often be accomplished by gearing a retransmitting potentiometer to the transmission-slide-wire-pointer drive mechanism. Commercial double-beam spectrophotometers must usually be modified for modulated reflectance studies owing to their use of low chopper frequencies and/or ac power supplies for lamps.

Single-beam monochromators offer the advantages of great flexibility, freedom from cell space restrictions, and building-block construction. In addition, they permit use of high-intensity sources (e.g., Xe arc lamps), and sophisticated detection systems such as lock-in amplifiers or signal averagers. Some double-pass monochromators discriminate against any dc scattered radiation emerging from the exit slit, by internally chopping the primary dispersed beam before the second pass. Light intensity fluctuations due to lamp aging or arc instability can present severe low-frequency drift problems when measuring differential reflection spectra. Such problems can be alleviated to some extent by use of a beam splitter and a ratio-recording system, or by employing a servo system which maintains the lamp intensity constant (220,221).

Single beam monochromators of low f-number are the instruments of choice for modulation spectroscopy studies. Measurements of the ratio $\Delta R/R$ are made using a single beam and one detector, so that this quantity is independent of spectral and temporal variations of the source intensity and photodetector response, as well as of the optical transfer characteristics of the other system elements.

Sensitive scanning reflectometers of high precision which combine the advantages of both double-beam and single-beam instruments have recently been described by Gerhardt and Rubloff (98), and Sell (253).

2. *Determination of Substrate Optical Constants*

In order to determine the optical properties of an electrode surface film as a function of wavelength, it is first necessary to ascertain the optical constants of the underlying substrate material. For metals whose reflectivities exhibit pronounced structural features (e.g., Cu, Ag, and Au), measurements must be made at closely spaced wavelength intervals throughout the spectral range of interest. The sensitivity of the measured optical constants to the surface treatment of bulk materials, the conditions of preparation of vacuum-evaporated or sputtered metal films, and the general problems of surface contamination and purity render it imperative that the optical constants characteristic of the particular method of electrode fabrication be determined, if meaningful quantitative results are to be obtained. An apparently formidable task thus confronts the research worker in this field, and methods must be sought which enable the desired properties to be determined with rapidity and ease.

Ideally, measurements of the optical constants of substrate materials should be made under ultrahigh vacuum conditions employing annealed, optically smooth, film-free surfaces in cells with windows free from strain birefringence. Examples of the elegant techniques employed in such studies have been described by Kruger (179), Detorre, Knorr, and Vaughan (63), and Archer (9). Much useful information can be obtained in combined optical-electrochemical investigations, however, by employing substrate surfaces prepared by the simpler, more practical techniques described in Section III.1B. It is possible to measure "apparent" optical constants (41,207) for such substrates that can subsequently be used in determining the optical properties of electrode-reaction products.

A wide variety of methods has been described in the optical literature for measuring the optical constants of metals and thin films (4,22,27,141,142,143,192,249,301). When the substrate material is opaque, as is usually the case for metallic electrodes, the useful methods are restricted to those techniques which measure the polarization state (amplitude and phase) of the *reflected* beam alone, relative to that of the incident beam. The methods can be classified as polarimetric and photometric. Here, we consider only those which appear to offer the greatest promise for determining optical constants over a wide spectral range. Greatest emphasis is placed on spectrophotometric methods, partly because of the orientation of this chapter but also because these methods are more rapid and convenient and require less-specialized apparatus. Owing to the sensitivity of the optical constants to the conditions of preparation, measurements which yield these parameters with an accuracy of ca. 1% are considered satisfactory.

A. Polarimetric Methods. The classical and still the most precise method for the determination of optical constants is ellipsometry. Initially plane-polarized radiation, when reflected from the surface of an absorbing material, becomes elliptically polarized. The ratio of the Fresnel reflection coefficients for parallel- and perpendicularly-polarized radiation is

$$\rho = \frac{r_{\parallel}}{r_{\perp}} = \frac{R_{\parallel}^{1/2}}{R_{\perp}^{1/2}} \exp(\delta_{\parallel}^{r} - \delta_{\perp}^{r}) = \tan \Psi e^{i\Delta} \tag{60}$$

Various ellipsometric configurations for measuring the azimuthal angle, Ψ, and the phase shift difference, Δ, are reviewed by Bennett and Bennett (22). The parameters Ψ and Δ are measured by appropriate adjustments of the polarizer, compensator, and analyzer. The optical constants are given by (68)

$$\varepsilon' = n^2 - k^2 = \sin^2 \phi_1 \left[1 + \frac{\tan^2 \phi_1 (\cos^2 2\Psi - \sin^2 2\Psi \sin^2 \Delta)}{(1 + \sin 2\Psi \cos \Delta)^2} \right] \tag{61a}$$

$$\varepsilon'' = 2nk = \frac{\sin^2 \phi_1 \tan^2 \phi_1 \sin 4\Psi \sin \Delta}{(1 + \sin 2\Psi \cos \Delta)^2} \tag{61b}$$

Owing to the requirement of successively setting the optical elements of the instrument, measurement of Ψ and Δ for a large number of wavelengths is extremely tedious if manual methods are employed. The recent advent of automatically balancing ellipsometers (49,109,154,182,218,293,298) has largely removed this difficulty, however. Methods have been developed for obtaining high accuracy in automated ellipsometric optical constant measurements that do not require use of a quarter-wave plate or compensator (10,150). The analyzer setting is first scanned at constant wavelength in the vicinity of the null, while the corresponding local null settings of the polarizer are recorded. The process is then reversed by scanning the polarizer and recording the local null settings of the analyzer. The loci of the null settings are straight lines of slope $\tan \Psi \cos \Delta$ and $\cos \Psi \cos \Delta$, respectively (195). From the measured slopes, the values of Ψ and Δ are found; the optical constants at each wavelength are then calculated from Eq. 61. This derivative technique is reported to eliminate constant offset errors due to residual ellipticities in the polarizer prisms and minimize errors due to surface imperfections and sample misalignment.

Since such automated instruments are not as yet commonly available, the ellipsometric method is generally restricted at present to measurements at a few isolated wavelengths. The accessible spectral range is limited by the transmission of calcite prism polarizers to ca. 2100 Å in the ultraviolet to 3 μ in the near infrared.

B. Photometric Methods. Photometric determination of optical constants involves measurement of the reflectivities, $R_{\parallel}$ and $R_{\perp}$, or their ratio, $R_{\parallel}/R_{\perp}$, for a collimated, monochromatic, obliquely incident light beam. The principal advantage of these methods lies in the facility with which measurements can be made while the wavelength is continuously scanned.

(a) External Reflection. A comparison of the external reflection methods available for measuring optical constants has been given by Humphreys-Owen (148). Those of greatest interest for the present purposes consist of measurement of:

(i) $R_{\parallel}$ at two angles of incidence.
(ii) $R_{\perp}$ at two angles of incidence.
(iii) $R_{\parallel}/R_{\perp}$ at two angles of incidence.
(iv) $R_{\parallel}$ and $R_{\perp}$ separately at a single angle of incidence.

Measurement of $R_{\perp}$ at two angles of incidence yields lower sensitivity to changes in n and k than does measurement of $R_{\parallel}$ (149). Avery (12) has pointed out the advantages of measuring the ratio, $R_{\parallel}/R_{\perp}$, rather than the absolute reflectivities, $R_{\parallel}$ and/or $R_{\perp}$, alone. Problems of small sample aperture, variation of the beam intensity over its cross section, variation of the photosensitivity of the detector across its surface, and source and detector drifts are minimized by ratio measurements. In some instances, these advantages may be offset by the convenience of the single-angle method (iv).

Until recently, analytical solutions of the reflectivity equations were not available. In the past, optical constants were commonly evaluated by graphical procedures (38,58,151,250,261,279), geometric construction (186), or numerical analysis (149,156). Greatest accuracy is usually obtained by employing angles of incidence on either side of the pseudo-Brewster angle.

Exact analytical solutions of the reflectivity equations for n and k are now available, however, for all but one of the above spectrophotometric methods. Fahrenfort and Visser (82) have given an explicit solution for determination of optical constants by method (ii) using an attenuated total reflection technique (81). Their solution is generally applicable to both internal *and* external reflection modes. Computational aspects of this method have been discussed by Hansen (123) and the preceding authors (83). An approximate numerical solution for Avery's ratio method (iii) was first given by Abelès (2); an exact solution has recently been obtained by Kolb (169). Solutions for the single-angle method (iv), which are valid for all angles except 0° and 45°, have been given by Abelès (3), Kudo (180), and most recently by Querry (239). The optimum angle of incidence for determining optical constants using method (iv) is 74° (209). For the two-angle method (i) with parallel polarization, no analytical solution appears to be available. Neglect of higher-order terms of a polynomial expansion permits a numerical solution to be obtained by computer (170). An error analysis of the reflectance versus angle of incidence methods (i) and (ii) has been given by Hunter (149).

The methods proposed by Humphreys-Owen (148), which involve measurement of the pseudo-Brewster angle ϕ_B', where $R_\parallel$ exhibits a minimum, are of doubtful attraction when readings must be taken at many wavelengths. Potter (233,234,235) has described an analytic method for the determination of the optical constants from measurements of the ratio $(R_\parallel/R_\perp)_{\phi_B'}$ and the angle ϕ_B' and has designed a reflectometer for such measurements. These methods were formerly of interest because explicit solutions were available (148). Now that exact analytical solutions are available for the much less tedious methods (i), (ii), and (iv) above, this advantage no longer persists.

To avoid the necessity for a standard reflector, it is possible to take advantage of the Abelès relation (1). This relation, valid for *isotropic* samples only, states that

$$R_{\parallel,45^\circ} = R^2_{\perp,45^\circ} \tag{62}$$

An instrumental calibration factor can thus be determined since $R_{\perp,45^\circ} = (R_\parallel/R_\perp)_{45^\circ}$.

(b) Internal Reflection. Internal reflection spectroscopy is a particularly attractive method for determining the optical constants of metallic substrates that are free from surface contamination. The sample material is deposited under high-vacuum conditions onto the clean inert surface of a transparent internal reflection element (IRE) whose index of refraction is greater than that of the ambient medium. Measurements of the reflectivity of the IRE-sample interface, which is effectively self-encapsulated, are made using one of the techniques previously described at angles of incidence greater than the critical angle. This method was employed by Schulz and

Tangherlini (252) to determine the optical constants of the metals Ag, Au, Cu, and Al. Their vacuum-evaporated films were carefully annealed and the results of reflectivity measurements made at a 45° angle of incidence were compared with the Abelès relation, Eq. 62, to ensure that the films were optically isotropic.

One of the principal advantages of this method is its high sensitivity to small changes in n and k, as compared to external reflection (22,122). A second advantage is that an uncoated area of the IRE can conveniently be employed as a standard reflector with $R \equiv 1.00$, since there is no power loss in the evanescent wave.

The results of internal reflection measurements on metal films can serve as a reference for comparison with reflectivity measurements on front-surface mirror electrodes deposited under the same conditions. The effects of surface films formed by exposure of the mirror to the atmosphere, and varying degrees of roughness of the external surface can then be evaluated.

(c) Kramers-Kronig Analysis. As a consequence of the principle of causality (53,181,277), the optical constants of a material are not independent of each other but are integrally related by the Kramers-Kronig dispersion relations (173,177). At normal incidence, the Fresnel reflection coefficient of the interface between air (or vacuum) and an absorbing medium with optical constants n, k is

$$r = |r|e^{i\delta^r} = \frac{n - ik - 1}{n - ik + 1} \tag{63}$$

In order to determine n and k, both the modulus and phase of r must be known. The value of $|r|$ can be determined from normal incidence reflectivity measurements since $R = |r|^2$. If R and, hence, $\ln |r|$ are known over the entire frequency spectrum, the phase $\delta(\omega_0)$ at a single frequency, ω_0, can be calculated by means of the Kramers-Kronig relation between the real and imaginary parts of the complex function $\ln r = \ln |r| + i\delta^r$. Thus

$$\delta^r(\omega_0) = \frac{2\omega_0}{\pi} \int_0^\infty \frac{\ln |r(\omega)|}{\omega^2 - \omega_0^2} \, d\omega \tag{64}$$

This relationship was first employed for the determination of optical constants from normal incidence reflection data by Robinson and Price (245,246). Although experimental measurements of R can only be made over a limited frequency range, the method is successful since the influence of remote regions of the spectrum on $\delta(\omega_0)$ tends to be negligibly small. This fact becomes apparent when Eq. 64 is expressed in the form (37)

$$\delta^r(\omega_0) = \frac{1}{\omega}\int_0^\infty \frac{d \ln |r(\omega)|}{d\omega} \ln \left|\frac{\omega + \omega_0}{\omega - \omega_0}\right| d\omega \qquad (65)$$

The weighting function ln $(\omega + \omega_0)/(\omega - \omega_0)$ peaks sharply at $\omega = \omega_0$, but becomes small and flat at remote frequencies. At low frequencies, the optical constants and reflectivity become frequency-independent and make no contributions to the integral. Similarly, in the case of an isolated reflectivity band which has constant-reflectivity wings, there is no contribution to $\delta^r(\omega)$ from the spectral region outside the band. When R is still varying at frequencies higher than the limit of measurement, it is necessary to extrapolate the measured reflectance to higher energies. Various procedures have been employed and are reviewed in detail elsewhere (5,53,110, 224,286,287). One method is to extrapolate $R(\omega)$ beyond ω_1 by assuming

$$R = R_1\left(\frac{\omega_1}{\omega}\right)^p \qquad (66)$$

where R_1 is the measured reflectivity at ω_1. The exponent $p \approx 4$ at very high frequencies where the primary contribution to the dielectric constant is due to free electrons. In general, the value of p is chosen to give a good fit of the reflectivity data at low energies. Empirical values of p ranging from 1.3 to 1.9 have been found to yield the best results for noble metals in the visible-near UV in this laboratory (170). Nilsson and Munkby (216) have recently developed an analytical model that eliminates errors from measured reflectance data and tests the reliability of the approximations used.

Once the phase angle $\delta^r(\omega)$ has been computed, the optical constants $n(\omega_0)$ and $k(\omega_0)$ are evaluated from the relations

$$n = \frac{1 - R}{1 - 2R^{1/2} \cos \delta^r + R} \qquad (67a)$$

$$k = \frac{2R^{1/2} \sin \delta^r}{1 - 2R^{1/2} \cos \delta^r + R} \qquad (67b)$$

The values of α, ε', ε'' are then obtained from Eqs. 6 and 9.

Figure 23 illustrates the frequency dependence of the real and imaginary components of the dielectric constant of a vacuum-evaporated Au film on a smooth quartz substrate, computed by

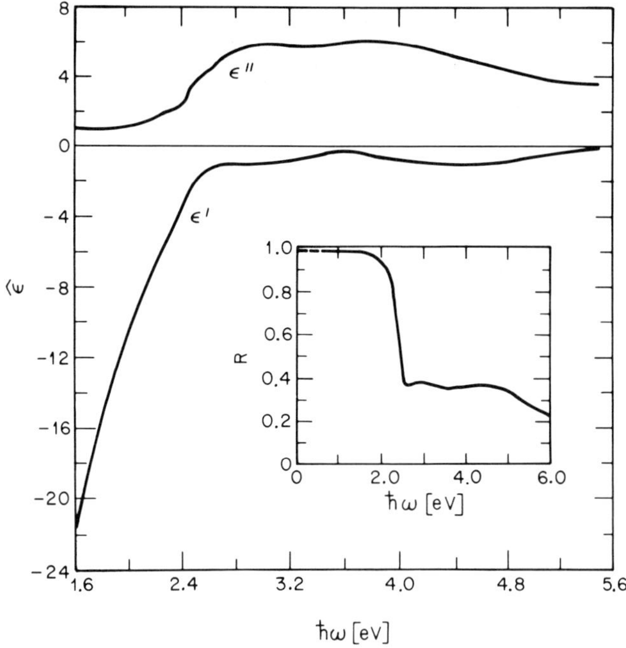

Fig. 23. Frequency dependence of the real and imaginary components of the dielectric constant of Au: $\hat{\varepsilon} = \varepsilon' - i\varepsilon''$. The normal-incidence reflectivity of Au (in air) is shown in the inset. After Kolb and McIntyre (171).

Kramers-Kronig analysis of the normal incidence reflectivity spectrum (in air) shown in the inset (171). Similar data for a sputtered Pt film are shown in Fig. 24.

3. *Determination of Surface Film Optical Constants*

In the past, the optical constants and thickness of a surface film on an absorbing substrate have been determined exclusively by ellipsometry. In general, three optical parameters must be determined for an absorbing thin film—n_2, k_2, and d. Since only two quantities, Δ and Ψ, can be measured polarimetrically, one of the three parameters must be determined by an independent measurement. The fundamental equation of ellipsometry for a three-phase system:

$$\tan \Psi e^{i\Delta} = \frac{r_{\parallel 12} + r_{\parallel 23} e^{-2i\beta}}{1 + r_{\parallel 12} r_{\parallel 23} e^{-2i\beta}} \cdot \frac{1 + r_{\perp 12} r_{\perp 23} e^{-2i\beta}}{r_{\perp 12} + r_{\perp 23} e^{-2i\beta}} \tag{68}$$

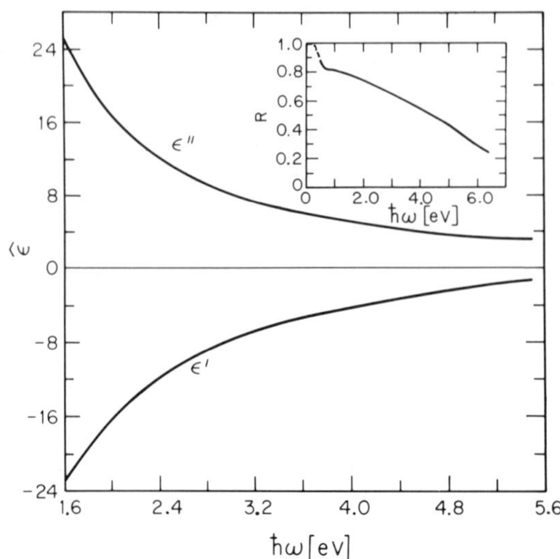

Fig. 24. Frequency dependence of the real and imaginary components of the dielectric constant of Pt. The normal-incidence reflectivity of Pt (in air) is shown in the inset. After McIntyre and Kolb (201).

cannot be inverted directly to yield the optical constants of the film; they must be found by numerical solution. Computational methods (26,27,195,196,197) and various approximations for thin films (7,8,69,70,137,183,285,300) are well documented in the optical literature. Of the various first-order approximations (7,8,69,70,285), that due to Archer (7,8) is the most accurate. Burge and Bennett (47) have given corrected versions of the equations.

A spectrophotometric procedure for determining the optical constants of very thin isotropic surface films has been developed by Kolb and McIntyre (171). The first-order Eqs. 23a and 23b for the normalized reflectivity changes for *s*- and *p*-polarized light can be combined and inverted to yield a cubic equation for the real part of the film dielectric constant, ε_2', as a function of $(\Delta R/R)_\perp$, $(\Delta R/R)_\parallel$, ε_1, $\hat{\varepsilon}_3$, ϕ_1, and d/λ. If measurements are carried out over a range of photon energies, it is generally obvious which of the three possible exact solutions of this equation is physically significant. Substitution of this value of ε_2' into Eq. 23a yields the corresponding value for the imaginary component, ε_2''. Again, there remains the problem of determining the three parameters, ε_2', ε_2'', and d, from two spectrophotometrically measured values, $(\Delta R/R)_\perp$ and $(\Delta R/R)_\parallel$. In studies of electrochemically generated surface films, it is often possible to estimate the mean film thickness from coulometric measurements of the charge passed during film formation. The monolayer thickness can be estimated by assuming a value for the film density or by summing atomic diameters. If the film is transparent, measurement of $(\Delta R/R)_{\perp}$ and $(\Delta R/R)_{\parallel}$ will yield the real refractive index and film thickness directly.

The above spectrophotometric method has been employed to determine the optical constants of a series of metallic and nonmetallic films, in the monolayer thickness range, adsorbed on metallic substrates (201). The results are in qualitative agreement with those of ellipsometric studies. Ellipsometric results are sparse, however, and usually confined to a single wavelength and angle of incidence, so that a proper comparison of the accuracy of the two methods cannot yet be made. At present, the spectrophotometric method appears to be the only one suitable for measuring the optical properties of adsorbed species such as reaction intermediates on a very short time scale ($t \leq 10^{-6}$ sec).

IV. Electroreflectance of Metals

Since their inception in the mid-1960s, modulated reflection spectroscopy techniques (electroreflectance, piezoreflectance, thermoreflectance, magnetoreflectance, wavelength modulation) have been employed with remarkable success in elucidating the optical properties and band structure of solids (54). The penetration depth of light in metals and semiconductors (in the energy region above the fundamental absorption edge) is very small, typically less than 0.1 μ. As a result, the use of conventional transmission spectroscopy is precluded unless extremely thin, uniform films of the sample material can be prepared (e.g., by vacuum deposition). The static reflectivity spectra (R versus $\hbar\omega$) of strongly absorbing solids generally consist of broad structural features superimposed on a constant or slowly varying background, a result of the breadth of the energy bands in solids and free-electron effects. Further, since the reflectivity is a function of *both* the real and imaginary components of the dielectric constant of the sample, direct interpretation of the static spectrum in terms of characteristic absorption losses is usually not possible. However, when the optical properties of a semiconductor are perturbed by the application of an external electric field, $\mathscr{E}$, sharp structure is observed (52,255,256,259) in narrow energy regions of the modulated reflectance spectrum, $R^{-1}\Delta R/\Delta\mathscr{E}$ versus

$\hbar\omega$. This structure is associated with "critical points" in the band structure of the solid, where the joint interband density-of-states function exhibits analytic singularities. It is the ability of modulation spectroscopy† to reveal these critical points and to define the type of absorption edge (225) that has made it such a powerful diagnostic tool in semiconductor band-structure analysis.

The Franz-Keldysh effect (90,158) is primarily responsible for the electroreflectance (ER) effect in semiconductors in the region of the fundamental edge. An applied electric field "bends the bands" in the space-charge region at the semiconductor surface and induces electron-tunneling below the conduction band or hole-tunneling above the valence band. The result is a shift of the absorption edge to longer wavelengths since photons of lower energy than the band gap, $E_g = E_c - E_v$, can excite an electron from the valence band to the conduction band. This field-assisted absorption process is an *even* function of the field strength and is thus independent of the sign of the field.

A detailed discussion of ER effects in semiconductors is beyond the scope of this article. Excellent reviews have recently been given by Seraphin (256,257,258) and Cardona (54). Here we wish to discuss the ER effect in metals, since this promises to be an important new tool for studying the electronic properties of metal surfaces in solution and extending our knowledge of electrosorption and electrocatalysis.

Although the space-charge region in a semiconductor in contact with an electrolyte solution extends approximately 1 μ inward from the surface, a distance comparable to the penetration depth of visible light, the corresponding lengths in a metal are much smaller and of different relative magnitude owing to the high free-electron concentration (ca. 10^{23} cm^{-3}) and the greater frequency dispersion of metallic optical properties. The Thomas-Fermi screening length in metals, analogous to the Debye-Hückel length in electrolytes, is given by (165)

$$l_{TF} = \left(\frac{E_F}{6\pi N e^2}\right)^{1/2} \tag{69}$$

where E_F is the maximum electronic kinetic energy in the degener-

† Electroreflectance is the most sensitive of the modulation techniques since low-field ER spectra (of semiconductors) are related to the *third* derivative of the linear dielectric function (11), whereas the other methods yield first-derivative spectra.

ate Fermi gas, N is the volume concentration of free electrons in the bulk metal, and e is the electronic charge. For noble metals such as Cu, Ag, and Au, $l_{TF} \approx 0.6$ Å; the static electric field due to a test charge at the surface of a metal is thus screened out within the *first* atomic layer. A low-frequency field should, therefore, have negligible effect on the band structure of the bulk metal in the region penetrated by visible-UV light, a layer of thickness $1/\alpha \approx 100$ Å. Nevertheless, Feinleib (88), employing the "electrolytic" method (52,259,297), was able to detect a strong ER effect for the metals Ag and Au by modulating the high field ($\sim 10^7$ V/cm) in the electrical double layer. Although the quantitative significance of his nonpotentiostatic experiments is now in doubt,† the phenomenon generated considerable interest among electrochemists because of its apparent potential for revealing details of the structure of the double layer. Since Feinleib's original investigation, several experimental studies of the ER effect in metals have been made. Typical modulated ER spectra for Ag and Au, measured under controlled electrochemical conditions (198,199), are illustrated in Figs. 25 and 26. Various mechanisms have been advanced to account for this optical effect; their essential features are described below.

1. Electrolyte Effects

A. Feinleib Model. Noting the symmetry in the normal-incidence reflectivity relation for a *two*-phase system

$$R = \left| \frac{\hat{\varepsilon}_M^{1/2} - \hat{\varepsilon}_E^{1/2}}{\hat{\varepsilon}_M^{1/2} + \hat{\varepsilon}_E^{1/2}} \right|^2 \tag{70}$$

where $\hat{\varepsilon}_M$ and $\hat{\varepsilon}_E$ represent the complex dielectric constants of the metal and electrolyte respectively, Feinleib proposed that the observed ER effect was due to the modulation of the dielectric constant of the electrolyte in the double-layer region by the applied low-frequency field. From Eq. 70

$$\frac{1}{R}\frac{\partial R}{\partial |\hat{\varepsilon}_E|} = -\frac{|\hat{\varepsilon}_M|}{|\hat{\varepsilon}_E|}\left(\frac{1}{R}\frac{\partial R}{\partial |\hat{\varepsilon}_M|}\right) \tag{71}$$

† In Feinleib's experiments, the metal mirror electrode and a Pt counter-electrode were immersed in aqueous KCl. A high field normal to the test electrode surface was applied by modulating the cell voltage with a 2 V peak-to-peak signal at 35 Hz. No precautions were taken to avoid the formation of surface films or soluble metal-chloride complexes.

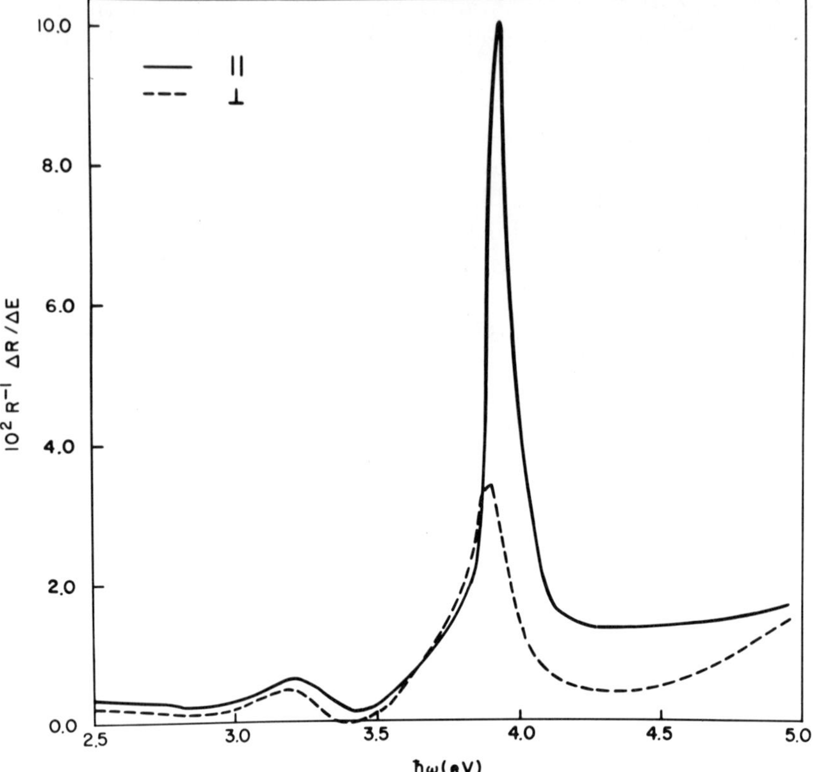

Fig. 25. Electroreflectance spectra of Ag in 1 *M* $NaClO_4$ (Ar-saturated) at $\phi_1 = 45°$. $E = -0.5$ V vs. S.C.E. $\Delta E = 100$ mV (r.m.s.) at 27 Hz. After McIntyre (198).

and it follows that in the transparent range of the electrolyte

$$\frac{1}{R}\frac{\Delta R}{\Delta \mathscr{E}} \approx -\frac{|\hat{\varepsilon}_{\mathrm{M}}|}{|\hat{\varepsilon}_{\mathrm{E}}|}\left(\frac{1}{R}\frac{\partial R}{\partial|\hat{\varepsilon}_{\mathrm{M}}|}\right)\frac{\partial|\hat{\varepsilon}_{\mathrm{E}}|}{\partial \mathscr{E}} \tag{72}$$

Since the frequency dispersion of $\hat{\varepsilon}_{\mathrm{E}} = n_{\mathrm{E}}^2$ in this photon energy range is small and monotonic, the primary origin of the sharp peaks in the experimental ER spectrum, $\Delta R/R$ versus $\hbar\omega$, was attributed to the pronounced structure in the optical constants of the metal (cf. Fig. 23). The reflectivity change predicted by this model is inaccurate, however, because it corresponds to a perturbation of the refractive index of the *whole* electrolyte, not just that of the

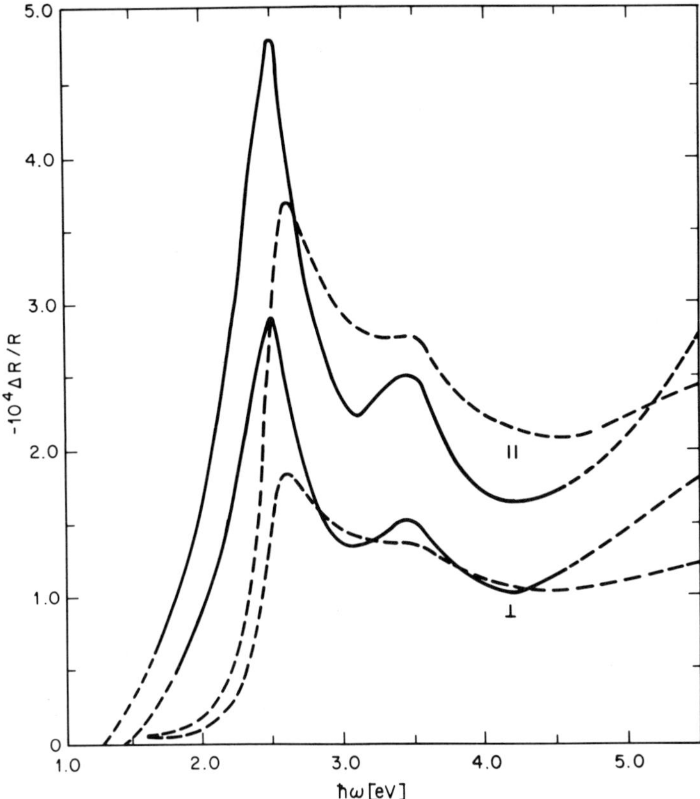

Fig. 26. Electroreflectance spectra of Au in 1 N $HClO_4$ (Ar-saturated) at $\phi_1 = 45°$. $E_H = 0.4$ V. $\Delta Q_m = +1.91$ μC cm^{-2} (r.m.s.) at 270 Hz. ——, $(\Delta R/R)_{expt.}$; – – –, $\Delta R/R$ calculated according to McIntyre-Aspnes model (202). After McIntyre (199).

ionic double layer. It cannot safely be concluded on the basis of Feinleib's analysis therefore, that the ER effect originates in the electrolyte rather than in the metal.

B. Stedman Model. Applying Gouy-Chapman theory and using a *three*-phase model for the electrode-electrolyte interfacial region, Stedman (266) computed the optical effects due to a perturbation of the refractive index in the diffuse layer, in both the absence and presence of specific anion adsorption. Figure 27 illustrates her results for the equivalent layer refractive index of aqueous NaF solutions. For a 0.5 V potential change on a Hg electrode in 0.1 *M*

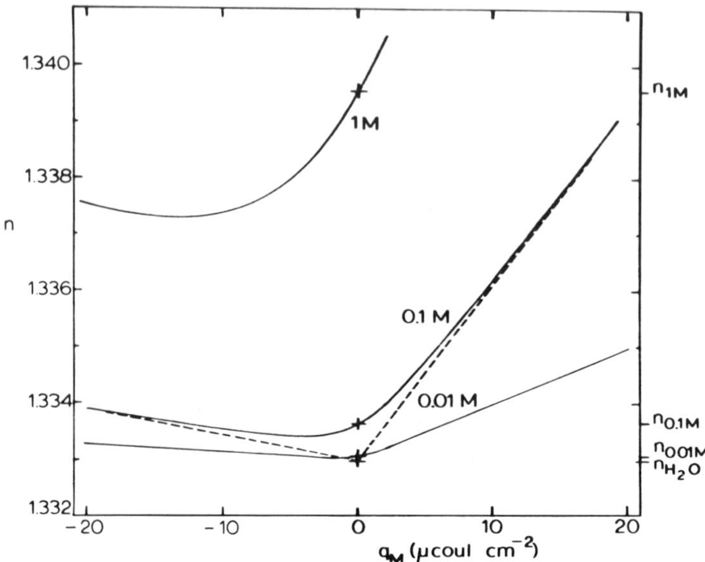

Fig. 27. Equivalent layer refractive index for aqueous NaF solutions; thickness of equivalent layer, 6.10×10^{-8} cm, 1.93×10^{-7} cm, 6.10×10^{-7} cm for $c = 1, 0.1, 0.01$ M, respectively. After Stedman (266).

aqueous KCl, the magnitude of $\Delta R/R$ at $\phi_1 = 0°$ was found to be ca. 10^{-5}. This value is much lower than that observed experimentally for the ER effect of other metals and suggests that modulation of the refractive index in the ionic double layer is only of secondary importance at low angles of incidence.

More recently, Stedman (267) has considered the optical effects arising from electrostrictive compression of the layer of oriented solvent dipoles which comprise the compact or inner double layer. According to Hills and Payne (145), an increase of the surface charge on Hg (in 0.1 M aqueous NaF) from -10 to $+20$ $\mu C/\text{cm}^2$ produces a compression of ca. 17% in the compact water layer. Stedman estimates the resulting refractive index of this layer to be 1.405. This large change yields $(\Delta R/R)_{0°} = 1.1 \times 10^{-4}$, $(\Delta R/R)_{\perp,70°} = 3.6 \times 10^{-5}$ and $(\Delta R/R)_{\parallel,70°} = 9.1 \times 10^{-4}$. Such an effect is not negligible, although it will be much smaller for the modulating voltages employed in differential ER experiments. In the absence of other complicating factors, the compressive effect $(\propto \mathscr{E}^2/8\pi)$ should become zero at a potential close to the point of zero charge.

Further diagnostic criteria for distinguishing electrolyte effects are discussed in Section IV.2B.

2. *Electronic Effects*

A. Hansen and Prostak Models. Employing a multilayer model of the type described in Section II.2, Prostak and Hansen (237) showed by computer calculations that electric-field modulation of the refractive index of the ionic double layer could not lead to an ER spectrum for gold of the shape reported by Feinleib and concluded that the effect was due rather to a perturbation of the optical constants of the *metal*. They proposed that the ER effect in metals could be accounted for by a field-induced rigid shift of the optical constants of a 0.5 Å thick metal surface layer to lower photon energies. Field-induced shifts of the fundamental absorption edge of semiconductors to lower energies have frequently been observed (40,93,117,211,296,297) and attributed to the Franz-Keldysh effect. For an exponential edge, the uniform displacement $\Delta\omega$ varies as $\mathscr{E}^2$ and is independent of the sign of $\mathscr{E}$. The Franz-Keldysh effect is not expected to be operative for metals, however, owing to the very short screening length, l_{TF}. Nevertheless, Prostak and Hansen were able to demonstrate by a simple rigid-shift model calculation, that a modulation of the optical constants of a thin surface layer of gold could account for the predominant peak at 2.35 eV, near the plasma edge of gold, in the ER spectrum reported by Feinleib. Their conclusion was subsequently verified experimentally by spectroscopic studies employing internal reflection (127,129), modulated ellipsometry (45,46), and specular reflection techniques (198).

The rigid-shift model yields an ER spectrum that is simply a normalized first-derivative of the reflectivity with respect to energy. For a metal such as Ag, which exhibits a pronounced reflectivity minimum at 3.86 eV, peaks of both signs should occur in the ER spectrum. However, for external reflection spectra measured under controlled electrochemical conditions (cf. Figs. 25 and 26), McIntyre showed that the optical signal $R^{-1}\,\partial R/\partial\mathscr{E}$ is everywhere negative and linearly proportional to the modulating voltage (198). Further, the experimental ER spectrum of Au exhibits strong modulation in energy regions well-removed from the main peak (cf. Fig. 26)—an effect not predicted by the Prostak-Hansen theory. The simple rigid-shift model is, therefore, inapplicable.

The above model was subsequently refined by Hansen and Prostak (127). The contribution made by *free* electrons to the dielectric response function of a metal is (165)

$$\begin{aligned} \hat{\varepsilon}_f(\omega) &= 1 - \frac{\omega_p^{\ 2}}{\omega^2 - i\omega/\tau} \\ &= \left(1 - \frac{\omega_p^{\ 2}}{\omega^2 + 1/\tau^2}\right) - i\left(\frac{\omega_p^{\ 2}}{\omega^2 + 1/\tau^2}\right)\frac{1}{\omega\tau} \\ &= \varepsilon_f' - i\varepsilon_f'' \end{aligned} \tag{73}$$

where τ is the relaxation time of the electrons, and ω_p, the plasma frequency of the metal, is given by

$$\omega_p = \left(\frac{4\pi Ne^2}{m^*}\right)^{1/2} \tag{74}$$

where m^* is the effective electronic mass. An increase of the free-electron concentration by an amount ΔN produces a plasma frequency shift, $\Delta\omega_p = \frac{1}{2}\omega_p \Delta N/N$, which displaces plots of ε_f' and ε_f'' against ω to *higher* frequencies. Hansen and Prostak proposed that the Fermi level, E_F, in the conduction band, is modulated by the field-induced concentration change ΔN, but that other bands involving bound electronic states are unaffected by the field. The photon energies required to excite electrons to or from the Fermi surface are thus altered, with the result that the interband component, $\hat{\varepsilon}_b$, of the dielectric constant of the metal is modulated. On the assumption that the components of $\hat{\varepsilon}_b$ corresponding to the allowed interband transitions are either constant with N or vary with N in the same way as $\hat{\varepsilon}_f$, they postulated that the optical constants of the metal are shifted along the frequency axis by the same fraction that ω_p is shifted.†

This semiempirical model has also proved to be unsatisfactory, however, since it predicts ER spectra for Ag and Au different in shape from those observed experimentally. Again it fails to account

† It should be noted that a plasma frequency shift does not lead to a *uniform* translation of the curves ε_f' and ε_f'' versus ω along the frequency axis. At high frequencies where $\omega\tau \gg 1$, $\varepsilon_f'' \approx 0$ and $\hat{\varepsilon}_f$ is nearly pure real. The frequency displacement of $\hat{\varepsilon}_f'$ is then: $\Delta\omega \approx \omega(\Delta\omega_p/\omega_p)$—i.e., the shift is not rigid but varies as the frequency. At lower frequencies, however, $\Delta\omega$ is a more complicated function of ω. The rigid frequency-shift model employed by Hansen and Prostak in their calculations (127), if valid, would be strictly applicable only over a very narrow frequency interval.

for the one-sided nature of external spectra, and is physically unrealistic because it assumes a low-frequency modulation of the Fermi level of the metals.

B. McIntyre and Aspnes Model. The optical properties of the electrolyte-metal interfacial region can be described in terms of two semi-infinite phases with uniform dielectric constants ε_1 and $\hat{\varepsilon}_3$, and a thin intermediate region of thickness $d \ll \lambda$, whose dielectric constant $\hat{\varepsilon}_2$ varies continuously between these values in a direction normal to the surface plane, as illustrated schematically in Fig. 28. Owing to the symmetry between ε_1 and $\hat{\varepsilon}_3$ in the linear approximation theory (cf. Eq. 24), the perturbation, $\Delta\hat{\varepsilon}(z)$, in this gradual transition region can be defined with reference to either bounding phase. A mean value, $\langle\Delta\hat{\varepsilon}\rangle$, is defined by averaging the local change $\Delta\hat{\varepsilon}(z)$ over the transition region, viz.

$$\langle\Delta\hat{\varepsilon}\rangle = \frac{1}{d}\int_{-d}^{0} \Delta\hat{\varepsilon}(z)\, dz \tag{75}$$

We, therefore, consider the results predicted by linear approximation theory for the two limiting cases: $\Delta\hat{\varepsilon} = \hat{\varepsilon}_2 - \varepsilon_1$ and $\Delta\hat{\varepsilon} = \hat{\varepsilon}_2 -$

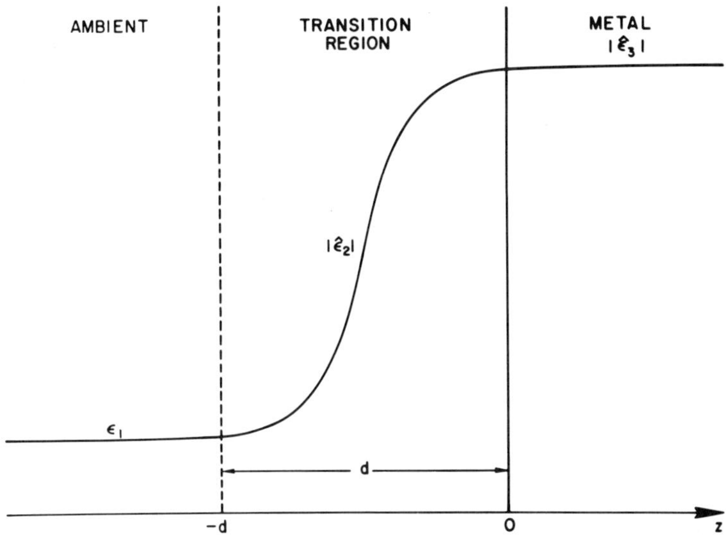

Fig. 28. Schematic illustration of the optical properties of the transition layer at a metal-electrolyte interface.

$\hat{\varepsilon}_3$, corresponding to field-induced perturbations of the optical constants of the electrolyte and metal, respectively. A remarkable simplification occurs for oblique-incidence spectra at $\phi_1 = 45°$.

(a) Modulation of the Optical Constants of the Electrolyte. As illustrated by Stedman's calculations in Fig. 27, the perturbation of the refractive index of the electrolyte in the ionic double layer by an applied field is very small and real. It follows immediately from Eqs. 23a and 23b that $(\Delta R/R)_{\perp,45°}$ is small and finite, but that $(\Delta R/R)_{\parallel,45°} \approx 0$. The physical basis for the disappearance of $(\Delta R/R)_{\parallel}$ is simply that the Brewster angle, $\phi_B = \tan^{-1}(n_2/n_1)$, for external reflection ($n_1 < n_2$) of p-polarized radiation from the interface of the transparent bulk solution and double-layer phases is very close to 45°. At the Brewster angle, it will be recalled, $R_{\parallel,\phi_B} = 0$ while $R_{\perp,\phi_B} > 0$. An analogous situation exists for internal reflection ($n_1 > n_2$) at the complementary principal angle. These predictions of the linear approximation theory are readily confirmed by examination of the rigorous reflectivity relation, Eq. 18.

Inspection of Figs. 25 and 26 reveals that such behavior is not observed experimentally. Modulation of the ionic double-layer refractive index is, therefore, not the primary mechanism responsible for the ER effect of the metals Ag and Au. It does become important at long wavelengths for p-polarization and near-grazing incidence (cf. Fig. 6), and may well account for the optical effects ascribed by Walker to absorption by solvated electrons (290,291).†

† The specular reflection cell employed by Walker for spectroscopic detection of solvated electrons generated in the hydrogen evolution reaction (her) is shown in Fig. 22b and described in Section III.1.A. In Walker's nonpotentiostatic experiments, the cell voltage was modulated with a 1.1 V (rms) sine wave at a frequency of 82.5 Hz, producing an ac cell current of 38 mA (rms). In this cell, the ratio of the area of the Ag-film electrode to that of the Pt-wire counter electrode is ca. 100:1 and the electrolyte resistance is ca. 1.3 ohm. Thus the major fraction of the cell voltage drop occurs at the Pt-wire/solution interface. Further, measurements in this laboratory have shown that under open-circuit conditions the stable rest potential of an Ag electrode in deoxygenated 0.25 *M* Na_2SO_4 is +0.16 V (SCE), a potential ca. 0.8 V *positive* of a reversible hydrogen electrode in this electrolyte. This potential is rapidly restored even after prolonged cathodic polarization of the Ag electrode in the H_2-evolution region.

A simple ac circuit analysis of this cell reveals that owing to the very high double-layer capacity of the Ag electrode (ca. 20,000 μF), its maximum potential excursion from rest cannot exceed 10 mV (rms) under the experimental conditions employed by Walker. At these potentials hydrogen evolution is thermodynamically impossible and the observed optical effects cannot, therefore, be ascribed to absorption by intermediate species such as solvated electrons

(b) Modulation of the Optical Constants of the Metal. If the optical constants of the strongly absorbing metal are perturbed in a very thin surface layer ($d \ll \lambda$), it can readily be shown from linear-approximation theory that $(\Delta R/R)_{\parallel,45°} \approx 2(\Delta R/R)_{\perp,45°}$. Such behavior is indeed observed experimentally for ER spectra measured with film-free electrodes in a potential range where no Faradaic reactions occur (cf. Figs. 25 and 26). These simple oblique-incidence tests† provide further convincing evidence that the ER effect of metals is principally due to a modulation of the optical constants of the metal by the high field in the electrical double layer.

In the above derivation, no specific model is assumed for the form of the field-induced shift of the optical constants of the metal. The effects of the field on the dielectric response function of the metal can thus be deduced by comparison of the calculated results for various assumed interaction models with the results of experiment. The essential features of a simple model (199,202) that accurately predicts the general form of the ER spectra of metals are discussed below.

(*i*) *Plasma-Frequency Effects.* The electron-density profile of a metal extends outward, beyond the surface layer of positive metal ions, and decays smoothly to zero within the inner Helmholtz layer. The primary optical effect due to the electric field results from the perturbation of the one-electron wave functions of the surface metal atoms. We now consider specific effects due to free and bound electrons.

The dielectric function of the metal can be resolved into two components (75,76)—the free-electron intraband transition term, $\hat{\varepsilon}_{3f}$, and the interband transition contribution, $\hat{\varepsilon}_{3b}$, such that

$$\hat{\varepsilon}_3 = \hat{\varepsilon}_{3f} + \hat{\varepsilon}_{3b} \tag{76}$$

---

generated in the her. The fact that the same optical effect was observed by Walker in his rectified ac experiments where hydrogen *was* evolved suggests that the phenomenon is of different origin. It was proposed by the author (198) that electroreflectance effects (cf. Fig. 25) were responsible. Subsequent calculations (204) have shown that the sign and approximate magnitude of the intensity changes observed by Walker can be accounted for by a combination of ionic and electronic ER effects.

† These simple criteria may also be useful in other surface studies for establishing whether an observed optical effect is due to: (i) adsorption of a transparent species on the substrate; (ii) a perturbation of the substrate optical constants in a thin surface layer arising from interaction with a transparent adsorbed species; (iii) formation of an absorbing surface film of chemical reaction products.

The form of the free-carrier term is given by Eq. 73. If it is assumed that $\hat{\varepsilon}_{3b}$ is *not* altered by the applied field,† then

$$\Delta\hat{\varepsilon}(z) = (\hat{\varepsilon}_{3f} - 1)\frac{\Delta N(z)}{N} \tag{77}$$

where $\hat{\varepsilon}_{3f}$ and N are, respectively, the dielectric constant and free-electron concentration of the *bulk* metal.

Averaging over the transition-layer thickness, we obtain

$$\langle\Delta\hat{\varepsilon}\rangle = \frac{(\hat{\varepsilon}_{3f} - 1)}{d}\frac{\Delta N_s}{N} = -\frac{(\hat{\varepsilon}_{3f} - 1)}{d}\frac{\bar{C}_{dl,E}E}{N} \tag{78}$$

where ΔN_s is the total excess number of free carriers per unit surface area required to shield the field, $\bar{C}_{dl,E}$ is the *integral* double-layer capacity and E is the electrode-solution potential difference referred to the point of zero charge (pzc). In this model, it is assumed implicitly that the Fermi level of the metal is *not* modulated by the applied field and that bound electron states are unaffected; only the plasma frequency of the free-electron gas in the transition region is shifted.

For normal incidence, this model yields the result

$$\frac{\Delta R}{R} = \frac{8\pi n_1}{\lambda}\left(\frac{\bar{C}_{dl,E}E}{N}\right)\operatorname{Im}\frac{\hat{\varepsilon}_{3f} - 1}{\hat{\varepsilon}_{3f} + \hat{\varepsilon}_{3b} - \varepsilon_1} \tag{79}$$

The prefactor is thus linearly dependent on both the applied field and the double-layer capacity. The remaining term determines the lineshapes of the spectrum and the sign of the ER effect. It is evident that the predicted effect does not exhibit a minimum at the pzc but passes smoothly through it, changing sign. Further, to first order, the value of $\Delta R/R$ is independent of the screening length and the transition-layer thickness. Thus, if $d \ll \lambda$, the ER effect is insensitive to the actual shape of the charge-distribution profile at the metal surface.

† This assumption is based on the facts (cf. Friedel, 92) that *d*-states have small orbits compared to sp-valence states of comparable energy, are localized and not strongly perturbed by the lattice potential, and are ineffective in screening the nuclear charge within the atom. Evidence that *d*-orbitals are not readily deformed is given by the low compressibility and high cohesive energy of the noble metals. Thus, since the *d*-orbitals are "stiff" and, further, are shielded from the applied field by the conduction-band electrons, it is reasonable to assume as a first approximation that $\Delta\hat{\varepsilon}_b \ll \Delta\hat{\varepsilon}_f$ and $\Delta\hat{\varepsilon}_b \approx 0$.

For modulation with small voltages, ΔE, about a fixed dc bias potential, E, it follows that

$$\left(\frac{\Delta R}{R}\right)_E = \frac{8\pi n_1 d}{\lambda} \operatorname{Im} \left(\frac{\langle\Delta\hat{\varepsilon}\rangle_{E+dE} - \langle\Delta\hat{\varepsilon}\rangle_E}{\varepsilon_1 - \hat{\varepsilon}_3}\right) \tag{80}$$

where it is assumed that d remains effectively constant. This result is general and *independent* of the origin of the optical modulation effect. Neglecting contributions to $\Delta\hat{\varepsilon}$ other than $\Delta\hat{\varepsilon}_{3f}$, we have to first order

$$\left(\frac{\Delta R}{R}\right)_E = \frac{8\pi n_1}{\lambda} \left(\frac{C_{dl,E}\Delta E}{N}\right) \operatorname{Im} \frac{\hat{\varepsilon}_{3f} - 1}{\hat{\varepsilon}_3 - \varepsilon_1} \tag{81}$$

where $C_{dl,E}$ is the *differential* double-layer capacity. The ER effect will thus be very sensitive to changes in $C_{dl,E}$ resulting from changes in double-layer structure with potential, specific adsorption of ions, surface-film formation, or impurity adsorption.

For illustrative purposes, it is useful to divide metals into two classes: (a) pure free-electron metals (e.g., the alkalis) whose dielectric function in the energy range of interest is determined solely by intraband transitions; (b) metals which undergo both intraband and interband transitions in the photon energy range of interest. In the latter case, Re $\hat{\varepsilon}_b$ extends to frequencies well below the threshold frequency for the onset of interband transitions since it corresponds to the sum of contributions from all polarizability mechanisms (other than intraband) at higher frequencies.

Pure free-electron metals: For this case, $\hat{\varepsilon}_{3b} = 0$, $\hat{\varepsilon}_3 = \hat{\varepsilon}_{3f}$, and the normalized reflectivity change is given by

$$\begin{aligned} \frac{\Delta R}{R} &= -\frac{8\pi n_1}{\lambda}\frac{\Delta N_s}{N} \operatorname{Im} \frac{\hat{\varepsilon}_{3f} - 1}{\hat{\varepsilon}_{3f} - \varepsilon_1} \\ &= -\frac{8\pi n_1}{\lambda}\frac{\Delta N_s}{N} \frac{\varepsilon_3''(\varepsilon_1 - 1)}{(\varepsilon_3' - \varepsilon_1)^2 + \varepsilon_3''^2} \end{aligned} \tag{82}$$

Since ε_3'' is positive and $\varepsilon_1 > 1$ always, the response $R^{-1}\, \partial R/\partial E$ is predicted to be very small and *positive*.† It is interesting to note that to first order in d/λ, a free-electron metal in an air or vacuum ambient ($n_1 = \sqrt{\varepsilon_1} = 1$) will *not* exhibit an ER effect. From Eq.

† This sign is opposite to what might be predicted intuitively from consideration of a two-phase system. Its origin in the three-phase case is made clear by linear-approximation theory. The effect is equivalent to a shift of the ε_f'' curve to lower energies by anodic polarization of the metal.

82, it is also apparent that $\Delta R/R$ is closely related to the dielectric loss function (76) of the metal, Im $\hat{\varepsilon}_3^{-1} = \varepsilon_3''/|\hat{\varepsilon}_3'|^2$.

Metals with interband transitions: The positive response predicted for a pure free-electron metal is opposite in sign to the ER effect observed experimentally for the noble metals Ag and Au. Computer calculations based on this simple plasma-frequency shift model and measured values of the optical constants of these metals also indicate that the normal incidence ER spectra, $R^{-1}\,\partial R/\partial E$ versus $\hbar\omega$, are *negative* over the entire visible-UV wavelength region. The origin of the sign reversal is due to the contribution of the interband component. The onset of interband transitions in Au near 2.0 eV is evident from the curve of ε_3'' versus $\hbar\omega$ in Fig. 23. For Ag, the threshold frequency for the transition $L_3 \rightarrow$ Fermi surface occurs at 3.9 eV. Even though $\hat{\varepsilon}_b$ is not modulated by the field according to this model, it remains as a constant term in Eq. 79. Expansion of this relation gives the general result

$$\frac{\Delta R}{R} = \frac{8\pi n_1}{\lambda}\frac{\Delta N_s}{N}\frac{\varepsilon_{3f}''[\varepsilon_{3b}' - (\varepsilon_1 - 1)] - \varepsilon_{3b}''(\varepsilon_{3f}' - 1)}{(\varepsilon_3' - \varepsilon_1)^2 + \varepsilon_3''^2} \tag{83}$$

Since $(\varepsilon_{3f}' - 1)$ is always negative, while ε_{3f}'' and ε_{3b}'' are inherently positive, it is evident that $R^{-1}\Delta R/\Delta N_s$ is always positive and $R^{-1}\,\partial R/\partial E$ is therefore always negative when Re $\hat{\varepsilon}_{3b} > (\varepsilon_1 - 1)$. Decomposition of the computed values of $\hat{\varepsilon}_3$ into the intra- and interband contributions according to the method of Ehrenreich and Philipp (75) reveals this condition is satisfied for the metals Ag and Au over the complete transparent wavelength range of the electrolyte. Figure 29 illustrates the relative magnitudes of the free and bound electron terms in $\hat{\varepsilon}_{Au}$.

Again, $\Delta R/R$ is closely related to the dielectric loss function of the metal at high frequencies where $\omega\tau \gg 1$, such that $\varepsilon_{3f}'' \approx 0$ and $\varepsilon_3'' \approx \varepsilon_{3b}''$. Equation 83 then reduces to

$$\frac{\Delta R}{R} \approx -\frac{8\pi n_1}{\lambda}\frac{\Delta N_s}{N}\frac{\varepsilon_3''(\varepsilon_{3f}' - 1)}{(\varepsilon_3' - \varepsilon_1)^2 + \varepsilon_3''^2} \tag{84}$$

which is similar in form to Eq. 82, since ε_{3f}' varies only slowly with ω in this region.

In the above treatment, no specific model has been assumed for the structure of the double-layer region. The optical modulation effects are assumed to be solely due to an alteration of the surface-charge density and to first order are independent of its actual distribution. Although obviously idealized, this model accounts for

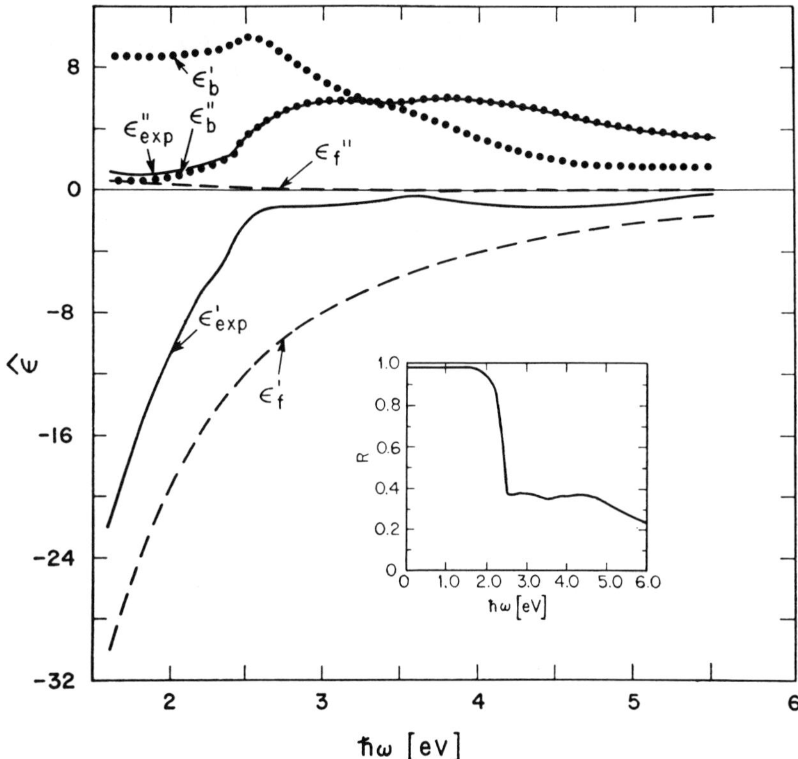

Fig. 29. Decomposition of experimental values of the dielectric constant of Au into free and bound electron contributions: $\hat{\varepsilon}_{\text{expt}} = \hat{\varepsilon}_f + \hat{\varepsilon}_b$; ——, $\hat{\varepsilon}_{\text{expt}} = \varepsilon'_{\text{expt}} - i\varepsilon''_{\text{expt}}$; – – –, $\hat{\varepsilon}_f = \varepsilon'_f - i\varepsilon''_f$; $\cdots$ $\hat{\varepsilon}_b = \hat{\varepsilon}'_b - i\varepsilon''_b$. After McIntyre (199).

the main features of the ER spectra of Ag and Au on a semiquantitative basis. The influences of interband transitions appear only indirectly, through their contribution to the bulk-metal dielectric constant, which is not modulated by the applied field.

(*ii*) *Field-Assisted Interband Transitions.* The effects of a field-induced perturbation of $\hat{\varepsilon}_b$ in the surface region can be seen from the normal-incidence relation

$$\frac{\Delta R}{R} = \frac{8\pi n_1 d}{\lambda}\left[\frac{-\varepsilon''_3\langle\Delta\varepsilon'_3\rangle + (\varepsilon'_3 - \varepsilon_1)\langle\Delta\varepsilon''_3\rangle}{(\varepsilon_1 - \varepsilon'_3)^2 + \varepsilon''^2_3}\right] \tag{85}$$

If such modulation occurred for the noble metals, anodic polarization would either: (i) lower the Fermi level (127), or (ii) raise the energy levels of *d*-band states in the surface atom layer. Simple calculations show (204) that in the absence of free-electron effects, a large peak of *positive* sign should appear in the $R^{-1}\,\partial R/\partial E$ spectrum of Ag near 4.0 eV, close to the threshold energy of the $4d \rightarrow 5s$ (Fermi surface) transition. The absence of such a peak in the experimental spectrum, which is everywhere negative, indicates that $\hat{\varepsilon}_b$ is not appreciably modulated by the applied field.

C. Surface States. Certain features of the ER effect in metals remain unexplained. With increasing anodic bias, the spectral peaks are observed to broaden and shift to lower energy (50,198). Figure 30 illustrates the effects observed for Au. Cahan, et al. (50) have suggested these effects may be associated with electromodulation of the energies of electronic surface states resulting from lattice truncation (intrinsic Tamm states) or specific chemical interactions of the metal surface atoms with electrolyte species (extrinsic states). They proposed that the rounding of the reflectivity curve of Au (cf. Fig. 29) is associated with excess absorption due to such states.

A number of difficulties are encountered with this model. Since there is no bandgap in metals and the densities of bulk electron states at the Fermi level and lower-lying d-electron states are both very high, metallic surface states are difficult to distinguish from

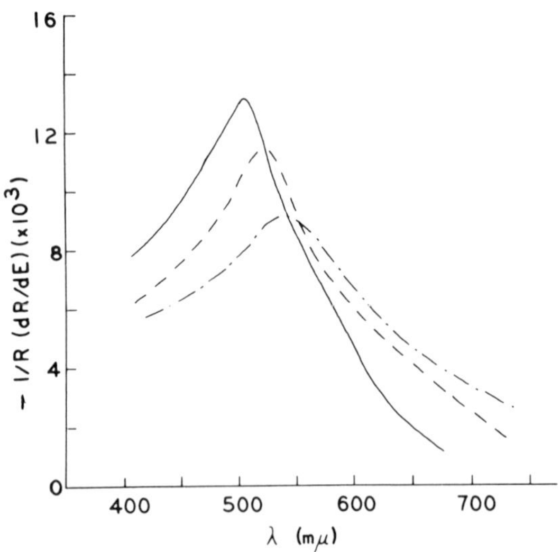

Fig. 30. Electroreflectance spectra of Au in 1 N $HClO_4$ for different bias potentials: ——, 0.2 V; – – –, 0.6 V; and –·–, 1.0 V. After Cahan et al. (50).

bulk states (62). In fact, no experimental evidence for *intrinsic* states has yet been found by either photoemission or ion-neutralization spectroscopy (74). For a metal in contact with a liquid, it is indeed doubtful that intrinsic states could exist.

Further, in the low-energy section of the pseudoplasma edge of Au ($\hbar\,\omega \leq 2.1$ eV), the shape of the reflectivity curve is controlled by the *free*-electron properties of the metal. This is a result of the fact (cf. Eq. 31) that the energy absorption in the metal, $A_m = 1 - R$, is proportional to $\varepsilon''\langle E^2 \rangle$. For Au, the quantities ε_f'' and ε_b' vary slowly with energy in this region, whereas ε_f' varies rapidly (cf. Fig. 29). This frequency dispersion in ε_f' is reflected in the mean-square field strength in the metal and hence in R. Any small structural features in $\hat{\varepsilon}_b$ in this long-wavelength region (e.g., those due to possible intrinsic surface states near the edge of the 5d band in Au) will be strongly masked by the rapid fall in ε_f' towards $-\infty$. As a result, such surface states will be very difficult, if not impossible, to detect by optical methods. Extrinsic surface states, formed by chemisorption of foreign atoms on the metal surface, may have unique optical properties of their own. In this case, the "surface compounds" formed may best be considered as a distinct phase, rather than as a perturbed layer of the metal.† The optical properties of such systems are discussed in Section V.

A number of surface-scattering mechanisms may be responsible for the broadening of the ER peaks. These include scattering by localized surface charges, scattering by lattice vibrations,‡ and scattering from surface roughness (111). Such effects produce a lowering of the surface conductivity and a decrease in the relaxation time of the electrons. Thus the value of τ estimated from the classical Lorentz-Sommerfeld relation

$$\tau = \frac{m^* \sigma_0}{Ne^2} \tag{86}$$

may be much too high for the very thin surface region where $\hat{\varepsilon}_m$

† The term "electroreflectance" has unfortunately come to have a much broader connotation than its original physical definition. It would seem desirable to restrict its use to pure field-induced modulation effects and to use a term such as "electrochemical modulation spectroscopy" for optical effects of chemical origin, e.g., adsorption.

‡ Additional surface electron-phonon coupling may result from interaction of the one-electron wave functions with the water molecules in the compact double layer.

is modulated by the applied low-frequency field. Microscopic surface roughness may also induce broadening through generation of surface plasmons, when the appropriate boundary conditions are satisfied (199), or through anomalous plasma resonance absorption (65). It is probable that the peak shift in the ER spectrum is caused by a distortion or shift of the free-electron density profile at the surface with change in dc bias or, equivalently, by polarization of microscopic surface bumps. A more detailed model of the metal-electrolyte interface must be developed to take such effects into account.

D. Piezoreflectance Effects. Finally, it may be noted that the shapes of the experimental ER spectra of Ag and Au are not consistent with a piezoreflectance effect induced by the Helmholtz stress ($\mathscr{E}^2/8\pi$) normal to the surface. Piezoreflectance spectra for these metals are derivativelike (97), in contrast to the one-sided ER spectra.

V. Optical Properties of Electrode Surface Films

Initial applications of specular reflection spectroscopy for characterization of the electrode-solution interphase have primarily emphasized the *speed* of this technique for *in situ* kinetic studies of the formation of adsorbed layers and ultra-thin surface films. A few studies, however, have demonstrated its utility for measuring the actual *spectroscopic* properties of these layers. In general, these spectro-electrochemical investigations have been confined to "known" systems, in order to gain an understanding of their optical behavior.

1. Anion Adsorption

The adsorption of halide ions on platinum and gold electrodes has recently been studied, using both reflectometric and ellipsometric techniques. Figure 31 illustrates the relative intensity changes observed by Barrett and Parsons (17) for Cl^- and Br^- ion adsorption on Pt in 0.5 M H_2SO_4. For Br^- adsorption, the surface coverage θ is independent of concentration over the range $[Br^-] = 10^{-4}$ to 10^{-2} mole/liter. For both Cl^- and Br^- adsorption, θ increases with potential. This conclusion is based on the assumption that a given coverage will result in the same optical effect at different potentials, without any complicating factors such as a change in the distance of closest approach of the anions or their polarization, or a change in state of the platinum surface. The assumptions are borne

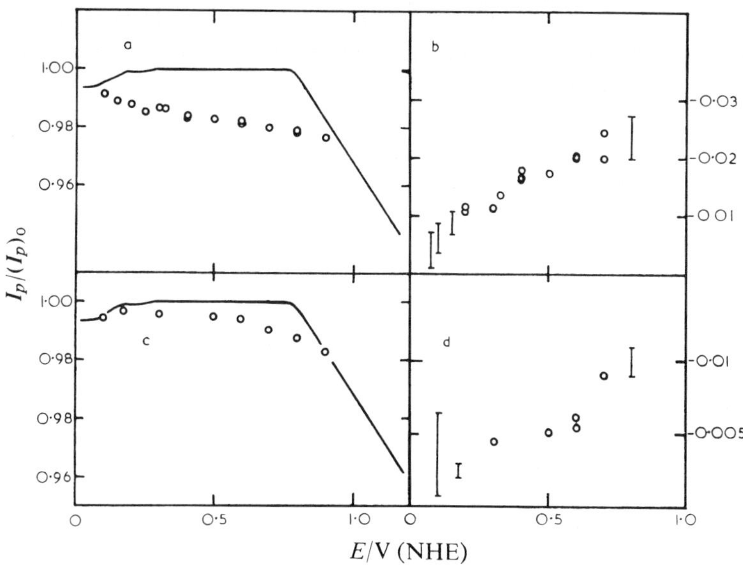

Fig. 31. Dependence of the reflectivity change on potential and halide-ion concentration for Pt in 0.5 M H_2SO_4. (a) ——, $[Br^-] = 0$; ∘ ∘ ∘, 10^{-4}, 10^{-3}, and 10^{-2} M KBr added; (b) lowering of I_p due to Br^- adsorption deduced from (a); (c) the same as (a) but 2.5×10^{-5} M KCl added; (d) lowering of I_p due to Cl^- adsorption deduced from (c). After Barrett and Parsons (17).

out by the (time)$^{1/2}$ dependence of $\Delta R/R$ following an anodic potential step, indicating that the adsorption process is diffusion-controlled and that θ is potential-dependent. The results of Barrett and Parsons for Cl^- adsorption are similar to those obtained by coulometric (14,102,103) and radiotracer studies (16). The recent ellipsometric results of Chiu and Genshaw (55,56) exhibit quite different concentration and potential dependences. The latter authors assumed the optical effects were entirely due to a change in refractive index of the ionic double layer.† It is now believed, however, that electroreflectance effects, arising from the change in electronic surface-charge density with change in potential, the alteration of the double-layer capacity due to specific adsorption of Cl^- ions, and the

† The Chiu-Genshaw model has been criticized by Barrett and Parsons on the basis that a *nonabsorbing* layer cannot account for the sign of the optical effect observed at low angles. Their criticism, however, is inconsistent with the predictions of Eqs. 23 and 30 if the transparent adsorbed layer has a refractive index higher than that of the bulk electrolyte, as is the case for anionic adsorption (266).

perturbations of the optical properties of the surface metal-atom layer by strong interactions with these chemisorbed species, are of greater importance.

The adsorption of halide ions on gold electrodes in 0.2 *M* $HClO_4$ has been studied by Takamura, et al. (273,274). The reflectivity change following an anodic potential step consists of two parts: (i) a fast component, which is the same as the total change in solutions without halide ions and corresponds to the rapid rearrangement of electronic surface charge and the relaxation of the ionic double layer; (ii) a slow component corresponding to the diffusion-controlled adsorption of halide ions. The reflectivity change accompanying cathodic desorption is virtually instantaneous, indicating that this process is not diffusion-controlled and, further, that the reflectivity is insensitive to the perturbation of the electrolyte refractive index by the excess halide-ion concentration in the ionic double layer and the Nernst diffusion layer.

In Fig. 32 is illustrated† the Cl^- concentration dependence of the

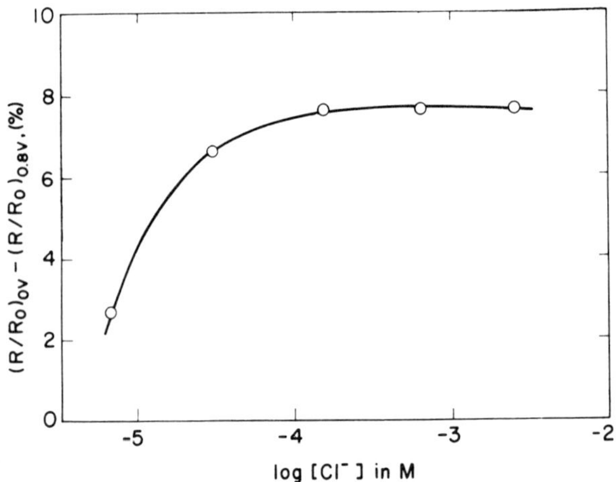

Fig. 32. Dependence of the reflectivity change of Au on Cl^- concentration in 0.2 *M* $HClO_4$. $\phi_1 = 50°$, $\lambda = 560$ nm, parallel-polarization, 13 reflections. After Takamura et al. (274).

† Owing to the effects of scattered light in the multiple-reflection cell used in these experiments, the absolute values of $\Delta R/R$ may be seriously in error. At a given wavelength, however, the relative dependence of $\Delta R/R$ on potential can still be used to monitor adsorption/desorption processes at the electrode surface (274).

reflectivity of Au in 0.2 M $HClO_4$, at a potential of 0.80 V versus Ag, AgCl/10^{-2} M Cl^- where voltammetric measurements indicate Cl^- is adsorbed. Assuming that the change in reflectivity is linearly proportional to the surface coverage with specifically adsorbed Cl^- (cf. Section II.4), this plot corresponds to the adsorption isotherm at constant potential. This assumption is substantiated by the observed linear variation of $\Delta R/R$ with the change in double-layer charge resulting from halide adsorption, as illustrated in Fig. 33. The linearity of these plots suggests that there is relatively little

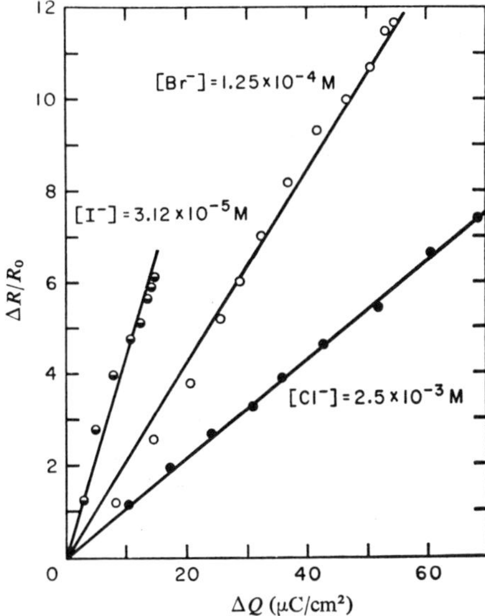

Fig. 33. Variation of reflectivity with change in double-layer charge due to halide adsorption. $\Delta Q = Q' - Q$, where Q' and Q are, respectively, the anodic coulombs required to charge the double layer in the presence and absence of halide; ΔR is defined analogously. After Takamura et al. (273).

variation in the optical-frequency dielectric constant of the adsorbed layer arising from dipole-dipole repulsion at the low surface coverages involved ($\theta \leq 0.3$).

In alkaline electrolytes, the reflectivity versus potential curve for Au exhibits a sharp break in the oxide-free region at -0.40 V versus S.C.E. (272). This potential corresponds to the point of zero charge (pzc) reported by Bode et al. (36). Specific adsorption of I^- shifts the pzc to more negative values; the shift is linear with log $[I^-]$ as expected on the basis of the Esin-Markov effect (80). Takamura et al. attribute the change in slope at the pzc to a change in the fraction of the applied dc electric field that penetrates the metal surface. The redistribution of potential is associated with the change in ionic constitution of the inner layer and solvent-dipole reorientation.

The optical effects due to the structure of the electrical double layer on the solution side of the interface have been discussed by Stedman (266,267). Parsons (223) has noted that, since specific ion adsorption is a rather common phenomenon, optical techniques for determining the pzc will not be generally useful.

2. *Adsorbed Oxygen and Oxide Layers*

One of the first demonstrations of the sensitivity of normal-incidence specular reflection spectroscopy for submonolayer film detection was the study by Koch, et al. (166,167) of *O* adsorption on Pt mirror electrodes. The results of these initial studies were analyzed as transmission spectra. Since this original work, a number of similar investigations have been carried out for Pt, Pd, Au, and Ni electrodes and analyzed according to the principles of thin film optics. The results are summarized below.

A. Platinum. The spectroscopic properties of the adsorbed oxygen or oxide layer formed anodically on Pt have recently been investigated using oblique-incidence techniques (17,201). Figure 34 illustrates the variation of the reflectivity change (normalized at $E_H = 0.4$ V) with potential for a Pt mirror electrode in 1.0 *M* $HClO_4$.† The apparent independence of the reflectivity change on the occurrence of a simultaneous Faradaic reaction demonstrates the utility of this technique for monitoring surface coverage. As frequently observed in such studies (146), the number of coulombs passed during the anodic formation of the O layer is greater than the number consumed during its cathodic reduction. The closure of the loop in the $\Delta R/R$ versus E curve, however, indicates the surface is restored to its initial state when the potential cycle is completed.

The spectral variation of the reflectivity change caused by O layer formation is shown in Fig. 35. The shift of the minimum of these curves to lower energy with increasing film thickness is due to a

† The potential at which the reflectivity begins to decrease rapidly during the anodic scan is a function of electrolyte composition and purity and electrode surface preparation, and can occur as low as 0.8 V (cf. Fig. 34). The "sudden" onset of a film near 0.95 V, attributed by Reddy et al. (242, 243) to "phase oxide" formation, has not been observed in recent ellipsometric studies (95, 108), which demonstrate a direct correspondence between optical effect and anodic charge.

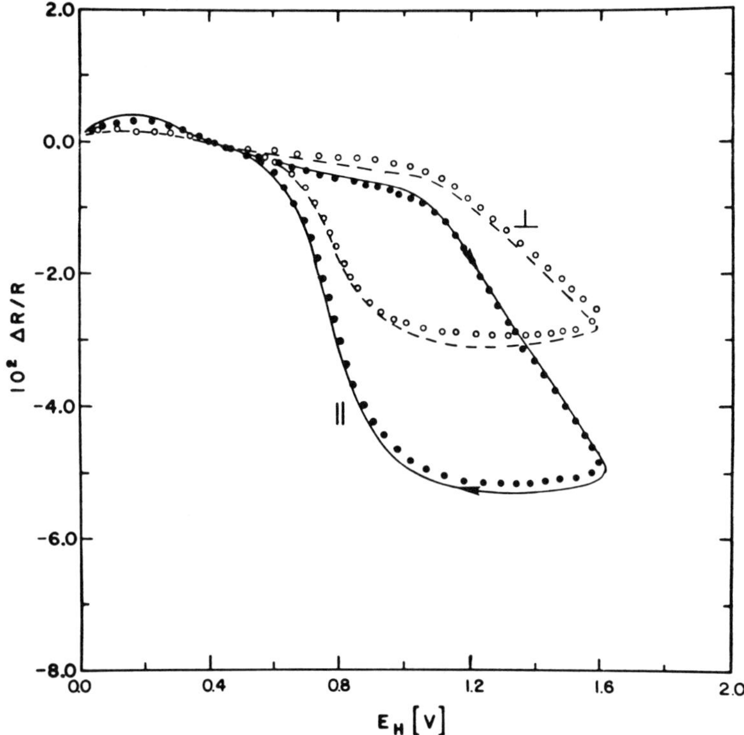

Fig. 34. Variation of normalized differential reflectance with potential for a Pt electrode in 1.0 M $HClO_4$ at 3000 Å. $\phi_1 = 45°$; scan rate $= 30$ mV sec^{-1}. —— Ar-saturated solution; $\cdots$, ∘∘∘, O_2-saturated solution. After McIntyre and Kolb (201).

change in the optical properties of the film. Analysis of oblique-incidence reflection spectra according to an idealized three-phase model (201) reveals the presence of two distinct bands in the absorption spectrum of the O layer at 3.4 and 4.5 eV, as shown in Fig. 36. The relative heights of these peaks vary with potential. Correlation with coulometric data (32) and electron spectroscopy (ESCA) results (159) suggests these bands are associated with the surface molecules PtO_2 and PtO, respectively. Conway and co-workers (59) have proposed, however, that the oxidation of the surface corresponds to successive coverage of Pt sites by "*O*H," then by "O," and finally by "phase oxide," formed by place exchange of Pt and O atoms at potentials more positive than 1.1 V.

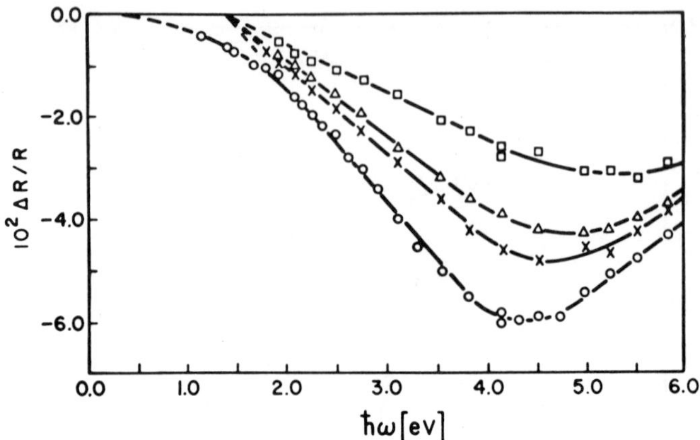

Fig. 35. Normal-incidence differential reflection spectra of a Pt electrode in 1.0 *M* $HClO_4$ (Ar-saturated) at a series of electrode potentials. □, 1.2 V; △, 1.4 V; ×, 1.5 V; ○, 1.8 V. After McIntyre and Kolb (201).

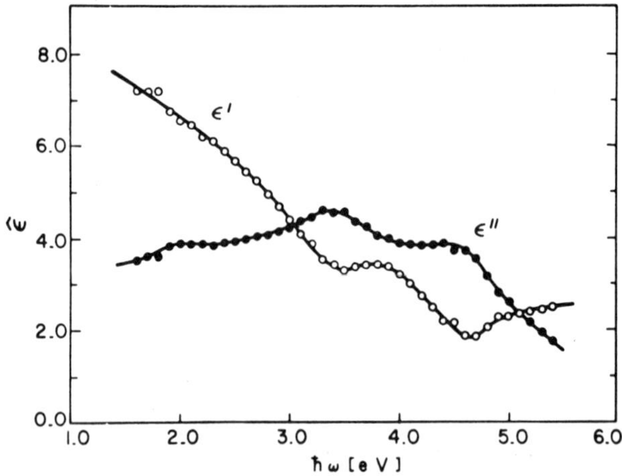

Fig. 36. Frequency dependence of the real and imaginary components of the dielectric constant, $\hat{\varepsilon} = \varepsilon' - i\varepsilon''$, of the O layer on Pt in 1.0 *M* $HClO_4$. $E_H = 1.5$ V. After McIntyre and Kolb (201).

Further work is required to identify the adsorbed species with certainty.

The optical distinction between a chemisorbed O layer and a "phase-oxide" layer is not a clear one. The high magnitude of the absorption coefficients (10^5–10^6/cm) suggests that photon-assisted charge-transfer transitions are responsible for the large reflectivity changes produced by monolayer-film formation. It has previously been proposed (243,244), however, that the optical properties of a chemisorbed O layer are not sufficiently different from those of the compact water layer to register an optical effect. This assumption is based on the observation (178) that thin films of physisorbed or condensed O_2 and H_2O are *transparent* and have refractive indices very close to those of bulk water.

Modern quantum-mechanical treatments of atoms *chemisorbed* on metal surfaces (73,95,96,106,114,116,120,251) now lead us to believe that monolayer films, less than ca. 5 Å in thickness, may have optical properties remarkably different from those of bulk material and may in fact exhibit *strong* absorption. When a foreign "impurity" atom (or ion) is chemisorbed on the surface of a metal, the discrete unperturbed ground-state levels of the valence electrons in the free atom are shifted and lifetime-broadened due to the interaction of the electrons and ion core with the continuum of metallic states. The absorption spectrum of the surface monolayer is, thus, characteristic of the electronic structure of the metal-adparticle *complex*. The latter may be ionic, polar-metallic, or metallic, depending upon the relative magnitudes of the ionization potential of the free atom, the work function of the metal, and the image potential and repulsion terms. Since the broad distribution of available energy levels of the virtual bound state of the adatom may either overlap or lie close to the Fermi level of the metal (either above or below), photon-induced charge-transfer transitions will be allowed over a broad energy range in the UV-visible region. This simple physical model can thus account for the apparent strong absorption exhibited by chemisorbed O layers on the metals Pt, Pd, and Au. The distinction between chemisorbed O and phase oxide layers then depends on the relative extent of orbital overlap between the metal surface atoms and the adsorbed species.

Anisotropy and submicroscopic surface roughness are additional possible sources of "apparent" strong absorption in monolayer surface films (65,66). These effects are discussed in Section V.5.

B. Gold. Anodic oxide formation on Au also produces a hysteresis loop in the $\Delta R/R$ versus E curve (cf. Fig. 41), analogous to that observed for Pt. In the anodic scan, the change in slope near 1.2 V (S.C.E.) ($\sim$500 μC/cm^2) is attributed to the completion of monolayer coverage (272). Figure 37*a* illustrates the polarized differential reflection spectra of the O layer formed by stepping the potential from 0.4 to 1.6 V (N.H.E.). Takamura et al. (272) proposed that the reflectivity change is associated with a shift of the threshold frequency of the $5d \rightarrow 6s$ interband transition, resulting from the interaction of the oxide ions with the surface orbitals of the metal or,

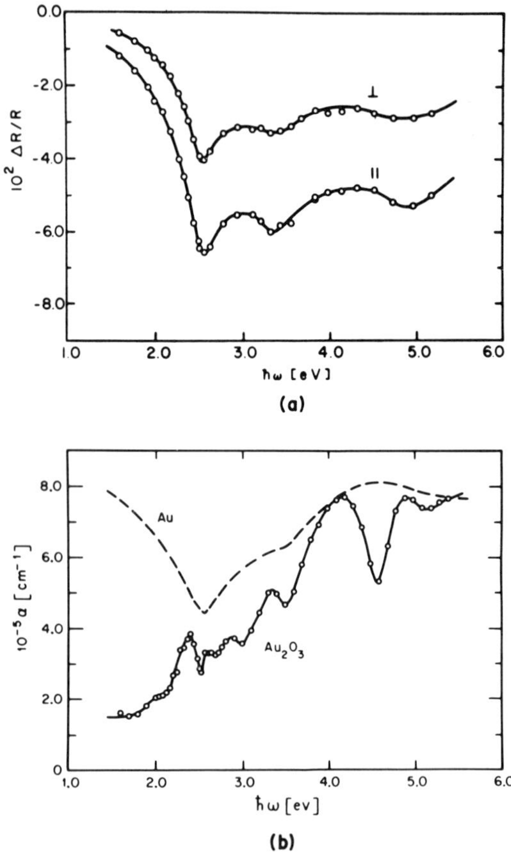

Fig. 37. (*a*) Differential reflection spectra ($\phi_1 = 45°$) of the anodic O layer formed on an Au electrode at $E_H = 1.6$ V in 1.0 *M* $HClO_4$ (Ar-saturated). After McIntyre and Kolb (201). (*b*) Absorption spectra of bulk Au and the anodic O layer on Au (assumed to be Au_2O_3). After Kolb and McIntyre (171).

alternatively, from a penetration of the local field of these ions into the metal phase.

As noted in Section II, the form of the differential reflection spectrum for moderately reflecting metals such as Au is determined by three factors: (i) absorption in the surface film; (ii) the change in absorption by the metal substrate; (iii) the normalization factor, R. Thus, it is exceedingly difficult to deduce the mechanism of the absorption loss directly from the $\Delta R/R$ versus $\hbar\omega$ spectrum. Figure 37*b* illustrates the actual absorption spectrum of the anodic O layer on Au, as calculated on the basis of an idealized isotropic model (203). Strong absorption bands at 4.2 and 4.9 eV are evident here but are concealed in the original differential reflection spectrum. Comparison of the absorption spectrum of the O layer with that of bulk Au (cf. Fig. 37*b*) indicates the surface layer has unique optical properties, which are more closely characteristic of a semiconducting surface compound than of a perturbed layer of metal.

It is interesting to note, however, that a rough parallelism exists between $\alpha_{Au_2O_3}$ and α_{Au}. Indeed, this is to be expected on the basis of the quantum-mechanical treatment outlined above, because the unperturbed wave functions of the free atom are mixed with those of the metal surface in forming the virtual bound state. As yet, however, the extent of the contributions made by surface roughness and/or film anisotropy is not clear, so that we cannot state with certainty whether the apparent strong absorption is real or an artifact of the over-simplified model used in computing the optical constants of the film.

C. Nickel. The differential reflection spectrum of the oxide layer NiOOH, formed by anodization of Ni electrodes at high potentials in alkaline solution, differs notably from the preceding cases. As shown in Fig. 38, both negative *and* positive values of $\Delta R/R$ are observed for this system and these reflectance changes are extremely large for a film ca. 13 Å thick. The absorption spectrum of Ni(III) oxide has two broad bands at 3.0 and 4.1 eV, close to the bands near 3.1 and 4.5 eV observed by McClure (194) and Tippins (276) for Ni^{+3} ions substitutionally dissolved in corundum single crystals. As the thickness of the oxide layer is substantially greater than that of a monolayer, it is probable that the main absorption occurs within the film itself in this case. The magnitudes of the absorption coefficients (10^5–10^6/cm) and the high oscillator strengths indicate that the absorption maxima are associated with charge-transfer transitions between the $O^{2-}(2p^6)$ bands and vacant $3d$ or $4s$ levels

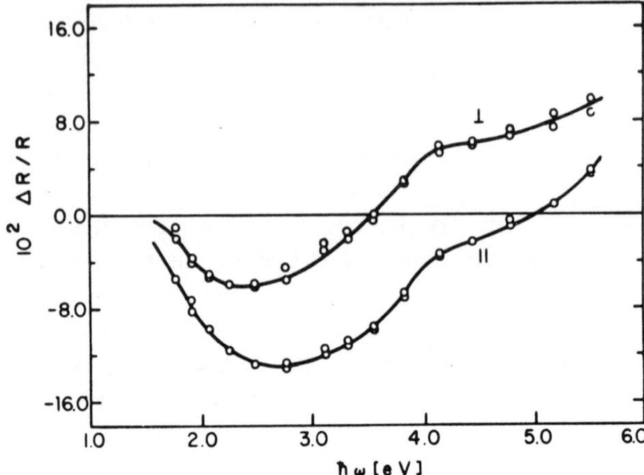

Fig. 38. Differential reflection spectra ($\phi_1 = 45°$) of the NiOOH layer formed on a Ni electrode in 0.1 *M* KOH (Ar-saturated) at $E_H = 1.5$ V. After McIntyre and Kolb (201).

of the Ni^{3+} ions. The real component of the dielectric function of NiOOH is *negative* for $\hbar\omega > 2.2$ eV, indicating *metallic* conductivity (201). Recent work (204) has shown that the dielectric function of the divalent oxide, $Ni(OH)_2$, is characteristic of a strongly absorbing insulator or semiconductor. It is this transition from low to high conductivity in the surface-oxide film which is responsible for the passivity of Ni.

3. *Adsorbed Hydrogen*

In studies of the gas-solid interface, molecular hydrogen (H_2) is found to be chemisorbed by Pt in two distinct adsorption states, which are clearly revealed by surface-potential measurements (208), field-emission microscopy (185,247), thin-film resistance measurements (231), and infrared spectroscopy (228). Mignolet (208) found that on a clean Pt surface at low temperatures (−190°C), a strongly bound electronegative layer is adsorbed first, causing the work function Φ to increase by 0.15 eV. At higher coverages ($\theta \geq 0.4$), a weakly bound electropositive component was added, leading to a net work-function decrease of $\Delta\Phi = -0.23$ eV at full coverage. Both forms remained on the surface when heated in an H_2 atmosphere to 20°C; evacuation removed only the weakly bound species. Recent field-emission measurements (185,247) indicate that desorption of the weakly bound electropositive component occurs at 200–250°K; the more strongly bound component desorbs in turn at 275–300°K and desorption is complete at 320°K.

The numerous models put forth to describe the adsorbed states have been reviewed by Eley, et al. (79). It is generally agreed that since the *s*-orbital of H is spherically symmetric, the strongest covalent bonding will occur when this atom is situated in an interstitial surface site, thereby allowing simultaneous interaction with several substrate atoms. Lewis and Gomer (185) have recently proposed a configuration of this type, analogous to the "*s*-state" discussed by Horiuti and Toya (147) but differing significantly in that the adsorbed layer has a *negative* surface dipole.† Lewis and Gomer postulate that the net electronic charge associated with the proton is greater than unity and that the center of gravity of this charge distribution lies outside the plane of electroneutrality. In accord with Mignolet, they conclude that the electropositive adsorbate is probably molecular H_2 bound by charge-transfer forces. Disagreement persists concerning the polarity of the Pt—H bond (60,94) and the nature of the adsorbed states.

Eley and co-workers (79) have recently deduced from infrared studies that hydrogen adsorption on supported platinum catalysts probably occurs at *platinum oxide patches*, rather than at atomic and interstitial sites on metallic Pt, as originally proposed by Pliskin and Eischens in their classic study (228). The latter's conclusion that the weakly bound hydrogen is not adsorbed as molecule-ions (H_2^+) at the gas-metal interface may now be open to question.

Electrochemical studies of hydrogen adsorption on Pt single crystal electrodes (15,295) also indicate that hydrogen adsorbs in two distinctly different binding states. These states are stabilized by the presence of electrolyte and continue to exist at high temperatures (350°K). The magnitudes of the heats of adsorption and kinetic considerations led Breiter (43) to conclude that *both* weakly and strongly bound forms of hydrogen are *atomically* adsorbed on Pt electrodes in aqueous electrolytes. Will (295) postulated that on single-crystal electrodes each of the nominal faces exposes several different planes in different proportions and tentatively concluded

† Horiuti and Toya identified the electronegative component as an "*r*-state," in which the H atom is covalently bonded to a single substrate atom and located directly above it. They apparently concluded incorrectly, however, that the *electropositive* component predominates at high temperatures, in opposition to results of work function (185, 247, 208) and resistance (231) measurements.

that the strongly and weakly bound forms correspond to atomic adsorption on {100} and {110} planes, respectively. However, in recent work by Bagotzky, et al. (15), the adsorption isotherms on different Pt single-crystal faces and on polycrystalline Pt were found to be practically identical. They concluded that the shape of the isotherms and the dependence of the heat of adsorption of hydrogen on surface coverage are governed not by surface structure but by the electronic properties of platinum.

Optical studies of hydrogen adsorption on Pt offer the possibility of resolving this controversy. Although the fractional reflectivity change associated with this process is very small, of the order of 0.1% (cf. Figs. 31 and 34), it can be highly resolved using modulation spectroscopy techniques, owing to the reversibility of the Volmer reaction

$$H_3O^+ + e \rightleftharpoons H_{ad} + H_2O$$

Phase-sensitive detection measurements with small modulating voltages reveal the presence of two distinct maxima in the plot of $(R^{-1}\Delta R/\Delta E)_\lambda$ versus E_H in the H-adsorption region (201), analogous to those observed in differential capacitance plots (248) and linear-sweep voltammograms (104,294). This follows, because when the adsorption pseudocapacitance, C_2, is much larger than the double-layer capacitance,

$$\frac{dR}{dE} = \frac{dR}{dQ}\frac{dQ}{dE} \approx C_2 \frac{dR}{dQ_{ad}} = i_{ad} \frac{dR/dQ_{ad}}{dE/dt} \tag{87}$$

where Q_{ad} and i_{ad} represent, respectively, the Faradaic charge and current density. Here, it is assumed that the optical effect is directly associated with the adsorbed layer and not due to an alteration of the double-layer capacitance and electroreflectance. This is supported by the fall of $R^{-1}\Delta R/\Delta E$ towards zero as the hydrogen coverage approaches saturation near $E_H = 0.0$ V. These measurements should be extended into the H_2-evolution region to verify this conclusion.

Although phase-sensitive detection methods are convenient for observing the wavelength variation of $\Delta R/R$, they are not as well suited for obtaining kinetic information (31).† If, for example, a

† Another significant disadvantage of using phase-sensitive detection with a slow potential scan is the high susceptibility of the electrode to impurity adsorption. This may account for the differences in behavior observed in initial modulation spectroscopy studies of H-adsorption (31, 201). Digital signal averagers facilitate use of potentiodynamic pulse techniques (105) for preparing a clean surface and studying the kinetics of adsorption processes.

potential step is applied to the electrode, a lock-in amplifier will reject all but the fundamental Fourier component of the ΔR signal. Signal-averaging or cross-correlation techniques retain the valuable high-frequency information. Figure 39 illustrates the signal-averaged time-dependence of the reflectivity observed by Bewick and Tuxford (31), when the potential of a Pt electrode in 1.0 M $HClO_4$ was modulated with a 30 Hz square wave. The transient behavior during the anodic period of Fig. 39*a* was attributed to a rapid ionization of weakly bound H_{ad}, followed by a slower ionization of the strongly bound adsorbate. During the cathodic half-cycle, both forms re-adsorb rapidly—the strongly bound component producing a small reflectivity increase, the weakly bound component producing a larger reflectivity decrease. The results of these optical studies are in agreement with previous voltammetric kinetic measurements by Breiter (42). The work of Bewick and Tuxford illustrates well that

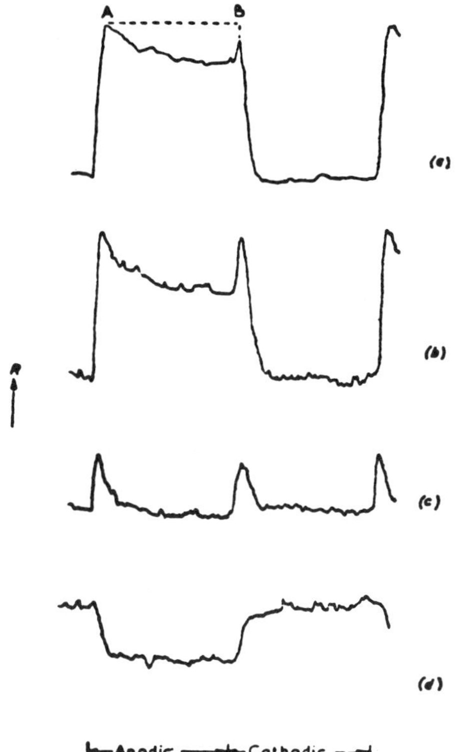

Fig. 39. The time dependence of the reflectance change at 435 nm on switching the potential of a Pt electrode in 1.0 M $HClO_4$ with a 30 Hz square wave from +320 mV (N.H.E.) to the potentials: (*a*) +20 mV; (*b*) +70 mV; (*c*) +120 mV; (*d*) +170 mV. After Bewick and Tuxford (31).

fast kinetic effects on electrode surfaces can be detected readily by modulation spectroscopy techniques. Owing to their wavelength selectivity, these optical methods are particularly useful for kinetic studies in potential regions where two or more species are coadsorbed [cf. (201)].

The origin of the minute reflectivity changes associated with H adsorption is not yet clear since accurate dielectric constant data have not been reported. Again, it is probable that the adsorbed species form strongly absorbing virtual bound states as discussed in Section V.2. Owing to the small size of the adatoms, however, the optical effect may simply be due to an alteration of the surface conductivity of the substrate due to covalent bond formation, diffuse-scattering of conduction electrons, or surface reconstitution, thus making it analogous to an electroreflectance phenomenon. Simultaneous measurements of film conductivity (cf. Ref. 260) may help to resolve this problem. When optical constant data become available, various physical models for the adsorbed states can be tested.

Adsorption of organic compounds on platinum has been studied optically by Barrett and Parsons (17). Methanol, formic acid, and formaldehyde are observed to displace hydrogen from the surface. The effects are too complicated to be reviewed adequately here. Determination of the spectroscopic properties of the adsorbed layers should prove of great value in elucidating the mechanisms of electrocatalysis.

4. Metal Atom Adsorption

Metal surfaces covered by a fractional monolayer of a foreign metal are of great technological importance in physical electronics devices because of their enhanced electron and ion emission properties. Adsorption of electropositive alkali and alkaline earth metals has been extensively studied using field-emission (229,251,271) and LEED (99) techniques, and significant advances in the quantum-mechanical description of the adsorbed state have recently been made (73,95,96). The ability of noble-metal electrodes such as Pt and Au immersed in aqueous electrolytes to adsorb monoatomic layers of foreign metals (e.g., Ag, Bi, Cd, Hg, Pb, Sn, Tl) at "underpotentials" (cf. Refs. cited in 201), has enabled the optical properties of metal adatom layers to be measured *in situ* by differential reflection spectroscopy. Such measurements are of considerable theoretical interest because it is possible to monitor the continuous variation in optical and electronic properties of the adsorbed layer as a function of surface coverage and to observe the transition from an atomic to a metallic state. The technique also offers a new means of studying the initial stages in metal electrodeposition.

Figure 40 illustrates the $\Delta R/R$ changes observed during the initial stages of Ag electrodeposition on a Pt substrate. Not unexpectedly, the optical properties of the Ag atomic layers initially deposited differ markedly from those of the bulk metal. A distinct minimum in the reflectivity occurs at a thickness corresponding to ca. one monolayer. With increasing film thickness and Ag atomic wavefunction overlap, the transition to metallic behavior is observed. This is a particularly interesting case for study because the dielectric function of the surface film varies rapidly as a function of $\bar{d}$ in the monolayer thickness range. At $\lambda = 4000$ Å and $d \approx 10$ Å, ε_2'' exhibits a broad minimum very close to zero, then rises and asymptotically approaches the value characteristic of bulk Ag (172).

This behavior can be interpreted qualitatively in terms of the

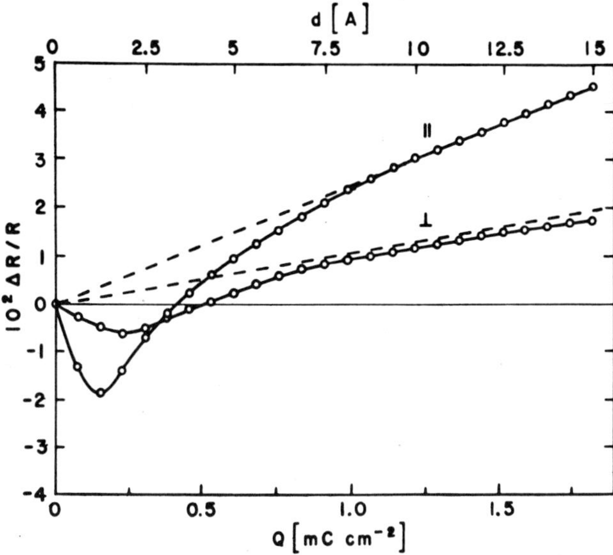

Fig. 40. Variation of normalized differential reflectance with cathodic charge passed during deposition of bulk Ag on a Pt electrode in 1.0 M $HClO_4$ (Ar-saturated). $E_H = 0.4$ V; $\lambda = 4000$ Å; $\phi_1 = 45°$. ——, experimental; – – –, computed. After McIntyre and Kolb (201).

quantum-mechanical model discussed in Section V.2. When $\theta \leq 1$, there is strong interaction of the valence electrons and ion cores of the Ag adatoms with the continuum of metallic states in the Pt substrate; the virtual bound states which are formed are strongly absorbing. At monolayer coverage, ε_2'' exhibits a broad absorption maximum near $\hbar\omega = 3.8$ eV (172). As the film thickness increases, interaction with the substrate is reduced and so is the probability of optically exciting an electron to (or from) its Fermi surface. The film becomes almost transparent since it cannot yet sustain metallic conductivity.† Finally, at a thickness of several monolayers, the film acquires metallic characteristics and becomes strongly absorbing once more. It is interesting to note that bulk metal behavior is approached in this system at a film thickness nearly an order of magnitude smaller than is the case for evaporated metal films on insulating substrates (cf. Ref. 249).

The Faradaic adsorption of Pb and Cd on Au electrodes at

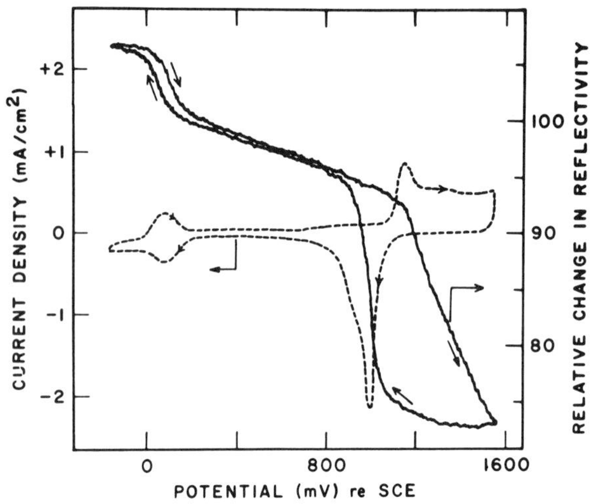

Fig. 41. Relative change in reflectivity-potential and current-potential curves for Au in the presence of Pb^{2+} ions. Electrolyte: 5×10^{-4} *M* Pb^{2+} + 0.2 *M* $HClO_4$. Potential sweep rate = 105 mV sec^{-1}; $\lambda = 5200$ Å; electrode area = 23 cm^2; $\phi_1 = 50°$; 18 reflections. After Takamura et al. (272).

† The initial Ag adatom layer forms uniformly on the surface (201), but further deposition may result in the formation of three-dimensional nuclei. Recall that the mean free path for electron-phonon collisions in a metal is ca. 10^3 Å (165).

underpotentials has been studied by Takamura, et al. (272). Figure 41 illustrates the variation of reflectivity and current for an Au electrode in 0.2 M $HClO_4$ containing 5×10^{-4} M Pb^{+2}. The peaks in the voltammetry curve at 0.1 V (S.C.E.) correspond to the cathodic adsorption and oxidative desorption of a single monolayer of Pb; similar behavior is observed for Cd. Oxide formation is evident at potentials more anodic than 0.9 V. The reflectivity versus potential curves exhibit marked changes in the metal adsorption region and are strikingly wavelength-dependent. As shown in Fig. 42, the relative change, $\Delta R/R$, alters sign when the wavelength is varied. A more detailed interpretation of these interesting results must await determination of the optical constants of the metal adatom layers.

In related studies of Ag adsorption at underpotentials on Au (172), the $\Delta R/R$ spectra indicate a gold-rich surface alloy is formed.

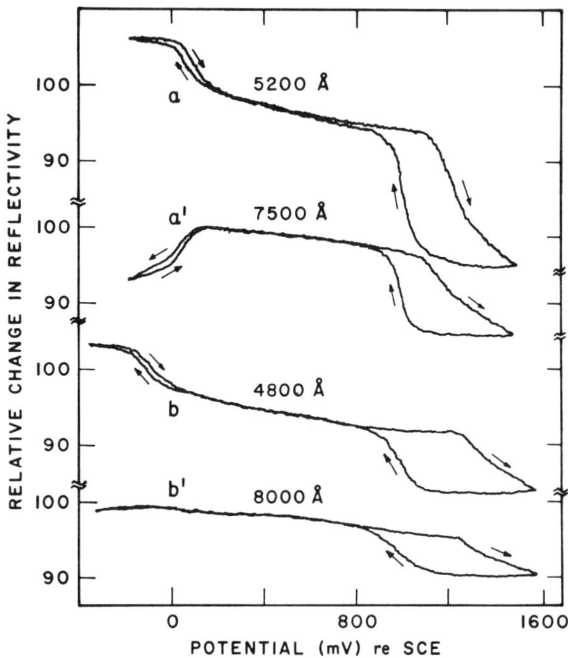

Fig. 42. Relative change in reflectivity-potential curves at various wavelengths in the presence of Pb^{2+} or Cd^{2+}. Curves a, a': 5×10^{-4} M Pb^{2+} + 0.2 M $HClO_4$. Curves b, b': 1×10^{-3} M Cd^{2+} + 0.2 M $HClO_4$. Other conditions the same as for Fig. 41. After Takamura et al. (272).

5. Anisotropy and Surface Roughness

In the preceding sections, we have treated the topic of monolayer spectroscopy from a highly idealized viewpoint, assuming: (i) the substrate surface is atomically smooth and optically flat and (ii) both substrate and adsorbate have uniform *isotropic* optical properties. Failure to meet these criteria may lead to severe problems in interpretation of experimental results.

The optical effects arising from anisotropy and surface roughness are very complicated to describe mathematically. Only a brief summary of the results can be given here.

A. Anisotropy. At low surface coverages, adsorbed molecules are expected to be preferentially oriented relative to the surface normal. If, in addition, they are randomly oriented with respect to rotation about this normal, the adsorbed film will behave as a *uniaxial* optical medium. Dignam and co-workers (66) have deduced that the "effective" optical constants, calculated on the assumption that the film is isotropic, are actually a function of the optical constants of all *three* phases. Thus a transparent uniaxial film on a metallic substrate will appear to be absorbing. Further, for specular reflectance measurements (but *not* ellipsometry), the dielectric constant of the film will appear to be angle-dependent. Apparent anisotropy of the O layer on Pt was in fact detected by McIntyre and Kolb (200,201).

The origin of the effect can be seen as follows. The complex dielectric constant of an absorbing uniaxial film is a second-rank tensor whose principal components normal and tangential to the surface are denoted as $\hat{\varepsilon}_n$ and $\hat{\varepsilon}_t$, respectively. Using first-order approximations for very thin films (analogous to those described in Section II.2B), Dignam, et al. showed that for perpendicularly polarized light

$$\operatorname{Im}\left[\frac{1 - \hat{\varepsilon}_2/\hat{\varepsilon}_3}{1 - \varepsilon_1/\hat{\varepsilon}_3}\right] = \operatorname{Im}\left[\frac{1 - \hat{\varepsilon}_t/\hat{\varepsilon}_3}{1 - \varepsilon_1/\hat{\varepsilon}_3}\right] \tag{88a}$$

whereas for parallel-polarized light

$$\operatorname{Im}\left[\frac{(1 - \hat{\varepsilon}_2/\hat{\varepsilon}_3)(1 - \varepsilon_1/\hat{\varepsilon}_2)}{(1 - \varepsilon_1/\hat{\varepsilon}_3)(\cot^2 \phi_1 - \varepsilon_1/\hat{\varepsilon}_3)}\right] = \operatorname{Im}\left[\frac{(1 - \hat{\varepsilon}_t/\hat{\varepsilon}_3)(1 - \varepsilon_1/\alpha_3\hat{\varepsilon}_3)}{(1 - \varepsilon_1/\hat{\varepsilon}_3)(\cot^2 \phi_1 - \varepsilon_1/\hat{\varepsilon}_3)}\right] \tag{88b}$$

where

$$\alpha_3 = \frac{\hat{\varepsilon}_n}{\hat{\varepsilon}_t}\left(\frac{\varepsilon_t - \hat{\varepsilon}_3}{\varepsilon_n - \hat{\varepsilon}_3}\right) \tag{89}$$

and where ε_1, $\hat{\varepsilon}_2$, $\hat{\varepsilon}_3$ have their usual isotropic significance. Thus, owing to the changes in the relative interactions of the X and Z fields with the film and/or substrate, the "effective" dielectric constant $\hat{\varepsilon}_2$ will vary with angle of incidence when the surface film is anisotropic.

As an example of the magnitude of the effect to be expected, numerical calculations for a thin, *transparent*, anisotropic film ($\varepsilon_t = 2.4$, $\varepsilon_n = 1.6$) on a Pt substrate yield the "effective" values: $k_2 \approx 0.03$ ($\phi_1 = 30°$) and $k_2 \approx 0.18$ ($\phi_1 = 70°$) in the visible wavelength region (66). The "effective" values of n_2 are close to 1.5 for both these angles of incidence. The film thus *appears* to acquire the optical properties of the substrate, a characteristic noted in the case of O layer formation on Au (cf. Fig. 37b). However, in the case of O layer formation on *both* Pt and Au, the measured absorption coefficients are more than an order of magnitude larger than those calculated according to the anisotropic model above. Thus it seems doubtful that film anisotropy can account entirely for the strong "apparent" absorption that is observed experimentally. More extensive calculations are required to establish the maximum "effective" values of k_2 that can be generated by this mechanism.

B. Surface Roughness. When light is incident on a surface with irregularities small in height relative to the wavelength, the reflected radiation comprises a coherent specularly reflected beam, and a diffuse coaxial cone of incoherently reflected rays that are uncorrelated in phase (22). While this scattering process will affect the accuracy of absolute reflectivity measurements, its effects will be largely cancelled in reflectivity ratio measurements. An effect of much greater importance is the anomalous plasma resonance absorption that can arise in a metal substrate when submicroscopic pits, domes, or foreign particles, a few angstroms in diameter, bestrew its surface. According to Berreman (28,29,30), the effect is due to electrically polar resonances in the metal in the neighborhood of the irregularities. The resonances are displaced from the bulk plasmon frequencies because of the electric fields induced around the highly curved surfaces of the bumps or pits. This coherent effect can cause marked changes in reflectance even when incoherent scattering is negligible.

A similar plasma resonance damping occurs in colloidal metal particles (174,175,176). The free electrons in these small particles can absorb only in a narrow band at the resonant frequency. The breadth of the band is determined by the radius of the spheres, owing to its limitation of the mean free path of the conduction electrons.

Dignam and Moskovits (65) have recently described an optical model that takes into account the effects of adsorption on slightly rough metal surfaces, approximating the influence of metal bumps by treating them as spheres, and assuming that the adsorbed layer is *transparent*. The reflectivity alteration is determined by four wavelength-independent parameters—the mean radius of the bumps, the volume fraction of the roughness layer that they occupy, the product of their polarizability and surface coverage, and a parameter accounting for the electron transfer that occurs on chemisorption. Excellent agreement was demonstrated between the optical effects predicted by the model and those observed experimentally on adsorption of methanol and oxygen from the gas phase onto 500 Å Ag films. The surface roughness region was assumed to consist of spherical bumps with a mean radius of ~ 10 Å and a packing fraction of ~ 0.3.

A similar effect is anticipated to occur in electrochemical systems. A change in the dielectric constant of the phase surrounding the metal bumps will result from adsorption, surface film formation, or simply a change in double-layer structure or constitution with potential. This will in turn alter the local electric field polarizing the bumps, and hence their free-electron concentration, conductivity, and resonance absorption properties. In some instances, the optical effect may be so large as to mask the inherent absorption characteristics of an adsorbed film *per se*! It is essential that the magnitude of this phenomenon be established as soon as possible.

VI. Future Applications

A wide variety of research problems in electrochemistry, surface chemistry, and surface physics is now open to attack by *in situ* specular reflection spectroscopy provided that accurate dielectric constant data for the interphase region can be measured. A major difficulty that beset previous optical studies of absorbing films on absorbing substrates has now been removed. Paik and Bockris (219) have recently developed a method for *uniquely* determining the optical constants and thickness of a surface film by combining *in situ* ellipsometric and reflectometric measurements. If the layer thickness can be determined accurately at one wavelength using this technique, fast kinetic studies can then be carried out at a series of wavelengths by using spectrophotometric methods (171) to measure the optical properties of the adsorbed layer.

Examples of chemical mechanistic applications include the use of modulation spectroscopy techniques for detection and identification of reaction intermediates and products, and determination of kinetic rate constants. Such studies should be of particular value for elucidating the mechanisms of electrocatalysis. In corrosion studies, the speed and high sensitivity of this method enable it to be used to monitor rapidly varying surface coverages. It can thus be employed effectively to distinguish the relative roles of adsorption and oxide films in establishing passivity and for determining the mechanism of surface film growth. Its spectral selectivity suggests its use for analysis of the chemical composition of passivating films. The optical and electronic properties of other compounds, which are difficult to prepare in an isolated state but which can be formed as stable surface films on electrodes, can also be studied in this way. In metal deposition studies, the size of nuclei formed in the growth process can be estimated.

The spectroscopic study of surface states has just begun. Preliminary results suggest it may be possible to locate the electronic energy levels of virtual bound states relative to the Fermi level of a metal substrate, thus yielding information of the same type as found by ion-neutralization and field-emission studies of the gas-solid interface.

Electroreflectance measurements enable detailed information to be gained about the electronic properties of the surface atomic layer of metal catalysts. Using high angles of incidence, information about the structure and dynamics of the ionic double layer can also be obtained. Electroreflectance measurements are also of potential interest in surface physics since they offer a convenient and very sensitive means of studying surface plasmon generation in metals. In addition, information may be gained about the shape of the electronic charge-density profile at metal surfaces and the effects of strong electric fields in assisting interband transitions. The electroreflectance effect in metals also offers the simplest practical system for studying the magnitude of the anomalous plasma resonance absorption caused by microscopic surface roughness. The results of these studies will be crucial in determining the future role of reflection spectroscopy in surface science.

References

1. Abelès, F., *Compt. Rend.*, **230**, 1942 (1950).
2. Abelès, F., *Proc. Phys. Soc.* (*London*), **B65**, 996 (1952).
3. Abelès, F., *Rev. d'Opt.*, **31**, 127 (1952).
4. Abelès, F., in *Progress in Optics*, Vol. II, E. Wolf, Ed., North-Holland, Amsterdam, 1963, p. 251.
5. Andermann, G., A. Caron, and D. A. Dows, *J. Opt. Soc. Amer.*, **55**, 1210 (1965).
6. Andreeva, V. V., *Corrosion*, **20**, 35t (1964).
7. Archer, R. J., *J. Electrochem. Soc.*, **104**, 619 (1957).
8. Archer, R. J., *J. Opt. Soc. Amer.*, **52**, 970 (1962).
9. Archer, R. J., in *Ellipsometry in the Measurement of Surfaces and Thin Films*, E. Passaglia, R. R. Stromberg, and J. Kruger, Eds., Natl. Bur. Stand. Misc. Publ. 256, U.S. Govt. Printing Office, Washington, 1964, p. 255.
10. Aspnes, D. E., *J. Opt. Soc. Amer.*, **61**, 1077 (1971).
11. Aspnes, D. E., *Phys. Rev. Lett.*, **28**, 168 (1972).
12. Avery, D. G., *Proc. Phys. Soc.* (*London*), **65B**, 425 (1952).
13. Axe, J. D., and R. Hammer, *Phys. Rev.*, **162**, 700 (1967).
14. Bagotsky, V. S., Yu. B. Vassilyev, J. Weber, and J. N. Pirtskhlava, *J. Electroanal. Chem.*, **27**, 31 (1970).
15. Bagotsky, V. S., Yu. B. Vassiliev, and I. I. Pyshnograeva, *Electrochim. Acta.*, **16**, 2141 (1971).
16. Balashova, N. A., and V. E. Kazarinov, in *Electroanalytical Chemistry*, Vol. 3, A. J. Bard, Ed., Marcel Dekker, Inc., New York, 1969, p. 135.
17. Barrett, M. A., and R. Parsons, *Symp. Faraday Soc.*, **4**, 72 (1970).
18. Bartell, L. S., and D. Churchill, *J. Phys. Chem.*, **65**, 2242 (1961).
19. Bashara, N. M., and D. W. Peterson, *J. Opt. Soc. Amer.*, **56**, 1320 (1966).
20. Baumann, R. P., *Absorption Spectroscopy*, Wiley, New York, 1962, p. 314.
21. Bennett, H. E., and J. O. Porteus, *J. Opt. Soc. Amer.*, **51**, 123 (1961).
22. Bennett, H. E., and J. M. Bennett, in *Physics of Thin Films*, Vol. 4, G. Hass and R. E. Thun, Eds., Academic Press, New York, 1967, p. 1.
23. Bennett, H. E., J. M. Bennett, E. J. Ashley, and R. M. Motyka, *Phys. Rev.*, **165**, 755 (1968).
24. Bennett, H. E., D. K. Burge, R. L. Peck, and J. M. Bennett, *J. Opt. Soc. Amer.*, **59**, 675 (1969).
25. Bennett, H. E., and J. M. Bennett, in *Handbook of Optics*, W. G. Driscoll, Ed., McGraw-Hill, New York, in press.
26. Berning, J. A., and P. M. Berning, *J. Opt. Soc. Amer.*, **50**, 813 (1960).
27. Berning, P. H., in *Physics of Thin Films*, Vol. 1, G. Hass and R. E. Thun, Eds., Academic Press, New York, 1963, p. 65.
28. Berreman, D. W., *Phys. Rev.*, **163**, 855 (1967).
29. Berreman, D. W., *J. Opt. Soc. Amer.*, **60**, 499 (1970).
30. Berreman, D. W., *Phys. Rev.*, **B 1**, 381 (1970).
31. Bewick, A., and A. M. Tuxford, *Symp. Faraday Soc.*, **4**, 114 (1970).

32. Biegler, T., D. A. J. Rand, and R. Woods, *J. Electroanal. Chem.*, **29**, 269 (1971).
33. Bockris, J. O'M., A. K. N. Reddy, and M. A. V. Devanathan, *Proc. Roy. Soc.*, **279A**, 327 (1964).
34. Bockris, J. O'M., A. K. N. Reddy, and B. Rao, *J. Electrochem. Soc.*, **113**, 1133 (1966).
35. Bockris, J. O'M., and A. K. N. Reddy, *Modern Electrochemistry*, Plenum Press, New York, 1970, p. 1141.
36. Bode, D. D., T. N. Anderson, and H. Eyring, *J. Phys. Chem.*, **71**, 792 (1967).
37. Bode, H., *Network Analysis and Feedback Amplifier Design*, van Nostrand, New York, 1945.
38. Boeckner, C., *J. Opt. Soc. Amer.*, **19**, 7 (1929).
39. de Boer, J. H., *Z. Phys. Chem. (Leipzig)*, **B18**, 49 (1932).
40. Böer, K. W., H. J. Mänsche, and U. Kümmel, *Z. Physik*, **155**, 170 (1959).
41. Bootsma, G. A., and F. Meyer, *Surf. Sci.*, **14**, 52 (1969).
42. Breiter, M. W., *Trans. Faraday Soc.*, **62**, 2887 (1966).
43. Breiter, M., *Electrochim. Acta*, **7**, 25 (1962).
44. Brown, W. F., in *Handbuch der Physik*, Vol. XVII, S. Flügge, Ed., Springer-Verlag, Berlin, 1956, p. 1.
45. Buckman, A. B., and N. M. Bashara, *Phys. Rev.*, **174**, 719 (1968).
46. Buckman, A. B., *Surf. Sci.*, **16**, 193 (1969).
47. Burge, D. K., and H. E. Bennett, *J. Opt. Soc. Amer.*, **54**, 1428 (1964).
48. Byers, H. G., *J. Amer. Chem. Soc.*, **30**, 1718 (1908).
49. Cahan, B. D., and R. F. Spanier, *Surf. Sci.*, **16**, 166 (1969).
50. Cahan, B. D., J. Horkans, and E. Yeager, *Symp. Faraday Soc.*, **4**, 36 (1970).
51. Cahan, B. D., private communication, 1971.
52. Cardona, M., K. L. Shaklee, and F. H. Pollak, *Phys. Rev.*, **154**, 696 (1967).
53. Cardona, M., in *Optical Properties of Solids*, S. Nudelman and S. S. Mitra, Eds., Plenum Press, New York, 1969, p. 137.
54. Cardona, M., *Modulation Spectroscopy*, Academic Press, New York, 1969.
55. Chiu, Ying-Chech, and M. A. Genshaw, *J. Phys. Chem.*, **72**, 4325 (1968).
56. Chiu, Ying-Chech, and M. A. Genshaw, *J. Phys. Chem.*, **73**, 3571 (1969).
57. Churchill, D., and L. S. Bartell, *J. Phys. Chem.*, **67**, 2518 (1963).
58. Collins, J. R., and R. O. Bock, *Rev. Sci. Instr.*, **14**, 135 (1943).
59. Conway, B. E., *Symp. Faraday Soc.*, **4**, 95 (1970).
60. *Ibid.*, p. 128.
61. Crawford, V., *Quart. Revs.*, **14**, 378 (1960).
62. Davison, S. G., and J. D. Levine, in *Solid State Physics*, Vol. 25, H. Ehrenreich, F. Seitz, and D. Turnbull, Eds., Academic Press, New York, 1970, p. 1.
63. Dettorre, J. F., T. G. Knorr, and D. A. Vaughan, in Ref. 9, p. 245.
64. Dietz, R. W., and J. M. Bennett, *Appl. Optics*, **5**, 881 (1966).
65. Dignam, M. J., and M. Moskovits, *Symp. Faraday Soc.*, **4**, 208 (1970).
66. Dignam, M. J., M. Moskovits, and R. W. Stobie, *Trans. Faraday Soc.*, **67**, 3306 (1971).
67. Dignam, M. J., B. Rao, M. Moskovits, and R. W. Stobie, *Can. J. Chem.*, **49**, 115 (1971).
68. Ditchburn, R. W., *J. Opt. Soc. Amer.*, **45**, 743 (1955).

69. Drude, P., *Ann. Phys. Chem. N.F.*, **36**, 532, 865 (1889).
70. Drude, P., *Ann. Phys. Chem. N.F.*, **39**, 481 (1890).
71. Drude, P., *Ann. Phys. Chem. N.F.*, **43**, 126 (1891).
72. Drude, P., *The Theory of Optics*, C. R. Mann and R. A. Millikan, transl., Longmans, Green, London, 1920, p. 287.
73. Duke, C. B., and M. E. Alferieff, *J. Chem. Phys.*, **46**, 923 (1967).
74. Eastman, D. E., *Phys. Rev.*, **B. 3**, 1769 (1971).
75. Ehrenreich, H., and H. R. Philipp, *Phys. Rev.*, **128**, 1622 (1962).
76. Ehrenreich, H., in *The Optical Properties of Solids*, J. Tauc, Ed., Academic Press, New York, 1966, p. 106.
77. Eischens, R. P., and W. A. Pliskin, in *Advances in Catalysis*, Vol. 10, D. D. Eley, W. G. Frankenburg, and V. I. Komaresky, Eds., Academic Press, New York, 1958, p. 1.
78. Eischens, R. P., *Science*, **146**, 486 (1964).
79. Eley, D. D., D. M. Moran, and C. H. Rochester, *Trans. Faraday Soc.*, **64**, 2168 (1969).
80. Esin, O. A., and B. F. Markov, *Acta Physicochim. URSS*, **10**, 353 (1939).
81. Fahrenfort, J., *Spectrochim. Acta*, **17**, 698 (1961).
82. Fahrenfort, J., and W. M. Visser, *Spectrochim. Acta*, **18**, 1103 (1962).
83. Fahrenfort, J., and W. M. Visser, *Spectrochim. Acta*, **21**, 1433 (1965).
84. Faraday, M., *Phil. Mag.*, **9**, 57, 122 (1836); **10**, 175 (1837).
85. Farnsworth, H. E., in *Advances in Catalysis*, Vol. 15, D. D. Eley, H. Pines, and P. B. Weisz, Eds., Academic Press, New York, 1964, p. 31.
86. Farnsworth, H. E., in *Experimental Methods of Catalytic Research*, R. B. Anderson, Ed., Academic Press, New York, 1968, p. 224.
87. Farnsworth, H. E., and M. Onchi, in *Molecular Processes on Solid Surfaces*, E. Drauglis, R. D. Gretz, and R. I. Jaffee, Eds., McGraw-Hill, New York, 1969, p. 31.
88. Feinleib, J., *Phys. Rev. Lett.*, **16**, 1200 (1966).
89. Francis, S. A., and A. H. Ellison, *J. Opt. Soc. Amer.*, **49**, 131 (1959).
90. Franz, W., *Z. Naturforsch.*, **13A**, 484 (1958).
91. Freundlich, H., H. Patscheke, and H. Zocher, *Z. Phys. Chem. (Leipzig)*, **130**, 289 (1927).
92. Friedel, J., in *Physics of Metals*, J. M. Ziman, Ed., Cambridge University Press, Cambridge, 1969, p. 340.
93. Frova, A., and P. Handler, *Proc. Int. Conf. Semicond., Paris*, 1964, Dunod Cie, Paris, 1964, p. 157.
94. Frumkin, A. N., in *Advances in Electrochemistry and Electrochemical Engineering*, Vol. 3, P. Delahay, Ed., Interscience, New York, 1963, p. 287.
95. Gadzuk, J. W., in *The Structure and Chemistry of Solid Surfaces*, G. A. Somorjai, Ed., Wiley, New York, 1969, p. 43–1.
96. Gadzuk, J. W., *Phys. Rev.*, **B 1**, 2110 (1970).
97. Garfinkel, M., J. J. Tiemann, and W. E. Engeler, *Phys. Rev.*, **148**, 695 (1966).
98. Gerhardt, U., and G. W. Rubloff, *Appl. Opt.*, **8**, 305 (1969).
99. Gerlach, R. L., and T. N. Rhodin, *Surf. Sci.*, **17**, 32 (1969).
100. Germer, L. H., in *Advances in Catalysis*, Vol. 13, D. D. Eley, P. W. Selwood, and P. B. Weisz, Eds., Academic Press, New York, 1962, p. 191.
101. Germer, L. H., *Annals. N. Y. Acad. Sci.*, **101**, 599 (1963).

102. Gilman, S., *J. Phys. Chem.*, **68**, 2098, 2112 (1964).
103. Gilman, S., *Electrochim. Acta*, **9**, 1025 (1964).
104. Gilman, S., *J. Electroanal. Chem.*, **7**, 382 (1964).
105. Gilman, S., in *Electroanalytical Chemistry*, Vol. 2, A. J. Bard, Ed., Marcel Dekker, New York, 1967, p. 111.
106. Gomer, R., and L. W. Swanson, *J. Chem. Phys.*, **38**, 1613 (1963).
107. Grahame, D. C., *Z. Elektrochem.*, **62**, 264 (1958).
108. Greef, R., *J. Chem. Phys.*, **51**, 3148 (1969).
109. Greef, R., *Rev. Sci. Instr.*, **41**, 532 (1970).
110. Greenaway, D. L., and G. Harbeke, *Optical Properties and Band Structure of Semiconductors*, Pergamon Press, Oxford, 1968.
111. Greene, R. F., in *Solid State Surface Science*, Vol. 1, M. Green, Ed., Marcel Dekker, New York, 1969, p. 87.
112. Greenler, R. G., *J. Chem. Phys.*, **44**, 310 (1966).
113. Greenler, R. G., *J. Chem. Phys.*, **50**, 1963 (1969).
114. Grimley, T. B., in *Advances in Catalysis*, Vol. XII, D. D. Eley, P. W. Selwood, P. B. Weisz, Eds., Academic Press, New York, 1960, p. 1.
115. Grubb, W. T., *Nature*, **198**, 883 (1963).
116. Gurney, R. W., *Phys. Rev.*, **47**, 479 (1935).
117. Gutsche, E., and H. Lange, *Proc. Int. Conf. Phys. Semicond., Paris*, 1964, Dunod Cie, Paris, 1964, p. 129.
118. Hagstrum, H. D., *Phys. Rev.*, **150**, 495 (1966).
119. Hagstrum, H. D., and G. E. Becker, *Phys. Rev.*, **159**, 572 (1967).
120. Hagstrum, H. D., and G. E. Becker, *J. Chem. Phys.*, **54**, 1015 (1971).
121. M. L. Hair, *Infrared Spectroscopy in Surface Chemistry*, Marcel Dekker, New York, 1967.
122. Hansen, W. N., *Inst. Soc. Amer. Trans.*, **4**, 263 (1965).
123. Hansen, W. N., *Spectrochim. Acta*, **21**, 209 (1965).
124. Hansen, W. N., R. A. Osteryoung, and T. Kuwana, *J. Amer. Chem. Soc.*, **88**, 1062 (1966).
125. Hansen, W. N., T. Kuwana, and R. A. Osteryoung, *Anal. Chem.*, **38**, 1810 (1966).
126. Hansen, W. N., in *Modern Aspects of Reflectance Spectroscopy*, W. W. Wendlandt, Ed., Plenum Press, New York, 1968, p. 182.
127. Hansen, W. N., and A. Prostrak, *Phys. Rev.*, **174**, 500 (1968).
128. Hansen, W. N., *J. Opt. Soc. Amer.*, **58**, 380 (1968).
129. Hansen, W. N., *Surf. Sci.*, **16**, 205 (1969).
130. Hansen, W. N., *Symp. Faraday Soc.*, **4**, 27 (1970).
131. Hansen, W. N., this volume.
132. Harrick, N. J., *J. Opt. Soc. Amer.*, **55**, 851 (1965).
133. Harrick, N. J., *Internal Reflection Spectroscopy*, Interscience, New York, 1967, p. 67.
134. Ref. 133, p. 139.
135. Harris, L. A., *J. Appl. Phys.*, **39**, 1419, 1428 (1968).
136. Hass, G., and L. Hadley, in *American Institute of Physics Handbook*, 2nd ed., D. E. Gray, Ed., McGraw-Hill, New York, 1963, p. 6–103.
137. Hauschild, H., *Ann. Physik*, **63**, 816 (1920).
138. Hayfield, P. C. S., in *First Int. Congr. Met. Corros.*, Butterworths, London, 1963, p. 663.

139. Hayfield, P. C. S., and G. W. T. White, in Ref. 9, p. 157.
140. Hayfield, P. C. S., *Werkst. Korros.*, **19**, 950 (1968).
141. Heavens, O. S., *Rep. Prog. Phys.*, **23**, 1 (1960).
142. Heavens, O. S., *Physics of Thin Films*, Vol. 2, G. Hass and R. E. Thun, Eds., Academic Press, New York, 1964, p. 193.
143. Heavens, O. S., *Optical Properties of Thin Solid Films*, Dover, New York, 1965.
144. Hedges, E. S., *Protective Films on Metals*, van Nostrand, New York, 1937, p. 105.
145. Hills, G. J., and R. Payne, *Trans. Faraday Soc.*, **61**, 326 (1965).
146. Hoare, J. P., *The Electrochemistry of Oxygen*, Interscience, New York, 1968, p. 24.
147. Horiuti, J., and T. Toya, in Ref. 111, p. 1.
148. Humphreys-Owen, S. P. F., *Proc. Phys. Soc.* (*London*), **77**, 949 (1961).
149. Hunter, W. R., *J. Opt. Soc. Amer.*, **55**, 1197 (1965).
150. Hunter, W. R., D. H. Eaton, and C. T. Sah, *Surf. Sci.*, **20**, 355 (1970).
151. Ishiguro, K., T. Sasaki, and S. Nomura, *Sci. Pap. Coll. Gen. Educ., Univ. Tokyo*, **10**, 207 (1960).
152. Jackson, J. D., *Classical Electrodynamics*, Wiley, New York, 1962, p. 189.
153. Jahoda, F. C., *Phys. Rev.*, **107**, 1261 (1957).
154. Jerrard, H. G., *Surf. Sci.*, **16**, 137 (1969).
155. Jørgensen, C. K., *Absorption Spectra and Chemical Bonding in Complexes*, Pergamon Press, Oxford, 1962.
156. Juenker, D. W., *J. Opt. Soc. Amer.*, **55**, 295 (1965).
157. *Organic Electronic Spectral Data*, M. J. Kamlet, and H. E. Ungnade, Eds., Interscience, New York, 1960.
158. Keldysh, L. V., *Soviet Phys.—JETP*, **34**, 788 (1958).
159. Kim, K. S., N. Winograd, and R. E. Davis, *J. Amer. Chem. Soc.*, **93**, 6296 (1971).
160. Kishi, K., S. Ikeda, and K. Hirota, *J. Phys. Chem.*, **71**, 4384 (1967).
161. Kishi, K., and S. Ikeda, *J. Phys. Chem.*, **73**, 15 (1969).
162. *Ibid.*, p. 729.
163. *Ibid.*, p. 2559.
164. Kissinger, P. T., and C. N. Reilley, *Anal. Chem.*, **42**, 12 (1970).
165. Kittel, C. *Introduction to Solid State Physics*, 3rd ed., Wiley, New York, 1968, Ch. 8, p. 225.
166. Koch, D. F. A., *Nature*, **202**, 387 (1964).
167. Koch, D. F. A., and D. E. Scaife, *J. Electrochem. Soc.*, **113**, 302 (1966).
168. Koenigsberger, J., and W. J. Muller, *Phys. Z.*, **6**, 847, 849 (1905); **12**, 606 (1911).
169. Kolb, D. M., *J. Opt. Soc. Amer.*, **62**, 599 (1972).
170. Kolb, D. M., private communication, 1971.
171. Kolb, D. M., and J. D. E. McIntyre, *Surf. Sci.*, **28**, 321 (1971).
172. Kolb, D. M., and J. D. E. McIntyre, to be published.
173. Kramers, H. A., *Atti Congr. Int. Fis., Sept. 1927*, Como-Pavia-Roma, Nicola Zanichelli, Bologna, **2**, 545 (1928).
174. Kreibig, U., and C. V. Fragstein, *Z. Physik*, **224**, 307 (1969).
175. Kreibig, U., and P. Zaccharias, *Z. Physik*, **231**, 128 (1970).
176. Kreibig, U., *Z. Physik*, **234**, 307 (1970).
177. de L. Kronig, R., *J. Opt. Soc. Amer.*, **12**, 547 (1926).

178. Kruger, J., and W. J. Ambs, *J. Opt. Soc. Amer.*, **49**, 1195 (1959).
179. Kruger, J., in Ref. 9, p. 131.
180. Kudo, K., *Sci. Light*, **13**, 11 (1964).
181. Landau, L., and E. M. Lifshitz, *Electrodynamics of Continuous Media*, Pergamon Press, New York, 1960, p. 256.
182. Layer, H. P., *Surf. Sci.*, **16**, 177 (1969).
183. Leberknight, C. E., and B. Lustman, *J. Opt. Soc. Amer.*, **29**, 59 (1939).
184. Leftin, H. P., and M. C. Hobson, Jr., in *Advances in Catalysis*, Vol. 14, D. D. Eley, H. Pines, and P. B. Weisz, Eds., Academic Press, New York, 1963, p. 115.
185. Lewis, R., and R. Gomer, *Surf. Sci.*, **17**, 333 (1969).
186. Lindquist, R. E., and A. W. Ewald, *J. Opt. Soc. Amer.*, **53**, 247 (1963).
187. Little, L. H., *Infrared Spectra of Adsorbed Species*, Academic Press, New York, 1966.
188. MacDonald, J. R., and C. A. Barlow, Jr., *J. Chem. Phys.*, **44**, 202 (1966).
189. MacDonald, J. R., and C. A. Barlow, Jr., *Surf. Sci.*, **4**, 381 (1966).
190. Mark, H. B., Jr., and B. S. Pons, *Anal. Chem.*, **38**, 119 (1966).
191. Mark, H. B., Jr., and E. N. Randall, *Symp. Faraday Soc.*, **4**, 157 (1970).
192. Mayer, H., *Physik Dünner Schichten*, Vol. 1, Wissenschaftliche Verlagsgesellschaft M.B.H., Stuttgart, 1950.
193. McBee, C. L., and J. Kruger, *Surf. Sci.*, **16**, 340 (1969).
194. McClure, D. S., *J. Chem. Phys.*, **36**, 2757 (1962).
195. McCrackin, F. L., E. Passaglia, R. R. Stromberg, and H. L. Steinberg, *J. Res. Natl. Bur. Std.*, **67A**, 363 (1963).
196. McCrackin, F. L., and C. P. Colson, in Ref. 9, p. 61.
197. McCrackin, F. L., *Natl. Bur. Std. (U.S.) Tech. Note 479*, U.S. Govt. Printing Office, Washington, D.C., 1969.
198. McIntyre, J. D. E., paper presented at the 135th National Meeting of The Electrochemical Society, New York, May, 1969 (Abstract No. 231).
199. McIntyre, J. D. E., *Symp. Faraday Soc.*, **4**, 50, 55, 61 (1970).
200. *Ibid.*, p. 126.
201. McIntyre, J. D. E., and D. M. Kolb, *Symp. Faraday Soc.*, **4**, 99 (1970).
202. McIntyre, J. D. E., and D. E. Aspnes, *Bull. Amer. Phys. Soc.*, **15**, 366 (1970).
203. McIntyre, J. D. E., and D. E. Aspnes, *Surf. Sci.*, **24**, 417 (1971).
204. McIntyre, J. D. E., and D. E. Aspnes, to be published.
205. McRae, E. G., in Ref. 87, p. 81.
206. Melmed, A. J., in *Experimental Methods in Materials Research*, H. Herman, Ed., Interscience, New York, 1967, p. 103.
207. Mertens, F. P., P. Theroux, and R. C. Plumb, *J. Opt. Soc. Amer.*, **53**, 788 (1963).
208. Mignolet, J. C. P., *J. Chim. Phys.*, **54**, 19 (1957).
209. Miller, R. F., A. J. Taylor, and L. S. Julien, *J. Phys.*, **D 3**, 1957 (1970).
210. Moss, T. S., *Optical Properties of Semiconductors*, Butterworths, London, 1959, p. 6.
211. Moss, T. S., *J. Appl. Phys.*, **32**, 2136 (1961).
212. Müller, E. W., in *Advances in Electronics and Electron Physics*, Vol. 13, L. Marton, Ed., Academic Press, New York, 1960, p. 83.

213. Müller, E. W., in Ref. 87, p. 67.
214. Muller, R. H., *Surf. Sci.*, **16**, 14 (1969).
215. Müller, W. J., and J. Koenigsberger, *Phys. Z.*, **5**, 413, 797 (1904).
216. Nilsson, P. O., and L. Munkby, *Phys. Kondens. Materi.*, **10**, 290 (1969).
217. Ord, J. L., and D. J. DeSmet, *J. Electrochem. Soc.*, **113**, 1258 (1966).
218. Ord, J. L., *Surf. Sci.*, **16**, 155 (1969).
219. Paik, Woon-Kie, and J. O'M. Bockris, *Surf. Sci.*, **28**, 61 (1971).
220. Pardue, H. L., and P. A. Rodriguez, *Anal. Chem.*, **39**, 901 (1967).
221. Pardue, H. L., and S. N. Deming, *Anal. Chem.*, **41**, 986 (1969).
222. Parsons, R., *Surf. Sci.*, **2**, 418 (1964).
223. Parsons, R., *Symp. Faraday Soc.*, **4**, 85 (1970).
224. Philipp, H. R., and E. A. Taft, *Phys. Rev.*, **113**, 1002 (1959).
225. Phillips, J. C., in Ref. 76, p. 316.
226. Pickering, H. L., and H. C. Eckstrom, *J. Phys. Chem.*, **63**, 512 (1959).
227. Plieth, W. N., *Symp. Faraday Soc.*, **4**, 137 (1971).
228. Pliskin, W. A., and R. P. Eischens, *Z. Phys. Chem. (Frankfurt) N.F.*, **24**, 11 (1960).
229. Plummer, E. W., and R. D. Young, *Phys. Rev.*, **B 1**, 2088 (1970).
230. Poling, G. W., *J. Coll. Interface Sci.*, **34**, 365 (1970).
231. Ponec, V., *J. Catal.*, **6**, 362 (1966).
232. Pons, B. S., L. O. Winstrom, J. S. Mattson, and H. B. Mark, Jr., *Anal. Chem.*, **39**, 685 (1967).
233. Potter, R. F., *J. Opt. Soc. Amer.*, **54**, 904 (1964).
234. Potter, R. F., *Appl. Opt.*, **4**, 53 (1965).
235. Potter, R. F., in Ref. 53, p. 489.
236. Powell, C. F., in *Vapor Deposition*, C. F. Powell, J. H. Oxley, J. M. Blocher, Jr., Eds., Wiley, New York, 1966, p. 21.
237. Prostak, A., and W. N. Hansen, *Phys. Rev.*, **160**, 600 (1967).
238. Prostak, A., H. B. Mark, Jr., and W. N. Hansen, *J. Phys. Chem.*, **72**, 2576 (1968).
239. Querry, M. R., *J. Opt. Soc. Amer.*, **59**, 876 (1969).
240. Reale, C., *Infrared Phys.*, **10**, 173 (1970).
241. Reddy, A. K. N., M. A. V. Devanathan, and J. O'M. Bockris, *J. Electroanal. Chem.*, **6**, 61 (1963).
242. Reddy, A. K. N., and J. O'M. Bockris, in Ref. 9, p. 229.
243. Reddy, A. K. N., M. A. Genshaw, and J. O'M. Bockris, *J. Electroanal. Chem.*, **8**, 406 (1964).
244. Reddy, A. K. N., M. A. Genshaw, and J. O'M. Bockris, *J. Chem. Phys.*, **48**, 671 (1968).
245. Robinson, T. S., *Proc. Phys. Soc. (London)*, **65B**, 910 (1952).
246. Robinson, T. S., and W. C. Price, *Proc. Phys. Soc. (London)*, **66B**, 969 (1953).
247. Rootsaert, W. J. M., L. L. van Reijen, and W. H. M. Sachtler, *J. Catal.*, **1**, 416 (1962).
248. Rosen, M., D. R. Flinn, and S. Schuldiner, *J. Electrochem. Soc.*, **116**, 1112 (1969).
249. Rouard, P., and P. Bousquet, in *Progress in Optics*, Vol. IV, E. Wolf, Ed., North-Holland, Amsterdam, 1965, p. 147.
250. Sasaki, T., and K. Ishiguro, *Jap. J. Appl. Phys.*, **2**, 289 (1963).

251. Schmidt, L., and R. Gomer, *J. Chem. Phys.*, **42**, 3573 (1965); **43**, 2055 (1965); **45**, 1605 (1966).
252. Schulz, L. G., and F. R. Tangherlini, *J. Opt. Soc. Amer.*, **44**, 362 (1954).
253. Sell, D. D., *Appl. Opt.*, **9**, 1926 (1970).
254. Seraphin, B. O., *Proc. Int. Conf. Phys. Semicond., Paris, 1964*, Dunod Cie, Paris, 1964, p. 165.
255. Seraphin, B. O., and R. B. Hess, *Phys. Rev. Lett.*, **14**, 138 (1965).
256. Seraphin, B. O., in Ref. 53, p. 153.
257. Seraphin, B. O., in *Semiconductors and Semimetals*, Vol. IX, R. K. Willardson and A. Beer, Eds., Academic Press, New York, 1972, p. 1.
258. Seraphin, B. O., in *Optical Properties of Solids*, F. Abelès, Ed., North-Holland, Amsterdam, 1972, p. 163.
259. Shaklee, K. L., F. H. Pollak, and M. Cardona, *Phys. Rev. Lett.*, **15**, 883 (1965).
260. Shimizu, H., *Electrochim. Acta*, **13**, 27 (1968).
261. Simon, I., *J. Opt. Soc. Amer.*, **41**, 336 (1951).
262. Sirohi, R. S., and M. A. Genshaw, *J. Electrochem. Soc.*, **116**, 911 (1969).
263. Sivukhin, D. V., *Soviet Phys.—JETP*, **3**, 269 (1956).
264. Smith, T., *J. Opt. Soc. Amer.*, **58**, 1069 (1968).
265. Srinivasan, S., H. Wroblowa, and J. O'M. Bockris, in *Advances in Catalysis*, Vol. 17, D. D. Eley, H. Pines, P. B. Weisz, Eds., Academic Press, New York, 1967, p. 351.
266. Stedman, M., *Chem. Phys. Lett.*, **2**, 457 (1968).
267. Stedman, M., *Symp. Faraday Soc.*, **4**, 64 (1970).
268. Strachan, C. S., *Proc. Cambridge Phil. Soc.*, **29**, 116 (1933).
269. Stratton, E. A., *Electromagnetic Theory*, McGraw-Hill, New York, 1941, p. 492.
270. *Ibid.*, p. 131.
271. Swanson, L. W., and R. W. Strayer, *J. Chem. Phys.*, **48**, 2421 (1968).
272. Takamura, T., K. Takamura, W. Nippe, and E. Yeager, *J. Electrochem. Soc.*, **117**, 626 (1970).
273. Takamura, T., K. Takamura, and E. Yeager, *Symp. Faraday Soc.*, **4**, 91 (1970).
274. Takamura, T., K. Takamura, and E. Yeager, *J. Electroanal. Chem.*, **29**, 279 (1971).
275. Terenin, A., in *Advances in Catalysis*, Vol. 15, D. D. Eley, H. Pines, P. B. Weisz, Eds., Academic Press, New York, 1964, p. 227.
276. Tippins, H. H., *Phys. Rev.*, **B 1**, 126 (1970).
277. Toll, J. S., *Phys. Rev.*, **104**, 1760 (1956).
278. Topping, J., *Proc. Roy. Soc. (London)*, **A114**, 67 (1927).
279. Tousey, R., *J. Opt. Soc. Amer.*, **29**, 235 (1939).
280. Tronstad, L., *Z. Phys. Chem. (Leipzig)*, **A142**, 241 (1929).
281. Tronstad, L., *Kgl. Norske Videnskab. Selskabs Skrifter* 1931, Nr. 1.
282. Tronstad, L., *Trans. Faraday Soc.*, **29**, 502 (1933).
283. Tronstad, L., and C. W. Borgmann, *Trans. Faraday Soc.*, **30**, 349 (1934).
284. Tronstad, L., and T. Höverstad, *Trans. Faraday Soc.*, **30**, 362 (1934).
285. Tronstad, L., *Trans. Faraday Soc.*, **31**, 1151 (1935).
286. Tulvinskii, V. B., and N. I. Terentev, *Opt. Spectry. (USSR)*, **28**, 484 (1970).

287. Velický, B., *Czech. J. Phys.*, **B 11**, 787 (1961).
288. Visscher, W., *Optik*, **26**, 402 (1968).
289. Walker, D. C., *Can. J. Chem.*, **44**, 2226 (1966).
290. Walker, D. C., *Can. J. Chem.*, **45**, 807 (1967).
291. Walker, D. C., *Anal. Chem.*, **39**, 896 (1967).
292. Weber, R. E., and W. T. Peria, *J. Appl. Phys.*, **38**, 4355 (1967).
293. Weingart, J. N., in Ref. 9, p. 113.
294. Will, F. G., and C. A. Knorr, *Z. Elektrochem.*, **64**, 258 (1960).
295. Will, F. G., *J. Electrochem. Soc.*, **112**, 459 (1965).
296. Williams, R., *Phys. Rev.*, **117**, 1487 (1960).
297. Williams, R., *Phys. Rev.*, **126**, 442 (1962).
298. Wilmanns, I., *Surf. Sci.*, **16**, 147 (1969).
299. Winograd, N., and T. Kuwana, *J. Electroanal. Chem.*, **23**, 333 (1969).
300. Winterbottom, A., *Kgl. Norske Videnskab. Selskabs Skrifter* 1955, Nr. 1.
301. Wolter, H., in *Handbuch der Physik*, Vol. XXIV, *Fundamentals of Optics*, S. Flügge, Ed., Springer-Verlag, Berlin, 1956, p. 461.
302. Young, L., *Anodic Oxide Films*, Academic Press, New York, 1961, p. 49.

Principles of Ellipsometry

ROLF H. MULLER

Inorganic Materials Research Division, Lawrence Berkeley Laboratory and Department of Chemical Engineering; University of California Berkeley, California

Contents

I. Introduction

Ellipsometry derives its name from the measurement of elliptically polarized light that results from optical reflection. More precisely, the *change* in the state of polarization due to reflection is measured and interpreted in terms of properties of the reflecting surface. The utility of ellipsometry for electrochemical purposes is largely due to its capability for examining surfaces in any optically transparent environment. Electrode surfaces can therefore often be observed *in situ*, without exposing them to conditions (e.g., vacuum, heat, electron impact) that might alter their properties. Previous reviews of ellipsometry (7,16,144) and proceedings from two conferences (13,98) are available.

Two parameters are measured in ellipsometry: the change in relative amplitude and relative phase of two orthogonal components of light due to reflection. The fact that relative, rather than absolute, measurements are made is one reason for the high resolution of ellipsometry. The technique employs a built-in reference, so to speak, that largely eliminates effects due to external fluctuations, such as those of the light source. The other reason for the high resolution lies in the fact that the measured quantities are usually azimuth angles (i.e., angles resulting from rotation around an optic axis), and angular measurements can easily be made with high resolution. For example, an angle of 0.01°, the resolution of currently available research instruments, represents 3×10^{-5} of a full turn, the largest measurement possible.

From the measured two quantities, change in relative amplitude and phase at constant wavelength, two parameters of the reflecting surface can be derived. For a bare surface, these can be the real and imaginary parts of the refractive index. For a surface covered with a transparent film, thickness and refractive index of the film can be determined, if the optical constants of the substrate are known.

If sufficiently narrow limits can be imposed on acceptable solutions, it is sometimes possible to determine more than two unknowns, e.g., thickness and complex refractive index of an absorbing film. Additional data have been generated by immersion in media of different refractive index and reflectivity measurements at normal incidence (79).

In many cases, an increase in film thickness by one wavelength results in an azimuth change of a full turn. Under these conditions, an angular resolution of 0.01° results in an average resolution in film thickness of 3×10^{-5} wavelengths or about 0.2 Å.

The availability of high-speed computers and comprehensive programs (84) has been an important factor for the renewed interest in ellipsometry, which has been in existence at least since 1888 (55). The exact classical equations, presently used for the interpretation of results, cannot be solved explicitly (130) and, despite their simple algebraic form, are too tedious for hand calculations in all but a few special cases. Approximations, introduced in earlier days, have more recently been found to be often not reliable (31).

In the following, a brief account of the classical theory of ellipsometry, its use and limitations are given. Detailed derivations of electromagnetic theory on which this theory is based exceed the scope of this chapter. Instead, an attempt is made to provide primarily a physical understanding of the optical phenomena involved and of the principles of ellipsometer measurements.

II. Polarized Light

1. Linear Polarization

The complete description of a monochromatic light wave, in addition to frequency, phase, propagation direction, and amplitude, has to include information on the orientation of electric and magnetic vectors in space. Since the two vectors are orthogonal, it is usually sufficient to consider the electric vector only. If the electric vector, along a wave in space, lies in a plane, the light is called linearly polarized. The connecting line between the end points of the vectors shows the sinusoidal electric field distribution along the propagation direction z at a fixed instant in time (Fig. 1). At a given point in space, the tip of the electric vector oscillates along a straight line as a function of time, hence the name. Polarization is a direct consequence of the transverse nature of light. In the present literature, the plane which contains the electric vectors in space is called

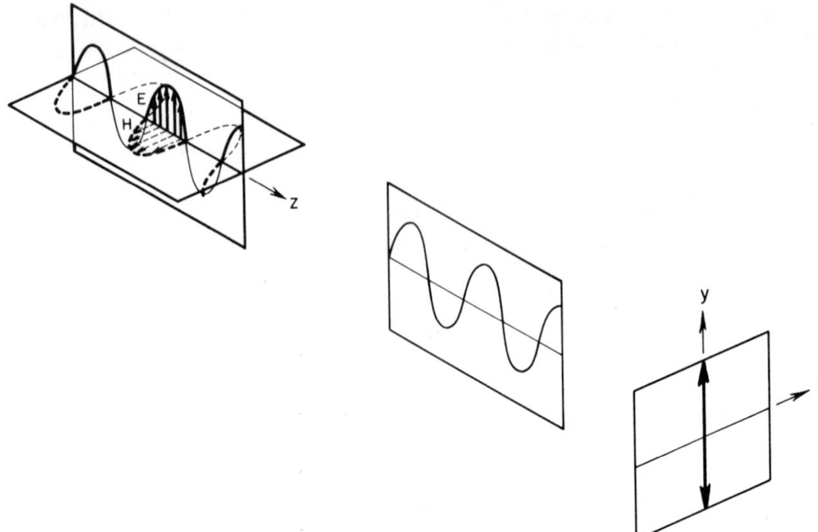

Fig. 1. Electric and magnetic vectors in a linearly polarized light wave. The plane of polarization contains the electric field vectors in space. At a fixed location, the tip of the electric vector traces a straight line as a function of time.

the *plane of polarization.* Superposition of two linearly polarized waves which are in phase results in another linear polarization. Linear polarization is the best known state of polarization of light and often is simply referred to as polarized light.

2. *Elliptic Polarization*

A. *s* and *p* Components. In analyzing the reflection of light, it is convenient to decompose incident and reflected waves into two orthogonal linear components. One of these components has its electric vector oriented parallel to the plane which contains incident and reflected beams (or wave normals), called the *plane of incidence.* This component is usually denoted by a subscript p. The other component, denoted by a subscript s, has its electric vector oriented normal to the plane of incidence (Fig. 2). Both components suffer different changes in phase and amplitude upon reflection.

B. Geometric and Physical Parameters. The superposition of linearly polarized s and p components of the same frequency, but different phase and amplitude, in general, results in elliptic polarization (96,111,122,141). The tip of the electric vector now describes a

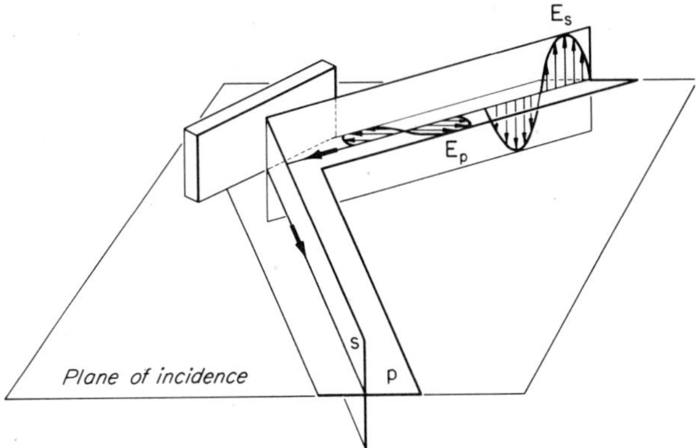

Fig. 2. Reflection of polarized light. Plane of incidence defined by incident and reflected beam. The *s* and *p* components indicated in incident light, with electric vector normal and parallel, respectively, to the plane of incidence.

helix in space or, at a given location, it traces an ellipse as a function of time (Fig. 3).

$$E_{tp} = |E_p| \cos(\omega t + \varepsilon_p) \tag{1}$$

$$E_{ts} = |E_s| \cos(\omega t + \varepsilon_s) \tag{2}$$

This ellipse also results from the projection of the helix on a plane normal to the propagation direction. The ellipse possesses a positive (counterclockwise) or negative (clockwise) sense of rotation, as seen looking into the beam, and is inscribed in a rectangle (if we limit these considerations to orthogonal components) with sides parallel to the planes of polarization of both components and lengths equal to twice their amplitudes (Fig. 3, top). A detailed review of the analysis of elliptic polarization, using matrix methods, has been given by Richartz and Hsue (104).

Elliptic polarization is the most general state of polarization, with circular and linear polarization being limiting cases. The state of polarization, independent of light intensity, can be characterized by either of two sets of two parameters each, which can be measured as azimuth angles with an ellipsometer.

The *physical parameters*, already mentioned, are the ratio of the electric field amplitudes $|E|$ of p and s components, expressed as the tangent of an angle ψ

$$\frac{|E_p|}{|E_s|} = \tan \psi \tag{3}$$

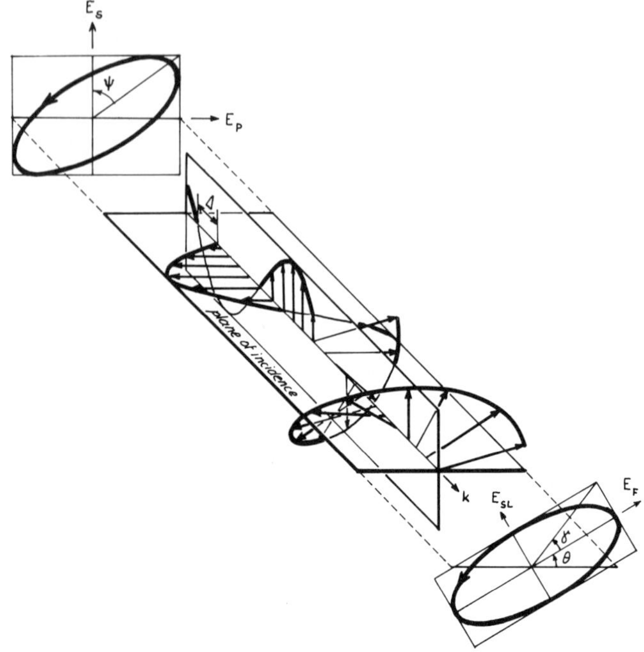

Fig. 3. Elliptic polarization resulting from the superposition of two linear components of different phase and amplitude. Representation as a helix in space or an ellipse in a plane normal to the propagation direction k. Physical parameters ψ, Δ, and geometric parameters θ, γ of the state of polarization. E_F, E_{SL} orthogonal fast and slow components ahead or behind each other in phase, respectively, by a quarter wave (90°).

and the difference Δ of the time-independent phase ε of the two components

$$\varepsilon_p - \varepsilon_s = \Delta \tag{4}$$

The quantities ψ and Δ are indicated in the upper part of Fig. 3. The range of values for the two physical parameters are

$$0° \leq \psi \leq 90° \tag{5}$$

$$0° \leq \Delta \leq 360° \tag{6}$$

A range of $\pm 180°$ for Δ is equivalent to the one given in Eq. 6. For reflection from bare surfaces the ranges of ψ and Δ values are further restricted (see Section III.1D).

The *geometric parameters* characterize elliptic polarization by the shape and orientation of the ellipse. They are the angle θ, which the major axis of the ellipse forms with the plane of incidence (measured counterclockwise)

$$\theta = \text{orientation of major axis} \tag{7}$$

and the eccentricity of the ellipse

$$\tan\gamma = \frac{\text{minor axis}}{\text{major axis}} \tag{8}$$

The angle γ is measured in the sense of the rotation of the ellipse and can therefore assume positive and negative values. The angles γ and θ are indicated in the lower part of Fig. 3. The ranges of the geometric parameters are

$$0^\circ \leq \theta < 180^\circ \tag{9}$$

$$-45^\circ \leq \gamma \leq 45^\circ \tag{10}$$

Physical and geometric parameters of elliptic polarization can be mutually converted either graphically by use of the Poincaré sphere (see below) or algebraically with the following expressions (27,47, 96,111,134), which can be derived by coordinate transformations.

$$\sin 2\gamma = \sin 2\psi \sin\Delta \tag{11}$$

$$\tan\Delta = \frac{\tan 2\gamma}{\sin 2\theta} \tag{12}$$

$$\tan 2\theta = -\tan 2\psi \cos\Delta \tag{13}$$

$$\cos 2\psi = -\cos 2\gamma \cos 2\theta \tag{14}$$

All the angles except the relative phase Δ appear doubled in the above relations.

Orientation and sense of rotation of the ellipse are determined by the phase difference Δ between both components (47,63). A few examples are given in Fig. 4, where the azimuth angles representing different parameters are also indicated. It can be seen that a phase difference of 0°, 180°, etc., produces linear polarization, while a phase difference of 90° (or a quarter wavelength), 270°, etc., results in the ellipse axes being oriented parallel to the planes of polarization of the two components. Fast and slow components, mutually out of phase by 90°, are indicated for a general elliptic polarization in Fig. 3. This fact will be used later. The orientations of the ellipse shown in Fig. 4 are valid only for the coordinate system indicated,

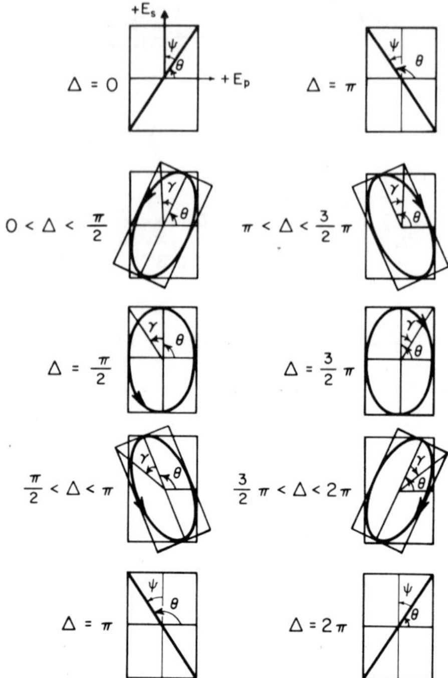

Fig. 4. Dependence of geometric parameters θ and γ of elliptic polarization on phase difference Δ between p and s components for constant relative amplitude tan ψ.

the relative phase Δ as defined in Eq. 4, and the phase ε as formulated in Eqs. 1 and 2. The same figures have been obtained with different definitions of Δ and ε (47).

Unpolarized light can be pictured as a superposition or rapid sequence of different states of polarization which may be linear or elliptic each (96).

C. Poincaré Sphere. The state of polarization of any light wave of unit intensity may be represented by a point E on the surface of a unit sphere, known as Poincaré sphere (18,64,82,112,128). The features of this representation can be derived from the Stokes parameters (27,82,85,122). A coordinate grid for the geometric parameters of polarization (Fig. 5, top) is established, with longitude 2θ, measured on the "equator" counterclockwise, as seen from pole L, from point H, which signifies the plane of incidence, and latitude 2γ. Points on the "northern" hemisphere (○), with positive γ

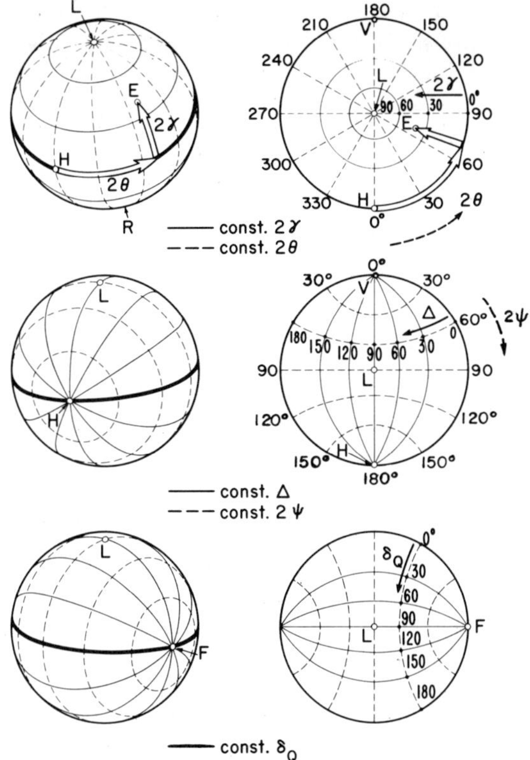

Fig. 5. Poincaré sphere (left) and its stereographic projection (right), as seen from the left-hand circular pole L. Reference point H denotes the plane of incidence. Top: grid for geometrical parameters, θ and γ; center: grid for physical parameters ψ and Δ; bottom: grid for retardation δ_Q of a compensator with fast axis F.

values, represent left-hand (positive) rotation of the ellipse, points on the "southern" hemisphere (×) right-hand rotation. The two poles represent left- and right-hand circular polarization, while points on the equator signify linear polarization states of different azimuth, θ, with pure p polarization represented by point H, pure s by point V.

A similar coordinate grid can be established to represent the physical parameters of polarization (Fig. 5, center). Point H forms one pole of this coordinate grid. Great circles represent lines of constant relative phase Δ, which is measured from the "equator" in a positive (counterclockwise) sense of rotation, as seen from point

H. Lines of constant parameter 2ψ are small circles with the origin at the pole V opposite to H. This pole represents the plane normal to the plane of incidence, or pure s polarization. Points on the circle of $2\psi = 90°$ (which is also a great circle through poles L and R) represent states of polarization of equal amplitude of p and s components.

It should be noted that all angles, except phase angles appear doubled on the Poincaré sphere, as in Eqs. 11 to 14, which can also be derived from the two coordinate grids by use of spherical trigonometry.

The equator of the Poincaré sphere is also used to indicate the orientation of polarizer and analyzer transmission axis P and A and the fast axis F of the compensator. Their azimuth angles p, a, and q (measured counterclockwise from the plane of incidence) are also doubled on the sphere and have the same value as the quantity θ.

A superposition of the coordinate grids for geometric and physical parameters provides a graphical conversion between the two. For this purpose, and other applications of the Poincaré sphere to follow, the use of a stereographic projection of the sphere, indicated in the right hand column of Fig. 5, is best suited (141). The stereographic projection provides an angle-true representation of the surface of a sphere. It is obtained by a central projection of a hemisphere from the opposite pole on a plane through the equator. Stereographic grids (polar and equatorial, Wulff's net) are commonly used in crystallography (11).

The third coordinate grid, indicated on the bottom of Fig. 5, represents the effect of a retardation plate. The orientation of the fast axis of this compensator plate, chosen to be of azimuth 45° in the figure, is represented by point F on the equator of the sphere. The retardation δ_Q of the plate results in a positive rotation on the sphere, as seen from F. This grid is identical to the one for Δ with the fast axis F taking the place of the plane of incidence H; its use will become clearer in some sections to follow.

III. Optical Reflection

1. *Reflection from Bare Surfaces*

A. Complex Amplitude. The complex notation provides a simple formalism for the description of oscillations and waves (47, 62,121). Its advantages are best seen in the addition of two waves of different phase which can be performed as an addition of two vectors in the complex plane without the use of tedious trigonometric formulas.

The instantaneous amplitude of a harmonic oscillation can be formulated trigonometrically (Eq. 1) as

$$E_t = |E| \cos (\omega t + \varepsilon) \tag{15}$$

According to the Gauss equation

$$e^{ix} = \cos x + i \sin x \tag{16}$$

the exponential formulation (Eq. 17) has the same real part as Eq. 15. (The imaginary unit $\sqrt{-1}$ is designated by the letter i.)

$$|E| e^{i(\omega t + \varepsilon)} = |E| [\cos (\omega t + \varepsilon) + i \sin (\omega t + \varepsilon)] \tag{17}$$

For this reason, it is used to represent the cosine formulation, with the understanding that the real part represents the oscillation.

$$E_t = |E| e^{i(\omega t + \varepsilon)} \tag{18}$$

The time-dependent factor in Eq. 18

$$e^{i\omega t}$$

is not particularly important and often is disregarded in considerations of monochromatic light. The time-independent part of Eq. 18 is called the complex amplitude E:

$$E = |E| e^{i\varepsilon} \tag{19}$$

Its modulus $|E|$ is the real amplitude, its argument ε the phase of the oscillation relative to a reference.

B. Fresnel Equations. The reflection of light on a dielectric (nonabsorbing) interface can be described by the Fresnel (amplitude) reflection coefficients r. These coefficients represent the ratio of reflected to incident electric field amplitude; they are different for s and p components

$$r_p = \frac{E''_p}{E_p} \tag{20}$$

$$r_s = \frac{E''_s}{E_s} \tag{21}$$

As indicated in Fig. 6, double-primed quantities refer to the reflected, unprimed to the incident wave. The Fresnel equations (27,63,126,

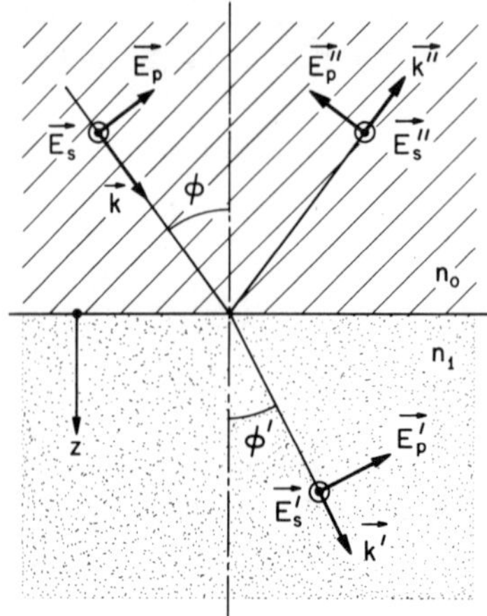

Fig. 6. Reflection and refraction at a dielectric interface. Definition of symbols and of positive coordinate directions for p and s components of electric field vector **E** in incident, reflected, and refracted waves. (Circles represent arrows pointing out of the plane of the drawing.) **k**—propagation direction of the three waves; ϕ—angle of incidence; ϕ'—angle of refraction; n_0—refractive index of incident medium; n_1—refractive index of reflecting medium; z—coordinate direction for amplitude attenuation if the reflecting medium is absorbing.

128,134,139) in their simplest form relate the amplitude reflection coefficients to the angles of incidence and refraction, indicated in Fig. 6.

$$r_p = \frac{\tan(\phi - \phi')}{\tan(\phi + \phi')} \tag{22}$$

$$r_s = -\frac{\sin(\phi - \phi')}{\sin(\phi + \phi')} \tag{23}$$

The angle of refraction ϕ' can be obtained from the angle of incidence ϕ and the refractive indices of both media at the interface by use of Snell's law

$$\sin\phi' = \frac{n_0}{n_1}\sin\phi \tag{24}$$

Fresnel coefficients of negative sign signify a phase change of 180° ($\exp i\pi = -1$) of the reflected with respect to the incident wave at the reflecting surface.

Application of electromagnetic theory (19,26,27,37,68,72,80,106, 123) shows that the Fresnel equations can be adapted to describe reflection from absorbing media by introduction of a complex refractive index n_c in place of n_1

$$n_c = n - ik \tag{25}$$

If n_c is defined here to be a material constant, independent of angle of incidence (41,90), application of Snell's law

$$n_0 \sin \phi = n_c \sin \phi_c' \tag{26}$$

results in a complex angle of refraction ϕ_c', which provides a valid formalism, but has no recognizable physical meaning.

The resulting complex reflection coefficients can be cast in exponential form

$$r_p = \frac{|E_p''|}{|E_p|} e^{i(\varepsilon_p'' - \varepsilon_p)} \tag{27}$$

$$r_s = \frac{|E_s''|}{|E_s|} e^{i(\varepsilon_s'' - \varepsilon_s)} \tag{28}$$

with the modulus representing the amplitude attentuation $|r|$

$$|r_p| = \frac{|E_p''|}{|E_p|} \tag{28}$$

$$|r_s| = \frac{|E_s''|}{|E_s|} \tag{30}$$

and the argument representing the (absolute) change in phase δ due to reflection

$$\delta_p = \varepsilon_p'' - \varepsilon_p \tag{31}$$

$$\delta_s = \varepsilon_s'' - \varepsilon_s \tag{32}$$

Thus the complex Fresnel reflection coefficients can also be expressed as

$$r_p = |r_p| e^{i\delta_p} \tag{33}$$

$$r_s = |r_s| e^{i\delta_s} \tag{34}$$

As shown by Koenig (68) the use of complex arithmetic to derive the above quantities can be avoided by the introduction of expressions involving only real quantities. For the present conventions and definitions these expressions are (51,88):

$$|r_s| = \sqrt{\frac{A^2 + B^2 - 2A \cos \phi + \cos^2 \phi}{A^2 + B^2 + 2A \cos \phi + \cos^2 \phi}} \tag{35}$$

$$|r_p| = |r_s| \sqrt{\frac{A^2 + B^2 - 2A \sin \phi \tan \phi + \sin^2 \phi \tan^2 \phi}{A^2 + B^2 + 2A \sin \phi \tan \phi + \sin^2 \phi \tan^2 \phi}} \tag{36}$$

$$\delta_s = \tan^{-1}\left[-\frac{2B\cos\phi}{A^2 + B^2 - \cos^2\phi}\right] \tag{37}$$

$$\delta_p = \tan^{-1}\left[\frac{2B\cos\phi(A^2 + B^2 - \sin^2\phi)}{A^2 + B^2 - (1/n_0^4)(n^2 + k^2)^2\cos^2\phi}\right] \tag{38}$$

with the intermediate variables A and B defined as

$$A = \sqrt{\frac{1}{2n_0^2}\left[\sqrt{(n^2 - k^2 - n_0^2\sin^2\phi)^2 + 4n^2k^2} + (n^2 - k^2 - n_0^2\sin^2\phi)\right]} \tag{39}$$

$$B = \sqrt{\frac{1}{2n_0^2}\left[\sqrt{(n^2 - k^2 - n_0^2\sin^2\phi)^2 + 4n^2k^2} - (n^2 - k^2 - n_0^2\sin^2\phi)\right]} \tag{40}$$

When the direction of propagation of the refracted wave (the physical, real angle of refraction ϕ_r') has to be known (e.g., for determining the penetration depth at nonnormal incidence), an alternate complex refractive index

$$n_c' = n' - ik' \tag{41}$$

which depends on the angle of incidence can be defined (41). The real part of this refractive index and the real angle of refraction ϕ_r' satisfy Snell's law

$$n_0 \sin\phi = n' \sin\phi_r' \tag{42}$$

The imaginary part describes the decay of the electric field amplitude with increasing distance z normal to the surface

$$E \sim e^{-(2\pi/\lambda_0)k'z} \tag{43}$$

where λ_0 is the vacuum wavelength.

The distance at which the square of the electric field amplitude (a measure of light intensity) decreases by a factor $1/e$ is the (intensity) penetration depth

$$z = \frac{\lambda_0}{4\pi k'} \tag{44}$$

Some penetration depth data are given in Fig. 7. For most metals,

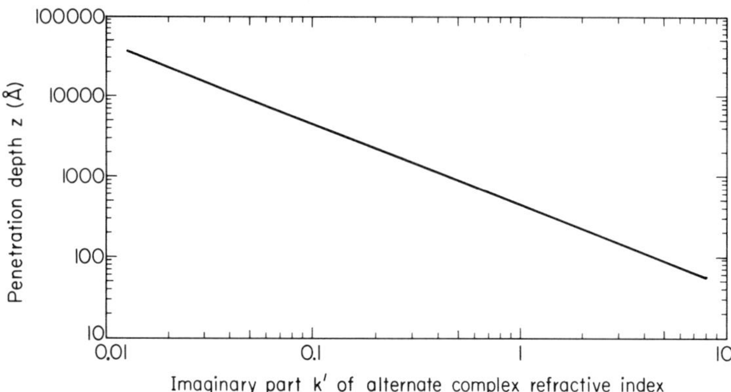

Fig. 7. Light penetration into absorbing media of alternate complex refractive index $n_c' = n' - ik'$. Wavelength 5.461×10^{-5} cm.

z is a few hundred angstrom units. The complex refractive index and its alternate are related (41) by

$$n^2 - k^2 = n'^2 - k'^2 \tag{45}$$

$$nk = n'k' \cos \phi_r' \tag{46}$$

The dependence of the alternate optical constants on the angle of incidence has been illustrated elsewhere (90). For many metals the change is small. At normal incidence the two quantities become identical. The angle-dependent optical constants have deliberately been used by Vašíček (133).

C. Equations of Ellipsometry. The ellipsometer determines the ratio ρ of the (complex) reflection coefficients for p and s components

$$\rho = \frac{r_p}{r_s} \tag{47}$$

The use of Eqs. 27 and 28 and rearranging of terms results in

$$\rho = \frac{\dfrac{|E_p''|}{|E_s''|}}{\dfrac{|E_p|}{|E_s|}} e^{i[(\varepsilon_p'' - \varepsilon_s'') - (\varepsilon_p - \varepsilon_s)]} \tag{48}$$

The modulus of this expression contains the relative amplitudes of

p and s components in reflected (subscript r) and incident (subscript i) waves, which can be formulated according to Eq. 3 as

$$\tan \psi_r = \frac{|E_p''|}{|E_s''|} \tag{49}$$

$$\tan \psi_i = \frac{|E_p|}{|E_s|} \tag{50}$$

The argument of Eq. 48 contains the relative phase of p and s components in reflected and incident waves. According to Eq. 4 they can be formulated as

$$\Delta_r = \varepsilon_p'' - \varepsilon_s'' \tag{51}$$

$$\Delta_i = \varepsilon_p - \varepsilon_s \tag{52}$$

Thus, Eq. 48 becomes the basic equation of ellipsometry

$$\rho = \frac{\tan \psi_r}{\tan \psi_i} e^{i(\Delta_r - \Delta_i)} \tag{53}$$

which is often given in a simplified form as

$$\rho = (\tan \psi) e^{i\Delta} \tag{54}$$

For the sake of simplicity, the present usage of ellipsometers employs equal amplitudes of p and s components in the incident light, so that $\tan \psi_i = 1$. The relative phase between the two components may be zero in the incident or reflected beam.

With the formulation of Eqs. 33 and 34, Eq. 47 can also be written as

$$\rho = \frac{|r_p|}{|r_s|} e^{i(\delta_p - \delta_s)} \tag{55}$$

Comparison with Eq. 54 leads to the definitions

$$\tan \psi = \frac{|r_p|}{|r_s|} \tag{56}$$

$$\Delta = \delta_p - \delta_s \tag{57}$$

Expressions for relative phase and amplitude change due to reflection from a bare surface using only real quantities (51,68,88) are:

$$\Delta = \tan^{-1}\left(-\frac{2B \sin \phi \tan \phi}{A^2 + B^2 - \sin^2 \phi \tan^2 \phi} \right) \tag{58}$$

$$\psi = \tan^{-1}\left\{\sqrt{\frac{A^2 + B^2 - 2A\sin\phi\tan\phi + \sin^2\phi\tan^2\phi}{A^2 + B^2 + 2A\sin\phi\tan\phi + \sin^2\phi\tan^2\phi}}\right\} \quad (59)$$

The intermediate variables A and B have been defined in Eqs. 39 and 40.

Conversely, the optical constants of a bare surface can be determined from the parameters ψ and Δ by

$$n = n_0\sqrt{\tfrac{1}{2}\{(G^2 - E^2 + \sin^2\phi) + \sqrt{(G^2 - E^2 + \sin^2\phi)^2 + 4G^2E^2}\}} \quad (60)$$

$$k = n_0\sqrt{\tfrac{1}{2}\{-(G^2 - E^2 + \sin^2\phi) + \sqrt{(G^2 - E^2 + \sin^2\phi)^2 + 4G^2E^2}\}} \quad (61)$$

With the intermediate variables G and E (equal to A and B in Eqs. 39 and 40) defined as

$$G = \frac{\sin\phi\tan\phi\cos 2\psi}{1 + \sin 2\psi\cos\Delta} \quad (62)$$

$$E = \frac{\sin\phi\tan\phi\sin 2\psi\sin\Delta}{1 + \sin 2\psi\cos\Delta} \quad (63)$$

D. Data for Reflection from Bare Surfaces. In the following, some results of numerical computations for reflections from bare surfaces (88) are presented in graphical form in order to provide a better idea of the relationships between the optical and ellipsometric quantities discussed.

Figure 8 illustrates the dependence of ellipsometer parameters and complex reflection coefficients on angle of incidence (40,63,126). It also shows the effect of an increased imaginary part of the refractive index, starting with dielectric reflection as a limiting case. For an incident dielectric medium of refractive index n_0, n and k/n have to be divided by n_0 (72,128). Similar data based on different conventions can be found in the literature (27,124). The sensitivity for determining metal optical constants greatly depends on angle of incidence (75,91).

The dependence of the same four quantities on the optical constants n and k/n of the reflecting surface is illustrated in Figs. 9–12 for angles of incidence 2°, 45°, and 75°. The data for 2° are a close approximation to normal incidence, where indeterminate values resulting from the computations would have to be resolved by a

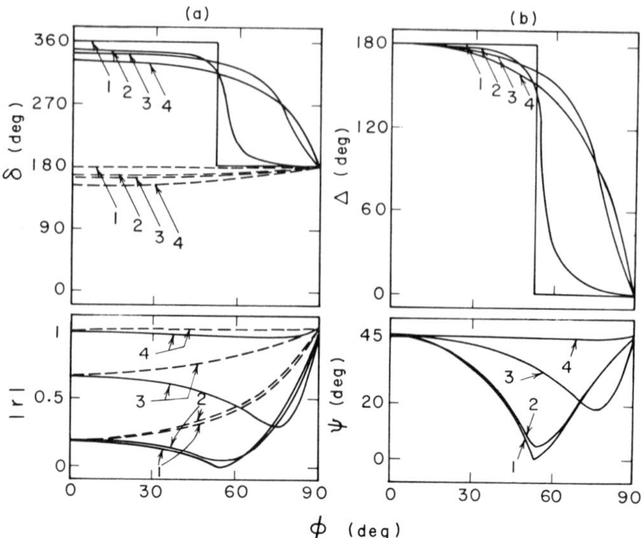

Fig. 8. Reflection from bare surfaces of complex refractive index n_c as a function of angle of incidence ϕ. (a) Argument δ (absolute phase) and modulus $|r|$ (amplitude) of reflection coefficient for p (——) and s (– – –) components. (b) Ellipsometer parameters Δ (relative phase change) and ψ (relative amplitude attenuation tan ψ). Identification of curves: 1. n_c = 1.5 (clear glass); 2. n_c = 1.5–0.15i (dark glass); 3. n_c = 3.3–2.31i (tantalum); 4. n_c = 0.2–4.0i (silver).

limiting process. At normal incidence, s and p components become indistinguishable. A difference in phase of 180° between s and p components is due to the choice of coordinate system (Fig. 6). Not all combinations of optical constants shown are physically realizable. Computed nets of n, k versus ψ, Δ values for bare surfaces (14,71,139) and detailed data for dielectric reflection (69) as well as the use of reflectance (19,43,113) can be found in the literature.

2. *Reflection from Film-Covered Surfaces*

A. Drude Equations. The classical theory of optical reflection from a film-covered surface assumes a planar substrate covered with a plano-parallel, homogeneous, isotropic film. Application of electromagnetic theory to this geometry provides solutions for waves traveling in incident medium, film, and substrate in the directions indicated in Fig. 13b. These waves are equivalent to the summation

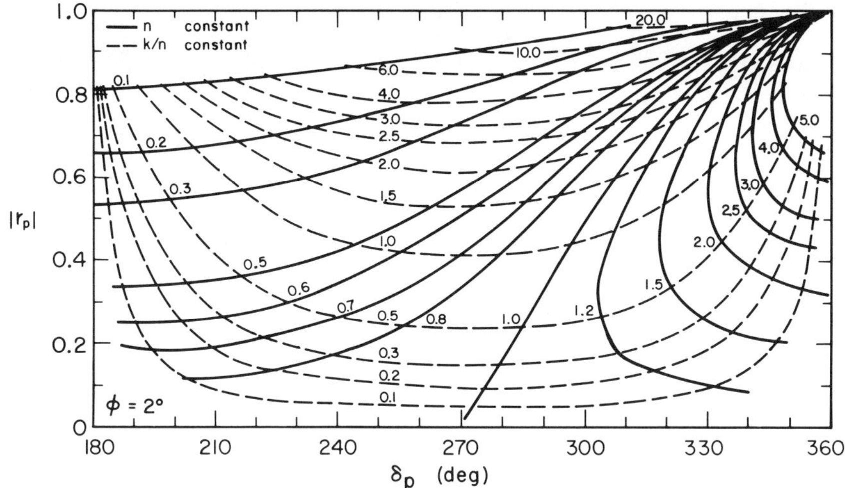

Fig. 9. Argument δ (absolute phase) and modulus $|r|$ (amplitude) of reflection coefficient for reflection from a bare metal surface of refractive index $n_c = n - ik$. Angle of incidence 2°, p-component. (For s-component the modulus is the same, the argument is 180° smaller.)

of multiple reflections shown in Fig. 13*a* (56,74,80,131). The resultant reflected wave is described in the (exact) Drude equations (Eqs. 64 and 65) by an overall, complex reflection coefficient r, which depends on the reflection coefficients r_1 and r_2 of the two interfaces and the phase delay D due to the travel of light in the film (42).

$$r_p = \frac{E_p''}{E_p} = \frac{r_{1p} + r_{2p}e^{-iD}}{1 + r_{1p}r_{2p}e^{-iD}} \tag{64}$$

$$r_s = \frac{E_s''}{E_s} = \frac{r_{1s} + r_{2s}e^{-iD}}{1 + r_{1s}r_{2s}e^{-iD}} \tag{65}$$

The delay in the film of refractive index n_{cf} and thickness L is

$$D = \frac{4\pi}{\lambda_0} L n_{cf} \cos \phi_{cf} \tag{66}$$

where λ_0 is the vacuum wavelength of the light and ϕ_{cf} the complex angle of refraction in the film (equal to ϕ_c' of Eq. 26). The reflection coefficients r_1 and r_2 are the Fresnel coefficients for the two interfaces

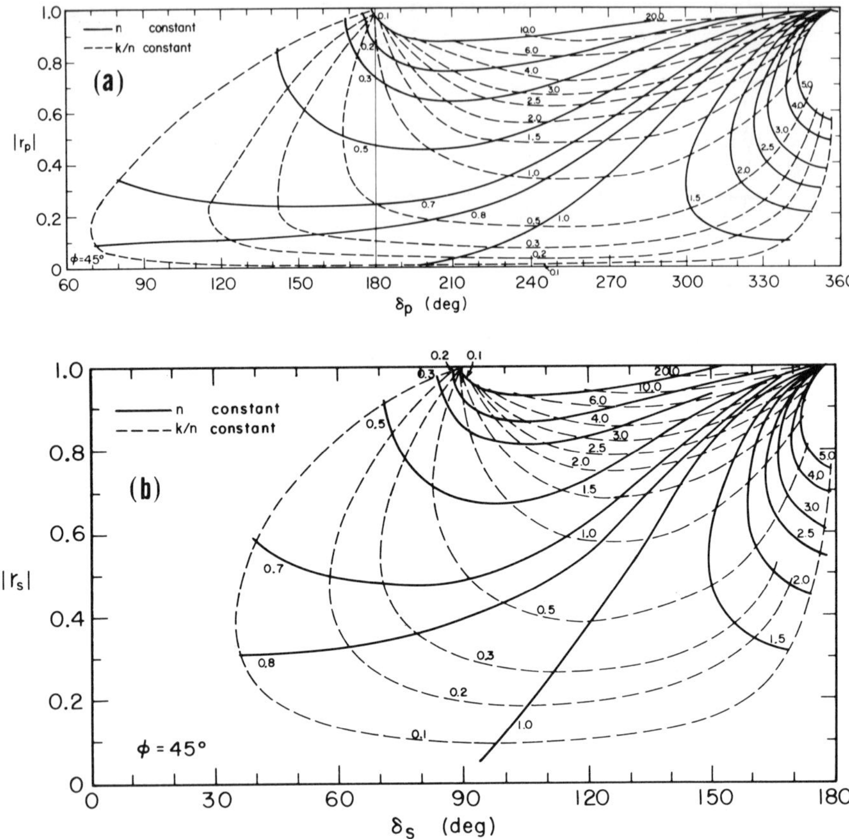

Fig. 10. As Fig. 9, angle of incidence 45°, (*a*) *p*-component, (*b*) *s*-component.

considered separately (with an infinitely thick film). With the electric field designations given in Fig. 14 they are defined as

$$r_1 = \frac{E_1}{E} \tag{67}$$

$$r_2 = \frac{E'''}{E'} \tag{68}$$

With Eqs. 22 and 23 they can be formulated as

$$r_{1p} = \frac{\tan\,(\phi - \phi_{cf})}{\tan\,(\phi + \phi_{cf})} \tag{69}$$

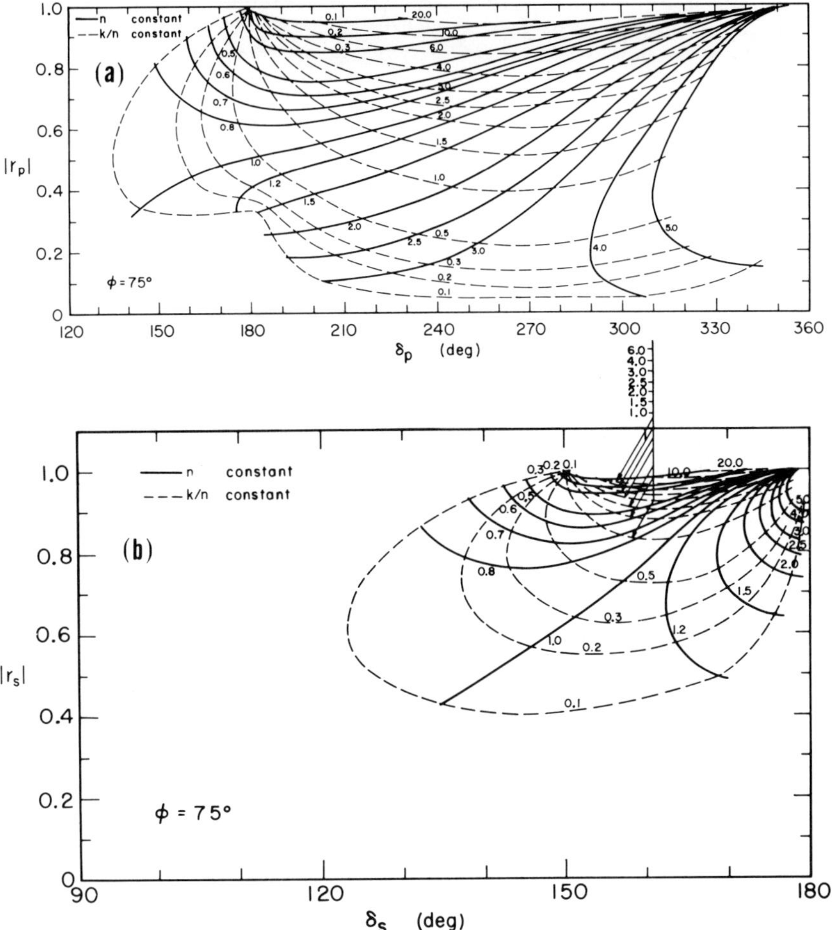

Fig. 11. As Fig. 9, angle of incidence 75°, (*a*) *p*-component, (*b*) *s*-component.

$$r_{1s} = -\frac{\sin(\phi - \phi_{cf})}{\sin(\phi + \phi_{cf})} \tag{70}$$

$$r_{2p} = \frac{\tan(\phi_{cf} - \phi_{cm})}{\tan(\phi_{cf} + \phi_{cm})} \tag{71}$$

$$r_{2s} = -\frac{\sin(\phi_{cf} - \phi_{cm})}{\sin(\phi_{cf} + \phi_{cm})} \tag{72}$$

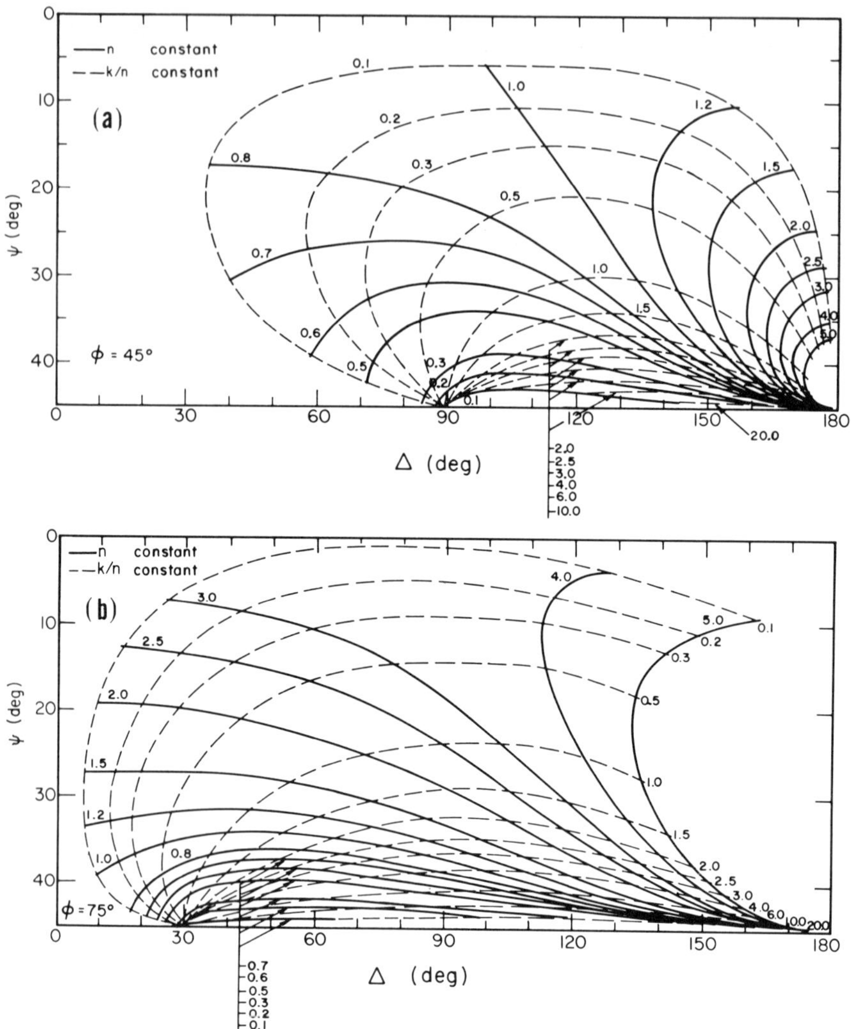

Fig. 12. Ellipsometer parameters Δ (relative phase change) and ψ (relative amplitude attenuation tan ψ) for reflection from a bare metal surface of refractive index $n - ik$. (*a*) Angle of incidence 45°, (*b*) angle of incidence 75°.

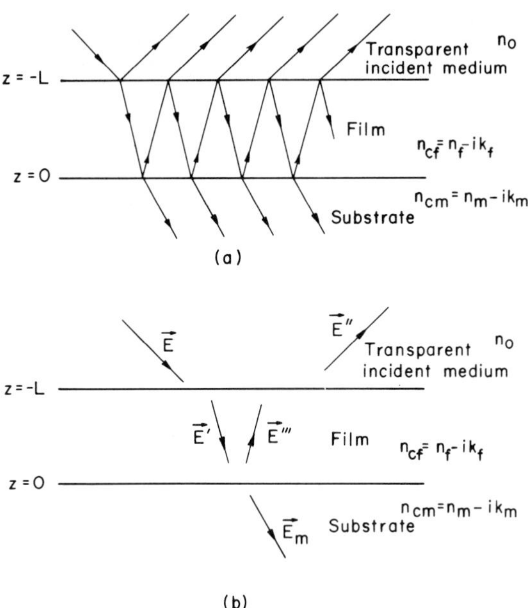

Fig. 13. Reflection from idealized film-covered surface (*a*) representation by multiple beam reflections, (*b*) representation by resultant waves.

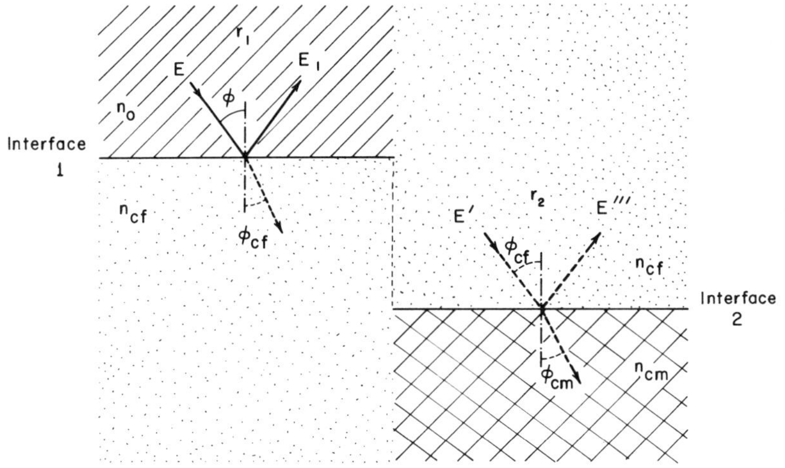

Fig. 14. Reflection coefficients r_1 and r_2 at both interfaces of a film-covered surface. Designation of refractive indices and electric fields. All the angles of propagation except ϕ are complex (indicated by dotted arches and lines) and cannot be interpreted geometrically.

where ϕ_{cf} and ϕ_{cm} are the complex angles of propagation in film and substrate, obtained by applying Snell's law (Eq. 26) to the two interfaces

$$n_0 \sin \phi = n_{cf} \sin \phi_{cf} \tag{73}$$

$$n_{cf} \sin \phi_{cf} = n_{cm} \sin \phi_{cm} \tag{74}$$

Drude has also derived a linearized form (31,42,80) of Eqs. 64 and 65 which are sometimes called *the* Drude equations (56,134). To avoid confusion, this nomenclature should be discontinued. Other approximations have been developed since (3,5,36,77). The exact equations are, however, easily handled by a computer and are now mostly used.

B. Computations. With the Drude equations for reflection from a film-covered surface, the basic equation of ellipsometry becomes

$$\rho = \frac{(r_{1p} + r_{2p}e^{-iD})(1 + r_{1s}r_{2s}e^{-iD})}{(r_{1s} + r_{2s}e^{-iD})(1 + r_{1p}r_{2p}e^{-iD})} = (\tan \psi)e^{i\Delta} \tag{75}$$

For given values of the optical constants of the three media involved (Fig. 14) and the thickness of the film, the ellipsometer parameters ψ and Δ can be computed by use of Eq. 75 (83). Examples for different classes of film-covered surfaces, with film thickness (in angstrom units) indicated as a parameter along the curve, are given in Figs. 15 and 16.

It is typical for dielectric films that the ellipsometer parameters ψ and Δ repeat themselves after an increment of one wavelength optical path (or an increment of 2π in phase delay D) in the film (Fig. 15). According to Eq. 66, one wavelength optical path usually requires a film thickness of about half a wavelength. A dielectric film on a metallic substrate (Fig. 15*b*) produces, in general, a greater change in Δ and ψ than a dielectric film on a dielectric substrate (Fig. 15*a*). In both cases, the curves are closed. Good agreement with measurements has been reported for the optical constants used in Fig. 15*b* (71).

With absorbing films, the effect of the substrate material is increasingly obscured, with increasing film thickness, by the light absorption in the film, and the ellipsometer parameters ψ and Δ approach the values representative of bulk film material. The ψ versus Δ curve is therefore not closed but runs from a point representing the bare substrate to a point representing the bare film material. Extremely thin absorbing films on a dielectric substrate

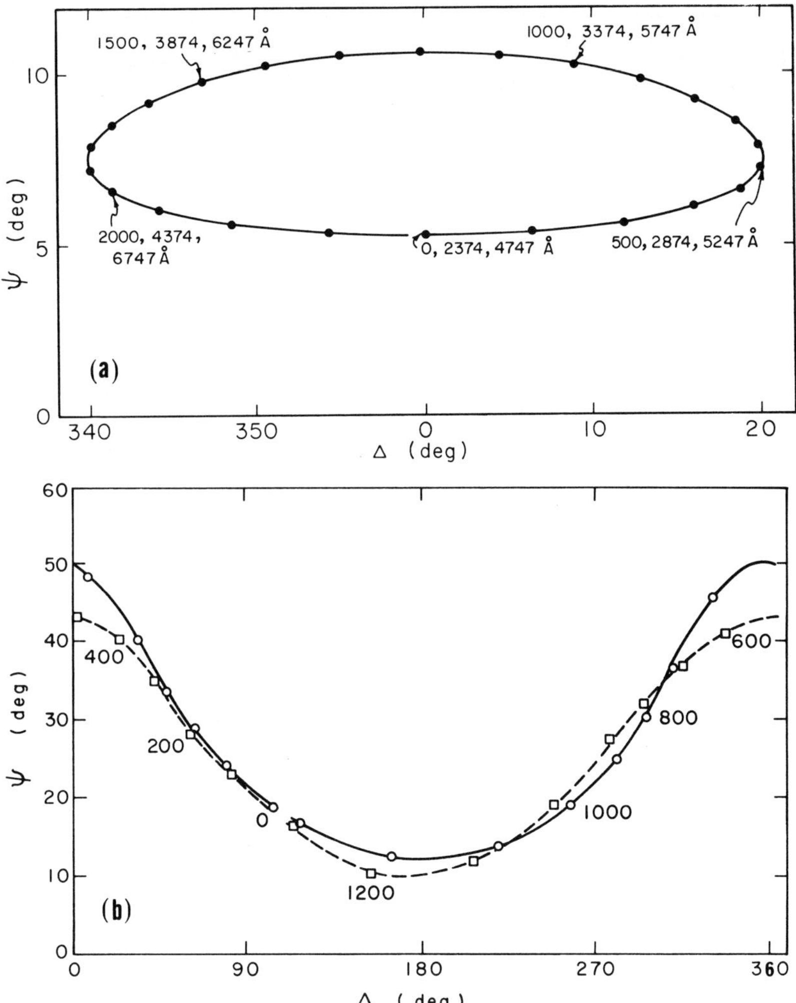

Fig. 15. Computed values of ψ and Δ (Eq. 75) for surfaces covered by a dielectric film. Incident medium vacuum ($n_0 = 1$), wavelength 5461 Å, film thickness in Å indicated along the curve. (*a*) Calcium fluoride ($n_f = 1.4339\text{–}0i$) on glass ($n_m = 1.519\text{–}0i$), angle of incidence 60°. (*b*) Tantalum oxide ($n_f = 2.26\text{–}0i$) on tantalum ($n_{cm} = 3.5\text{–}2.4i$), angle of incidence 75°. Solid curve—exact computations; dashed curve—graphical computation with double beam model (Section 2C).

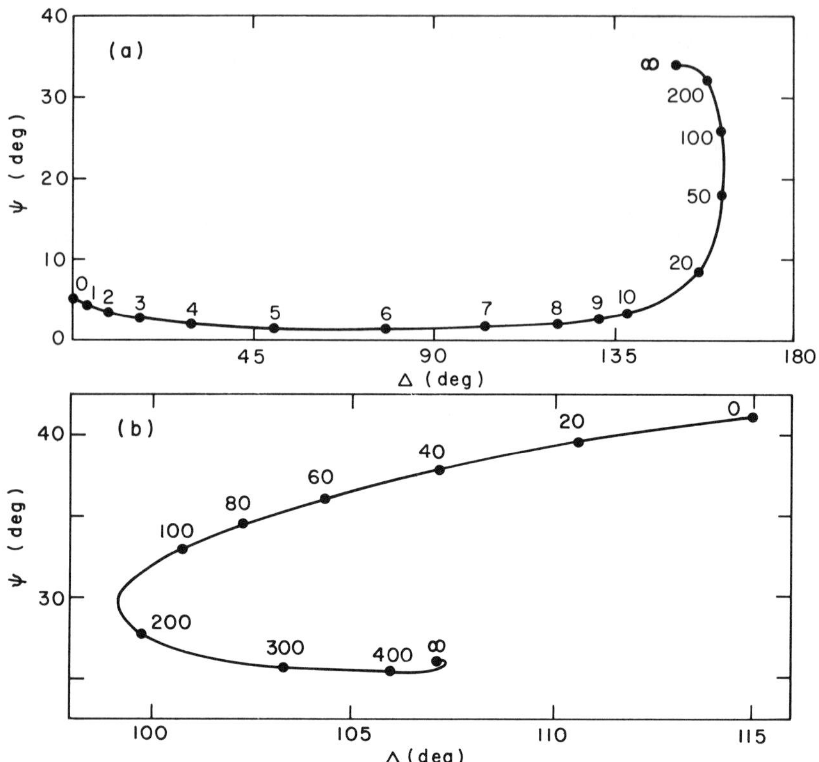

Fig. 16. Computed values of ψ and Δ for surfaces covered by an absorbing film. Incident medium vacuum ($n_0 = 1$), wavelength 5461 Å, film thickness in Å indicated along curve. (*a*) Chromium ($n_{cf} = 2.96–3.45i$) on glass ($n_{cm} = 1.519–0i$), angle of incidence 60°. (*b*) Chromium ($n_{cf} = 2.96–3.45i$) on nickel ($n_{cm} = 1.4–2.52i$), angle of incidence 75°.

(Fig. 16*a*) can be expected to produce particularly large changes in ψ and Δ, because of the large difference in optical constants of the two materials. Since the optical properties of a metal film and a metal substrate are more similar to each other, the ellipsometer parameters are much less sentitive to film thickness for such a combination (Fig. 16*b*).

The effect of film thickness (in the 100 Å range) on changes in ψ and Δ depends on the angle of incidence for each film-substrate combination and may even change sign. The optimum angle of incidence lies usually in the range of 60–80° (7,91,117,139).

In practice, one is often interested in determining the thickness

L and refractive index n_{cf} of a film on a substrate of known properties from measured values of ψ and Δ. The above procedure for calculating ψ and Δ for various combinations of n_f, k_f, and L can be used, until satisfactory agreement with measurement is found. Computed parametric curves (7,15,16,23,82,99) or computer-generated visual displays (92) can assist in this process. If an appreciable range of each of the three variables has to be covered in the search for a solution, the number of computations soon becomes too large to evaluate manually.

A more efficient determination of film thickness and refractive index from ellipsometer measurements (82,84) makes use of the fact that Eq. 75 can be solved explicitly for L as a function of n_{cf}, ψ, and Δ (this is, however, not possible for n_{cf}). This solution involves a quadratic equation and therefore provides two values of L which should be a real quantity. In practice, the solution with the smaller imaginary part is chosen to represent the physical situation.

C. Double Beam Model for Transparent Films. Since s and p components of reflected light are delayed to the *same* extent by the optical path in a film (Eq. 66), it is not obvious why the resulting state of polarization should depend on film thickness. The following double beam approximation to reflection from a transparent film on a metal substrate is intended to illustrate the physical basis of the ellipsometry of film-covered surfaces. The present approach is more easily understandable than a graphical computation (139) and a multiple beam model (44) employed before.

As an example for this approach, a tantalum oxide film on a tantalum surface will be considered. The multiple beam model of reflection (Fig. 17), which involves the summation of an infinite series of beams, will be approximated by the sum of the zero order reflection r_0, which involves only a dielectric reflection and r_∞, the resultant of all higher order reflections, which involve at least one metallic reflection.

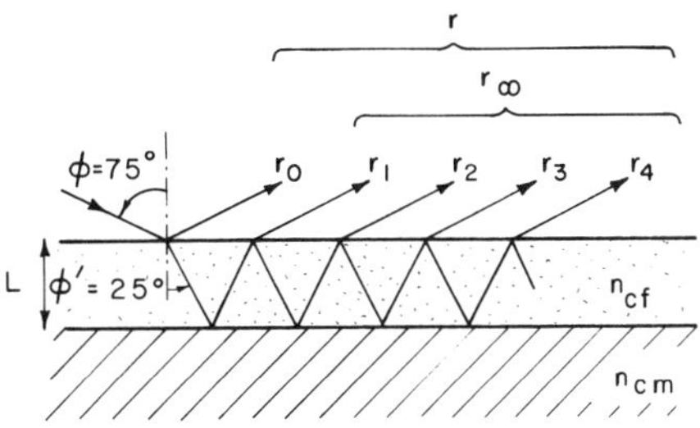

Fig. 17. Multiple beam model $r_0 + r_1 + r_2 + r_3$, etc., and double beam approximation $r_0 + r_\infty$ of reflection r from a dielectric film (Ta_2O_5, $n_{cf} = 2.26–0i$) on an absorbing substrate (Ta, $n_{cm} = 3.5–2.4i$). Angle of incidence 75°, incident medium vacuum ($n_0 = 1$).

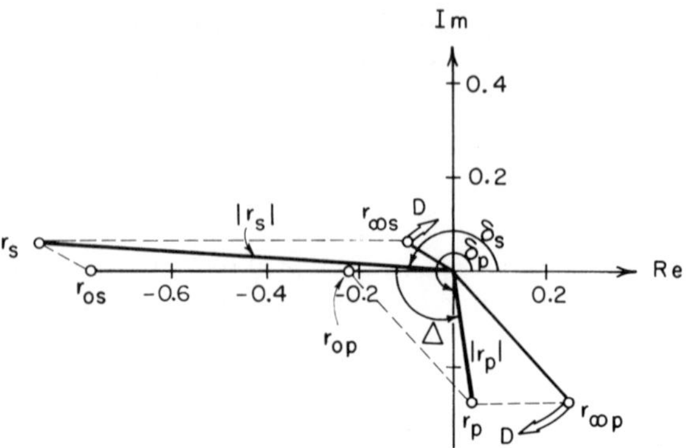

Fig. 18. Double beam model of ellipsometry. Addition of complex amplitudes r_0 and r_∞ for s and p components, Ta_2O_5 on Ta, data as in Fig. 17, negligible film thickness. D is the direction of phase change due to film growth.

For an incident wave of unit amplitude, r_0 and r_∞ are complex reflected amplitudes which can be added vectorially in the complex plane to form the resultant r. This process is carried out in Fig. 18, for s and p components for a film of negligible thickness. With the known Fresnel coefficient r_0, for the film material, r_∞ is chosen in such a way that the resultant r agrees with the known Fresnel coefficient for the bare substrate surface. The realtive phase Δ between p and s components is directly visible in the figure, the relative amplitude attenuation tan ψ has to be formed from the ratio of the lengths of the resultant p and s vectors.

An increase in film thickness results in an increased delay D, or a decrease in phase of r_∞ indicated by the double arrows in Fig. 18. This delay is *equal* for s and p components. (It will be assumed that the amplitude of r_∞ remains constant.) Phase and amplitude of r_0, on the other hand, are independent of film thickness. The sequence in Fig. 19 illustrates how the resultant s and p components are affected *differently* by the increase in film thickness and, thus, ψ and Δ values change. A comparison of results obtained graphically by this method with exact computations (Fig. 15*b*) indicates that the double beam model provides a reasonable approximation to reality.

D. Multiple Films. The optical effect of several different layers of film stacked up on a substrate can be computed by recursion formulas (38,84) or matrix methods (97,132,139). Inhomogeneous films with continuously varying optical properties in the direction

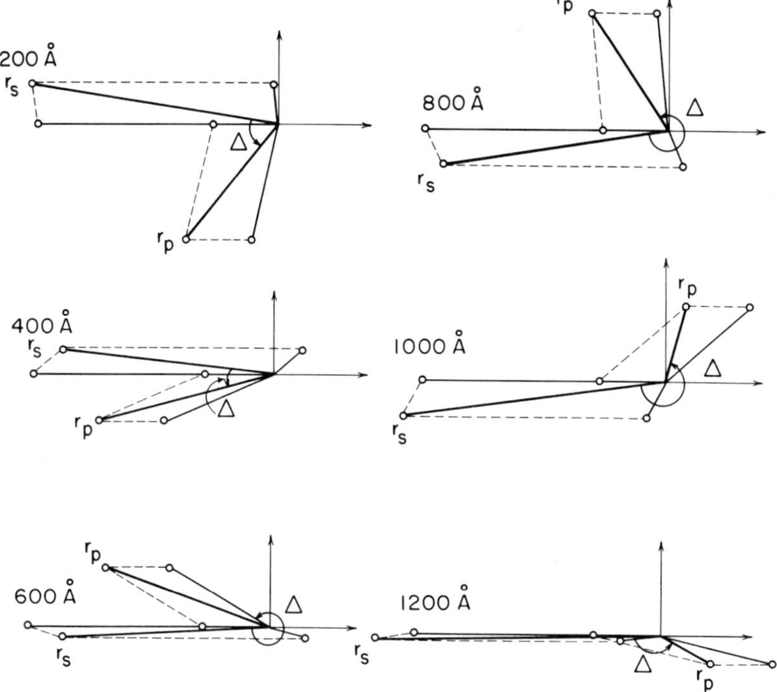

Fig. 19. Double beam model of ellipsometry. Effect of increasing film thickness on ellipsometer parameters Δ and $|r_p|/|r_s| = \tan \psi$, Ta_2O_5 film on Ta, data as in Fig. 17.

normal to the substrate surface (1,132) can be represented by multiple films (83). Evidence for the existence of multiple or inhomogeneous films cannot be obtained by ellipsometry alone, and considerable leeway usually exists in the choice of the additional parameters describing such films.

3. Conventions and Definitions

Numerical values of most of the parameters that appear in the theory of ellipsometry, or result from measurements, depend on the choice of arbitrary conventions and definitions. The effect of 9 twofold choices (resulting in 512 possible combinations) and their past usage has been discussed elsewhere (90).

Depending on the choices made, reflection from any given surface can be represented by eight numerically different combinations of ellipsometer parameters ψ and Δ. Conversely, from any measured set of ψ and Δ values, 16 different combinations of optical constants of a bare surface can be derived.

The preferred usage, determined at the International Ellipsometry Conference in Nebraska and used throughout this chapter is:

(1) Definition of the relative amplitude parameter ψ

$$\tan \psi = \frac{|r_p|}{|r_s|} \quad \left(\text{alternative: } \frac{|r_s|}{|r_p|}\right)$$

(2) Definition of the relative phase parameter Δ

$$\Delta = \delta_p - \delta_s \quad (\text{alternative: } \delta_s - \delta_p)$$

(3) Definition of complex relative amplitude attentuation

$$\rho = (\tan \psi)e^{i\Delta} \quad (\text{alternative: } (\tan \psi)e^{-i\Delta})$$

(4) Choice of time-dependence factor

$$E \sim e^{i\omega t} \quad (\text{alternative: } e^{-i\omega t})$$

(5) Definition of absolute phase

$$E \sim e^{i(\omega t + \delta)} \quad (\text{alternative: } e^{i(\omega t - \delta)})$$

(6) Positive coordinate directions for p component as shown in

Fig. 6 (alternative: E_p'' of opposite direction).

(7) Positive coordinate directions for s component as shown in

Fig. 6 (alternative: E_s'' of opposite direction).

(8) Formulation of complex refractive index

$$n_c = n - ik \quad (\text{alternative: } n_c = n(1 - i\kappa))$$

(9) Definition of complex refractive index

$$n_c = n - ik \quad (\text{alternative: } n_c' = n' - ik')$$

Azimuth angles of polarizers and analyzers are measured counterclockwise from the plane of incidence to the transmission direction for the electric field vector. Azimuth angles of compensators are measured counterclockwise from the plane of incidence to the fast axis of the compensator. In both cases the observer looks into the beam (toward the source).

IV. Ellipsometer Arrangements

1. Measurement With or Without Compensation

Elliptically polarized light (Fig. 3) can be analyzed by measuring the intensity transmitted by a rotating linear polarizer (analyzer) as a function of its azimuth angle (30). The azimuth of the intensity maximum represents the orientation θ of the major ellipse axis, the ratio of the square roots of intensity minimum to maximum represents the eccentricity $\tan \gamma$ of the ellipse (Fig. 20). This measurement without compensation requires absolute intensity determinations and therefore does not take full advantage of some of the inherent capabilities of ellipsometry.

The measurement of elliptic polarization with compensation is based on the restoration of linear polarization by the introduction of a known phase difference between two orthogonal components of the elliptic polarization. This phase difference is generated by a compensator (wave plate) and can have a fixed or variable value. The restored linear polarization is recognized by the *extinction* with a (linear) analyzer. In practice, a *minimum* in the transmitted light intensity is sought.

2. Fixed and Variable Compensators

Conceptually the simplest ellipsometer arrangement, although not normally used in practice, employs linearly polarized light from a polarizer P of azimuth $a = 45°$ incident on the reflecting surface. Thus, s and p components of the incident light are in phase and of equal amplitude. (In the basic ellipsometer equation (Eq. 53)

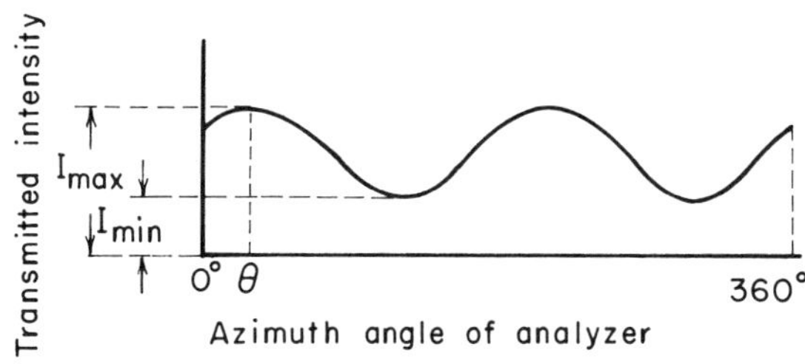

Fig. 20. Analysis of elliptic polarization with a rotating analyzer. Analyzer azimuth for, and intensity of, transmission minima and maxima.

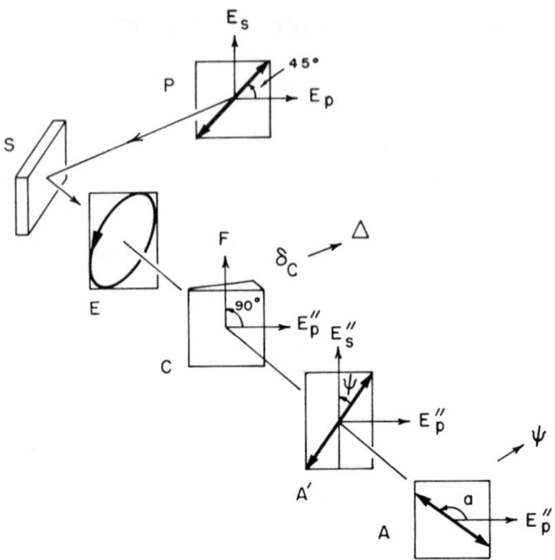

Fig. 21. Arrangement of components and states of polarization in an ellipsometer with linear polarization incident on the specimen. Use of a compensator of variable retardation. *P*—polarizer with transmitted electric field direction in fixed 45° azimuth. *S*—reflecting specimen surface. *E*—elliptic state of polarization after reflection. *C*—variable compensator of fixed azimuth (slow and fast axes parallel to *p* and *s* directions). Retardation at extinction is a measure of Δ. *A′*—restored linear state of polarization. *A*—Analyzer with transmitted electric field direction. Azimuth *a* at extinction is a measure of ψ.

$\tan \psi_i = 1$ and $\Delta_i = 0$.) As indicated in Fig. 21, reflection of this light results in elliptic polarization, due to a change in relative phase and amplitude of *p* and *s* components, e.g., the *p* component may now be ahead of the *s* component and be of smaller amplitude. (For the sake of simplicity, Δ is assumed to have a value between 0 and 90°, and ψ between 0 and 45° in Figs. 21–24. With bare metal surfaces this case is realized at angles of incidence above the principal angle.)

The phase difference between the two components can be restored by a compensator of *variable retardation* and fixed azimuth (Babinet-Soleil compensator, symbolized by a wedge *C* in Fig. 21), which introduces a relative phase change δ_C between *p* and *s* components which is equal but of opposite sign to that due to reflection (15,16). For this purpose, the fast axis *F* of the compensator is aligned with the component which has been retarded in reflection (e.g., the *s* component) and a retardation is chosen such that linear polarization is restored. The plane of this polarization forms the diagonal of a rectangle, in contrast to the diagonal of a square shown for the incident polarization, because p and s components, have been attenuated differently during reflection. Thus, the change in relative phase Δ due to reflection is obtained from the retardation δ_C of the compensator, while the change in relative amplitude, $\tan \psi$, is obtained from the azimuth of the restored linear polarization. In practice the latter is derived from the azimuth a of the analyzer A at extinction, as indicated in Fig. 21.

The elliptic polarization that results from the reflection of linearly polarized light can also be compensated with a compensator of *fixed retardation* and variable azimuth, as illustrated in Fig. 22. The elliptic polarization, in this case, is decomposed into components parallel to the major and minor axis of the ellipse. As had been pointed out in the discussion of Fig. 4, these two components are mutually out of phase by 90° or a quarter wavelength. Linear polarization can therefore be restored by the introduction of a quarter wave phase difference between these components by means of a *quarter wave plate* (Sénarmont compensator Q). Thus, the azimuth

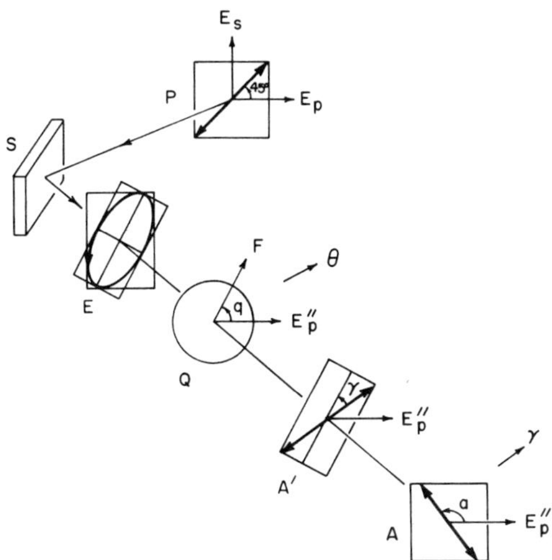

Fig. 22. Ellipsometer with linear polarization incident and compensator of fixed retardation. P, S, E, A'—as in Fig. 21. Q—fixed compensator (quarter wave plate) of variable azimuth q (at extinction, fast axis F parallel to major axis of ellipse, $q = \theta$). A—analyzer with transmitted electric field direction. Azimuth a at extinction is a measure of γ.

q of the quarter wave plate at extinction is a measure of the orientation θ of the major ellipse axis, and the azimuth of the restored linear polarization is a measure of the eccentricity tan γ of the ellipse. In contrast to the first arrangement, which provided a direct measurement of the physical parameters of elliptic polarization, this arrangement therefore provides the geometrical parameters.

Presently available compensators of variable retardation allow one to determine retardation only with a resolution of approximately 1°, because the measurement involves a translatory movement, in contrast to the rotatory movement employed with a compensator of fixed retardation. If a compensator of variable retardation is used in practice (e.g., for spectral work), it is therefore preferable not to change its retardation during a measurement, but to set it to quarter wave and rotate it like a compensator of fixed retardation. Optical imperfections are more prevalent in variable than in fixed compensators and may also affect the accuracy of such measurements. If reliable variable compensators of high resolution become available (e.g., electro-optic devices) some of the presently unused ellipsometer arrangements, such as the one in Fig. 23, may become of practical interest.

3. *Linear and Elliptic Polarization Incident*

Both ellipsometer arrangements described above employ *linearly* polarized light incident on the specimen and the elliptic polarization resulting from the reflection is converted back to linear polarization. A reversal of the sequence of optical elements results in *elliptically* polarized light incident on the specimen. The elliptic polarization is then chosen so that reflection results in linear polarization. If s and p components of the incident light are again chosen to be of equal amplitude, the use of compensators with variable or fixed retardation results in the following ellipsometer arrangements.

A compensator of variable retardation, shown in Fig. 23, introduces a phase difference δ_C between p and s components before reflection. This phase difference is of equal magnitude but opposite sign compared to the one due to reflection on the specimen. Linear polarization is therefore restored by reflection. As with the arrangement of Fig. 21, the parameters Δ and ψ, characteristic of the reflecting surface, are derived from the retardation δ_C of the compensator and the azimuth a of the analyzer.

The use of a quarter wave compensator with elliptic incidence is

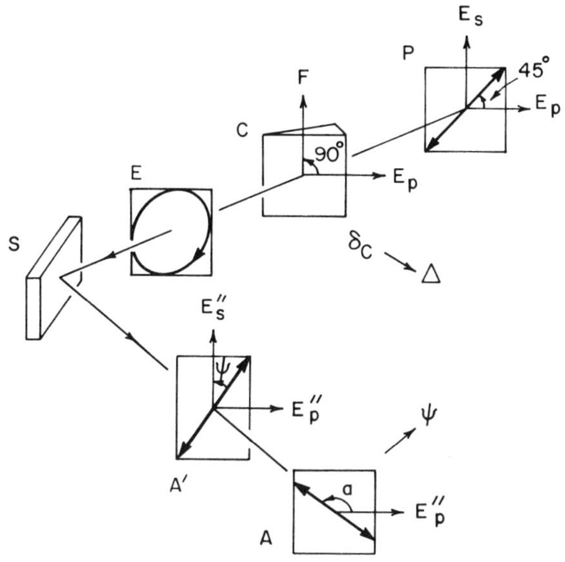

Fig. 23. Ellipsometer with elliptic polarization incident and compensator of variable retardation. *P*, *C*, *S*, *A′*, and *A* as in Fig. 21. *E*—elliptic state of polarization before reflection.

illustrated in Fig. 24. This arrangement is presently the preferred one for practical use. The 45° azimuth q of the quarter wave plate results in equal amplitudes of p and s components incident on the specimen. At extinction, the polarizer azimuth p is a measure of Δ, the analyzer azimuth a a measure of ψ due to the reflection. These features will be more apparent by use of the Poincaré sphere (Fig. 26*b*).

4. Application of Poincaré Sphere

A. Linear Polarization Incident. The state of polarization can easily be traced on the Poincaré sphere (34) through the elements of an ellipsometer (140) operated in the four modes shown above. The stereographic projection of the sphere, looking down on the "north" (L) pole, will be used for this purpose. The effect of reflection is represented by two separate steps, one affecting only Δ, the other only ψ. The following symbols are used:

H—plane of incidence (origin of azimuth measurements), p component.

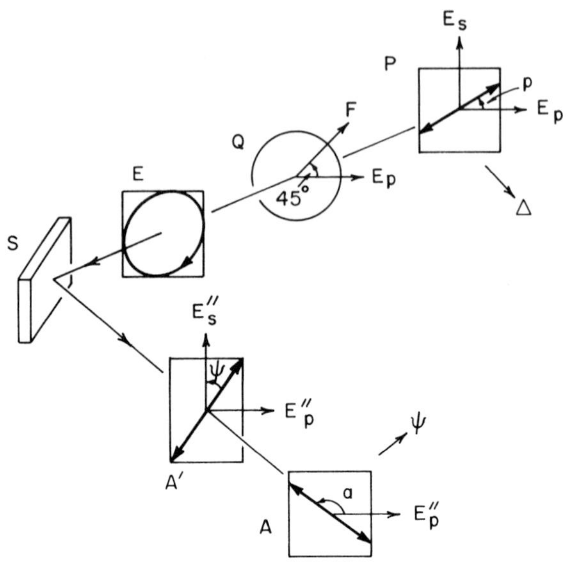

Fig. 24. Ellipsometer with elliptic polarization incident and compensator of fixed retardation. *P*—Polarizer with transmitted electric field direction. Azimuth *p* at extinction is a measure of Δ. *Q*—Fixed compensator (quarter wave plate) of fixed azimuth (fast axis *F* at 45° to plane of incidence). *E*—Elliptic state of polarization before reflection. *S*—Reflecting specimen surface. *A'*—Restored linear state of polarization. *A*—Analyzer with transmitted electric field direction. Azimuth *a* at extinction is a measure of ψ.

V—plane of *s* component.
L—left-hand circular pole of sphere.
P—polarizer transmission axis, azimuth *p* (angle to plane of incidence).
Δ—change in relative phase introduced by reflection between *p* and *s* components.
E—elliptic state of polarization before or after reflection.
(ψ)—change in relative amplitude ψ due to reflection.
δ_C—change in relative phase between fast and slow axis, introduced by a compensator of variable retardation.
δ_Q—change in relative phase introduced by a quarter wave plate.
F—compensator fast axis, azimuth *q*.
A'—restored linear polarization, azimuth *a'*.
A—analyzer transmission axis, azimuth *a* (at extinction).

——O—circles and points on the upper hemisphere (left-hand polarization).
---×—circles and points on the lower hemisphere (right-hand polarization).

With linear polarization incident and a variable compensator, the state of polarization through an ellipsometer, arranged as shown in Fig. 21 is given in Fig. 25*a*. The linear polarization produced by the polarizer of fixed 45° azimuth is represented by point P on the equator (angle from plane of incidence H doubled). Reflection on the specimen results in a change of relative phase Δ and relative amplitude ($\tan \psi$). (As in Figs. 21–24, Δ is assumed to have a value between 0 and 90° and ψ is assumed to lie between 0 and 45°.) Point E on the sphere represents the elliptic state of polarization after reflection. Introduction of the phase change δ_C (equal but opposite to Δ), between fast and slow axis of the variable compensator of fixed orientation of its fast axis F, results in the restored linear polarization A', which forms the angle ψ (doubled on the sphere) with the direction V of the s component (coincides here with F). In practice, ψ is derived from the azimuth a of analyzer A ($2a$ measured counterclockwise from H to A).

Figure 25*b* similarly traces the state of polarization for linear

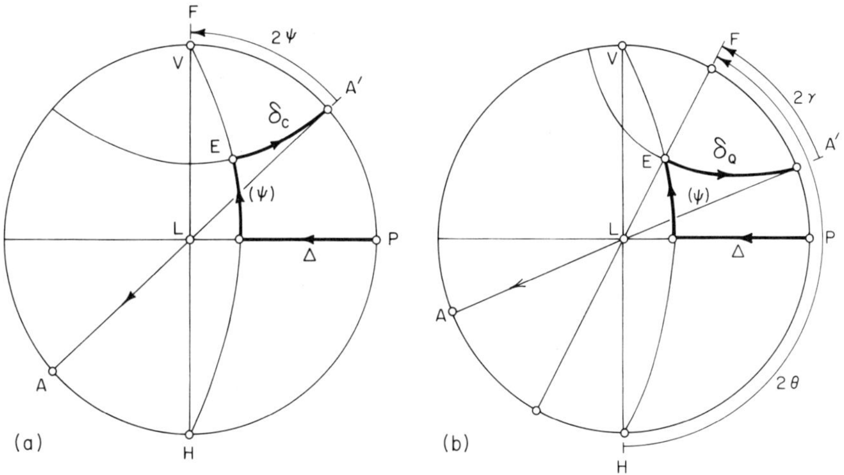

Fig. 25. Poincaré representations of the changes in the state of polarization through the components of an ellipsometer with linear polarization incident on the specimen. Stereographic projection, symbols listed in text. (*a*) Ellipsometer arrangement of Fig. 21; (*b*) ellipsometer arrangement of Fig.22.

incident polarization and a compensator of fixed retardation (140). In this case, the elliptic polarization E has to be converted to a linear polarization with a compensator of $\delta_Q = 90°$ retardation (quarter wave plate, MacCullagh method of compensation, Ref. 105). This is accomplished by choosing F on a great circle through L and E (parallel to a principal ellipse radius of azimuth θ). A 90° rotation in the positive direction, as seen from F then brings point E to A'. The difference in azimuth q between the fast axis F of the compensator and the azimuth a' of the restored linear polarization A' is the geometric ellipse parameter γ.

B. Elliptic Polarization Incident. Elliptic incident polarization, produced by a compensator of variable retardation, is illustrated in Fig. 26*a*. The polarizer P is again in a 45° azimuth (90° from point H on the sphere). The effect δ_C of the compensator (positive rotation as seen from F) results in right-hand elliptic polarization incident on the specimen. Point E, representing this state of polarization, lies on the lower half of the Poincaré sphere and is represented by a cross in Fig. 26*a*. The phase change Δ due to reflection is equal and opposite to δ_C, the unequal amplitude attenuation in reflection results in the linear polarization state A'. The angle ψ is derived from the azimuth of analyzer A.

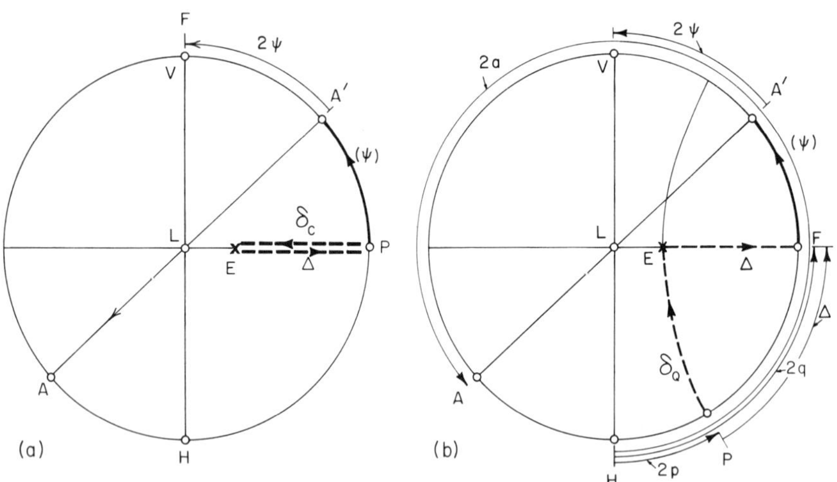

Fig. 26. Poincaré representations with elliptic polarization incident on the specimen. (*a*) Ellipsometer arrangement of Fig. 23; (*b*) ellipsometer arrangement of Fig. 24.

Elliptic incident polarization, produced with a quarter wave plate, is illustrated in Fig. 26*b* (140). This is the most important ellipsometer arrangement in practice. In this case, the incident elliptic state of polarization E, identical to the one in Fig. 26*a*, is obtained from the linear state P (determined by the polarizer azimuth p) by the action of the quarter wave plate with fast axis F of azimuth $q = 45°$ (90° from H on the sphere). The effect of the quarter wave plate on the Poincaré sphere is a 90° positive (counterclockwise) rotation on a small circle around F. As in Fig. 26*a*, phase and amplitude changes during reflection result in the restored linear state of polarization A', which is extinguished by the analyzer A of azimuth a.

The above argument assumes ideal, perfectly aligned optical components. In particular, the quarter wave plate is assumed to introduce a retardation of exactly 90° and not to change the relative amplitude. Deviations from these conditions can be determined (60) and employed in the computations (82,84). Averaging measurements in "four zones" (to be discussed) also serves to reduce the effect of some errors (82).

C. Error Propagation. The operation of an ellipsometer with a quarter wave compensator requires the rotation in azimuth of two components to reach extinction. For precise measurements, it is important that small deviations in azimuth from the correct (final) setting of one component do not affect the azimuth of the other component for minimum transmitted light intensity.

The ellipsometer arrangements of Figs. 22 and 24 differ in this respect. With incident linear polarization (Fig. 27*a*) any deviation in the quarter wave plate azimuth from the correct value q_1 results in a proportionate displacement of the analyzer azimuth a for intensity minimum, and vice versa. Compensator and analyzer azimuth therefore hunt each other in search for the deepest intensity minimum, which is not very well defined and often is affected by noise in the photodetector circuit.

With incident elliptic polarization, on the other hand, the analyzer azimuth a for minimum intensity is not affected by deviations in the polarizer azimuth from the final value p_1; the minima are not as low, but occur at the same angle (Fig. 27*b*). The same is true for intensity as a function of polarizer azimuth with the analyzer in different orientations. The lack of propagation of errors from one azimuth reading to the other is the reason for the preferred usage of the arrangement shown in Figs. 24 and 26*b*.

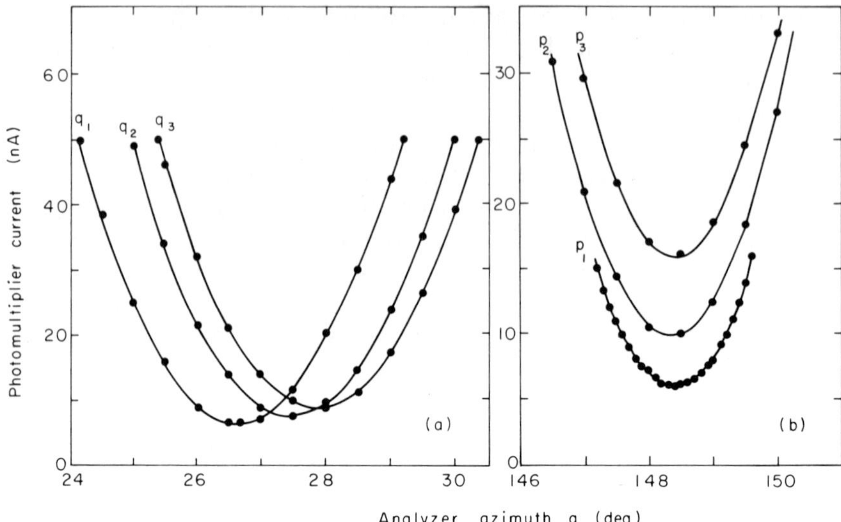

Fig. 27. Propagation of azimuth errors with quarter wave compensator. Transmitted light intensity as a function of analyzer azimuth. Stainless steel surface in air, angle of incidence 75°, Photomultiplier RCA 931A, operated at –800V. (*a*) Linear polarization incident, polarizer azimuth 135°, compensator azimuths $q_1 = 96.87°$, $q_2 = 95.87°$, $q_3 = 95.37°$; (*b*) elliptic polarization incident, compensator azimuth 45°, polarizer azimuths $p_1 = 0.33°$, $p_2 = 1.33°$, $p_3 = 1.83°$.

The difference in error propagation can be demonstrated on the Poincaré sphere. With linear polarization incident (Fig. 28*a*), a compensator of fast axis F_2, which is different from the correct location F_1, results in a restored polarization A_2', (indicated by ×) that is slightly elliptical. The major axis of the ellipse is indicated by a great circle through × and L, which appears as a diameter of the circle in the projection. A minimum in transmitted intensity is reached with the transmission direction of the analyzer in orientation A_2, parallel to the minor axis of the ellipse. Thus in first approximation, the difference in compensator azimuths between F_1 and F_2 is shown to be transmitted to the analyzer azimuths of A_1 and A_2, as found by experiment (Fig. 27).

The situation with elliptic polarization incident is illustrated in Fig. 28*b*. A polarizer P_2 of an azimuth that is different from the correct setting P_1 results, in first approximation, in a restored slightly elliptic polarization A_2' with major axis parallel to the correctly

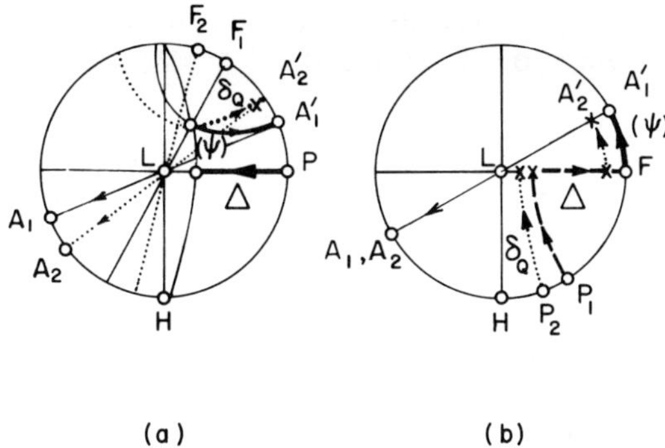

Fig. 28. Propagation of azimuth errors with quarter wave compensator, Poincaré representation. (*a*) Linear polarization incident; (*b*) elliptic polarization incident.

restored linear polarization A_1'. Thus the analyzer azimuths A_1 and A_2 for minimum transmitted intensity are shown to be the same, in agreement with experiment.

A similar analysis for the variable compensators, used as shown in Figs. 21 and 23, shows that errors in retardation δ_C do not propagate to the azimuth of the analyzer A.

For a more general analysis of the amplitude transmitted by an analyzer (82,103,139,141), a right triangle can be inscribed in the Poincaré sphere, with the three corners representing transmission and absorption directions of the analyzer and the state of polarization incident on it. The length of the side of the triangle opposite to the transmission of the analyzer (a chord on the sphere) is then a measure of the transmitted amplitude (82).

5. *Other Ellipsometer Arrangements*

The four ellipsometer arrangements discussed employ equal amplitudes of p and s components incident on the specimen (state of polarization represented by a point on the horizontal diameter of the circle, showing a great circle through pole L, in Figs. 25 and 26). In principle, any two optical elements can be adjusted to reach extinction, with the others remaining unchanged, and many more modes of operation are feasible. The derivation of results from the measured quantities is, however, somewhat more complicated if incident p and s components are not of equal amplitude. Arrangements with incident linear polarization and quarter wave plate of fixed azimuth (65) as well as the use of an optimum compensator azimuth (109) have been described.

6. Modern Instrumentation

A. Self-Compensating Ellipsometers. The manual operation of an ellipsometer is quite slow. To find extinction with good precision, several minutes are required for a measurement in one zone. A considerable fraction of this time is used for the repeated reading of azimuth circles. Mechanizing this process, e.g., by use of digital angle position encoders, seems to be a promising route which has not yet been taken to speed up the response of manually operated ellipsometers. With optical modulation techniques and mechanized azimuth readout, manual operation with response times of a few seconds appears feasible. A manually operated instrument with azimuth and phase modulation by use of Faraday and piezo-optic cells has been described by Wilmanns (138).

Self-compensating ellipsometers, for the most part, simulate manual operation at a greater speed and maintain the precision inherent in compensating measurements. Two kinds of mechanically operated self-balancing instruments have been described. One, by Ord (93, 94), is controlled by the intensity of the photodetector output. Compensation is reached in about a second. Another one, by Takasaki (127), is controlled by the phase of the photodetector output. A modulation in the state of polarization is introduced by two electro-optic elements (Pockels cells). Compensation takes several seconds.

Instead of by mechanical rotation of a polarizing prism, the plane of polarization of light can also be rotated electronically by a magneto-optic element (Faraday cell). This approach has been used by Layer (73) for an automatic ellipsometer, designed to respond within approximately 10^{-3} sec. A slew rate of 4°/sec has been demonstrated. Unfortunately, no final analysis of the performance of this instrument is available. The use of Faraday and Pockels cells for automatic ellipsometers has been suggested by several authors (137,140).

Several new designs of self-compensating ellipsometers would be made possible by the availability of a reliable electronic compensator of measurably variable retardation. Future forms of electro-optic or piezo-optic (61,66,95) devices may be useful for this purpose.

B. Noncompensating Ellipsometers. Off-balance intensity changes of the transmitted light can be used as a measure of changes in ellipsometer parameters (70,102,118). Depending on the azimuth settings chosen, the output is primarily due to changes in ψ or Δ, provided that the changes are of limited extent. Time resolution of less than 10^{-3} sec should be realizable by this approach.

As explained with Fig. 20, an elliptic state of polarization can also be determined from the transmitted light intensity as a function of the analyzer azimuth (103). An instrument that employs a rotating analyzer has been reported by Cahan and Spanier (33).

It must be remembered that the use of light intensity measurements in ellipsometry, common to all noncompensating techniques, to some extent negates one of the basic advantages of the technique: its use of the state of polarization, rather than the intensity of reflected light. To what extent this fact affects the reliability of results obtained from such measurements has to be analyzed for each specific case.

V. Experimental Aspects

1. Four Zones

The following discussion is restricted to the use of a compensator of fixed retardation, a quarter wave plate, and elliptic incident polarization, which are at present most commonly employed. Similar classifications can be derived for other modes of operation.

A given pair of values ψ and Δ, due to reflection, can be determined by 32 different combinations of polarizer, analyzer, and compensator azimuth settings. These combinations can be organized in four groups or "zones" (82), which are identified by the quadrant on the "equator" of the Poincaré sphere (with origin H), in which the polarizer is located (Table I).

TABLE I
Definition of Zone Numbering

Zones	Azimuth p of polarizer transmission axis
1	0°–45°
2	45°–90°
3	90°–135°
4	135°–180°

The algebraic expressions which relate the azimuth readings, p, q, and a of polarizer, quarter wave plate and analyzer, with the desired quantities ψ and Δ are slightly different for different ranges of the value of Δ. It is therefore convenient to distinguish four groups of zones *A* to *D* (Table II).

TABLE II
Definition of Zone Groups

Zone groups	Δ
A	0°–90°
B	90°–180°
C	180°–270°
D	270°–360°

The reason for the existence of four zones is most easily shown on the Poincaré sphere with a small value of Δ (group *A*) and is illustrated in Fig. 29. Zone *A*-1, with the fast axis *F* of the quarter wave plate in a 45° azimuth, is identical to the arrangement discussed in Fig. 26*b*. The same state of incident polarization is produced in zone *A*-2 with the quarter wave plate of 135° (or −45°) azimuth and the polarizer in a position symmetrical to the first one around the 45° azimuth. In both cases, the analyzer azimuth at extinction is the

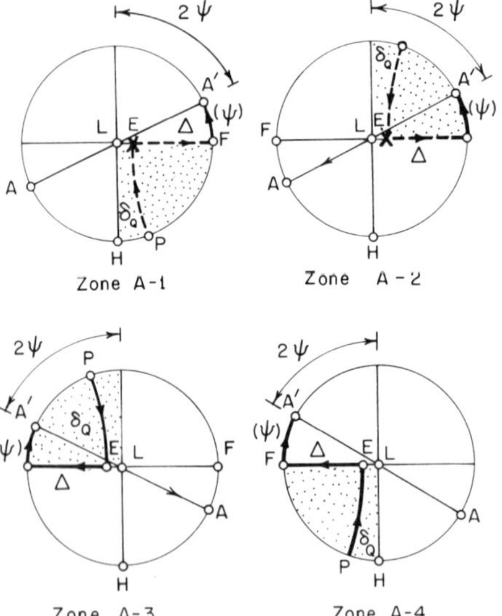

Fig. 29. Four zones for elliptic incident polarization and quarter wave compensator. Zone group *A* (Δ = 0 to 90°). Measurement of the same values ψ and Δ due to reflection. Stereographic projection of Poincaré sphere.

same. Zones 3 and 4 are related to each other like zones 1 and 2. The incident state of polarization is now represented by a point on the upper hemisphere (circle).

Because any two azimuth positions of polarizer, quarter wave plate, and analyzer 180° apart are optically identical, a measurement in each zone can be obtained by 8 different combinations of azimuth circle readings, or 32 combinations for all four zones. To avoid this additional multiplicity, all azimuth readings should be restricted (or converted) to values in the range of 0–180°.

2. *Sixteen Zones*

The derivation of ellipsometer parameters ψ and Δ from the measured azimuth angles in zone *A*-1 can be seen in Fig. 26*b* and may be applied to the other zones in group *A*. If we denote the azimuth angles (measured counterclockwise from the plane of incidence *H*) of polarizer, quarter wave plate, and analyzer by *p*, *q*, and *a*, the algebraic expressions listed in Table III result. The expressions for zone groups *B* to *D* can be derived similarly, as indicated in Fig. 30 for zone 1. All azimuth readings are assumed to lie in the range 0 to 180°.

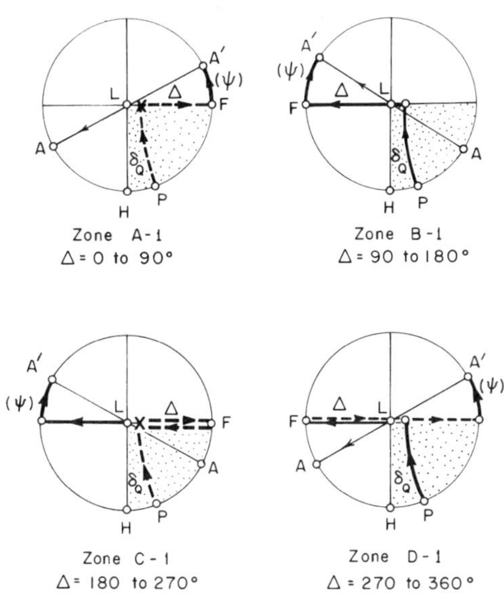

Fig. 30. Four zone groups for elliptic incident polarization and quarter wave compensator. Zone 1 ($p = 0$ to 45°). Stereographic projection of Poincaré sphere.

Azimuth circles in ellipsometers often have their origin in the plane of the s, rather than the p component and increase clockwise, rather than counterclockwise. Readings from such circles first must be converted to the azimuth angles defined here, before the expressions for ψ and Δ given in Table III can be used. Alternatively, a new set of expressions for ψ and Δ, which permits the use of instrument readings that do not conform to the definitions, can be derived. In practice, the use of Table III first involves a search for the correct zone, based on the values of polarizer, compensator, and analyzer azimuths. Then the appropriate algebraic expressions for deriving ψ and Δ can be used.

TABLE III

Derivation of ψ and Δ from ellipsometer azimuths of polarizer, quarter wave plate, and analyzer (all angles in degrees).
($\psi = 0°$ to $90°$, $\Delta = 0$ to $360°$)

Zone	Range of polarizer transmission azimuth p	Compensator fast axis azimuth q	Range of analyzer transmission azimuth a	ψ	Δ
A-1	0–45	45	90–180	180 − a	90 − 2p
A-2	45–90	135	90–180	180 − a	2p − 90
A-3	90–135	45	0–90	a	270 − 2p
A-4	135–180	135	0–90	a	2p − 270
B-1	0–45	135	0–90	a	90 + 2p
B-2	45–90	45	0–90	a	270 − 2p
B-3	90–135	135	90–180	180 − a	2p − 90
B-4	135–180	45	90–180	180 − a	450 − $2p$
C-1	0–45	45	0–90	a	270 − 2p
C-2	45–90	135	0–90	a	90 + 2p
C-3	90–135	45	90–180	180 − a	450 − 2p
C-4	135–180	135	90–180	180 − a	2p − 90
D-1	0–45	135	90–180	180 − a	270 + 2p
D-2	45–90	45	90–180	180 − a	450 − 2p
D-3	90–135	135	0–90	a	90 + 2p
D-4	135–180	45	0–90	a	630 − 2p

3. *Zone Averaging*

With ideal, perfectly aligned ellipsometer components, the algebraic expressions in Table III provide identical ψ and Δ values for a given surface measured in all four zones. In practice, however, the results are different (78,116). If the effect of different errors is investigated on the Poincaré sphere, it can be shown that by using the arithmetic average of results from four zones, errors due to different causes are reduced to different degrees:

Errors due to deviations in polarizer and analyzer circle alignment are closely eliminated.

Errors due to deviations of the compensator retardation from quarter wave and due to unequal transmission of fast and slow components are partially eliminated.

Errors due to deviations in compensator azimuth are not reduced.

Of the optical elements in an ellipsometer, the quarter wave plate usually deviates most from the ideal state. In addition to deviations from 90° retardation (50,52,107) differences in transmittance between fast and slow directions, due to absorption or interference (6,136,142), should be considered. A detailed theory of retardation plates (59) and their use and calibration (60) has been given by Holmes and Feucht. Corrections based on measurements in two zones are contained in the computer program by McCrackin (84) and corrections of measurements from one, two, or four zones have been analyzed by Azzam and Bashara (9).

4. Alignment and Calibration

The mechanical adjustment of azimuth circles (to read zero for the plane of incidence) has been described in detail (82). The procedure is based on the fact that pure s or p polarization remains linearly polarized in reflection from a metal surface, and linear polarization parallel to fast or slow axis of the compensator remains linear upon transmission through it. Dielectric reflection can also be used (8). Most instruments do not provide for a precise adjustment of azimuth circles. One, therefore, has to establish corrections, which are then applied to all the readings.

Less information is available on the alignment of the optical axes of polarizer, analyzer, and compensator (51). The first two elements are often built into collimator and telescope (polarizer and analyzer telescopes), respectively, the axes of which should lie in the same plane and intersect in the surface of the specimen (Fig. 31). If collimator or telescope can be rotated to vary the angle of incidence, the two axes are most easily made parallel to each other by aligning both telescopes with respect to each other in the straight-through position. The axes can then be made to intersect in the specimen

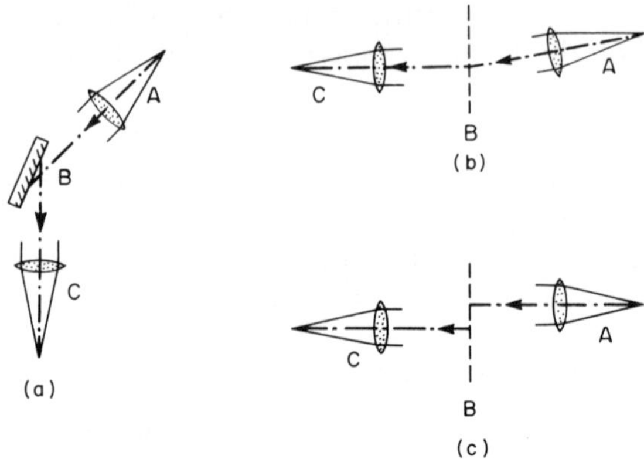

Fig. 31. Errors in alignment of axes of collimator *A* and telescope *C*. (*a*) Plan view with intersection of axes in front of specimen surface *B*; (*b*) side view with axes intersecting in specimen but not parallel to each other; (*c*) side view with axes parallel but not in the same plane.

surface by use of a mirror with a target on its surface. Iris diaphragms centered on the axes of collimator and telescope (*E* and *J* in Fig. 32) are of great help in this operation. The point of intersection also has to coincide with the axis of rotation of the movable telescope (138). Autocollimation techniques can be used to align the axis of rotation of the specimen table normal to the plane of incidence. The plane of incidence, defined by the two optical axes must be parallel to the plane of rotation of the telescope.

With ellipsometers of fixed angle of incidence, (89) a mirror stand with calibrated angle of rotation or a prism of the desired angle between reflecting faces has to be used for the alignment of collimator and telescope.

Adjustment of the angle of incidence circle, similar to the alignment of optical axes, is also accomplished in the straight-through position for instruments with variable angle of incidence and requires the use of a mirror with calibrated rotation or a prism with reflecting faces at the desired angle for instruments with fixed angle of incidence (51).

In order to employ a narrowly defined angle of incidence in ellipsometer measurements, it is important that the light incident on the specimen is well collimated (a parallel beam). For the use of most light sources, a small diaphragm (pinhole), precisely positioned in the focal plane of the collimator lens, provides the parallel beam.

The location of the focal plane can best be found by autocollimation with a cross hair in place of the pinhole: In the correct position the cross hair and its reflected image appear sharp and without parallax. The diameter of the pinhole has to be chosen to result in an acceptable divergence in the collimated beam and a sufficient output of the photodetector. Effects of angle at incidence errors have been discussed (75,110,119).

All adjustments and alignments should be checked periodically. Such complete checks can be quite time-consuming, but it would often be satisfactory to check the overall reproducibility of an ellipsometer, if a reproducible reflecting surface were available. A totally reflecting glass prism, with faces oriented normal to entering and exiting beams, has been found very suitable for this purpose. To render the reflecting surface independent of the surrounding medium, it has been coated with an opaque aluminum film, which in turn was protected by a layer of silicon monoxide. Except for some small changes due to aging during the first two weeks after preparation, this surface provided a stable reference. No changes could be detected when the environment of the prism was altered from vacuum to saturated vapors of several solvents.

5. *Other Sources of Errors*

Cell windows act as a source of errors in several different ways: Reflection adds "parasitic" beams that are reflected by other components into the photodetector. These parasitic beams are not extinguished together with the main beam and result in a broadening of the intensity minimum and possibly a shift in the azimuth at which the minimum is reached. The application of antireflective coatings (which must not be polarizing or birefringent) to all reflecting surfaces reduces the intensity of parasitic beams (144), but further reductions are often desired. For this purpose, a pinhole in the focal plane of the analyzer telescope can be used (138). Because of (inadvertent or purposeful) slight misalignment of reflecting surfaces from the axis, parasitic beams are in general not focused on the axis, as the main beam is aligned to be. The pinhole can therefore prevent parasitic beams from reaching the photodetector (Fig. 32).

A pinhole on the axis in the focal plane of the telescope also serves to better define the angle of incidence on the specimen. Light that is not reflected parallel to the axis, or under an angle of reflection

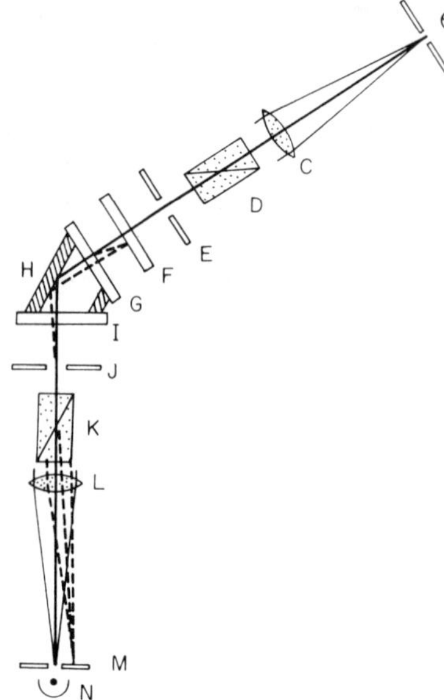

Fig. 32. Effect of pinhole in analyzer telescope. Ellipsometer for elliptic incident polarization with quarter wave compensator. *A*—light source; *B*—collimator pinhole; *C*—collimator lens; *D*—polarizer prism; *E, J*—iris diaphragms; *F*—quarter wave plate; *G, I*—cell windows; *H*—reflecting specimen; *K*—analyzer prism; *L*—telescope lens; *M*—telescope pinhole; *N*—photodetector; solid lines—main beam; dashed lines—parasitic beam.

equal to the angle of incidence (e.g., due to surface roughness), is not accepted for measurement. For this reason alone, the pinhole should be routinely employed. The diameter of the pinhole determines the range of angles of incidence that contribute to the measurement. To avoid unnecessary signal attenuation, the telescope pinhole should be slightly larger than the collimator pinhole (if both lenses are of the same focal length). The image of the collimator pinhole must be centered on the telescope pinhole for proper operation. This alignment can easily be achieved by fine-adjusting the reflecting surface to maximize the photodetector output with polarizer and analyzer set off extinction. Figure 32 also shows two iris diaphragms *E* and *J* that control the diameter of the light beam and are used in the alignment procedure. Not shown is a condenser lens, possibly inserted between light source and collimator pinhole, as well as color filters, usually positioned in the same area.

Birefrigence in cell windows, usually caused by stress, is a serious source of error (10,100) that often goes unnoticed, because under most circumstances, it can only be detected by use of a reference surface in place of the specimen. In all-glass equipment, carefully annealed, fused quartz windows have been used successfully (82). Satisfactory constructions for windows on bakeable stainless vacuum chambers still need to be developed. Glass with a low stress-optical coefficient (67) may find a use there.

Even a perfectly isotropic cell window changes the state of polarization of transmitted light that is not incident normal to the surface of the window (91). Measurements shown in Fig. 33 indicate that a one degree deviation from normal incidence, which is more than the accuracy usually achieved in construction and alignment of cells, can be tolerated. The use of substantial angles ϕ would require, however, that their value be known to a higher degree of accuracy than is usually possible, in order to satisfactorily compute the effect of such cell windows and subtract it from measurements (22,23,103). The optical properties of the window material, including the presence of surface layers, would also have to be known precisely for such computations.

Most of the errors discussed need not be considered if they remain

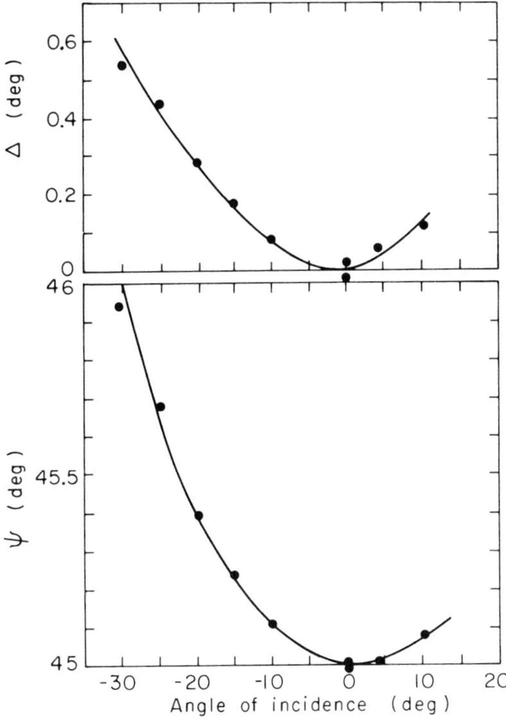

Fig. 33. Effect of nonnormal incidence on a 3/8 in. thick cell window on the state of polarization of transmitted light.

constant during an experiment and only "relative" measurements are made, in which a *change* in the optical properties of the surface, e.g., due to the formation of a film, is of interest. The interpretation of this change is usually little affected by substantial errors in the absolute magnitude of the measurements (120). "Absolute" measurements, on the other hand (e.g., bare-surface optical constants), require careful consideration of all sources of error (9,53,65,110,115). The availability of an absolute reference surface would be highly desirable for such purposes.

6. *Manual Ellipsometer Operation*

The use of electronic photodetectors (photomultipliers, phototransistors) for establishing minimum transmitted light intensity (108) has largely displaced half-shade devices (15,104,129) employed earlier for visual operation, although visual observation continues to be indispensable for detecting surface inhomogeneities. As shown in Fig. 27*b*, the minimum in the transmitted light intensity as a function of azimuth setting is flat and therefore not well defined. For precise measurements, advantage is taken of the symmetrical nature of the intensity curve, and the azimuth of minimum intensity is found as the center between the azimuth of two off-minimum readings of equal intensity (4). This technique provides the highest resolution possible with manual operation, but several minutes are required for establishing polarizer and analyzer azimuths in one zone. [Off-minimum readings are also employed in an automatic ellipsometer (93) and the same effect is achieved by modulation of the state of polarization (73,127,138).]

Noise due to the laboratory environment, the light source, the photodetector, and its instrumentation often limit the resolution obtainable. The use of a phase-sensitive detector and beam modulator (61,91) may be necessary to realize the $0.01°$ resolution in azimuth reading commonly offered by research-grade ellipsometers.

The physical appearance of a manually operated ellipsometer is shown in Fig. 34. A commercially available instrument (Gaertner L-119) has been extensively modified to provide space for large specimens, facilitate the exchange of light sources, and the conversion from elliptic to linear incident polarization. Other additions are iris diaphragms on collimator and telescope, the illumination of circles, a motor-driven, vertically movable specimen table, and a light chopper.

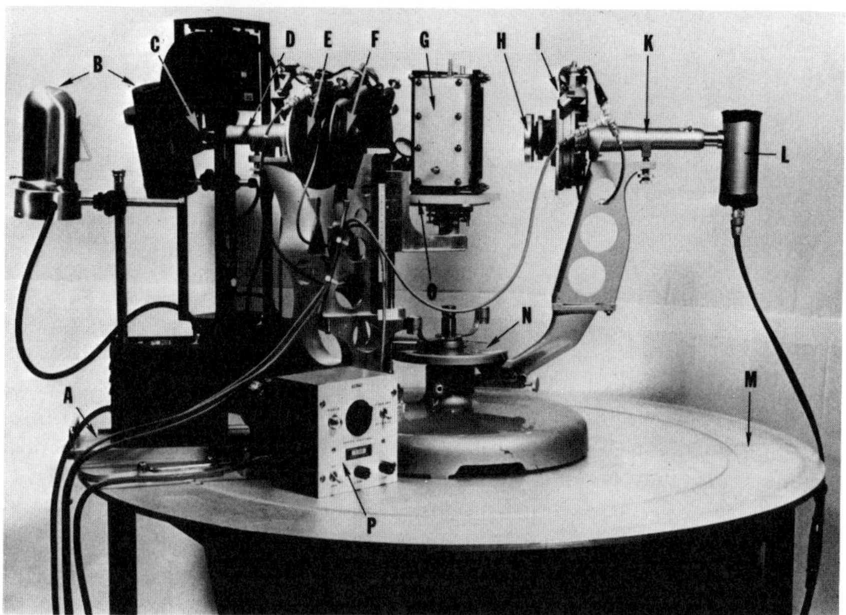

Fig. 34. Manual ellipsometer with incident elliptic polarization. *A*—movable carriage for light sources; *B*—exchangeable monochromatic light sources; *C*—light chopper and generator of reference signal for phase-sensitive detector; *D*—collimator tube with pinhole entrance; *E*—polarizer circle with illuminated scale; *G*—cell with specimen surface; *H*—iris diaphragm for control of observed specimen area; *I*—polarizer circle with illuminated scale; *K*—telescope tube with pinhole exit; *L*—photomultiplier; *M*—base plate for exchange of light sources and conversion to linear polarization incident; *N*—angle of incidence circle; *O*—specimen table with vertical movement; *P*—motor drive control unit for specimen table.

VI. Outlook

1. Theoretical Problems

The theory of optical reflection, on which the present theory of ellipsometry is based, employs the macroscopic Maxwell equations (123). Although extrapolations of this theory to atomic dimensions have in many cases been surprisingly successful (see also chapter by Kruger), and good arguments can be made that the ellipsometer measures the *amount* of material present on a surface (5), a general validity of the theory in this regime can hardly be expected.

A more realistic model for optical reflection from bare surfaces,

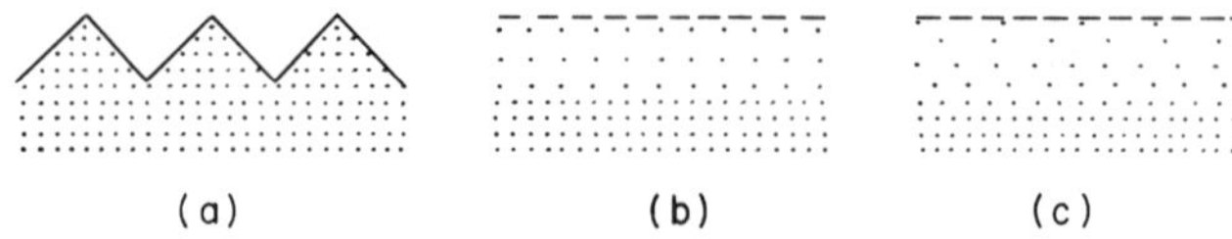

Fig. 35. Models for optical reflection from a rough surface. (*a*) Triangular ridge model of surface profile; (*b*) peak to peak variations in surface profile represented by a homogeneous transitional layer; (*c*) roughness represented by an inhomogeneous transitional layer with optical properties varying in the vertical direction.

in place of an ideally flat, continuous interface, will have to consider the effect of surface roughness (17), which is found to be present on most real surfaces, usually on a scale much larger than atomic dimensions (Fig. 35). The representation of roughness as an inhomogeneous surface layer, which is then treated by the present macroscopic theory, has been proposed by McCrackin (45). The validity of the assumptions inherent in this model still need to be tested experimentally. A new theoretical approach by Berreman (21) is based on optical scattering from submicroscopic spherical surface irregularities. Computational applications or experimental checks of this theory are also not available yet. Exact solutions for optical reflection from a plug-type model of surface roughness have been obtained by Deriugin (39).

Further information for the prediction of transitional layers on clean surfaces (25), due to the different environment of surface atoms as compared to bulk atoms (dangling bonds) will have to be developed.

The classical theory for ellipsometry of film-covered surfaces is also in need of refinement. Instead of the plano-parallel films of submolecular thickness assumed so far, and often successfully interpreted as equivalent films of equal mass (12,20,28,32,87,118), approaches that consider arrays of individual oscillators (114,125) may have to be extended to account for the discrete distribution of atoms in films of molecular coverage. A comparison of different models now available for submonolayers with measurements has shown a nearly linear relationship between the change in Δ and degree of coverage (24,25).

The effect of submicroscopic patchwise film coverage, or local agglomeration of material, found in many thin films (20,57,58,143), on ellipsometer measurements needs to be analyzed. It may be possible to identify patch size by use of variable lateral coherence in

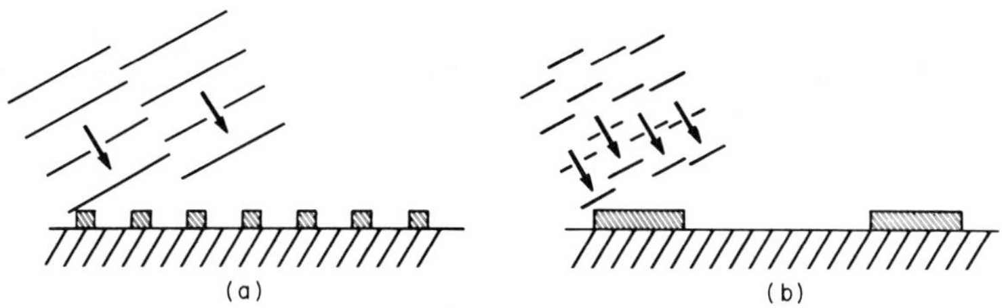

Fig. 36. Reflection from patch-wise distributed films, variation of lateral coherence (indicated by width of wavefronts) in incident light. (*a*) Coherent superposition of polarization states due to reflection on bare and film-covered surface elements; (*b*) incoherent superposition of polarization states.

the incident light, because coherent and incoherent superposition of light (54) reflected from bare and film-covered surface elements can be expected to result in different states of polarization (Fig. 36).

The availability of more explicit procedures for the ellipsometry of anisotropic films and substrates (49,139) would be desirable.

Of immediate electrochemical interest is a better understanding of the optical properties of electrolytic double layers (35) and mass transfer boundary layers which are present in most electrochemical situations. Procedures to account for the latter are in principle available but have not been sufficiently used or compared to results from other techniques yet.

2. *Instrumental Problems*

Automated ellipsometers should soon improve the time-resolution of measurements greatly. Response-times in the order of 10^{-3} sec appear well realizable. Some forms of automated operation should also make it possible to increase azimuth resolution to $10^{-3°}$ or better. Increased sensitivity will be necessary for some of the data interpretations suggested above, such as in the study of patchwise film coverage and surface roughness.

Ellipsometry, as presently constituted, provides information only on one surface element at the time. Size and shape of this element are determined by the beam cross section. The development of holographic ellipsometry (29,46,76) offers the capability for simultaneously observing many different surface elements of extended objects. No usable instrument of this kind seems to have been operated yet.

Some highly desirable instrumental developments concern the

convenient scan in wavelength of the light used in an ellipsometer, and very promising leads in the field of ellipsometric spectroscopy (81) have been developed.

3. Conclusions

Ellipsometry offers some unique possibilities for the study of surfaces and thin films. Some of these possibilities are particularly attractive for electrochemical applications. For the satisfactory use of ellipsometry, some of the principles outlined in this chapter must be considered. Except for precisely repetitive applications, an ellipsometer can therefore not normally be considered a routine analytical tool. Since ellipsometry, in general, provides only two measured parameters, the characterization of complex systems often requires a combination with other techniques (86,135).

Many applications of the ellipsometer in its present form remain to be explored and new possibilities will be opened by instrumental developments presently in progress.

Acknowledgment

This work was conducted in part under the auspices of the U.S. Atomic Energy Commission.

References

1. Abelès, F., in Ref. 98, p. 41.
2. Archer, R. J., *Phys. Rev.*, **110**, 354 (1958).
3. Archer, R. J., *J. Electrochem. Soc.*, **104**, 619 (1957).
4. Archer, R. J., *J. Opt. Soc. Am.*, **52**, 970 (1962).
5. Archer, R. J., and G. W. Gobeli, *J. Phys. Chem. Solids*, **26**, 343 (1965).
6. Archer, R. J., and C. V. Shank, *J. Opt. Soc. Am.*, **57**, 191 (1967).
7. Archer, R. J., *Manual on Ellipsometry*, Gaertner Scientific Corp., Chicago, 1968.
8. Aspnes, D. E., and A. A. Studna, *Appl. Opt.*, **10**, 1024 (1971).
9. Azzam, R. M. A., and N. M. Bashara, *J. Opt. Soc. Am.*, **61**, 600 (1971).
10. Azzam, R. M. A., and N. M. Bashara, *J. Opt. Soc. Am.*, **61**, 773 (1971).
11. Barrett, C. S., *Structure of Metals*, McGraw-Hill, New York, 1943, Ch. II.
12. Bartell, L. S., and J. F. Betts, *J. Phys. Chem.*, **64**, 1075 (1960).
13. Bashara, N. M., A. B. Buckman, and A. C. Hall, Eds., "Recent Developments in Ellipsometry," *Symposium Proceedings*, *University of Nebraska*, North-Holland, Amsterdam, 1968. Printed as *Surf. Sci.*, **16** (1969).
14. Beattie, J. R., and G. K. T. Conn, *Phil. Mag.*, **46**, 222 (1955).
15. Beckman, K. H., *Ber. Bunsenges.*, **70**, 842 (1966).

16. Beckmann, K. H., *Philips Techn. Rev.*, **29**, 129 (1968).
17. Beckmann, P., and A. Spizzichino, *The Scattering of Electromagnetic Waves from Rough Surfaces*, Pergamon Press, New York, 1963.
18. Becquerel, J., *Z. Physik*, **52**, 342 (1928).
19. Bell, E. E., in *Encyclopedia of Physics*, Vol. XXV/2a, S. Fluegge and L Genzel, Eds., Springer-Verlag, Berlin, 1967, p 1.
20. Bennett, H. E., D. K. Burge, R. L. Peck, and J. M. Bennett, *J. Opt. Soc. Am.*, **59**, 675 (1969).
21. Berreman, D. W., *J. Opt. Soc. Am.*, **60**, 499 (1970).
22. Bockris, J. O'M., M. A. V. Devanathan, and A. K. N. Reddy, *Proc. Roy. Soc. [A]*, **279**, 327 (1964)
23. Bockris, J. O'M., A. K. N. Reddy, and B. Rao, *J. Electrochem. Soc.*, **113,** 1133 (1966).
24. Bootsma, G. A., and F. Meyer, *Surf. Sci.*, **13**, 110 (1969).
25. Bootsma, G. A., and F. Meyer, *Surf. Sci.*, **14**, 52 (1969).
26. Born, M., *Optik*, Springer-Verlag, Berlin, 1965, 2nd ed., Ch. 1, 6.
27. Born, M., and E. Wolf, *Principles of Optics*, Macmillan, New York, 1964, 2nd ed., Ch. I, XIII.
28. Bornong, B. J., *Surf. Sci.*, **16**, 321 (1969).
29. Bryngdahl, O., *J. Opt. Soc. Am.*, **57**, 545 (1967).
30. Budde, W., *Appl. Opt.*, **1**, 201 (1962).
31 Burge, D. K., and H. E. Bennett, *J. Opt. Soc. Am.*, **54**, 1428 (1964).
32. Burge, D. K., J. M. Bennett, R. L. Peck, and H. E. Bennett, *Surf. Sci.*, **16**, 303 (1969).
33. Cahan, B. D., and R. F. Spanier, *Surf. Sci.*, **16**, 166 (1969).
34. Chaumont, M. L., *Ann. Phys.*, [9], **4**, 101 (1915).
35. Chiu, Y.-C., and M. A. Genshaw, *J. Phys. Chem.*, **73**, 3571 (1969).
36. Claussen, B. H., *J. Electrochem. Soc.*, **111**, 646 (1964).
37. Condon, E. U., in *Handbook of Physics*, E. U. Condon and H. Odishaw, Eds., McGraw-Hill, New York, 1958, Ch. 1.
38. Cook, A. W., *J. Opt. Soc. Am.*, **38**, 954 (1948).
39. Deriugin, L. N., *Radiotekhnika*, **15** [2], 15 (1960); **15** [5], 9 (1960).
40. Ditchburn, R. W., *J. Opt. Soc. Am.*, **45**, 743 (1955).
41. Ditchburn, R. W., *Light*, Interscience, New York, 1964, vol. II, p. 590.
42. Drude, P., *Ann. Phys. Chem., N.F.*, **36**, 865 (1889).
43. Engelsrath, A., and E. V. Loewenstein, *Appl. Opt.*, **5**, 565 (1966).
44. Faucher, J. A., G. M. McManus, and H. J. Trurnit, *J. Opt. Soc. Am.*, **48**, 51 (1958).
45. Fenstermaker, C. A., and F. L. McCrackin, *Surf. Sci.*, **16**, 85 (1969).
46. Fourney, M. E., A. P. Waggoner, and K. V. Mate, *J. Opt. Soc. Am.*, **58**, 701 (1968).
47. Françon, M., in *Encyclopedia of Physics*, Vol. XXIV, S. Fluegge, Ed. Springer-Verlag, Berlin, 1956, p. 171.
48. Fry, T. C., *J. Opt. Soc. Am.*, **16**, 1 (1928), **22**, 307 (1932).
49. Graves, R. H. W., *J Opt Soc. Am.*, **59**, 1225 (1969).
50. Grunstra, B. R., and H. B. Perkins, *Appl. Opt.*, **5**, 585 (1966).
51. Gu, H., M. S. Thesis, Department of Chemical Engineering, University of California, Berkeley, 1971 (LBL 165).
52. Hall, A. C., *J. Opt. Soc. Am.*, **53**, 801 (1963).

53. Hall, A. C., *J. Opt. Soc. Am.*, **55**, 911 (1965).
54. Hall, A. C., *J. Phys. Chem.*, **70**, 1702 (1966).
55. Hall, A. C., *Surf. Sci.*, **16**, 1 (1969).
56. Heavens, O. S., *Optical Properties of Thin Solid Films*, Dover, New York, 1956, Ch. 4 and 5.
57. *Ibid.*, Ch. 3
58. Henderson, G., and C. Weaver, *J. Opt. Soc. Am.*, **56**, 1551 (1966).
59. Holmes, D. A., *J. Opt. Soc. Am.*, **54**, 1115 (1964).
60. Holmes, D. A., and D. L. Feucht, *J. Opt. Soc. Am.*, **57**, 466 (1967).
61. Jasperson, S. N., and S. E. Schnatterly, *Rev. Sci. Instr.*, **40**, 761 (1969).
62. Jenkins, F. A., and H. E. White, *Fundamentals of Optics*, McGraw-Hill, New York, 1957, 3rd ed., Ch. 14.
63. *Ibid.*, Ch. 25.
64. Jerrard, H. G., *J. Opt. Soc. Am.*, **44**, 634 (1954).
65. Jerrard, H. G., *Surf. Sci.*, **16**, 137 (1969).
66. Kemp, J. C., *J. Opt. Soc. Am.*, **59**, 950 (1969).
67. King, R. J., and M. J. Downs, *Surf. Sci.*, **16**, 288 (1969).
68. Koenig, W., in *Handbuch der Physik*, Vol. XX, H. Geiger and K. Scheel, Eds., Springer-Verlag, Berlin, 1928, p. 141, p. 240.
69. Komrska, J., *Opt. Acta*, **15**, 389 (1968).
70. Kruger, J., *Corrosion*, **22**, 88 (1966).
71. Kumagai, S., and L. Young, *J. Electrochem. Soc.*, **111**, 1411 (1964).
72. Laue, M. v., in *Handbuch der Experimentalphysik* Vol. 18, W. Wien and F. Harms, Eds., Akademische Verlagsges., Leipzig, 1928.
73. Layer, H. P., *Surf. Sci.*, **16**, 177 (1969).
74. Leberknight, C. E., and B. Lustman, *J. Opt. Soc. Am.*, **29**, 59 (1939).
75. Loescher, D. H., *Appl. Opt.*, **10**, 1031 (1971).
76. Lohmann, A. W., *Appl. Opt.*, **4**, 1667 (1965).
77. Lucy, F. A., *J. Chem. Phys.*, **16**, 167 (1948).
78. Lukeš, F., *Surf. Sci.*, **16**, 74 (1969).
79. Lukeš, F., W. H. Knausenberger, and K. Vedam, *Surf. Sci.*, **16**, 112 (1969).
80. Mayer, H., *Physik duenner Schichten*, Wiss. Verlagsges., Stuttgart, 1950, Ch. IV, 1.
81. McBee, C. L., and J. Kruger, *Surf. Sci.*, **16**, 340 (1969).
82. McCrackin, F. L., E. Passaglia, R. R. Stromberg, and H. L. Steinberg, *J. Res. Natl. Bur. Stand.*, **67A**, 363 (1963).
83. McCrackin, F. L., and J. P. Colson, in Ref. 98, p. 61.
84. McCrackin, F. L., "A Fortran Program for Analysis of Ellipsometer Measurements," Nat. Bur. Stand., Tech. Note 479, April 1969.
85. McMaster, W. H., *Am. J. Phys.*, **22**, 531 (1954).
86. Melmed, A. J., H. P. Layer, and J. Kruger, *Surf. Sci.*, **9**, 476 (1968).
87. Miller, J. R., and J. E. Berger, *J. Phys. Chem.*, **70**, 3070 (1966).
88. Mowat, J. R., and R. H. Muller, *Reflection of Polarized Light From Absorbing Media*, University of California, Lawrence Radiation Laboratory, UCRL-11813, 1966.
89. Muller, R. H., *Rev. Sci. Instr.*, **39**, 1593 (1968).
90. Muller, R. H., *Surf. Sci.*, **16**, 14 (1969).
91. Muller, R. H., R. F. Steiger, G. A. Somorjai, and J. M. Morabito, *Surf. Sci.*, **16**, 234 (1969).

92. Oldham, W. G., *Surf. Sci.*, **16**, 97 (1969).
93. Ord, J. L., and B. L. Wills, *Appl. Opt.*, **6**, 1673 (1967).
94. Ord, J. L., *Surf. Sci.*, **16**, 155 (1969).
95. Palmer, A. D. F., *Phys. Rev.*, **17**, 409 (1921).
96. Partington, J. R., *An Advanced Treatise on Physical Chemistry*, Vol. 4. Longmans, Green, London, 1953.
97. Partovi, E., *J. Opt. Soc. Am.*, **52**, 918 (1962).
98. Passaglia, E., R. R. Stromberg, and J. Kruger, Eds., "Ellipsometry in the Measurement of Surfaces and Thin Films," *Proc. Symposium Washington* 1963, Nat. Bur. Stand. Misc. Publ. 256, 1964.
99. Passaglia, E., and R. R. Stromberg, *J. Res. Nat. Bur. Stand.*, **68A**, 601 (1964).
100. Primak, W., in Ref. 98, p. 154.
101. Ramachandran, G. N., and S. Ramaseshan, in *Encyclopedia of Physics*, Vol. XXV/1, S. Fluegge, Ed., Springer-Verlag, Berlin, 1961, p. 1.
102. Reddy, A. K. N., M. A. V. Devanathan, and J. O'M. Bockris, *J. Electroanal. Chem.*, **6**, 61 (1963).
103. Reddy, A. K. N., and J. O'M. Bockris, in Ref. 98, p. 229.
104. Richartz, M., and H.-Y. Hsue, *J. Opt. Soc. Am.*, **39**, 136 (1949).
105. Richartz, M., *J. Opt. Soc. Am.*, **56**, 198 (1966).
106. Rossi, B., *Optics*, Addison-Wesley, Reading, Mass., 1959, 2nd ed., Ch. 7, 8.
107. Rothen, A., and M. Hanson, *Rev. Sci. Instr.*, **19**, 839 (1948).
108. Rothen, A., *Rev. Sci. Instr.*, **28**, 283 (1957).
109. Schmidt, E., *J. Opt. Soc. Am.*, **60**, 490 (1970)
110. Schueler, D. G., *Surf. Sci.*, **16**, 104 (1969).
111. Shubnikov, A. V., *Principles of Optical Crystallography*, Consultants Bureau, New York, 1960, p. 145.
112. Shurcliff, W. A., *Polarized Light*, Harvard University Press, 1962, Ch. 2, 7.
113. Šimon, I., *J. Opt. Soc. Am.*, **41**, 336 (1951).
114. Sivukhin, D. V., *Sov. Phys. JETP*, **3**, 269 (1956).
115. Smith, L. E., and R. R. Stromberg, *J. Opt. Soc. Am.*, **56**, 1539 (1966).
116. Smith, P. H., *Surf. Sci.*, **16**, 34 (1969).
117. Smith, R. C., and M. Hacskaylo, in Ref. 98, p. 83.
118. Smith, T., *J. Opt. Soc. Am.*, **58**, 1069 (1968).
119. So, S. S., W. H. Knausenberger, and K. Vedam, *J. Opt. Soc. Am.*, **61**, 124 (1971).
120. Steiger, R. F. J. M. Morabito, G. A. Somorjai, and R. H. Muller, *Surf. Sci.*, **14**, 279 (1969).
121. Stone, J. M., *Radiation and Optics*, McGraw-Hill, New York, 1963, Ch. 2.
122. *Ibid.*, Ch. 13.
123. *Ibid.*, Ch. 15, 16.
124. *Ibid.*, Ch. 16.
125. Strachan, C., *Proc. Cambridge Phil. Soc.*, **29**, 116 (1933).
126. Strong, J., *Concepts of Classical Optics*, Freeman, San Francisco, 1958, Ch. III, VI.
127. Takasaki, H., *Appl. Opt.* **5**, 759 (1966).
128. Tronstad, L., *Det Kgl Norske Videnskabers Selskabs Skrifter* (*1931*) No. 1, F. Bruns, Trondheim, 1931.
129. Tronstad, L., *J. Sci. Instr.*, **11**, 144 (1934).

130. Vašíček, A., *J. Opt. Soc. Am.*, **37**, 145 (1947).
131. Vašíček, A., *Optics of Thin Films*, North-Holland, Amsterdam, 1960, Ch. 2, 3.
132. *Ibid.*, Ch. 4.
133. *Ibid.*, Ch. 5.
134. Vašíček, A., in Ref. 98, p. 25.
135. Vedam, K., W. Knausenberger, and F. Lukes, *J. Opt. Soc. Am.*, **59**, 64 (1969).
136. Weinberger, H., and J. Harris, *J. Opt. Soc. Am.*, **54**, 552 (1964).
137. Weingart, J. M., and A. R. Johnson, in Ref. 98, p. 113.
138. Wilmanns, I., *Surf. Sci.*, **16**, 147 (1969).
139. Winterbottom, A. B., *Det Kongelige Norske Videnskabers Selskabs Skrifter* (*1955*) Nr. 1, F. Bruns, Trondheim, 1956.
140. Winterbottom, A. B., in Ref. 98, p. 97.
141. Wright, F. E., *J. Opt. Soc. Am.*, **20**, 529 (1930).
142. Yolken, H. T., R. M. Waxler, and J. Kruger, *J. Opt. Soc. Am.*, **57**, 283 (1967).
143. Yoshida, S., T. Yamaguchi, and A. Kinbara, *J. Opt. Soc. Am.*, **61**, 463 (1971).
144. Zaininger, K. H., and A. G. Revesz, *RCA Review*, March 1964, p. 85.

Application of Ellipsometry to Electrochemistry†

J. KRUGER
Institute for Materials Research
National Bureau of Standards
Washington, D.C.

Contents

During many electrochemical processes, the formation or removal of a film is involved, or the topography of a bare surface is changed. These events are usually of crucial importance in determining the course of the process being studied, and studying them gives us an insight into the nature of the process. The ellipsometer is an experimental tool that allows us to do this. With it, we can look at the partial monolayer level of surface coverage up to a coverage of thousands of angstroms thickness while simultaneously making conventional electrochemical measurements. The pioneer applications of this optical technique to electrochemical studies were made by Tronstad (96,97) nearly 40 years ago and by Winterbottom (102), a student of his. In recent years, an ever increasing number of workers in electrochemistry have turned to ellipsometry to gain additional insight to that provided by other electrochemical techniques.

This chapter describes the ways in which ellipsometry can be applied to study the phenomena of adsorption, film formation and dissolution, passivation and corrosion, electrodeposition and electropolishing. It discusses the determination of optical properties of bare surfaces, an essential step in ellipsometric studies, and also discusses gaseous adsorption, which has an indirect but important relevance to electrochemical phenomena. No discussion of the theoretical or experimental aspects of ellipsometry per se is given since this is covered in the preceding chapter by Muller and elsewhere (10,78, 102). The symbols and equations used in this chapter correspond to the ones introduced in that chapter.

I. Bare Surfaces

In order to determine the optical properties and thickness of a film forming on a surface whose electrochemistry is being studied, it is necessary to know the optical constants for that surface. This is clearly pointed out in the preceding chapter on the theoretical aspects of ellipsometry. This chapter points out ways in which these necessary constants can be obtained. It also discusses surface characterization that is intimately bound up with optical constant determination, because the nature of the surface determines the values of the constants.

Besides a number of compilations of optical constants in the literature (43,83), there also exist samples of recent redeterminations of the optical constants of many metals and semiconductors (3,22,49, 86,105). While many of the values from these sources may be quite reliable, the danger always exists that the surface being studied electrochemically may not be quite the same as that whose constants are available from the literature. This is so because the optical constants of a surface [$\bar{n}_3 \equiv n_3(1 - i\kappa_3) \equiv n_3 - ik_3$] are strongly dependent on the presence of impurities in the material used to make the surface, impurities on the surface (how clean it is), and finally the structure of the surface. The last item involves such metallurgical considerations as crystal orientation, grain size, strained (cold-worked) layers, as well as surface topography. Strain and surface roughness can strongly affect the value for the optical constants measured on a surface. A theoretical study by Fenstermaker and McCrackin (30) of the effects of roughness suggested that for surface asperities of 500 Å certain material can exhibit values of n_3 and κ_3 in error by 200 to 300% or more. It is, therefore, necessary to be quite sure that the surface whose electrochemistry is being studied has been prepared under exactly the same conditions as that surface upon which the literature optical constants were obtained. Failure to do this will result in meaningless values for the thickness and optical constants of the film being measured. In some instances, the literature values, as compared to those for the actual surface being studied, may be so far off that negative values are obtained for the film thickness calculated.

A much more satisfactory approach than reliance on values from the literature, when possible, is to produce a bare surface *in situ*, measure the optical constants of the surface so produced, initiate the process to be studied, and follow it ellipsometrically. When this procedure is possible, the state of purity of the surface, the existence of strain, and roughness become less important from an ellipsometric standpoint, although they obviously play a role in affecting the electrochemistry. The ellipsometric measurement, however, is less affected because the optical constants for the "bare" surface are those for the actual surface (with all its imperfections) being studied. When the surface is rough, even though one can obtain apparent optical constants for this surface, the possibility exists, as pointed out by Archer (4), that errors in film thickness can result. On the basis of a highly simplified model, he predicted that surface irregularities could lead to calculated thicknesses that are smaller than the true values. Therefore, *in situ* cleaning should lead to smooth surfaces [tens of atom layers (4)] if subsequent thickness measurements are to be relied upon.

The production of bare surfaces in electrochemical experiments, for the purpose of measuring $\bar{n}_3$, can be accomplished in two general ways: (1) use of ultrahigh vacuum techniques to prepare a bare surface with measurement of its optical constants prior to the introduction of an electrolyte, and (2) electrochemical techniques such as cathodic reduction of existing films or anodic dissolution to remove several outer layers of the surface.

An example of how it is possible to carry out electrochemical

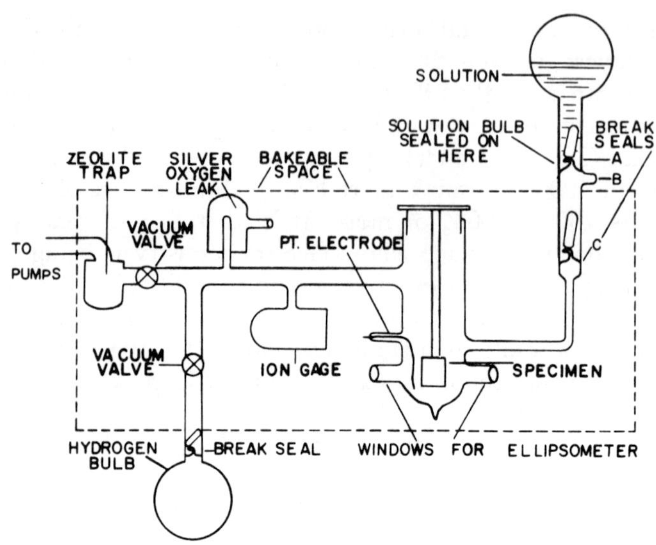

Fig. 1. Ultrahigh vacuum apparatus for studying the optical properties and the electrochemical potential of a metal surface immersed in a solution. From Kruger (51).

experiments using ultrahigh vacuum procedures to produce bare surfaces was given by Kruger (51). His apparatus, shown in Fig. 1, consisted of a conventional ultrahigh vacuum system capable of producing pressures less than 10^{-9} torr, with a means provided for the introduction of electrolyte after the surface under study had been cleaned and its $\bar{n}_3$ measured. A bare surface can be produced in a number of ways in an ultrahigh vacuum system: (*1*) heating in vacuum in the presence of reducing gases (51), (*2*) ion bombardment (90), (*3*) cleavage (5), and (*4*) vapor deposition (66).

During the cleaning procedure, measurements can be made with the ellipsometer using the windows, as shown in Fig. 1. These windows should be as strain-free as possible. Thick quartz optical flats offer the best means for achieving this end. A guide that is frequently used to determine the point in the cleaning procedure at which the surface can be judged clean is that repeated applications of the cleaning procedure do not change the values measured for Δ and ψ. A better way to judge whether the Δ and ψ values are truly characteristic of a bare surface is to measure the surface by another technique simultaneously under the same conditions. One way to do this is now possible by arrangements which combine ellipsometry with low energy electron diffraction (LEED). Such a combination has been successfully used (65,90) and does enable an independent judgment of the cleanliness of a surface to be made if the surface studied is a single crystal. Another useful combination would be ellipsometry and Auger spectroscopy. The latter technique can determine both the nature and quantity of impurities.

In the arrangement shown in Fig. 1, once a bare surface has been produced and its $\bar{n}_3$ measured, solution can be rapidly introduced by means of break seals. The main problem in incorporating a suitable reference electrode in such an arrangement is that it must be capable of withstanding the bakeout temperature ($\sim$300–400°C). The procedures using vacuum techniques are slow and tedious but sometimes are the only way that can be employed if the experiment requires an accurate knowledge of the $\bar{n}_3$ value of the electrode surface.

For some metals and semiconductors, however, it is possible to use electrochemical procedures to produce clean surfaces without affecting their topography. For some metals (such as copper, iron, tin) and semiconductors, cathodic reduction can be reasonably effective in reducing films present on this surface. McBee and Kruger (62) have shown how effective this can be. In Fig. 2 they show the optical constants n_3, κ_3 for an iron surface produced by ultrahigh vacuum techniques (105) compared to those measured on a cathodically reduced iron surface immersed in a neutral borate buffer solution. While there is not an exact correspondence at all wavelengths between the vacuum and solution values, the greatest divergence is less than 5%. Depending on the requirements of a given experimental objective, this could be well within tolerable limits. It should be mentioned, however, that a disadvantage of the production of bare surfaces by cathodic reduction, is that reduction products, while perhaps not on the surface when $\bar{n}_3$ is being measured, may affect electrochemical processes taking place in the solution.

Cleaning can also be accomplished electrochemically by anodic dissolution. It is necessary here, however, to do this in a manner that does not roughen the electrode surface, and does not produce anodic films. A single crystal surface whose orientation corresponds to the equilibrium plane developed in the environment in which dissolution takes place is the best candidate for carrying out this procedure for producing a bare surface.

In addition to the above direct techniques for producing bare

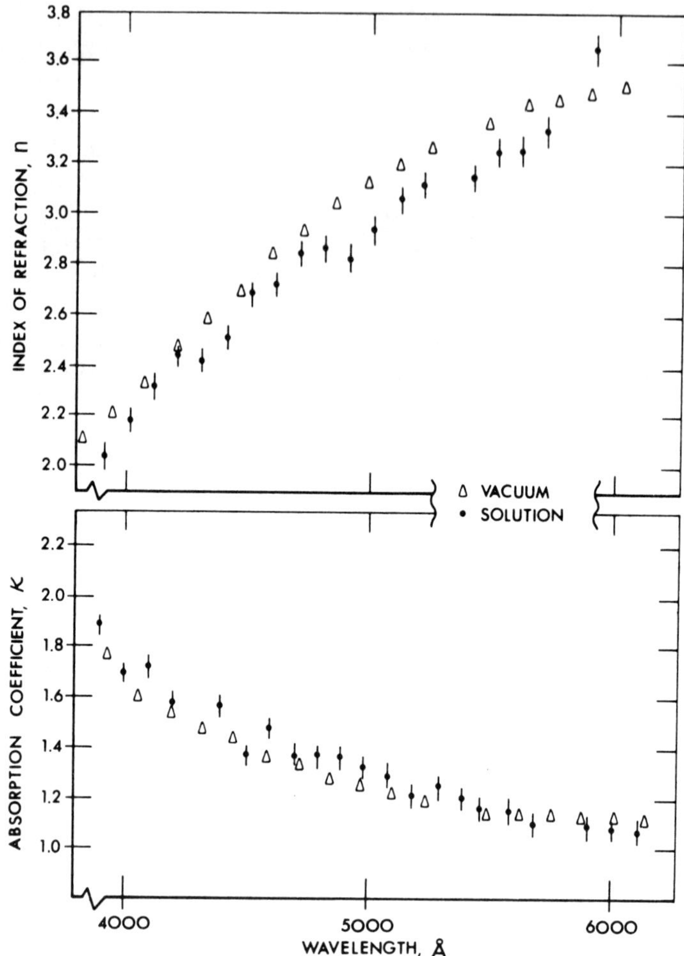

Fig. 2. Comparison of vacuum-produced clean Fe surface [data from (105)] and surface reduced cathodically in sodium borate-boric acid solution. The values for both n and $\kappa(= k/n)$ are shown. Error bars are given for the solution data, based on the uncertainty in measuring P $\pm$ 0.1° and A $\pm$ 0.1°. From McBee and Kruger (62).

surfaces upon which $\bar{n}_3$ can be determined, there are indirect methods. These methods must be applied when it is difficult or is not feasible to remove films from a metal surface.

One technique, which does not require the removal of an existing film, described by Barrett and Winterbottom (7) for aluminum, involves varying the refractive index of the surrounding medium, n_1. The rationale for this procedure is based on the fact that if n_1 is equal to n_2, the refractive index of the film, the film is eliminated optically. Under these circumstances $\bar{n}_3$, the refractive index of the substrate, can be determined directly as if it were film-free. The method suffers from the restriction that the film on the substrate must have a real $\bar{n}_2$ ($\kappa_2 = 0$) and that one must know its value.

Another indirect method described by Vedam, Knausenberger, and Lukes (99) combines a determination of the reflectance, at normal incidence, R, with ellipsometric measurements to determine $\bar{n}_3$. The method depends on the assumption that the reflectance of the clean surface is the same within $\pm 0.1\%$ of that of a surface with a film a few hundred angstroms thick for materials with low κ_3 values, and within $\pm 1\%$ for materials with high κ_3 values. By making measurements of n_3', κ_3', and R for surfaces with different film thicknesses, it is possible to relate R to n_3' and κ_3'. (The primed values are calculated from Δ and ψ values measured on filmed surfaces assuming no film is present. They are pseudooptical constants.) The true R is determined by extrapolation to zero film thickness and the theoretical relationship between the possible sets of n_3's and κ_3's for this R are determined for different thicknesses. The technique then determines the bare n_3 and κ_3 values by assuming that only one set of n_3 and κ_3 will give the right R and one n_2 (the film refractive index) will fit the Δ and ψ values measured for many film thicknesses. This technique, as was the case with the previously described one, requires a nonabsorbing film and depends on a reasonably close knowledge of its refractive index.

A variation of this technique, which involves immersion in different liquids, has also been developed by the same authors (59).

II. Adsorbed Surface Layers

Once a bare surface in the environment of interest is characterized ellipsometrically, it is then possible to follow the adsorption of various sorts of species from the environment onto the surface. Since the ellipsometer is capable of doing this in a way that enables other measurements, e.g., electrochemical ones, to be carried out simultaneously, it allows one to assess directly the role of the adsorbate in electrode processes. Ideally, the ellipsometer can measure the amount of coverage and possibly the nature of the adsorbate. We must then ask the question: How small a fraction of a monolayer of a species adsorbed on an electrode surface can be detected by the ellipsometer and what sort of reliance can we place in the answers? The first question, an experimental one, depends on the sensitivity of the ellipsometer.† Most modern instruments give measurements that allow calculation of Δ to $\pm 0.02°$ and ψ to $\pm 0.01°$. For a given system with specified values of n_2 and n_3, the values of $d\Delta/dt$ and $d\psi/dt$, where t is film thickness, depend strongly on the angle of incidence (103). At a sensitive angle of incidence, Archer and Gobeli (5) estimate that they can detect ± 0.02 monolayer. This is probably a reasonable estimate, but it depends on the development of a satisfactory theory relating ellipsometric measurements of submonolayer films and is related to the second part of the question just posed. Unfortunately, there is now no completely satisfactory answer to that part of the question, especially for surfaces immersed in solutions, which is considered later. They suffer from problems introduced by the double layer.

To avoid the complications present when the surface is in a solution, it is useful first to consider experiments where adsorption of gases introduced into ultrahigh vacuum systems is studied. These have the advantage that one can characterize the bare surface better and observe the adsorption by additional techniques such as LEED. From an ellipsometric standpoint, however, the conclusions derived from these experiments in terms of the questions just posed should be as applicable to immersed surfaces as they are to surfaces in vacuum or gaseous atmospheres. Before considering these experiments, however, we will make an examination of theoretical aspects of the ellipsometry of the partial monolayer.

The basic problem in a theoretical treatment of the ellipsometry of partial monolayers is to develop realistic two-dimensional substitutes for the three-dimensional quantities, index of refraction, and film thickness, and, once having chosen these, to relate them in a meaningful way to the Δ and ψ obtained ellipsometrically. Archer (4) has made initial attempts at such a treatment by utilizing the theory of Strachan (91) for the Fresnel coefficients of monolayers. Strachan

† See Hayfield (47) for a comparison of the sensitivity of ellipsometry to reflectivity measurements.

proposed that a monolayer could be represented by a two-dimensional distribution of scattering centers. These centers are characterized by three scattering indices σ_1, σ_2, and σ_3, each corresponding to a component of the electric vector of a light wave interacting with a surface.

Strachan then derived an expression for the ratio of the parallel to the normal Fresnel coefficients ($\tan \psi e^{-i\Delta}$) which replaced the refractive index of the film by σ_1, σ_2, and σ_3. Using these concepts of Strachan, Archer postulated that Δ and ψ could be related to the scattering indices by the equations

$$\Delta = \bar{\Delta} - \sum \alpha_i \sigma_i \tag{1}$$

$$\psi = \bar{\psi} + \sum \beta_i \sigma_i \tag{2}$$

where $\bar{\Delta}$ and $\bar{\psi}$ are the values of the quantities for the bare surface, α_i and β_i are constants that are functions of the angle of incidence, the refractive index of the medium, the refractive index of the substrate, and the wavelength of the light used. According to Strachan σ_i defines the strength of a Hertzian oscillator. Archer assumed that σ_i would be proportional to coverage, θ, and he could therefore replace σ_i by $\theta\sigma_i'$ in Eqs. 1 and 2 to relate Δ and ψ to θ as follows:

$$\Delta = \bar{\Delta} - \theta \sum \alpha_i' \sigma_i' \tag{3}$$

$$\psi = \bar{\psi} + \theta \sum \beta_i' \sigma_i' \tag{4}$$

Thus in this formulation Δ and ψ would be proportional to coverage.

Hall (39) criticized Archer's assumption that the σ_i is proportional to θ because σ_i defines an amplitude rather than an intensity. Hence if an absorbed molecule constitutes an oscillator, the scattered intensity would be related to the number of molecules per unit area. Hall concludes that θ is proportional to σ_i^2, and this leads to Δ and ψ being proportional to $\theta^{1/2}$. It is difficult to see, however, on physical grounds why intensities should be added, as Hall's treatment requires.

Another approach to relating coverage to the ellipsometric parameters has been used by Meyer and Bootsma (68). They calculate an effective refractive index for the adsorbed layer by using the Lorentz-Lorenz equation (70)

$$\frac{n^2 - 1}{n^2 + 2} = \frac{4\pi\alpha}{3V} \tag{5}$$

which relates the refractive index of the adsorbed molecule, n, to the volume occupied by one adsorbed molecule, V, and to α, the polarizability of a given molecule calculated from the sum of atomic polarizabilities. The effective refractive index is then calculated from the value obtained from Eq. 5, averaged over a unit surface area for a given θ. This value of θ can then be related to Δ and ψ by using the n calculated for this θ in the macroscopic Drude equation.

It can be seen from this brief description that the theory of ellipsometric measurements of partial monolayer films is still in an elementary stage. The treatments described above assume atomically flat surfaces. As crystal models and the results of LEED studies (33, 34) show, there exists on perfect single crystal surfaces an inherent "roughness" with hill and valleys of one- or two-atom dimensions. When one is talking about partial monolayer coverage, the interpenetration of the adsorbed molecules with these surface variations must be taken into consideration. The next section goes into some studies involved in the adsorption of gases on surfaces that illustrate these problems. Some of these provide insights that may be of value in the interpretation of future electrochemical ellipsometric experiments.

1. Gas Adsorption

A number of experiments have been carried out involving ellipsometric measurements of surfaces with monolayer or submonolayer coverage (5,21,28,68,90). In a study of the chemisorption of oxygen on freshly cleaved silicon by Archer and Gobeli (5), the two-dimensional theory of Archer described above was applied ($\Delta \propto \theta$) to estimate the sticking coefficients at submonolayer level. A precision of ± 0.02 monolayers was claimed. They estimated θ from the expression

$$\theta = \frac{\delta\Delta}{\delta\Delta_0} \tag{6}$$

where $\delta\Delta_0$ was defined as the change in Δ corresponding to complete monolayer coverage. The value of $\delta\Delta_0$ was estimated from $\delta\Delta$ versus time curves by assuming that it corresponded to the point at which the rate of change of $\delta\Delta$ markedly decreases. From the θ so determined, the sticking coefficient could be calculated.

Meyer and Bootsma (68) studied the same system as Archer and Gobeli and obtained the same results for the variation in Δ but also observed significant changes in ψ, which the latter did not report. Meyer and Bootsma found that these changes in $\psi(\delta\psi)$ were roughly the same for different gases on silicon while the $\delta\Delta$'s were widely different. From this fact they attributed the $\delta\psi$ to be due to changes in the substrate while $\delta\Delta$ was related to the adsorbed layer. They speculated that the change in the substrate comes about because the outer layer of the silicon they studied has a different $\bar{n}_3$ than bulk silicon since it consists of uncompensated bonds at the surface. Upon undergoing chemisorption, the bonds are compensated and the surface layer shifts to bulk values.

Whether this picture is strictly accurate or not, the observation of Meyer and Bootsma that $\delta\psi$ is not to be neglected is important. This is even more strongly underscored in a study by Carroll and Melmed (21) on the interaction of oxygen with (011) tungsten where $\delta\psi$ and $\delta\Delta$ were measured continuously for low oxygen exposures ($<0.1 \times 10^{-6}$ torr-sec). They used a semiautomatic recording technique that enabled the rapid determination of $\delta\Delta$ and $\delta\psi$ to ± 0.02 and 0.01, respectively, while carrying out simultaneous LEED measurements. Figure 3 shows the complicated sets of events that can be observed when following $\delta\Delta$, $\delta\psi$, and the intensity of a specific LEED spot. Of most interest is the fact that $\delta\Delta$ changes direction at a definite oxygen exposure. When hydrogen was used in place of oxygen, the peaking behavior for Δ disappeared, but $\delta\psi$ increased continuously rather than decreasing as was observed for oxygen. Since LEED shows that hydrogen does not lead to surface rearrangements of (011) tungsten, and oxygen does, they tentatively concluded that they were observing the more complicated of the two oxygen-tungsten rearrangements proposed by Germer and May (33). They were not able, however, to find a quantitative fit for the Δ and ψ data using the Drude equations over a wide range of possible complex refractive indices for the film. For oxygen on tungsten this peaking behavior rules out both Archer's and Hall's theories, which require that Δ be proportional to θ or $\theta^{1/2}$. More recent work by Melmed and Carroll (67) utilizes the directions that Δ and ψ take with an increase in coverage or film thickness to serve as a further constraint in calculating $\bar{n}_2$. Thus in Fig. 3 during the initial part of the process Δ increases and ψ decreases, i.e., $d\Delta/dt$ is positive and $d\psi/dt$ is negative. Requiring Δ and ψ to vary in this manner reduces the number of possible values one can choose for n_2. Using such a constraint, they then calculate an effective $\bar{n}_2$ following the procedure used by Fenstermaker and McCrackin (30) for rough surfaces. This effective $\bar{n}_2$ is calculated from the theory of Garnett (31). Carroll and Melmed consider the roughness as being the atomic configura-

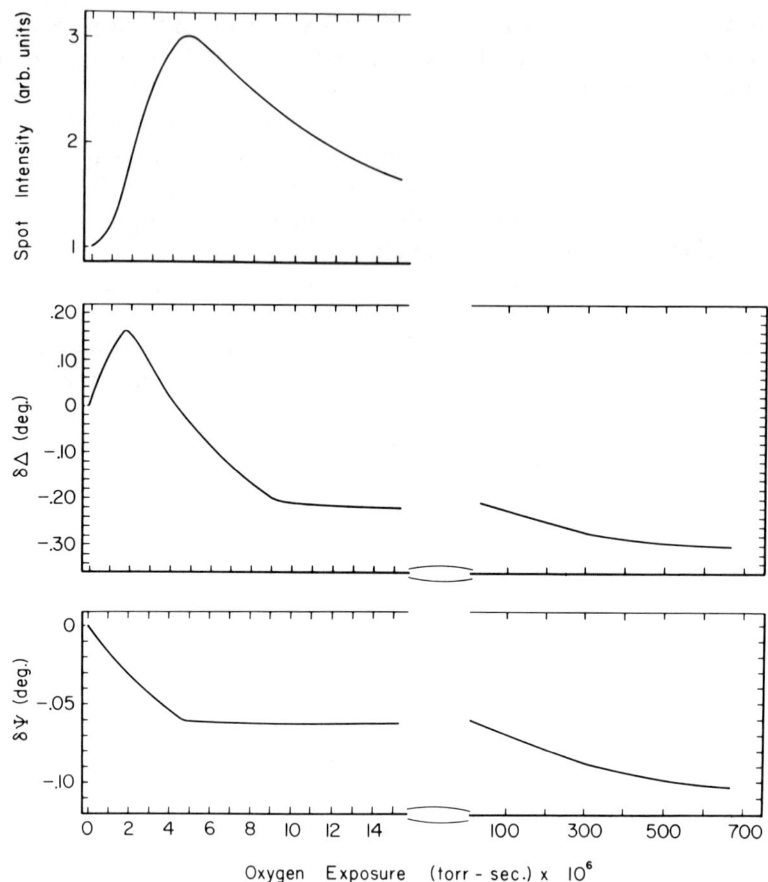

Fig. 3. The results from combined LEED-ellipsometry measurements for oxygen adsorption on clean (011) W. The intensity of LEED 1/2, 3/2 spot as well as the changes in Δ and ψ ($\delta\Delta$, $\delta\psi$) vs. oxygen exposure. LEED data taken at 118V. Ellipsometry data taken at $\lambda = 4800$ Å. From Carroll and Melmed (21).

tion of the surface, and thus their effective $\bar{n}_2$ involves substrate atoms and gas atoms. By means of this approach they were able, for example, to find values of $\bar{n}_2$ that gave a $d\psi/dt$ that was positive during the first part of the process and then became negative during the second part when the atomic configuration, as judged by LEED, changed, while $d\psi/dt$ remained negative for the $\bar{n}_2$'s found for both parts of the adsorption process.

Whether any of the described approaches is valid, or whether one, indeed, is more valid than the other, remains to be more conclusively demonstrated. It is, however, obvious from these gas adsorption experiments that ellipsometry is an extremely sensitive tool for the detection of the adsorption of less than monolayer quantities of a species on a surface. As such it is valuable in detecting adsorption qualitatively. For example, Steiger et al. (90) demonstrated for the physical adsorption of a number of gases on silver that ellipsometry could detect fractional monolayers in many instances where LEED was insensitive to their presence up to monolayer coverage. For chemisorbed layers Carroll and Melmed (21) also observed the superior sensitivity of the ellipsometer over LEED. Thus, while a firm quantitative basis for measurement of coverage is still lacking, ellipsometry nevertheless offers a sensitive tool to study adsorption.

2. *Ion Adsorption from Solution*

The concepts and some of the experimental and calculational techniques just outlined for gas adsorption can be applied to the adsorption of ions from solution. There are, however, two possible complicating factors that must be pointed out before we can deal with ion adsorption. These are electromodulation effects and double layer effects.

A. Electromodulation Effects. The first of these complicating factors is the possibility that the optical constants of the metal interface will change under the influence of the applied external fields existing at electrode interfaces in many electrochemical experiments when the potential is changed. This effect—called electromodulation—has been studied ellipsometrically by Buckman (18) and using a multiple reflection technique by Hansen (41). By impressing a potential that varied sinusoidally from 0 to -1.3 V with respect to the counterelectrode, Buckman indicates he could observe for Au and Ag samples in 1 *M* KCl changes of Δ and ψ of the order of 6×10^{-6} degrees using differential lock-in amplifier techniques. He carried out these measurements at different wavelengths in the energy range 1.9–4.1 eV and was able to locate the interband transitions. Hansen showed that these changes in Δ and ψ could be maximized at the pseudo-Brewster angle, and he calculated expected $\delta\Delta$ and $\delta\psi$ of the order of $0.1°$.

The usual concern with these electromodulation measurements is to be certain that the changes observed are not due to film formation or metal dissolution leading to roughening, rather than actual changes in the optical constants of the metal substrate. Hansen (41) gives arguments supporting the optical change in the metal itself based on his demonstration that all observed data are consistent with his model and inconsistent with other models suggested, but it is difficult to see how it is possible to rule out completely other surface changes in aggressive environments, such as chloride solutions, which can lead to changes in Δ and ψ.

B. Double Layer Effects. The second complicating factor is the possible optical effect of the electrical double layer at electrode-electrolyte interfaces. Stedman (89) has analyzed the optical effects resulting from the fact that this almost universally present double layer region contains a solution whose composition differs from that of the bulk. She calculates these changes in Δ and ψ by making the following reasonable assumptions: (1) The surface excess charge of each species in the diffuse layer can be predicted by the Gouy-Chapman theory; (2) The material in the double layer region can be rearranged so that it can be considered as optically equivalent to a single homogeneous layer; (3) The interaction of specifically adsorbed ions with the substrate does not affect the contribution of the adions to the layer's refractive index; (4) The Lorentz-Lorenz equation is valid for calculating the refractive index of nonneutral ionic solutions; and (5) The refractive index of the substrate and the solvent are not changed by the electric field in the double layer. (It should be noted that the electromodulation studies just described would consider assumption 5 as invalid.)

Using these assumptions, Stedman then calculates from the Lorentz-Lorenz equation the effective refractive index, n_2, for the double layer as

$$n_2 = \left[\left(\frac{\bar{M} + 2\,d\bar{R}}{\bar{M} - d\bar{R}}\right)\right]^{1/2} \tag{7}$$

where d is the density predicted from molar ion volumes, and where $\bar{M}$, the mean molecular weight of the ions in the layer, and $\bar{R}$, their mean molar refractivity, are given by $\bar{M} = \sum_i (x_i M_i)$ and $\bar{R} = \sum_i (x_i R_i)$. The n_2 thus calculated is then inserted into the Drude equation along with a double layer thickness taken as $2/\delta$ where δ is the Debye-Hückel reciprocal length, and the relationship between the ellipsometric parameters and double layer parameters can thereby be determined.

By the above approach Stedman calculated changes of Δ of the order of 0.005° for a 1 M NaF solution on mercury at negative

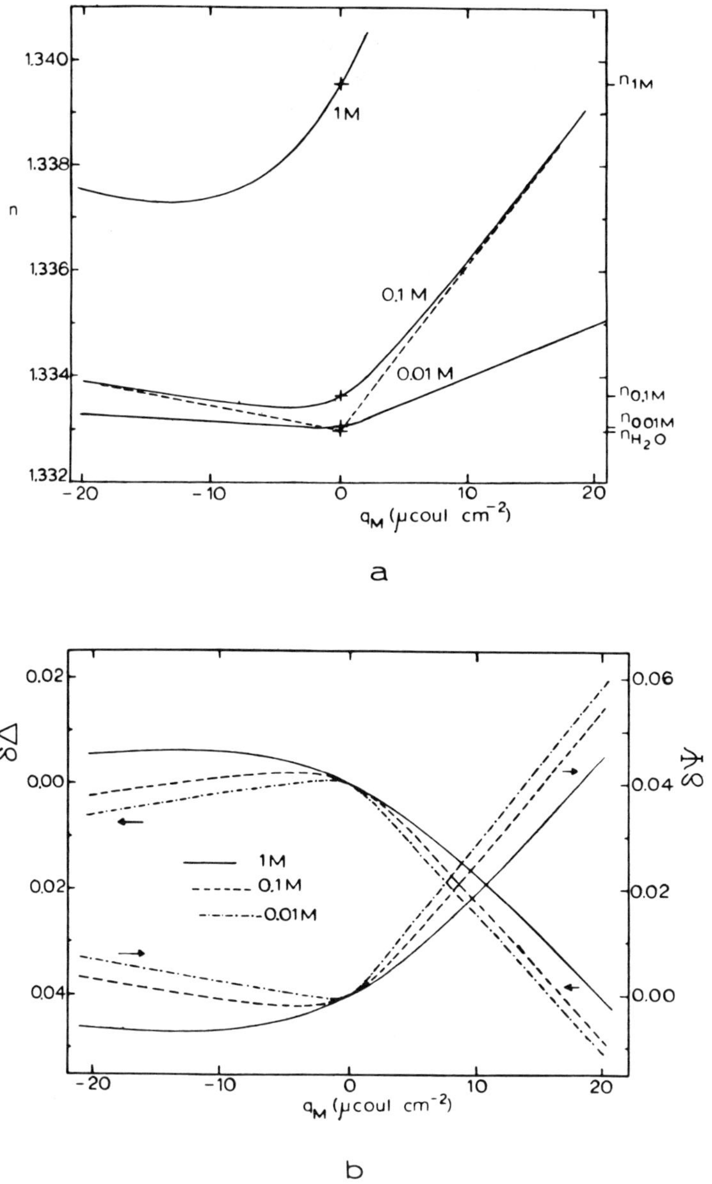

Fig. 4. (*a*) The variation of the refractive index, n, of the equivalent layer with the excess charge density on the electrode, q_m. Thickness of the equivalent layer, 6.10×10^{-8}, 1.93×10^{-7}, 6.10×10^{-7} cm for three NaF aqueous solutions of concentrations 1, 0.1, and 0.01 M respectively. (*b*) Change in Δ and ψ with q_m for the three NaF solutions shown in (*a*). From Stedman (89).

surface charges and 0.04° for a surface charge of 20 $\mu C/cm^2$. The changes in ψ for the same system are of the same magnitude, but the large effect is for negative surface charge whereas the small effect occurs for positive charge. Figure 4 shows how such changes in Δ and ψ affect the layer's refractive index, and it also shows the effect of specific adsorption. If these effects can be demonstrated to be valid experimentally, Stedman points out that they may be used to determine points of zero charge and obtain absolute values of ion refractivity.

C. Specific Adsorption of Ions. In spite of its obvious value to electrochemists, there have been few ellipsometric studies of specific adsorption of ions. This is so because the changes in Δ and ψ involved are very small and difficult to measure, especially for partial monolayer coverage. However, because of the development of more sensitive techniques that can detect changes less than 0.01° such experiments become feasible. Examples of these techniques are those of Carroll and Melmed (21), a similar technique by Brusic, Genshaw, and Cahan (16) and multiple reflection techniques (48,93).

Thus far, the very little work done in ion adsorption has been primarily that at the University of Pennsylvania (6,24,25). The earliest attempts by Chiu and Genshaw on Hg and Pt (24,25) were based on considerations similar to those described by Meyer and Bootsma (68). Thus they used the Lorentz-Lorenz equation to calculate the refractive index of the ion, n_i. This value was then used to calculate the effective refractive index of the film, n_2, using the equation

$$n_2 = n_w(1 - \theta) + n_i\theta \tag{8}$$

where n_w is the refractive index of water, 1.33, and θ is the surface coverage of the ion. By using n_2 in the usual Drude equation a relationship between Δ and ψ and θ can be obtained. Chiu and Genshaw (24), studying Br^- and CNS^- on mercury, found from a comparison between ellipsometric and electrocapillary methods for determining θ that the two techniques gave similar results within experimental limits. In their study of the adsorption of various ions on Pt, Chiu and Genshaw (25) also compare the ellipsometric determination of coverage, θ, to that obtained by other methods, radiotracer (6) and electrochemical methods (35) (effects of adsorption on the kinetics of oxygen deposition). The results for such a comparison for Cl^- are shown in Table I. It can be seen that at the lower potentials where the coverage is low, the agreement is not

TABLE I
Comparison of Ellipsometric Determination of Coverage of Chloride Ions on Pt to Other Techniques (After Chiu and Genshaw (25))

Potential V vs. S.H.E.	Coverage		
	Ellipsometric	Radiotracer (6)	Electrochemical (35)
0.177		0.00	
0.377	0.00	0.018	0.27
0.577	0.01	0.07	0.35
0.777	0.05	0.101	0.50
0.977	0.11	0.12	

good. It improves considerably at higher coverages, when comparing the radiotracer technique to ellipsometry, but the electrochemical technique is always higher. Besides their having to assume a roughness factor to calculate θ from the radiotracer data, Chiu and Genshaw attributed these differences to two causes: (1) the possibility that the adhering solution may cause too high values in the radiotracer technique, and (2) the ellipsometric values were relative to a chosen potential rather than to the zero point of adsorption. Thus, because of the second reason, the θ determined ellipsometrically may not include those adions thought to be present on the surface at the chosen potential. In an attempt to correct for this by first determining the change in Δ at a constant potential in the presence and absence of the adsorbing ion, Chiu and Genshaw did succeed in increasing the ellipsometrically determined θ. Their values for I^- were, at the highest coverages measured, still 25% lower than that determined by radiotracers. As Table I shows for Cl^-, however, not using this correction, the differences between the two techniques are less at highest coverage. Results similar to Cl^- were also obtained for Br^-. There is, therefore, no straightforward way to eliminate the differences or decide which technique more accurately measures θ.

The above approach was spelled out in more detail by Paik, Genshaw, and Bockris (75) who included in their formulation for n_2 the area of the electrode covered by the ion S_i and thickness of adsorbed layer, τ_i. Thus their form of the Lorentz-Lorenz equation for an adion coverage of θ is

$$\frac{\theta R_i}{S_i \tau_i N_a} + (1 - \theta)\frac{R_w}{V_w} = \frac{n_2{}^2 - 1}{n_2{}^2 + 2} \tag{9}$$

where R_i and R_w are the molar refractions for the ion and water respectively, V_w is the molar volume of water, and N_a is Avogadro's number. They define the coverage where anions are adsorbed as $q_-/q_{-,\mathrm{sat}}$ where q_- is the amount of adsorbed anions in units of charge and $q_{-,\mathrm{sat}}$ is q_- at $\theta = 1$. They assume that, since the ellipsometer readings are determined by the total amount of surface excess of ions, regardless of their degree of separation from the electrode surface, q_- will include the amounts of ions adsorbed on the surface and those in the diffuse layer. Both this treatment and Stedman's (which it resembles closely) are based on the arguments of McCrackin and Colson (64) that for an inhomogeneous film, Δ and ψ are related to $\int \Delta n\, dt$ where Δn is the difference between the refractive index in the thickness increment dt and the refractive index of the surrounding medium. Paik et al. assume that Δc, the excess concentration of ions (those ions not in direct contact with the surface), is proportional to Δn. Thus in this treatment Δ and ψ are not determined by the distribution of ions perpendicular to the electrode surface but by either assuming all ions are in contact with the surface or by using the surface concentration. The Paik, Genshaw, Bockris treatment focuses on the adions in a layer where thickness, τ_i, is approximately two ionic radii, while Stedman is concerned with the whole diffuse layer of a thickness $2/\delta$.

In addition to the foregoing considerations, the Paik et al. treatment has modified other approaches to the ellipsometry of adions by taking into consideration the effects of electromodulation mentioned earlier. Using Hansen and Prostak's (40) treatment of this phenomenon, which considers changes of electron density in the surface layer of a metal, they found the change in Δ due to electromodulation, $\delta\Delta_{\mathrm{mod}}$, to be proportional to q_{m}, the total charge on the metal. The effect on the observed Δ, $\delta\Delta_{\mathrm{obs}}$, is given by

$$\delta\Delta_{\mathrm{obs}} = \delta\Delta_{\mathrm{ads}} + \delta\Delta_{\mathrm{mod}} = \delta\Delta_{\mathrm{ads}} + b\,\delta q_{\mathrm{m}} \tag{10}$$

where $\delta\Delta_{\mathrm{ads}}$, a function of q_-, is the change in Δ due to ion adsorption and b is a proportionality factor. They calculated b for light at a wavelength of 5500 Å for the metals they studied and found it to be -0.0042 deg μC^{-1} cm^2 for Au, -0.0029 for Ag, -0.0010 for Pt, 0.0017 for Ni, 0.0017 for Rh, and 0.0021 for Hg.

The above described treatment was applied by Paik et al. to studies of the adsorption of Cl^-, Br^-, ClO_4^-, SO_4^{-2}, and OH^- on Ag, Au, Rh, and Ni. They were able to determine q_- as a function

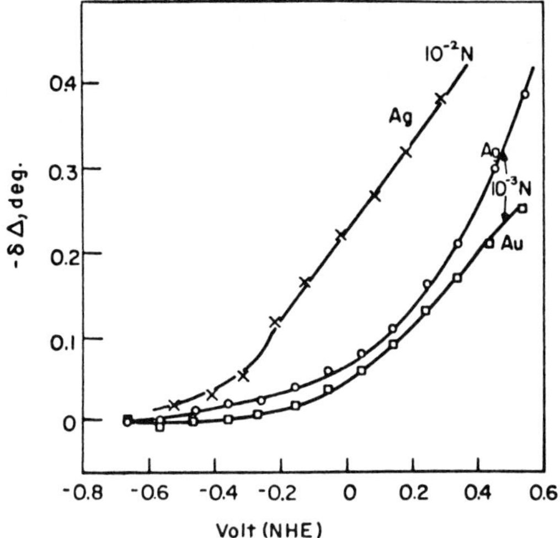

Fig. 5. The extent of change in Δ, δΔ, brought about by the adsorption of OH^- on Ag and Au electrodes as a function of potential. From Paik, Genshaw, and Bockris (75).

of potential and to demonstrate that the adsorption of the above anions increased in the order $ClO_4^- < SO_4^{-2} < Cl^- < Br^-$. The effects of adsorption of $\delta\Delta$ for various potentials are shown in Fig. 5. A later paper by Paik and Bockris (76) showed that for Au the contribution of electromodulation $\delta\Delta_{mod}$ was about equal to that of ion adsorption $\delta\Delta_{ads}$ for the adsorption of Br^- on Au.

3. Adsorption of Organic Molecules

Unlike the studies of ionic adsorption, most, if not all, of the ellipsometric studies of the adsorption of organic molecules from solution were not made under conditions involving current flow. In fact, the electrochemical aspects of the process were not considered by any of the studies found in the literature. In spite of this, it is important to treat briefly the application of ellipsometry to organic molecule adsorption for two reasons. First, this process has relevance to such electrochemical interests as electro-organic chemistry and corrosion (organic inhibitors). Secondly, some of the ellipsometric problems and their solutions in studying organic adsorption can possibly be applied to other systems.

Two types of organic film adsorbed from a liquid phase on metal or semiconductor surfaces have been studied: polar organic molecules adsorbed from solution or deposited as Langmuir-Blodgett films, and polymer molecules adsorbed directly from solution.

Studies by Bartell and Betts (9), Miller and Berger (69), and Bornung and Martin (14) dealing with adsorbed polar molecules were directed towards the question, as posed by Miller and Berger, "Is ellipsometry a reliable method for studying monolayer adsorption of polar molecules on metal surfaces?" The first two authors compare radiotracer techniques to ellipsometry; the second two use contact potential measurements. In those experiments involving radiotracers, the lineal thickness found using ellipsometry is compared with that determined by measurements of surface concentration using radiotracer techniques. The ready availability of C^{14} tagged organic molecules and the knowledge of the dimensions of the organic molecules used (capric acid, benzoic acid, stearic acid, *n*-octadecylamine) make it easy to correlate the two techniques. Both studies found good correlation. For example, Miller and Berger's results for tagged capric acid films formed by vapor deposition on nickel measure a film thickness of 33.0 Å by the ellipsometer and 32.1 Å by their radiotracer technique. The radiotracer values were calculated assuming a tightly packed capric acid monolayer 11 Å thick with a roughness factor of 1.3 chosen for nickel. While there is a certain amount of arbitrary choice of parameters (roughness factors, molecular size) in making the correlation, the values chosen are reasonable. It can be seen that the problems associated with ellipsometric studies of the adsorption of long chain polar molecules is less severe than those associated with the adsorption of ions discussed in the last section.

Another study that goes even more deeply into the question of the reliability of ellipsometric measurements of adsorbed organic molecules is that of Smith (88). He looked at monolayers and partial monolayers of adsorbed molecules of known dimensions on mercury (where the roughness factor = 1) in a Langmuir trough and combined the ellipsometric measurements with surface tension and contact-potential determinations. By using the approach of Archer (4) described earlier, which assumes a change in $\bar{\Delta} - \Delta$ proportional to the coverage, θ, he showed that the index of refraction of the film, n_2, is a function of the surface density. Since he was working with molecules of known dimensions and could easily vary the packing of molecules or coverage in the Langmuir trough by varying the surface pressure, he was in a position to compare carefully the ellipsometric values for thickness with the expected values from dimensions and packing. For rodlike molecules the correlation was excellent. He showed that the equation

$$\bar{\Delta} - \Delta = \alpha t_0 \theta, \tag{11}$$

a modification of Eq. 3, in which t_0 is the film thickness, $\theta = 1$ holds experimentally, and α is a constant described earlier in the discussion of Archers' theory, can be calculated using the index of refraction of the bulk material. The experimentally determined α's, using surface tension and contact potential measurements to obtain θ, were used to calculate the refractive index of the film, n_2. As Table II shows these compare very favorably with the bulk values, n_D, for rod-shaped molecules but not for a planar molecule.

TABLE II
Comparison of Calculated Values of Film Index of Refraction, n_2, with Bulk Values, n_D at Monolayer Coverage (After Smith (88))

Molecule	No. C- Atoms	Shape	α Degrees/Å	n_2	n_D
Pentane	5	Rod	0.127 ± 0.008	1.36	1.36
Caproic acid	6	Rod	0.129 ± 0.013	1.42	1.42
Lannic acid	12	Rod	0.140 ± 0.003	1.43	1.43
Stearic acid	18	Rod	0.180 ± 0.010	1.66	1.68
α-bromonaphthalene	10	Planar	0.22 ± 0.04	2.02	1.66

Whereas the dimensions of the organic molecules thus far described are easy to characterize and use in the ellipsometry of adsorbed organic films, the dimensions of polymeric species are by their very nature variable and cannot be so simply used. Even if the dimensions were known, equations like Eqs. 11 or 3 could not be used because they require that Δ be proportional to thickness and be the only parameter sensitive to film thickness. Thus variations in ψ with thickness that are related to β in Eq. 4 must be very small. These requirements only hold for very thin films (<50 Å). Therefore, films of adsorbed polymeric molecules which are usually thicker than 50 Å cannot be treated using Eqs. 3 or 11, and the complete exact Drude equations must be utilized. The major work on the adsorption of polymeric molecules on metals has been done by Stromberg and co-workers (92,94,95). They have been concerned with both the amount (thickness) and properties of the films. Since an adsorbed polymer chain has many possibilities available to it on adsorption because various segments can attach themselves to the surface in different ways, the concept of a "monolayer," indeed, the meaning of thickness, is somewhat uncertain. Moreover, since ellipsometric calculations require a homogeneous film with discrete boundaries, the "average" thickness, t_{av}, and refractive index, n_{av}, obtained assuming homogeneity from the Drude equation must be related to the very inhomogeneous polymeric film formed by polymeric molecules interspersed with solvent molecules. This problem was attacked by McCrackin and Colson (64) who substituted for the inhomogeneous film many much thinner homogeneous films whose refractive indices were to vary as the distribution of the refractive index throughout the total thickness of the inhomogeneous film being studied. The t_{av} could then be related to the distribution of refractive index of the heterogeneous film, $(n_2 - n_1)$, by the equation

$$\int_0^\infty (n_2 - n_1)\, dt = (n_{av} - n_1)t_{av} \tag{12}$$

where n_2 is the refractive index of the film at thickness t and n_1 is the refractive index of the solvent.† In the adsorption studies of Stromberg and colleagues (94) it was found more meaningful to use another sort of average, the root-mean-square average, defined by

$$t^2{}_{rms} = \frac{\int_0^\infty (n_2 - n_1)t^2\, dt}{\int_0^\infty (n_2 - n_1)\, dt} \tag{13}$$

Assuming that the distribution $(n_2 - n_1)$ is an exponential function of thickness as suggested by other theoretical studies (94), they calculate and use t_{rms} by dividing t_{av} (determined in the usual way from measured Δ and ψ values and the Drude equation) by 1.5. This factor was determined by McCrackin and Colson (64) for an exponential distribution. They found t_{av}/t_{rms} to be 1.75 for a linear distribution and 1.73 for a gaussian.

By using such an approach or other ones to determine the extension of the adsorbed polymer molecule normal to the surface, and by measurements of the fraction of chain segments attached to a surface by infrared techniques (95), information on the configuration

† Some of the special experimental problems associated with measuring t_{av} for polymeric films result from the fact that they are swollen with solvent so that their refractive index is nearly the same value as that of the solvent. These problems are discussed by Stromberg, Passaglia, and Tutas (92).

of the adsorbed polymers can be obtained. Examples of such studies are described in a review by Stromberg (95).

Before leaving the application of ellipsometry to the adsorption of large polymeric molecules, mention should be made, for the sake of possible future applications to bioelectrochemistry, of studies concerned with the adsorption of the large molecules of biological interest, e.g., blood proteins. Ellipsometric studies of adsorption of these has been made by a number of workers (61,87,100,101). Of special interest to electrochemistry is that by Mathot and Rothen (61) who carried out electroadsorption studies.

III. Reaction Layers

Ellipsometry has been applied more extensively to studies of reaction layers than to the partial monolayer adsorbed films described in the preceding section because the thicker reaction layer films are easier to measure and, more importantly, the classical optical theory can be more confidently applied in the interpretation of the measurements. This section deals with these reaction layers by first concentrating on anodic films which have received the most attention in ellipsometric-electrochemical studies. Special attention is paid also to passive films, some of which are formed by anodic polarization.

Thus this section deals with an area in which the concepts developed in the discussion on anodic and passive films are especially applicable—corrosion. This area of applied electrochemistry has used ellipsometry to a greater extent than any other area. Finally, applications of ellipsometry to electropolishing and electrodeposition are considered. These phenomena, in contrast to corrosion, have been virtually neglected by those investigators applying ellipsometry to electrochemical studies and can thus be discussed only briefly.

1. Anodic and Passive Films

This section considers the three kinds of information the ellipsometer can provide about anodic and passive films, namely, their thickness, their kinetics of formation, and their properties. The kinetics of film dissolution are dealt with in the next section on corrosion.

A. Thickness. The problems associated with precise measurements of reaction layer thickness are related to whether accurate values of the optical constants of the film, n_2 and k_2, are known.

The determination of these constants is discussed in a later section. If the film and substrate constants (see Section I) are known, the thickness of the film can be accurately determined from the Δ and ψ values measured on the surface studied in an electrochemical experiment. Unfortunately, the same Δ and ψ values are used to provide both the thickness, t, n_2, and k_2 at the same time. For absorbing films where $k_2 \neq 0$, this becomes a problem of using two measured parameters, Δ and ψ, to determine three unknowns, namely, n_2, k_2, and t. Fortunately, when the objective is mainly a thickness determination of compound films, i.e., "three-dimensional films" (usually oxides when anodic or passive films are being studied), a useful value of the thickness of films less than around 50 Å can be determined without knowing accurate values of n_2 and k_2. Therefore, if reasonable guesses are made as to the values of these parameters, e.g., the values of the bulk oxide, the error in the thickness determined is not too large. Figure 6 shows the extent of the error introduced when there is wide uncertainty in our knowledge of n_2 and k_2 for a hypothetical but realistic oxide film. It shows that for such a film, with a thickness of 40 Å, a maximum error of 11% in n_2 and 67% in k_2 produces a maximum error of roughly 3 Å (7.5%). For thicknesses less than 40 Å, the thickness error limits collapse onto the true value even though the percent of error may be as large or larger than that for thicker films. Whereas this situation makes

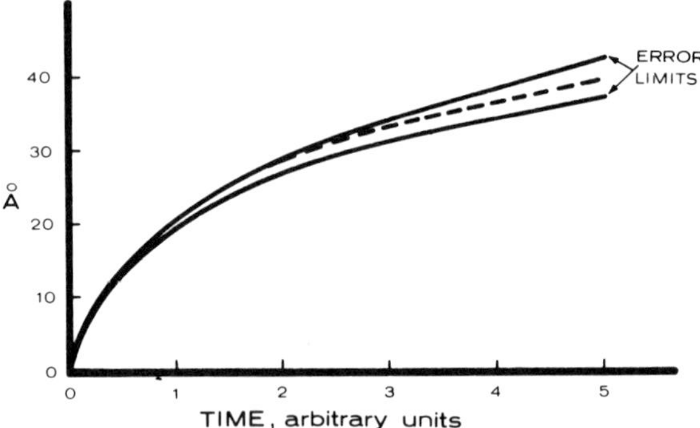

Fig. 6. Thickness-time dependence for a hypothetical film with $\bar{n}_2 = 2.6-.52i$ ($\bar{n}_3 = 3.28-4.038i$, $n_1 = 1.335$, $\phi_1 = 70.4°$, $\lambda = 5500$ Å). The variation from the correct time dependence (dashed curve) is shown when incorrect values for $\bar{n}_2$ are chosen with n_2 values ranging from 2.5 to 2.9 and k_2 from .25 to .87. Δ-values, the more sensitive to thickness changes in this range of optical constants, were used for the comparison.

thickness determination feasible, it makes the measurement of the optical constants for the very thin films, whose $k_2 \neq 0$, much more uncertain.

Another important aspect pointed out by Winterbottom (102) for anodic or passive oxide films is that for films of around 40 Å or less, Δ is the parameter that is most sensitive to variations in thickness and usually varies linearly with thickness. Thus for thin anodic films, if necessary, one can look at thickness variations and even measure thickness by measuring only Δ. The problems encountered in only using Δ for extremely thin films (less than a monolayer) are, however, pointed out in Section II. Moreover, below about 5 Å in thickness the kind of problems and approaches discussed in Section II become important.

There are some examples in the literature of electrochemical studies involving thin films at the lower range of Fig. 6 (<10 Å). These involve noble metals (Au, Pt) or metals that form very thin passive films (Cr). For both Au and Pt it has been shown (61–63) by ellipsometry that there exist definite potentials during anodic polarization where $-d\Delta/dV$ changes from a very low value and becomes appreciable (see Fig. 7). Since $-\Delta$ is directly proportional to thickness for very thin films, the Δ variation shown in Fig. 7 can be considered as thickness changes. The ψ variations were all less than 0.1°. This sort of behavior was interpreted by Reid and Kruger (81) and Sirohi and Genshaw (84) to be a transition from a chemisorbed layer below the potential where $-d\Delta/dV$ changes abruptly

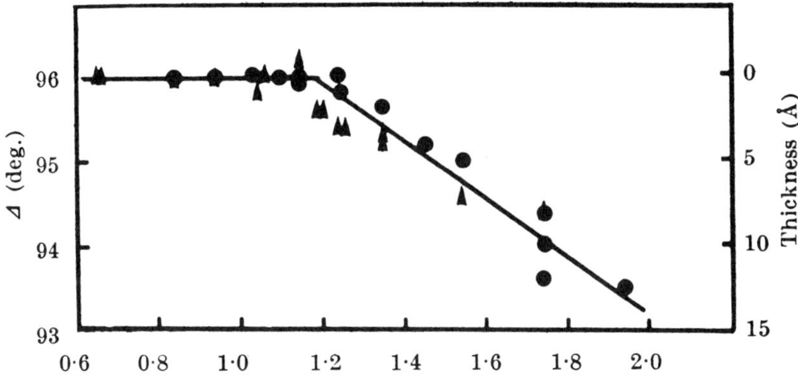

Fig. 7. Change in Δ and thickness values as a function of potential for gold in 1 M sulfuric acid (triangles) and in 1 M sulfuric acid and 4g/1 $Na_2C_2O_4$ (circles). From Reid and Kruger (81).

to the formation of a "three-dimensional" oxide above this potential. The Δ and ψ variations point to the formation of an oxide on the gold that is absorbing ($k_2 \neq 0$). Where a chemisorbed layer may be present, Sirohi and Genshaw ascribed the very small variation in Δ with potential to an electromodulation effect. By such ellipsometric experiments it is thus possible to answer the question as to whether or not a phase oxide can form on gold. Purely electrochemical measurements (see 81 for references), namely, charging curves, could not unequivocally resolve this question. Work by Reddy et al. (80) found results similar to that for gold in their thickness measurements on Pt.

A metal said to form an especially thin passive film is Cr (98). Ellipsometric measurements of this film by Genshaw and Sirohi show that the thickness required for passivation corresponds to a change in Δ of 0.25° which corresponds to a thickness of 0.8 Å for the optical constant they assumed ($n_2 = 2.54$, $k_2 = 0$). One problem that must always be considered in accepting such values for thickness for the very thin films on metals such as Cr, Pt, Au, etc., is that they are determined from a change in Δ that is obtained by assuming that a bare surface was present before film growth was initiated. Only when one is assured that the $\bar{\psi}$ and $\bar{\Delta}$ used in the calculation were indeed measured on a bare surface, can one say, for example, that 0.8 Å of a film is required to initiate passivity. The changes in Δ do, however, give minimum thicknesses.

For compound films >50 Å that absorb ($k_2 \neq 0$) it is perhaps more difficult to determine the film thickness because, as Fig. 6 shows, one needs to know the refractive index more precisely to arrive at more accurate values of thickness. One approach to overcome this problem is to compare the thickness obtained ellipsometrically, using a particular value for the refractive index of the film, to the thickness measured in other ways in order to see if the value chosen is reasonable. Coulometry offers such an opportunity for studies made in aqueous solutions. One area where such a comparison has been made is in investigations of the passive film formed on iron by anodic polarization. A number of workers have made such comparisons (13,53,58) between coulometric and ellipsometric thickness. The discrepancies usually found between the coulometric and ellipsometric thicknesses are explained in a number of ways. For example, Kruger and Calvert (53) studying the passive film on iron, explained the discrepancy between their optical and electrochemical determinations of thickness as being due to a roughness factor for their surface of 1.3 instead of 1.0. A discrepancy of 15% between coulometric measurements and ellipsometry for the same film studied by Kruger and Calvert was found by Kudo et al. (58). They attributed this discrepancy to the presence of water in the film. They used the ellipsometric thickness of the film to give them the dimensions of the film in terms of the number of oxygen ions making up its lattice. Coulometry measured the iron in the film and a radiotracer technique measured water in the film. A similar study by Sato, Kudo, and Noda (82) combined ellipsometric thickness measurements with chemical analysis of reductively dissolved Fe^{2+} to determine the amount of water in the passive film. It must be pointed out that the discrepancies between coulometric and ellipsometric thickness values, besides being due to a lack of precise knowledge of film optical constants, can also be attributed to poor knowledge of the film's density and the surface roughness factor, which are difficult to determine precisely and upon which accurate coulometry depends. There is also the problem of being sure that all the current measured goes into film formation (or dissolution) and not, for example, hydrogen oxidation. Such experiments, however, illustrate that ellipsometric thickness determinations can serve as a useful adjunct and check to coulometric studies.

Another example of a different way in which thickness measurements contribute to electrochemical studies is by Ord and deSmet (72). They measured the changes in the ellipsometrically determined thickness of what they called an electrically limiting outer layer as a function of potential to conclude that a field-dependent conduction mechanism for film formation was operative.

When one considers absorbing films thicker than 50 Å, the sensitivity of thickness to refractive index becomes greater, and more exact knowledge of the refractive index often becomes more imperative than for films < 50 Å. For films whose thickness is in the hundreds of angstroms range, this sensitivity usually allows one to choose a reasonable value of the refractive index if one can plot Δ versus ψ measured for a number of films of different thicknesses. This can be seen from Fig. 8, which shows such a plot for various values of n_2 and k_2. Cathcart and Petersen (23), however, found that for films in the 50–200 Å range that the assumption of a uniform refractive index can lead to erroneous thickness values. They were able to conclude this because the films they measured ellipsometrically (cuprous oxide films on {111} copper single crystal surfaces) could also be measured accurately by an x-ray technique. As Table III

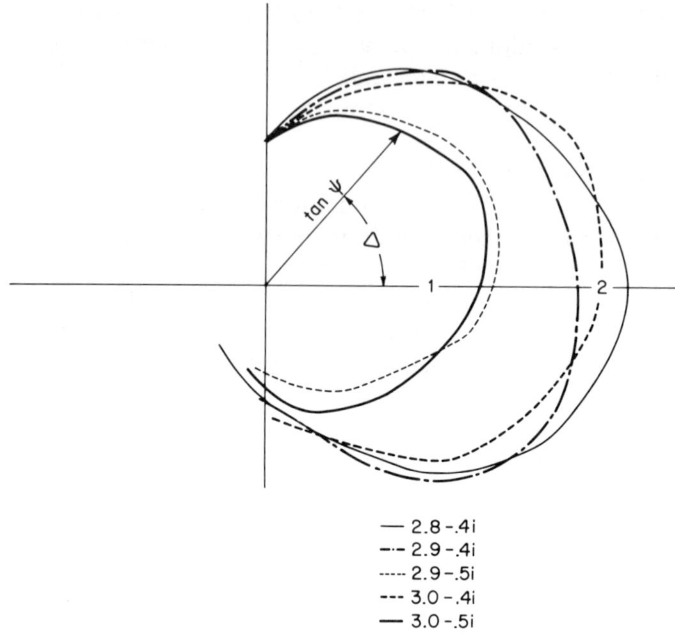

Fig. 8. Theoretical polar Δ-tan ψ plots for absorbing films of the $\bar{n}_2$ values indicated to thicknesses up to 1000 Å ($n_1 = 1.333$, $\bar{n}_3 = .92–2.94i$, $\phi_1 = 60°$, $\lambda = 5461$ Å).

shows, values calculated assuming a uniform film (isotropic model) are larger than the x-ray values. If a varying refractive index is assumed (gradient model), a better correspondence is obtained. Since the gradient model uses many adjustable parameters (the optical constants of the many layers making up a varying film), no real significance should necessarily be attached to the values of the refractive indices used to produce the good fit. However, if one can rely on the x-ray determination of thickness used, the procedure gives a good calibration curve for ellipsometric measurements.

Such problems as those just discussed for absorbing films are present to a much lesser extent for nonabsorbing films. The most widespread example of these are the anodic films formed on valve metals. The reason that the film thickness for nonabsorbing films can be determined is that $k_2 = 0$ and therefore both n_2 and t can be calculated (64). Problems can arise; the film can be nonuniform and its optical constants can change gradually. This is discussed

TABLE III
Comparison of Thickness Values Obtained by X-ray and Ellipsometer Measurements (From Cathcart and Petersen (23)) (Å)

X-ray	Ellipsometer	
	Gradient model	Isotropic model
76	92	126
130	125	169
148	141	197
205	205	296

later in the section concerned with film properties. Assuming homogeneous films, there are a number of examples of determinations of film thickness of thick anodic films (7,8,26,42,74,106).

B. Kinetics of Formation. This application involves a determination of the time dependence of the thickness measurements just described. In these measurements, while accurate knowledge of thickness is desirable, it is not indispensable. The rate of change in the optical parameters measured, Δ and ψ, their direction of change (increasing or decreasing), and the rate law governing these changes are the focus of interest. In many situations involving films of oxides, Δ is proportional to thickness for films < 50 Å. Thus a plot of Δ versus time can yield the form of a rate law. An exact knowledge of the refractive index of the film becomes less necessary, and consequently kinetic studies are both the easiest to apply and perhaps the most reliable application of ellipsometry. Figure 6 shows the thickness range of most interest to a large number of electrochemical systems (0–40 Å) where inexact knowledge of the refractive index of the film does not appreciably mislead one with regard to the time dependence of thickness change. Of course, if accurate rate constants must be determined, it is necessary to know the thickness to the limit of accuracy desired.

There are, however, many instances where one can obtain valuable information about a process without determining the thickness to a high degree of accuracy. An interesting example of this was developed by Budrys (19) who compared the change in Δ and/or ψ with time to the change in current density during the formation of a film by anodic polarization. By so doing, a diagnosis of how much of the current is involved in growing or dissolving the film can be made. To accomplish this, Budrys uses a semiautomatic intensity recording technique (79) and feeds the signal coming from a photomultiplier tube into a device that takes the time derivative of the intensity. The resulting signal (proportional to the time derivative of the thickness) thus becomes directly proportional to the current involved in film growth or dissolution. By then comparing this signal to the current, he can determine during potentiodynamic sweeps involving growth and decay, whether the current measured is only involved in forming or dissolving the film measured by the ellipsometer, or if it is involved in other reactions not seen by the ellipsometer, e.g., those taking place in solution.

The most widely used application of ellipsometry is to determine the form of a rate law governing film formation. Not only can this be done, but sometimes it can be determined at what point in time there is a change from one rate law to another. Figure 9 shows such a change from one rate law to another at different stages in the growth of the passive film on iron. This plot shows three possible stages and reveals the rate laws governing each stage. Thus for the last two stages, the thickness varies as the log of time. The first stage, if replotted, gives a linear dependence of thickness on the square root of time. These results can be obtained by using intensity transient techniques (16,21) or rapid automatic ellipsometry (20,73,104). Similar determinations of rate laws have been made for Fe (72,104), Fe-Cr-Ni alloys (36), Ni (50), Cr (32), gold (84), Pt (80), and Cu (80).

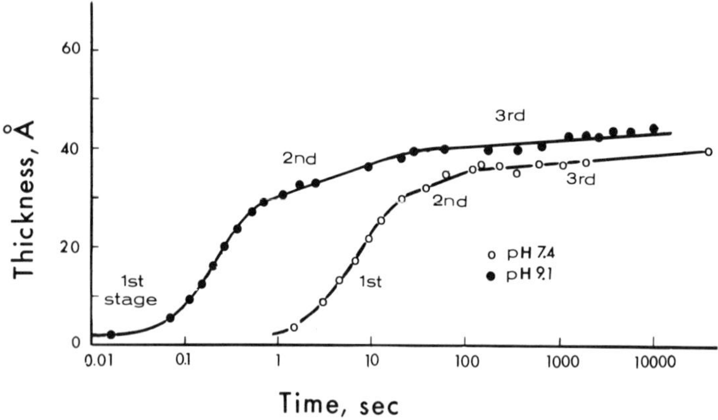

Fig. 9. Plot shows the three stages of passive film growth on iron in solutions of pH 7.4 (open circles) and 9.1 (closed circles) for potentiostatic growth at 0.8 V (S.C.E.). From Kruger and Calvert (53).

All the work discussed above deals with the kinetics of formation of rather thin films or slowly forming thick films. Special mention should be made of a study that measured the kinetics of a very rapidly forming thick film. This was done in the work of Green, Mengelberg, and Yolken (37) who looked at the tarnishing of brasses in ammoniacal solutions where in some cases 900 Å of film formed in 1–2 sec. They accomplished this by presetting the ellipsometer at Δ and ψ values corresponding to a particular film thickness. As the Δ–ψ polar spiral plot shown in Fig. 8 indicates, the Δ values are cyclic as one goes around the spiral with increasing thickness. Thus the time to reach the preset thickness can be measured by recording the point at which the intensity of light reaches a minimum, as recorded by a photodetector. As the thickness continues to increase, another minimum can be measured when Δ again becomes the preset value. By measuring the time to reach these minima and calculating the thickness at each minimum, a rough estimate for the rate of film formation can be obtained and a rate law derived.

In addition to determining rate laws and the point in a process where these laws change value or form, we can determine at what point in time a change in the film properties, e.g., formation of a new phase, occurs. An abrupt reversal in the direction of change of Δ and ψ values may signal such an event. An example of this kind of application can be found in the studies of copper by Kruger (50). He found a reversal in the direction of change of Δ when the film that initially formed as Cu_2O started to form as CuO. Also, one can determine induction times required for a film to start growing on a bare substrate. Here no changes are observed ellipsometrically until the induction time τ is reached. At τ an abrupt change in Δ results. While the thickness measurement may be qualitative, τ can be quantitatively determined as was done, for example, in the galvanostatic study of the formation of calomel films on mercury by Bockris, Devanathan, and Reddy (11). Thus the mechanism of this process can be determined by ellipsometric measurements of τ. Similar studies can be carried out potentiostatically. For example, Harvey and Kruger (47) were able to determine that passive film formation of GaAs took place at the point in time during a triangular potential sweep when the current reached its peak value in the active region (the passivation current). Figure 10 shows the relationships between the potential, the current, and the ellipsometer intensity signals (proportional to thickness) which allowed them to decide that this was so.

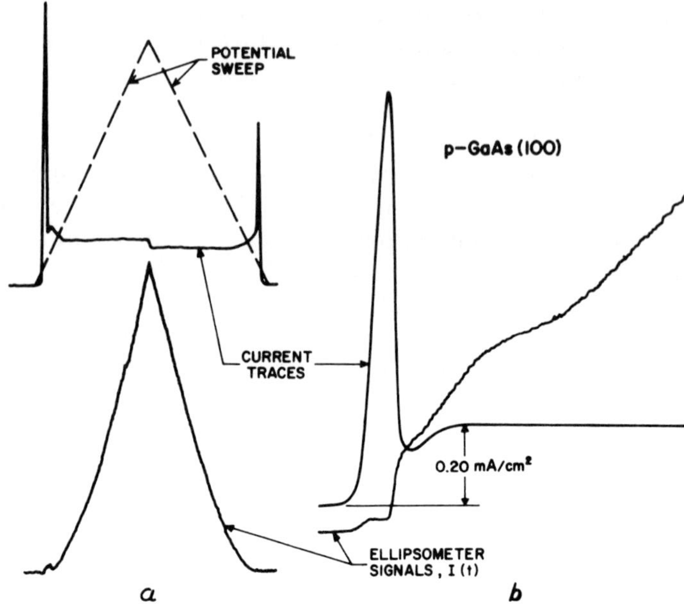

Fig. 10. Anodic polarization of *p*-GaAs (100) crystallographic orientation in borate buffer solution pH 7.4. Variation of anodic current and ellipsometer signal during a triangular potential sweep. (*a*) Ellipsometer nulled at the rest potential. (*b*) Expanded time base; ellipsometer initially offset from null. From Harvey and Kruger (42).

C. Film Properties. Determining the nature of the films formed at an electrode is potentially the most valuable contribution that ellipsometry can make to electrochemistry. Unfortunately, it is also the most difficult to achieve. The origin of this difficulty lies in the fact mentioned earlier in the discussion of thickness measurements, namely, that for absorbing films the two experimentally determined parameters, Δ and ψ, must be used to measure three parameters, t (film thickness), n_2, and k_2 (the real and imaginary parts of the film's refractive index). It is by the determination of these optical parameters that information about the nature of the film can be obtained.† Fortunately, for some metals the films formed are non-

† A review by Kruger and Hayfield (55) discusses the interpretation of optical parameters measured at different wavelengths and describes how insights may be gained from these measurements into the solid state properties of reaction layer films.

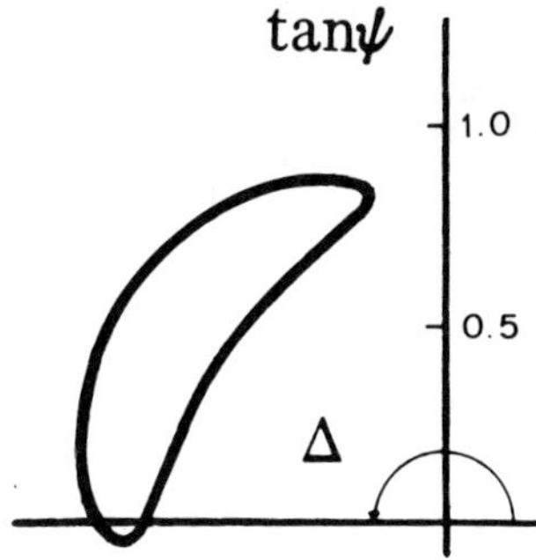

Fig. 11. Theoretical polar Δ-tan ψ plot for a nonabsorbing film, $n_2 = 1.24$ ($n_1 = 1.0$, $\bar{n}_3 = 0.77–2.08i$, $\phi_1 = 60°$, and $\lambda = 5461$ Å).

absorbing, i.e., have real refractive indices. For these metals, e.g., aluminum, the determination of the film's optical properties is much more tractable, although the measurement is not without problems.

Most of the metals forming films with real refractive indices are the valve metals, e.g., Ta, Nb, Al, and Ti. There are a number of examples of studies of these metals aimed at determining the properties of their anodic films (8,19,26,42,74,106). There have also been some studies of anodic films on semiconductors (3,42,85). The first question that comes to mind is how to know whether the film does indeed have a real rather than a complex refractive index. Many years ago Winterbottom (102) pointed out that a polar plot of Δ versus tan ψ yielded a closed curve of a "kidney bean" shape, if the film formed on a metal had a real refractive index. Figure 11 shows such a curve. In order to obtain such a closed curve, films of many thicknesses have to be measured, with the closure occurring when a thickness equivalent to one-half the wavelength of the light used in the experiment is reached. Unfortunately, many systems of interest to electrochemistry involve films that are much thinner. Therefore, the results from studies of thick films can only be used to infer that the films of interest to electrochemistry have real and not complex refractive indices. The work of Dell'Oca and Young (74) is an example of such a study. They have shown that for tantalum in phosphoric acid solutions it is possible, by means of ellipsometry, to demonstrate that the real film that formed consisted of two individually homogeneous layers and that the indices of refraction and thicknesses of each layer could be determined by curve fitting. While other interpretations (i.e., a linear gradient) of n could equally well fit the curve obtained, Dell'Oca and Young pointed out that nonoptical evidence ruled out such a gradient model.

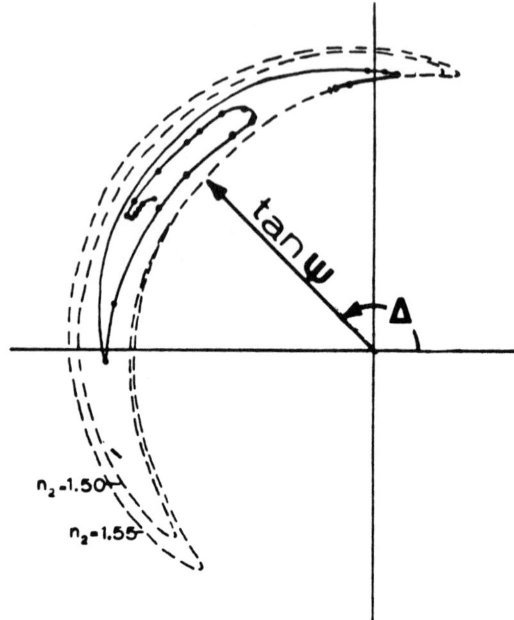

Fig. 12. Typical curve for changes in polar plots of Δ and tan ψ during corrosion of aluminum in hot water compared to theoretical curves (dashed lines) for films with refractive indices 1.50 and 1.55. From Barrett (8).

These results illustrate one problem that usually precludes unequivocal determinations of the optical constants of real films—the possibility of nonuniformity. Barrett (26) studying the films formed on aluminum in hot water showed how difficult this problem can be. In Fig. 12 it can be seen that the actual curve of Δ versus tan ψ departs markedly from the "kidney bean" theoretical curves. Two or more layer models can yield reasonable fits between experimental and theoretical, but ambiguities remain.

The above discussion has thus far dealt only with thick real films. With very thin real films, good fits can be obtained between experimental and theoretical curves. This does not always serve to determine the optical constants for very thin (< 10 Å) films. The reason lies in the fact that Δ and ψ approach the same values regardless of n as the thickness approaches zero. Thus the variations of Δ and ψ for the very thin films with n_2 may not be large enough to determine which of a number of n_2 values the data fit best. Harvey and Kruger (42) showed still another complication for real films in their study

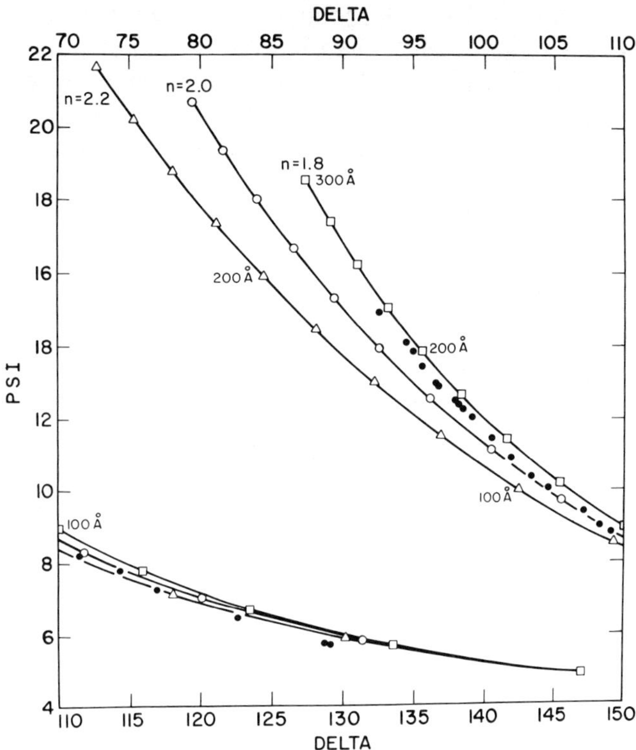

Fig. 13. Comparison of measured ellipsometric angles (black dots) with curves computed for nonabsorbing films on GaAs, $\phi = 70.3°$. Thicknesses on computed curves are indicated at 25 Å intervals. From Harvey and Kruger (42).

of passive films on GaAs. Here for films of less than 200 Å thickness, the n_2 of the real film increased in value as the film got thinner and Δ and ψ approached the bare substrate values (see Fig. 13). They found they could produce a better fit than that in Fig. 13 if they assumed that the film had two layers and that the first 25 Å had a complex refractive index, i.e., it was absorbing. Thus this points to the possibility of films that, while mostly nonabsorbing, can be duplex films, with one of the layers having a $k_2 \neq 0$.

The great difficulty that is present when we have to determine exactly the optical properties of absorbing (complex refractive index) films is that an additional parameter, k_2 must be determined. Thus even disregarding the problems of uniformity and working in a thickness range where all values of the parameters do not converge, problems of ambiguity remain. McBee and Kruger (62) showed how this can come about for the conditions that would exist in an actual electrochemical experiment assuming a reasonable error limit of accuracy (not precision) of $\pm 0.1°$ in polarizer or analyzer readings. It can be seen in Fig. 14 where the values of Δ and ψ corresponding to selected values of n_2 and κ_2 ($\kappa_2 = k_2/n_2$) are plotted for a film with a fixed thickness of 20 Å. If an experimental measurement of this film gives the values of Δ and ψ represented by point A, a narrow error limit will give a κ of 0.85 only, while the larger, more realistic limit shown, allows another κ of 0.25, along with the former κ value for an n of 2.1. At point B, two different values of κ are also possible but with two different values of n. This situation differs from the case where experimental error for a real film can yield two n_2 values. For the real film, the two n_2's are close to each other in value. For absorbing films, as Fig. 14 shows, the overlap is for widely differing values, especially for κ. Thus it is difficult to select the correct value.

Some studies of passivity minimize this very serious problem. An example is that of Bockris, Reddy, and Rao (12), who looked at passive film formation on Ni and ascribed the changes in Δ and ψ encountered in going from active to passive potentials to definite changes in n_2 and k_2, allowing them to make definite statements about the nature of the passive film. For example, they conclude that because the value of k_2 goes up at passive potentials, the film on Ni becomes an electronically conducting compound when passivity is achieved. Besides the very real possibility of multiple values of n_2 and k_2 as shown by Fig. 14, there is also the possibility of a large effect on the ellipsometric measurements that can be brought about by surface roughening that probably occurs while the Ni is at potentials in the active region. Thus facile conclusions cannot be drawn about film properties in going from active to passive potentials. A more recent similar study (13) on iron recognized the roughness problem but did not consider fully the more serious multiple solution problem.

What are the strategies for overcoming this great problem? McCrackin and Colson (64) examined this question by comparing the values of n_2 and k_2 obtained by computation to those assumed for a hypothetical absorbing film when they varied the following: (1) the thickness of the film, (2) the surrounding medium, (3) the substrate, and (4) the angle of incidence. They hoped by varying these parameters that they could arrive at unambiguous values for

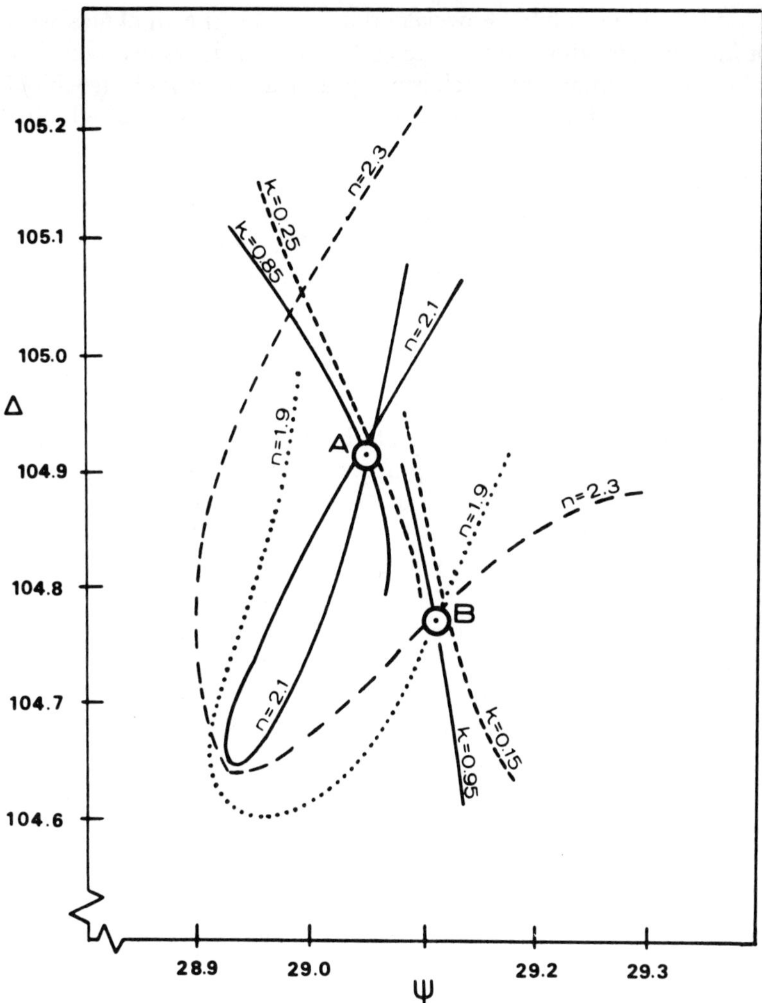

Fig. 14. An example of the variation in computed Δ and ψ values, for chosen n and κ ($\kappa = k/n$) values. The values of λ, 4700 Å; angle of incidence, 70.4°; thickness, 20 Å and substrate index, 2.77–1.39i are fixed. Experimental points are shown at A and B. The diameters of the circles indicate the larger error allowance. From McBee and Kruger (62).

n_2 and k_2 They found, however, that varying the thickness worked for films greater than 100 Å. Varying items 2 and 3 gave unambiguous values, but such an approach is not practical in most electrochemical studies. Using different angles of incidence (item 4) worked only for one particular thickness.

What sort of compromise has been arrived at in studying the properties of films formed under conditions of interest to electrochemistry? The popular approach has been to choose a reasonable estimate of the value of the film's optical constants as a starting point for determining its value. This guess can be based on the bulk refractive index of the material presumed to compose the film. This, of course, assumes that the bulk values are the same as that of the thin film, an assumption that may not be warranted. It also assumes knowledge of the composition of the film possibly obtainable by diffraction techniques. Lacking any bulk values from the literature, an estimate can be made for the real part of the refractive index using the Lorentz-Lorenz approximation (70). (This approach, however, does not give values for k_2.) By this "starting with a near answer" approach, good fits to experimental data can be obtained and a value for the refractive index estimated. A number of authors (12,13,57,72) have used this approach. As stressed above, however, and as Fig. 14 emphatically shows, other solutions to the Drude equation are ignored because this approach essentially assumes the answer in order to determine it.

Another scheme is to determine the film thickness by nonoptical means, e.g., coulometry, and then to determine n_2 and k_2 using this thickness value. Again Fig. 14, which is based on an assumed knowledge of thickness, shows that this strategy can involve ambiguities. A very recent approach to overcome these ambiguities by Paik and Bockris (77) measures an additional parameter, the reflectivity of the filmed surface. The reflectivity measurements are made by the ellipsometric optics without the refinements necessary to obtain precise reflectivity values. Thus the errors in the values obtained may still admit to ambiguities.

Even failing to arrive at reliable values for the optical constants of the films studied for the reasons just discussed, ellipsometry can provide valuable qualitative insights into film properties. An example that shows this is the study by Kruger (51), which addressed itself to the question of whether the passive film on iron was an adsorbed monolayer of oxygen or a "three-dimensional" reaction layer. Figure 15 shows that for the formation of an adsorbed film whose

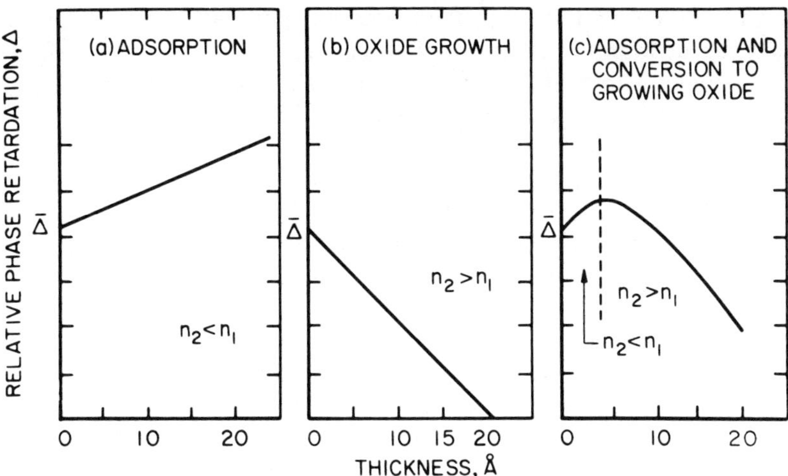

Fig. 15. Theoretically derived curves for the variation of Δ with film thickness for: (*a*) a film whose refractive index is less than the solution ($n_2 < n_1$); (*b*) a film whose refractive index is greater than the solution ($n_2 > n_1$); (*c*) a film where initially $n_2 < n_1$, but later n_2 becomes greater than n_1. From Kruger (51).

refractive index, n_2, would be undoubtedly less than the refractive index, n_1, of an aqueous solution, Δ would increase as a film formed, Δ becoming larger than that for the bare surface, $\bar{\Delta}$. If the film were an absorbing one where a reasonable n_2 for an iron oxide would be larger than n_1, Δ would decrease from the initial $\bar{\Delta}$. Finally, a combination of both an adsorbed film and then conversion to a reaction layer would cause Δ first to be larger than $\bar{\Delta}$ and then become smaller. The second kind of behavior (Fig. 15*b*) indicated a three-dimensional passive film on iron. Thus simply watching the directions of change of Δ and ψ as a film grows gives a very valuable qualitative indication of the nature of a film on an electrode.

All the difficulties just described have been for thin absorbing films. Problems also exist for thicker ones, but to a much lesser extent. This is so, because as one goes to greater thicknesses (100–1000 Å) the variation in the Δ–ψ plots with n_2 and k_2 over the whole thickness range becomes much more sensitive to changes in the film's optical constants (see Fig. 8). While ambiguities can still exist, the chances that they are present for the whole thickness range are much less. Also changes in film properties at some point in the film formation process can be readily detected.

An example of the sort of information one can obtain on thick films can be found in the work of Hayfield and White (46) on copper and copper alloys in seawater. They showed that, initially, an unprotective film was formed that was highly absorbing (k_2 = 1–1.5). The conversion of this film later on in the process to a protective one could be detected by a change in shape of the Δ–tan ψ curve (a Fig. 8 type curve). Reviews by Hayfield and colleagues (44,45, 46) give other instances of similar experiments that provide useful information on the properties of thick films.

2. *Corrosion*

Ellipsometry has been applied more extensively to studies of corrosion than to any of the other areas to which it is applicable. This is so because one of the central problems in studies of corrosion is an understanding of the role played by films on metal surfaces. The ellipsometer's main function is to provide such information. Moreover, while allowing us to measure the amount or thickness of these films, their rate of growth or disappearance and their optical properties, it can usually be combined without too much difficulty with other techniques, and in aqueous corrosion the other techniques are almost always electrochemical ones. This area was pioneered by Tronstad (43,83) who showed how effectively the optical and electrochemical techniques could work together. There exist a number of reviews on the applications of ellipsometry to corrosion (44,45,52,55).

A. Film and Substrate Dissolution. An important step in many corrosion processes is film dissolution especially because this may lead to the most elementary of all corrosion processes—substrate dissolution. In studying such phenomena with the ellipsometer, no new problems are introduced when only film dissolution occurs. This is simply the reverse of film growth and all the considerations discussed in the preceding section apply; the parameters are simply changing in the opposite direction with time. An example of such a situation is found when a metal surface bearing a passive film is polarized to a potential lower than the potential at which the film was formed, but at a potential still in the passive region. This dissolution of the film results in an increase in Δ for most films on metals. Complications arise, however, when both film and substrate dissolution occur if the dissolution results in surface roughening.

Roughening of a metal surface can give changes in Δ and ψ that are indistinguishable from the changes resulting from film growth.

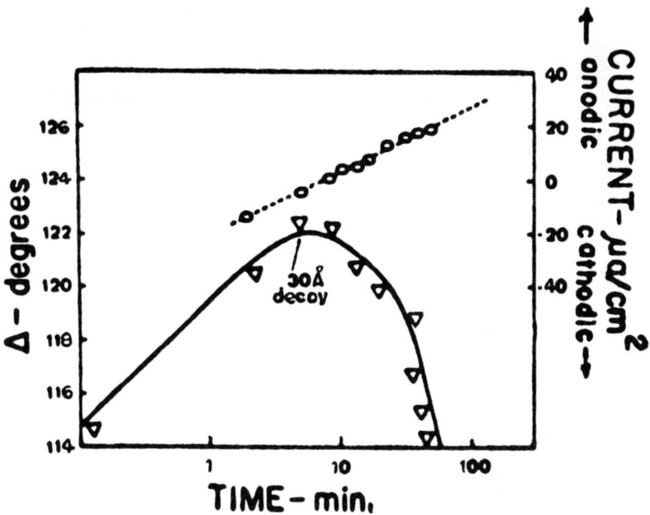

Fig. 16. The change in the relative phase retardation when a film formed potentiostatically at +0.8 V (S.C.E.) in a sodium borate-boric acid solution (pH 8.1) is dissolved by changing the potential to a value where passivity breaks down. From Kruger (52).

Kruger (52) showed an example of this effect in studying the dissolution of the passive film formed on iron at a passive potential. When the filmed surface is brought to a potential in the active region, the values of Δ initially increase with a decrease in film thickness. Figure 16 shows that negative current is measured while this is occurring. When the current becomes positive however, Δ starts to decrease, indicating either film growth or surface roughening. The latter can be detected if subsequent measurements on a cathodically reduced bare surface indicate a change in the substrate optical constants. Usually, the most significant change is that values of Δ are lower than those measured on a bare surface that has not been held for a significant time at potentials where active dissolution takes place.

For illustrative purposes, Fig. 17 shows why roughening results in a lowering of Δ and ψ values. Essentially, what this simplified picture assumes is that the roughened but bare surface can be considered as a smooth surface with a film made up of metal islands and solution-filled spaces. The index of refraction of this hypothetical film can be calculated for different degrees of roughening by assuming it to be made up of different percentages of metal and solution,

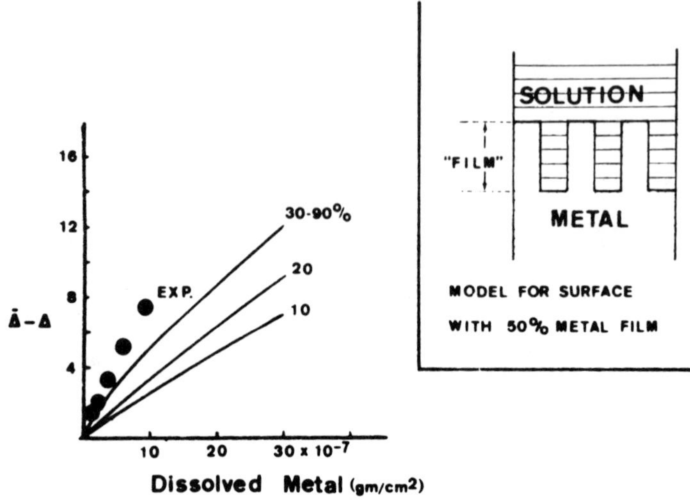

Fig. 17. A comparison between the experimental effect of roughening by iron dissolution on the departure of Δ from the smooth filmfree value $\bar{\Delta}$ and that calculated using the model shown in the upper right-hand corner. From Kruger (52).

knowing the index of refraction of the metal and the solution. The thickness of the film is, as shown in Fig. 17, the height of the metal solution region. A somewhat similar treatment has also been recently applied to electrochemical studies by Brusic, Genshaw, and Bockris (17) who showed that the conclusion of the above simplified picture —that roughening leads to optical effects similar to film growth—is reasonable.

One can use the approach illustrated in Fig. 17 to learn about other aspects of the corrosion process. For example, it can serve as a starting point to make ellipsometric estimates of the amount of metal going into solution during the roughening of a metal surface by corrosion. As Fig. 17 shows, if one assumes a film made up of any percentage greater than 30% metal (the surface is approaching localized roughening with an increase in percentage of metal), the calculated variation of $\bar{\Delta}-\Delta$ with amount of metal lost is independent of the percentage of metal making up the hypothetical film. Since the sensitivity of $\bar{\psi}-\psi$ to roughening is much less than $\bar{\Delta}-\Delta$, it is not plotted in Fig. 17. When the film is made up of less than 30% metal, the dissolution process is producing less roughening and the amount of metal going into solution affects $\bar{\Delta}$–Δ less sensitively. Here the calculated variation of $\bar{\Delta}$–Δ with the amount of metal going into solution does, in contrast to films of $>30\%$ metal, depend on the amount of metal assumed to be making up the film.

When an estimate of the amount of metal going into solution is made from current measurements and plotted versus $\bar{\Delta}$–Δ, using the data of Fig. 16, it can be seen that the experimental points do not follow the curves derived from the crude model too well. It is, however, closer to the $>30\%$ metal curve and thus serves to indicate that perhaps the dissolution is localized, possibly at grain boundaries. Recent work (54) using single crystals where the grain boundaries were eliminated showed that the assumption had merit because Δ did not change direction as in Fig. 16.

B. Pitting. For studies of pitting, the ellipsometer is more applicable to an investigation of the processes leading to film breakdown and pitting than it is for studying the progress of pitting. Once a film has broken down and pitting is actively taking place, the formation of corrosion products obscures the pitting process and one is simply looking at film formation. Thus the ellipsometer is best employed in studying prepitting rather than pitting.

An example of this application can be found in a study by Ambrose and Kruger (1) on the breakdown of passive films on iron by chloride ions. They formed a passive film by potentiostatic anodic polarization and introduced chloride ions holding the potential at the same value as that at which the film was formed. The results shown in Fig. 18 indicate that two events related to the initiation of pitting can be detected. These two events were associated with two induction times. The first, τ_1, signaled by a rapid rise in current but by no ellipsometric changes, indicated breakdown of the film. At a later time, Fig. 18 shows the onset of ellipsometric changes indicating the deposition of corrosion products at the sites of incipient pits. Figure 18 shows one of the factors that influences τ_1 and τ_2, chloride concentration. Thus ellipsometry gives details about the pitting process that can be missed by using only electrochemical techniques.

Moreover, one can examine even further details of the pitting process by using ellipsometry at many wavelengths. For example, it would be valuable in studying the mechanism of pitting to know whether any changes are taking place in the passive film before the first induction time τ_1 is complete. An ellipsometric spectroscopy study addressed to this question by McBee and Kruger (63) showed that if one worked at certain selected wavelengths one could observe

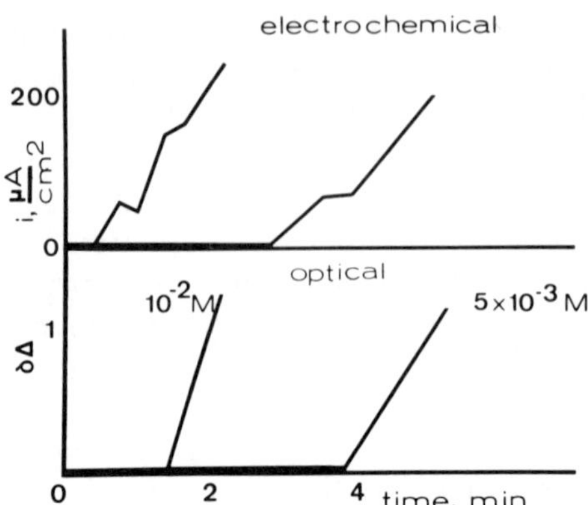

Fig. 18. The effect of chloride ion concentration on breakdown after introduction of chloride into a boric acid-sodium tetraborate buffer solution (pH; 8.4). Films were grown for 100 minutes. The breakdown was observed by simultaneously following changes in phase retardation ($\delta\Delta$) and current, potential being held constant at + 1.04 V SHE. From Ambrose and Kruger (1).

ellipsometric changes in the film well before breakdown, as indicated by a rise in current, occurred. They found that at most of the wavelengths in the visible spectrum, such as the one used in most ellipsometric studies (5461 Å) and used by Ambrose and Kruger above, no changes could be detected. At three of the wavelengths tried, however, significant effects were observed. By working at one of these wavelengths, as Fig. 19 shows, they found that chloride produced changes in the passive film prior to breakdown and that these changes could be reversed if chloride were removed from the solution. Thus ellipsometry is capable of showing that the initiation of pitting involves subtle and reversible changes in a passive film that are not detected by electrochemical measurements alone.

For films thicker than the passive films on iron where optical constants can be characterized with more certainty, ellipsometry can sometimes be used to characterize pitting tendency. Hayfield (as described in a review in Ref. 55) found that different cuprous oxide films with different k_2 values produced by varying conditions of formation or impurity levels exhibited different potentials and various pitting tendencies. Thus those films with the highest k_2 values (produced by the addition of iron impurities) had more noble potentials and were more susceptible to pitting attack than noncontaminated oxide films.

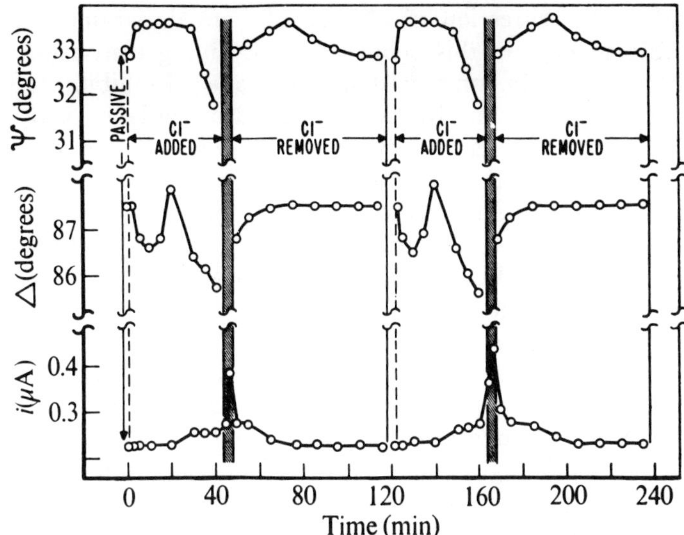

Fig. 19. Changes occurring in Δ and ψ for a passive film on iron in the presence of Cl^- prior to breakdown. These changes could be observed at a wavelength of 4100 Å but not at all wavelengths studied. As shown, the removal of Cl^- from solution causes the parameters to go back to their original values. From McBee and Kruger (63).

C. Inhibition. The ellipsometric aspects of the mechanism by which inorganic and organic inhibitors act have already been discussed in previous sections—the inorganic by means of passive film formation, the organic by adsorption. In this section, a few brief examples are given of the application of ellipsometry to the details of corrosion inhibitor action.

For those inorganic inhibitors called anion passivators, such as chromates, nitrites, etc., where metals immersed in solutions of these anions form passive films, no details need to be given, because the films form in the same manner as that for anodic films (see Ref. 51). A different kind of inorganic inhibitor is the cation inhibitor, e.g., stannous ions. An ellipsometric study of the inhibition of iron in sulfuric acid by Kudo and Okamoto (56) investigated the action of these. They found that the film formed on anodic polarization in the absence of stannous ions changed when stannous ions were added to the solution. Accompanying a change in potential to more noble values, which reached a maximum and dropped off, was a reversal in the direction of change of optical parameters. They were able to explain this behavior by assuming conversion of the film into an iron-tin compound acting as a good inhibitor.

Organic inhibitors act by forming an adsorbed layer or by modifying the formation of a protective oxide film in a manner somewhat like that just pointed out in describing the action of cations. For example, Mansfeld, Smith, and Parry (60) found that benzotriazole, an excellent inhibitor of copper corrosion, reduced the thickness of the oxide film formed by over one-third in a 5% NaCl solution. They attributed this to the barrier provided by the adsorbed inhibitor. However, since some films formed in the presence of the inhibitor were as thick as 60 Å it is unlikely that the benzotriazole stopped oxidation but instead promoted the formation of a more protective film. Work by Bornung (15), examined the effect of organic inhibitors on anodic film growth. He found that some inhibitors desorb at more positive potentials where anodic film growth occurs, whereas others, e.g., stearic acid, react with steel and any oxide on its surface.

D. Stress Corrosion. Ellipsometry has been applied to stress corrosion studies because some proposed mechanisms of stress corrosion involve a film as an integral part of the stress corrosion process. One mechanism considers stress corrosion as the formation and fracture of a brittle film. The study of the kinetics of film formation on brasses in ammoniacal solutions by Green et al. (37) described in an earlier section was aimed at examining the validity of such a mechanism. Another mechanism relates stress corrosion susceptibility to the rate of repassivation of a passive film fractured when a metal is stressed. A technique recently developed by Ambrose and Kruger (2) called "tribo-ellipsometry" applies ellipsometry to a measurement of this rate of repassivation. The apparatus shown in Fig. 20 allows one to simulate the production of bare surface that results from film fracture by abrading the metal surface studied *in situ* while maintaining the potential at a value relevant to stress corrosion conditions and then recording on an oscilloscope the ellipsometric and current transients taking place during repassivation. Combining the ellipsometric measurements with the current measurements is especially valuable because one can determine how much of the total current passed during repassivation produces film formation and how much produces metal dissolution.

Ellipsometry has also been applied to the determination of the ductility and mechanical properties of oxide films (29), which may

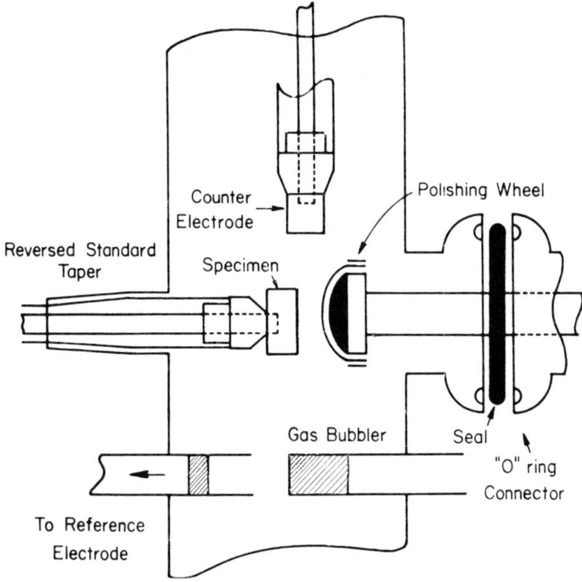

Fig. 20. Apparatus for studying repassivation kinetics by removing the film on an electrode by the rotating abrasion wheel shown, springing the wheel back rapidly and measuring the rate of film formation by transient ellipsometry. From Ambrose and Kruger (2).

have relevance to the film rupture mechanism of stress corrosion, and to studies of the effect of film thickness on susceptibility (38).

3. *Electrodeposition and Electropolishing*

Very little application of ellipsometry has been made to electrodeposition. It is quite possible to study the early stages of electrodeposition by ellipsometry but no such application could be found in the literature. That it is possible to study this process by ellipsometry can be seen by considering studies on the vapor deposition of metal on a metal substrate. There should be no difference between this process and electrodeposition as far as ellipsometry is concerned. An example of an ellipsometric study of metal deposition is that of Melmed and Carroll (66). They vapor deposited iron on a tungsten single crystal substrate and were able to measure changes in the refractive index n and the absorption coefficient κ ($\kappa = k/n$) of the iron-tungsten substrate as a function of number of doses of iron vapor. The thickness of the iron film could be determined up to the point where n and κ approached the values for bulk iron—about 400 Å. This result shows one of the limitations for the application of ellipsometry to electrodeposition. One can only study the thin layers that form in the beginning of the process. Within this limitation, though, the kinetics of electrodeposition and the optical properties of the film can be studied in the same manner as anodic film formation.

The surface properties of thick deposited films can be studied if one knows the optical constants for the bulk deposited metal. Figure 21 from Melmed and Carroll's work shows that the as-deposited layers do not have the same optical constants as the bulk metal. Only by annealing does the deposited layer approach bulk values. Therefore, ellipsometry offers an opportunity to measure the dif-

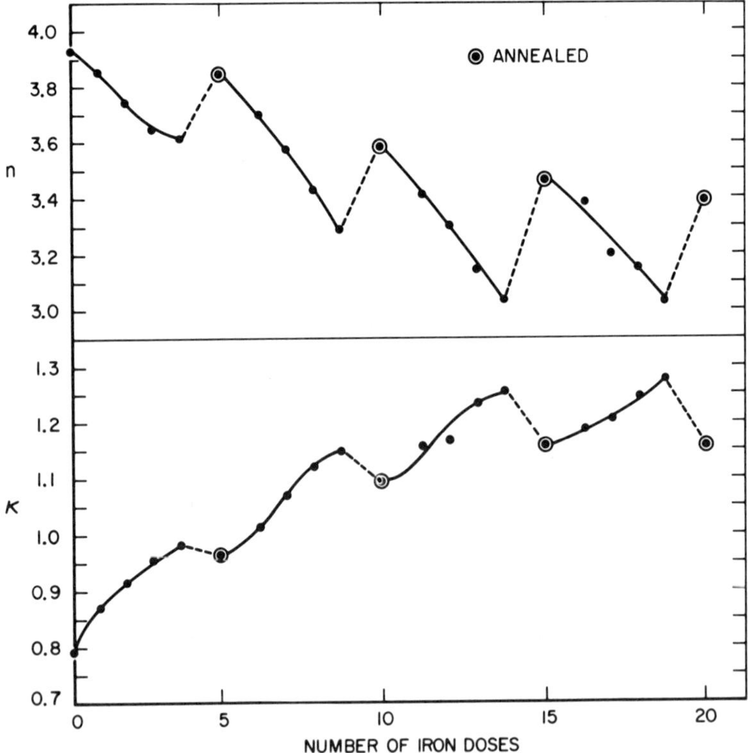

Fig. 21. Measured index of refraction (n) and absorption coefficient (κ) at $\lambda = 5461$ Å as a function of iron doses. The specimen was annealed, as shown, after each 5 doses. From Melmed and Carroll (66).

ference between the properties of a deposited metal surface and those of another surface of the same bulk metal. Thus the surface properties of even thick electrodeposits can be studied by ellipsometry as long as they remain specularly reflecting. If they do not remain smooth, it may be possible to measure topographical changes during electrodeposition by using the theory that treats the effect of roughening on ellipsometry measurements (30).

Topographical changes are also a concern of electropolishing. As with electrodeposition, very little attention has been paid to the application of ellipsometry to a study of this process. A notable exception is the study of the electropolishing of copper by Novak, Reddy, and Wroblowa (71). This study was directed at determining whether the visible viscous layer produced in the electropolishing of copper is in contact with a film-free metal surface, a surface with an adsorbed layer of oxygen or ions, or a surface with a thick reaction layer of oxide or salt. Prior to the application of ellipsometry none of the *in situ* methods used to attack this problem were capable of providing direct evidence for the existence of the film sandwiched between the viscous layer and the metal substrate or its thickness and nature. In order to apply ellipsometry to study a surface being electropolished, Novak et al. had to overcome the difficulty of determining whether their measurements were looking at a single or a duplex film.

They accomplished this objective by displacing the visible viscous film with glycerol. Prior to this displacement, Δ and ψ values were obtained under steady state polishing conditions. The glycerol was introduced while maintaining the potential at the polishing value with a potentiostat. While the anodic current ceased, potential control was maintained and ellipsometric measurements were made on the electrode in contact with the organic liquid, with its surface protected from the atmosphere.

If the assumption that no changes occurred on the electrode after removal of the viscous layer is correct, such a technique can determine whether a film is between the viscous layer and the metal. They found a solid film of around 40 Å and were able to estimate that the viscous layer was 2000–3500 Å thick. Moreover, they were able to establish that the solid film was completely or partially destroyed when the electrode was removed, washed, and dried. Thus only an *in situ* method such as ellipsometry would be capable of detecting the solid film sandwiched between the viscous layer and the electrode surface.

IV. Concluding Remarks

It is evident from the foregoing discussion that ellipsometry has the valuable advantage of being able to make completely different, independent measurements on an electrode surface while the more usual electrochemical parameters are measured. Thus, even as a qualitative tool, as an adjunct to electrochemical measurements, it can contribute greatly to the understanding of electrochemical processes. It is, of course, more. As detailed in what has gone before, it can in many instances make quantitative measurements of film thickness, rate of film formation, and optical properties. Its limitations, however, also lie in the quantitative realm. For thin absorbing films especially, one must be extremely careful not to attach too much quantitative significance to ellipsometric determinations of film properties. As stressed repeatedly, the Drude equations can lead to ambiguous results. Moreover, for monolayer and less than monolayer films it is even doubtful that the Drude equations, which consider films as continua, are even applicable. Many workers in ellipsometry, rightfully impressed by its superb sensitivity to atomic events on a surface, sometimes ignore these limitations and extract numbers whose validity may not always be substantial. As with every technique, if its limitations are recognized and borne in mind in interpreting data, ellipsometry can provide powerful insights into many electrochemical processes.

Moreover, future developments can even further enhance its promise. Possible directions for the future, some well on their way, will certainly include the following: (1) further development of the optical theory of monolayers, "atomic Drude equations," (2) greater use of ellipsometry at many wavelengths, "ellipsometric-spectroscopy," (3) ellipsometry of small features on an electrode, "micro-ellipsometry," and (4) fully automatic high speed ellipsometry, "transient-ellipsometry." The bringing to fruition of these and other future developments will further amplify the already considerable contributions of ellipsometry to electrochemistry.

References

1. Ambrose, J. R., and J. Kruger, 1969, *Proc. Fourth Intern. Cong. Met. Corros.*, N.A.C.E., Houston, Texas, 1972, p. 698.
2. Ambrose, J. R., and J. Kruger, *Corrosion*, **28**, 30 (1972).
3. Archer, R. J., *J. Opt. Soc. Amer.*, **52**, 970 (1962).
4. Archer, R. J., "Ellipsometry in the Measurement of Surfaces and Thin

Films," *Symposium Proceedings*, E. Passaglia, R. R. Stromberg, and J. Kruger, Eds., Nat. Bur. Stand. Misc. Pub. 256, 1964, p. 255.
5. Archer, R. J., and G. W. Gobeli, *J. Phys. Chem. Solids*, **26**, 343 (1965).
6. Balashava, N. A., and V. E. Kasarinov, *Electrokhimya*, **1**, 512 (1965).
7. Barrett, M. A., and A. B. Winterbottom, *First Intern. Cong. Met. Corros., London*, Butterworths, London, 1962, p. 657.
8. Barrett, M. A., "Ellipsometry in the Measurement of Surfaces and Thin Films," *Symposium Proceedings*, E. Passaglia, R. R. Stromberg, and J. Kruger, Nat. Bur. Stand. Misc. Pub. 256, 1964, p. 213.
9. Bartell, L. S., and J. F. Betts, *J. Phys. Chem.*, **64**, 1075 (1960).
10. Bashara, N. M., A. B. Buckman, and A. C. Hall, eds., "Recent Developments in Ellipsometry," *Symposium Proceedings, University of Nebraska*, North-Holland, Amsterdam, 1969. Also printed in *Surf. Sci.*, **16**, (1969).
11. Bockris, J. O'M., M. A. V. Devanathan, and A. K. N. Reddy, *Proc. Roy. Soc., A*, **279**, 327 (1964).
12. Bockris, J. O'M., A. K. N. Reddy, and B. Rao, *J. Electrochem. Soc.*, **113,** 1133 (1966).
13. Bockris, J. O'M., M. A. Genshaw, and V. Brusic, *Symposia Faraday Soc.*, **4**, 177 (1970).
14. Bornung, R. J., and P. Martin, *J. Phys. Chem.*, **71**, 3731 (1967).
15. Bornung, R. J., Tech. Rept.—12 RE TR 70–129, U.S. Army Weapons Command Science and Technology Laboratory, March, 1970.
16. Brusic, V., M. A. Genshaw, and B. D. Cahan, *Appl. Opt.*, **9**, 1634 (1970).
17. Brusic, V., M. A. Genshaw, and J. O'M. Bockris, submitted to *Surf. Sci.*
18. Buckman, A. B., "Recent Developments in Ellipsometry," Symposium Proceedings, University of Nebraska, N. M. Bashara, A. B. Buckman, and A. C. Hall, Eds., North-Holland, Amsterdam, 1969. Also printed in *Surf. Sci.*, **16**, 193 (1969).
19. Budrys, R., Submitted to *J. Electrochem. Soc.*
20. Cahan, B. D., and R. F. Spanier, "Recent Developments in Ellipsometry," *Symposium Proceedings, University of Nebraska*, N. M. Bashara, A. B. Buckman, and A. C. Hall, eds., North-Holland, Amsterdam, 1969. Also printed in *Surf. Sci.*, **16**, 166 (1969).
21. Carroll, J. J., and A. J. Melmed, *Surf. Sci.*, **16**, 251 (1969).
22. Carroll, J. J., and A. J. Melmed, *J. Opt. Soc. Amer.*, **61**, 470 (1971).
23. Cathcart, J. J., and C. F. Petersen, "Ellipsometry in the Measurement of Surfaces and Thin Films," *Symposium Proceedings*, E. Passaglia, R. R. Stromberg, and J. Kruger, Eds., Nat. Bur. Stand. Misc. Pub. 256, 1964, p. 201.
24. Chiu, Y. C., and M. A. Genshaw, *J. Phys. Chem.*, **72**, 4325 (1968).
25. Chiu, Y. C., and M. A. Genshaw, *J. Phys. Chem.*, **73**, 3571 (1969).
26. Dell'Oca, C. J., and L. Young, "Recent Developments in Ellipsometry," *Symposium Proceedings, University of Nebraska*, N. M. Bashara, A. B. Buckman, and A. C. Hall, Eds., North-Holland Amsterdam, 1969. Also printed in *Surf. Sci.*, **16**, 331 (1969).
27. Dell'Oca, C. J., G. Yan, and L. Young, *J. Electrochem. Soc.*, **118**, 89 (1971).
28. Dettorre, J. F., T. G. Knorr, and D. A. Vaughan, "Ellipsometry in the Measurement of Surfaces and Thin Films," *Symposium Proceedings*, E. Passaglia, R. R. Stromberg, and J. Kruger, Eds., Nat. Bur. Stand. Misc. Publ. 256, 1964, p. 245.

29. Escalante, E., and J. Kruger, Nat. Bur. Stand. Rept. 10594, 1971.
30. Fenstermaker, C. A., and F. L. McCrackin, "Recent Developments in Ellipsometry," *Symposium Proceedings, University of Nebraska*, N. M. Bashara, A. B. Buckman, and A. C. Hall, Eds., North-Holland, Amsterdam, 1969. Also printed in *Surf. Sci.*, **16**, 85 (1969).
31. Garnett, M., *Phil. Trans. Roy. Soc. London*, **203**, 385 (1904); **205**, 237 (1906). See also O. S. Heavens, *Optical Properties of Thin Solid Films*, Butterworths, London, 1955, p. 177.
32. Genshaw, M. A., and R. S. Sirohi, *J. Electrochem. Soc.*, **118**, 1558 (1971).
33. Germer, L. H., R. M. Stern, and A. U. MacRae, *Metal Surfaces; Structure, Energetics, and Kinetics*, ASM and Met. Soc. AIME, 1963, p. 287.
34. Germer, L. H., and J. W. May, *Surf. Sci.*, **4**, 452 (1966).
35. Gilman, S., *J. Phys. Chem.*, **68**, 2112 (1964).
36. Goswami, K. N., and R. W. Staehle, *Electrochim. Acta.*, **16**, 1895 (1971).
37. Green, J. A. S., H. D. Mengelberg, and H. T. Yolken, *J. Electrochem. Soc.*, **117**, 433 (1970).
38. Green, J. A. S., and A. J. Sedriks, *Metallurgical Transactions*, **2**, 1807 (1971).
39. Hall, A. C., *J. Phys. Chem.*, **70**, 1702 (1966).
40. Hansen, W. N., and A. Prostak, *Phys. Rev.*, **174**, 500 (1968).
41. Hansen, W. N., "Recent Developments in Ellipsometry," *Symposium Proceedings, University of Nebraska*, N. M. Bashara, A. B. Buckman, and A. C. Hall, Eds., North-Holland, Amsterdam, 1969. Also printed in *Surf. Sci.* **16**, 205 (1969).
42. Harvey, W. W., and J. Kruger, *Electrochim. Acta.*, **16**, 2017 (1971).
43. Hass, G., and L. Hadley, *American Institute of Physics Handbook*, McGraw-Hill, New York, 1963, pp. 6–103–6–122.
44. Hayfield, P. C. S., *First Intern. Congr. Met. Corros., Lond.*, Butterworths, London, 1962, p. 663.
45. Hayfield, P. C. S., *American Institute of Physics Handbook*, McGraw-Hill, New York, 1963, p. 370.
46. Hayfield, P. C. S., and G. W. T. White, "Ellipsometry in the Measurement of Surfaces and Thin Films," *Symposium Proceedings*, E. Passaglia, R. R. Stromberg, and J. Kruger, Eds., Nat. Bur. Stand. Misc. Pub. 256, 1964, p. 157.
47. Hayfield, P. C. S., *Surface Phenomena of Metals*, Society of Chemical Industry Monograph No. 28, 1968, p. 128.
48. Hayfield, P. C. S., "Recent Developments in Ellipsometry," *Symposium Proceedings, University of Nebraska*, N. M. Bashara, A. B. Buckman, and A. C. Hall, Eds., North-Holland, Amsterdam, 1969. Also printed in *Surf. Sci.*, **16**, 126 (1969).
49. Juenker, D. W., L. J. LeBlanc, and C. R. Martin, *J. Opt. Soc. Amer.*, **58**, 164 (1968).
50. Kruger, J., *J. Electrochem. Soc.*, **108**, 504 (1961).
51. Kruger, J., *J. Electrochem. Soc.*, **110**, 654 (1963).
52. Kruger, J., *Corrosion*, **22**, 88 (1966).
53. Kruger, J., and J. P. Calvert, *J. Electrochem. Soc.*, **114**, 43 (1967).
54. Kruger, J., and J. P. Calvert, unpublished work.
55. Kruger, J., and P. C. S. Hayfield, *Handbook on Corrosion Testing and Evaluation*, W. H. Ailor, Ed., Wiley, New York, 1971, p. 783.

56. Kudo, K., G. Okamoto, *Corr. Eng.* **17**, (3) (1968).
57. Kudo, K., N. Sato, and G. Okamoto, *Bull. Fac. Eng. Hokkaido Univ.*, **47**, 141 (1968).
58. Kudo, K., T. Shibata, G. Okamoto, and N. Sato, *Corros. Sci.*, **8**, 809 (1968).
59. Lukes, F., W. Knausenberger, and K. Vedam, *Surf. Sci.*, **16**, 112 (1969).
60. Mansfeld, F., T. Smith, and E. P. Parry, *Corrosion*, **27**, 289 (1971).
61. Mathot, C., and A. Rothen, "Recent Developments in Ellipsometry," *Symposium Proceedings, University of Nebraska*, N. M. Bashara, A. B. Buckman, and A. C. Hall, Eds., North-Holland, Amsterdam, 1969. Also printed in *Surf. Sci.*, **16**, 428 (1969).
62. McBee, C. L., and J. Kruger, "Recent Development in Ellipsometry," *Symposium Proceedings, University of Nebraska*, N. M. Bashara, A. B. Buckman, and A. C. Hall, Eds., North-Holland, Amsterdam, 1969. Also printed in *Surf. Sci.*, **16**, 340 (1969).
63. McBee, C. L., and J. Kruger, *Nat., Phys. Sci.*, **230**, 194 (1971).
64. McCrackin, F. L , and J. P. Colson, "Recent Developments in Ellipsometry," *Symposium Proceedings, University of Nebraska*, N. M. Bashara, A. B. Buckman, and A. C. Hall, Eds., North-Holland, Amsterdam, 1969. Also printed in *Surf. Sci.*, **16**, 61 (1969).
65. Melmed, A. J., H. P. Layer, and J. Kruger, *Surf. Sci.*, **9**, 476 (1968).
66. Melmed, A. J., and J. J. Carroll, *Surf. Sci.*, **19**, 243 (1970).
67. Melmed, A. J., and J. J. Carroll, To be published.
68. Meyer, F., and G. A. Bootsma, "Recent Developments in Ellipsometry," *Symposium Proceedings, University of Nebraska*, N. M. Bashara, A. B. Buckman, and A. C. Hall, Eds., North-Holland, Amsterdam, 1969. Also printed in *Surf. Sci.*, **16**, 221 (1969).
69. Miller, J. R., and J. E. Berger, *J. Phys. Chem.*, **70**, 3070 (1966).
70. Moelwyn-Hughes, E. A., *Physical Chemistry*, Pergamon Press, New York, 1961, p. 382.
71. Novak, M., A. K. N. Reddy, and H. Wroblowa, *J. Electrochem. Soc.*, **117**, 733 (1970).
72. Ord, J. L., and D. J. DeSmet, *J. Electrochem. Soc.*, **113**, 1258 (1966).
73. Ord, J. L., "Recent Developments in Ellipsometry," *Symposium Proceedings, University of Nebraska*, N. M. Bashara, A. B. Buckman, and A. C. Hall, Eds., North-Holland, Amsterdam, 1969. Also printed in *Surf. Sci.*, **16**, 147 (1969).
74. Ord, J. L., and D. J. DeSmet, *J. Electrochem. Soc.*, **118**, 206 (1971).
75. Paik, W., M. A. Genshaw, and J. O'M. Bockris, *J. Phys. Chem.*, **74**, 4266 (1970).
76. Paik, W., and J. O'M. Bockris, *Surf. Sci.*, **27**, 191 (1971).
77. Paik, W., and J. O'M. Bockris, *Surf. Sci.*, **28**, 61 (1971).
78. Passaglia, E., R. R. Stromberg, and J. Kruger, Eds., "Ellipsometry in the Measurement of Surfaces and Thin Films," *Symposium Proceedings*, Nat. Bur. Stand. Misc Pub. 256, p. 229, 1964.
79. Reddy, A. K. N., and J. O'M. Bockris, "Ellipsometry in the Measurement of Surfaces and Thin Films," *Symposium Proceedings*, E. Passaglia, R. R. Stromberg and J. Kruger, Eds., Nat. Bur. Stand. Misc. Publ. 256, 1964, p. 229.

80. Reddy, A. K. N., M. A. Genshaw, and J. O'M. Bockris, *J. Chem. Phys.*, **48**, 671 (1968).
81. Reid, W. E., and J. Kruger, *Nature*, **203**, 4943, 402 (1964).
82. Sato, N., K. Kudo, and T. Noda, *Corros. Sci.*, **10**, 785 (1970).
83. Schopper, H., in Landolt-Bornstein, Zahlenwerte und Funktionen aus Physik, Chem., Astron., Geophys., Tech. (6) **2**, 8, 1–2 to 1–36 (1962).
84. Sirohi, R. S., and M. A. Genshaw, *J. Electrochem. Soc.*, **116**, 910 (1969).
85. Sladkova, J., *Czech. J. Phys.*, **B18**, 801 (1968).
86. Smith, L. E., and R. R. Stromberg, *J. Opt. Soc. Amer.*, **56**, 1539 (1966).
87. Smith, L. E., C. A. Fenstermaker, and R. R. Stromberg, Amer. Chem. Soc., Papers Presented at Chicago 160th Meeting, **30**, 527 (1970).
88. Smith, T., *J. Opt. Soc. Amer.*, **58**, 1069 (1968).
89. Stedman, M., *Chem. Phys. Lett.*, **2**, 457 (1968).
90. Steiger, R. F., J. M. Morabito, Jr., G. A. Somorjai, and R. H. Muller, *Surf. Sci.*, **14**, 279 (1969).
91. Strachan, C., *Proc. Cambridge Phil. Soc.*, **29**, 116 (1933).
92. Stromberg, R. R., E. Passaglia, and D. J. Tutas, *J. Res. NBS*, **67A**, 431 (1963).
93. Stromberg, R. R., E. Passaglia, and D. J. Tutas, "Ellipsometry in the Measurement of Surfaces and Thin Films," *Symposium Proceedings*, E. Passaglia, R. R. Stromberg, and J. Kruger, Eds. Nat. Bur. Stand. Misc. Pub. 256, 1964, p. 281.
94. Stromberg, R. R., D. J. Tutas, and E. Passaglia, *J. Phys. Chem.*, **69**, 3955 (1965).
95. Stromberg, R. R., *Interface Conversion for Polymer Coatings*, P. Weiss and G. D. Cheever, Eds., American Elsevier, New York, 1969, pp. 321–337.
96. Tronstad, L., "Optische Untersuchungen zur Frage der Passivitat des Eisens und Stahls," Kgl. Norske Videnskab Selsk. Skr., **1**, (1931).
97. Tronstad, L., *Trans. Faraday Soc.*, **29**, 502 (1933).
98. Uhlig, H. H., Corrosion and Corrosion Control, Wiley, New York, 1963, p. 58.
99. Vedam, K., W. Knausenberger, and F. Lukes, *J. Opt. Soc. Amer.*, **59**, 64 (1969).
100. Vroman, L., "Ellipsometry in the Measurement of Surfaces and Thin Films," *Symposium Proceedings*, E. Passaglia, R. R. Stromberg, and J. Kruger, Eds. Nat. Bur. Stand. Misc. Pub. 256, 1964, p. 335.
101. Vroman, L., and A. L. Adams, "Recent Developments in Ellipsometry," *Symposium Proceedings*, *University of Nebraska*, N. M. Bashara, A. B. Buckman, and A. C. Hall, Eds., North-Holland, Amsterdam, 1969. Also printed in *Surf. Sci.*, **16**, 438 (1969).
102. Winterbottom, A. B., *Optical Studies of Metal Surfaces*, Kgl. Norske Videnskab. Selskabs Skrifter, 1, F. Bruns Bokhandel, Trondheim 1955.
103. *Ibid.*, p. 46.
104. Winterbottom, A. B., "Ellipsometry in the Measurement of Surfaces and Thin Films," *Symposium Proceedings*, E. Passaglia, R. R. Stromberg, and J. Kruger, Eds., Nat. Bur. Stand. Misc. Pub. 256, 1964, p. 97.
105. Yolken, H. T., and J. Kruger, *J. Opt. Soc. Amer.*, **55**, 842 (1965).
106. Young, L., and F. G. R. Zobel, *J. Electrochem. Soc.*, **113**, 277 (1966).

Double Beam Interferometry for Electrochemical Studies

ROLF H. MULLER

Inorganic Materials Research Division, Lawrence Berkeley Laboratory and Department of Chemical Engineering; University of California Berkeley, California

Contents

I. Introduction

1. Principles of Interferometry

Interferometry provides an elegant and sensitive tool for the study of refractive index fields. Local refractive index variations of electrochemical interest are due to variations in composition and concentration. Light is particularly well suited as a probe for the study of concentration fields, since it permits continuous observation of an object with a minimum of disturbance, and offers good geometrical resolution. Interferometer measurements usually result in pictures that are intuitively understandable, at least in a semiquantitative way.

The refractive index of a solution depends on its composition and is defined as the ratio of the phase velocity c of light in vacuum to the velocity v in the medium

$$n = \frac{c}{v} \tag{1}$$

Since refractive index of a medium is affected by all its constituents, as well as its temperature (and pressure), it is desirable to restrict the use of interferometry of concentration fields to one-component, isothermal solutions.

A plane wave front of light, entering a medium with locally variable refractive index, will not remain plane, because light travels more slowly in regions of higher refractive index. The resulting local variation in phase is proportional to the refractive index variation Δn and the geometrical path length d over which it exists. It is, therefore, convenient to define the product of geometical path length d and refractive index n as the optical path length p

$$p = nd \tag{2}$$

For fixed geometrical path length, the optical path difference is

$$\Delta p = (\Delta n)d \tag{3}$$

The optical path difference can also be expressed in number of wavelengths by use of the wavelength λ_0 of light in vacuum

$$\Delta z = \frac{\Delta p}{\lambda_0} \tag{4}$$

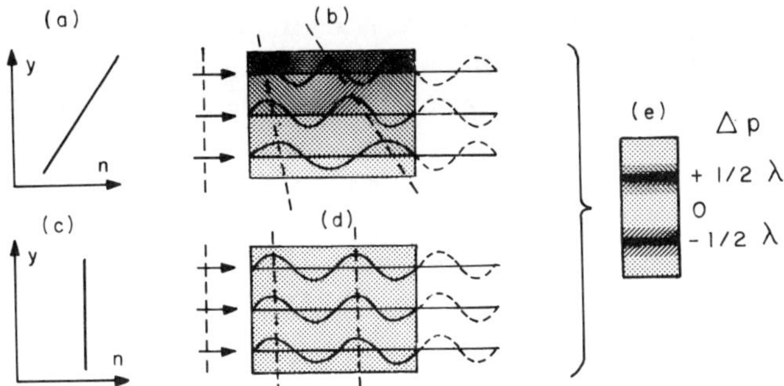

Fig. 1. Principle of double beam interferometry: (*a*) variable refractive index in specimen cell, (*b*) specimen cell with different parts of a plane light wave which enter the cell in phase and get increasingly out of phase during propagation through the cell, (*c*) uniform refractive index in reference cell, (*d*) reference cell with different parts of a plane wave which stay in phase during propagation, (*e*) interference pattern due to the superposition of specimen and reference waves; local optical path difference in wavelengths.

Similarly, a corresponding phase difference $\Delta\phi$ is formulated as

$$\Delta\phi = (\Delta z)2\pi \tag{5}$$

Figure 1 illustrates schematically the effect of a one-dimensional refractive index field (*a*), in a stratified specimen tank (*b*), on the propagation of light through it. Due to the local variation in wavelength, different parts of a wave entering the cell in phase, increasingly get out of phase, as they propagate through the cell, while a similar wave, propagating through a reference tank (*d*) of uniform refractive index (*c*), remains in phase. In Fig. 1, it is assumed that the optical path length in the center of both tanks is the same and that they differ by plus and minus half a wavelength at both edges. A suitable superposition of the light beams issuing from both tanks would result in alternate dark and bright interference fringes, with constructive interference occurring in the center and destructive interference at the edges, as indicated in (*e*). It is the purpose of a double beam interferometer to provide the two light beams, which possess a stationary point-by-point phase relationship, and to superimpose the exiting beams in such a way that both interference phenomenon and object are imaged. Thus local variations of phase due to differences between object and reference are made visible as intensity variations (usually in the form of interference fringe displacements), and can be quantitatively evaluated.

The principle illustrated above is characteristic of double beam interferometry, where two light beams are brought to interference. Typically, one beam has traversed the specimen, the other serves as a reference with a simple (usually plane) wave front. In general, the specimen beam is spacially separated from the reference beam and traverses the specimen once (occasionally twice). Due to the simple optical principles of double beam interferometry, measurements derived by this technique are more amenable to the analysis and correction of optical errors than measurements obtained by multiple beam interferometry, where interference is due to the superposition of an infinite series of waves with multiple passes through the specimen. The instrumentation required for double beam interferometry is, however, distinctly more complex.

It is characteristic of optical techniques that the propagation direction of the light is not equivalent to the two other dimensions of space. As a consequence, it is desirable that the light encounters over its entire path in the specimen the same value of the variable to be measured. The compound effect that results if areas of different properties are successively traversed can often not be resolved satisfactorily. Thus Interferometry is best suited for the observation of one- or two-dimensional refractive index fields.

As can be seen in Fig. 1, light propagating through a medium with refractive index variations across the beam suffers a tilt in wave front (dashed lines connecting crests of waves). Since the propagation direction is oriented normal to the wave front, the beam is deflected from its original propagation direction. This effect is utilized in Schlieren-optical devices.

An optical path difference, in addition to being due to a refractive index variation, can also be caused by a change in geometrical path, with constant refractive index of the medium. Thus double beam interferometry with reflected light can be used for determining the microtopography of electrode surfaces. In this case, local phase variations are introduced by the different geometrical position of reflecting surface elements.

Holographic interferometry is a special case of double beam interferometry. It can be applied to the observation of refractive index fields or surface topographies and is discussed in a separate chapter.

It is the purpose of this chapter to discuss the optical principles and to analyze some of the potentials and limitations of double beam interferometry for the mapping of refractive index fields and, to a lesser degree, surface topographies of electrochemical interest.

Examples of instrumentation and application have been chosen on the basis of their interest for future development rather than historical completeness.

2. Electrochemical Refractive Index Fields

Pure diffusion in solutions provides some of the best-defined refractive index fields for observation by interferometry. For binary solutions, the concentration is, in general, uniquely determined by the refractive index, which facilitates the derivation of diffusion coefficients from interferometer data.

In zone electrophoresis, the complete or partial separation of constituents results in refractive index fields that can be observed by interferometry. Components can be identified by their mobility and their amount is determined from the refractive index change.

The study of mass transfer boundary layers due to convective diffusion and migration near electrode surfaces offers some particularly rewarding uses of interferometry. Some interferometers for the study of diffusion layers, together with other techniques, have been discussed by Ibl (67). Electrochemical reactions across a solid-liquid interface, proceeding at a finite rate, require the transport of reactants and products to and from the interface. As a result, mass transfer boundary layers, i.e., regions in which the composition of the fluid phase is different from that in the bulk, are formed along the interface. A typical dimension of such mass transfer boundary layers would be 0.1 mm. Since mass transfer is often the limiting factor for the specific output of electrochemical reactors, a detailed understanding of mass transfer boundary layers is essential to the design and operation of many electrochemical devices. Local mass transfer boundary layers can be observed by interferometric techniques under conditions that are not restricted to operation at limiting current. The sensitivity to concentration changes typically is in the order of 10^{-4} moles/liter, the geometrical resolution can be in the range of microns. The observation of electrochemical mass transfer boundary layers presents a special challenge to the use of interferometry, because of the small dimensions and large gradients typically involved. Some of these problems have been of particular interest to the author. Mass transfer boundary layers are, therefore, referred to most often in the discussions to follow. However, most of the optical analysis applies equally to other electrochemical situations and, with appropriate modifications, even to applications in the areas of heat transfer or gas dynamics.

II. Refractive Index and Composition

1. Empirical Correlations

Variations in the refractive index of a solution, determined by interferometry, need to be interpreted as variations in composition. For this purpose, it is necessary to know the relation between the two variables. This relation is most firmly established by refractive index measurement of solutions of known composition. Algebraic expressions can then be developed to represent the data.

The concentration of most isothermal solutions of a single solute, such as copper sulfate in water (Fig. 2*a*), can be unequivocally determined from the refractive index. For some solutions, such as sulfuric acid (Fig. 2*b*), this relationship may not be unique over certain concentration ranges. The composition of multicomponent solutions, such as aqueous copper sulfate and sulfuric acid (Fig. 3*a*), cannot be determined from a single refractive index measurement. The number of measurements necessary to establish refractive index correlations of multicomponent solutions at different temperatures requires a considerable effort that is multiplied if data for different

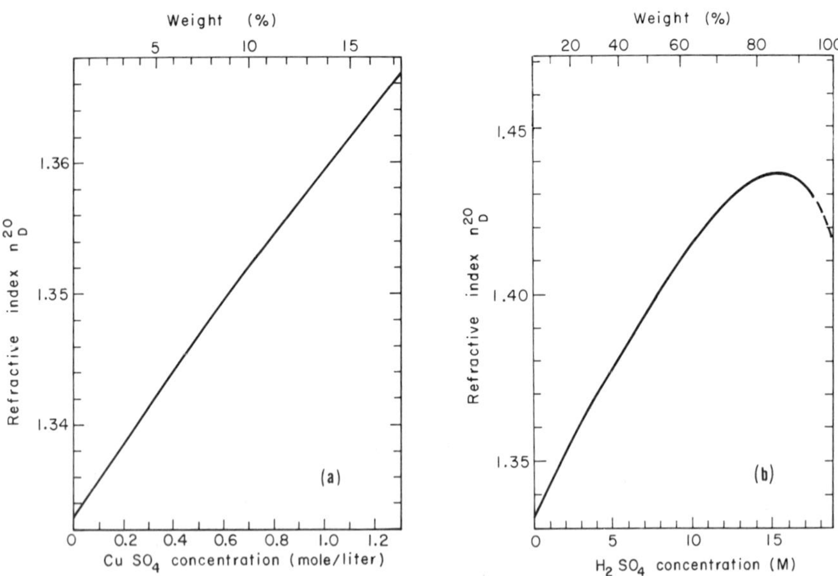

Fig. 2. Concentration-dependence of the refractive index of aqueous solutions: (*a*) copper sulfate at 20°C (116), (*b*) sulfuric acid at 20°C (116).

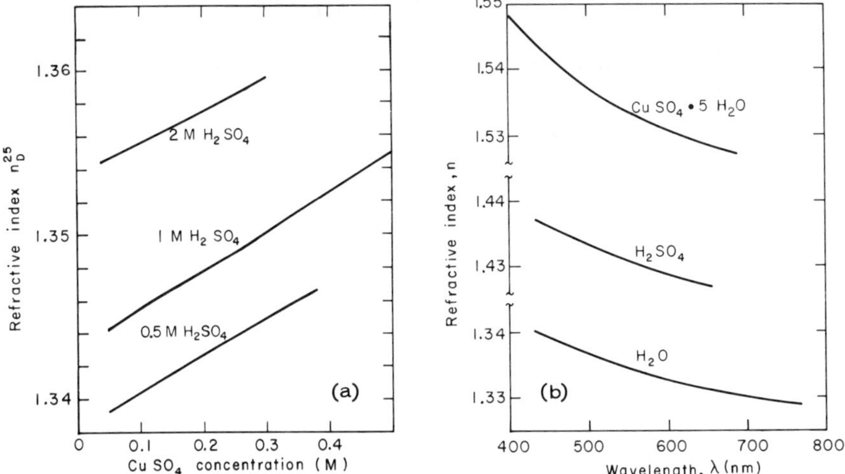

Fig. 3. (*a*) Refractive index of mixtures of sulfuric acid and copper sulfate (124). (*b*) Refractive indices at different wavelengths, literature data for H_2SO_4, 99.5%, 23°C (116); $CuSO_4 \cdot 5H_2O$ (101); H_2O (116).

wavelengths are desired. Due to inevitable deviations in composition from the desired values, the results have to be correlated by multiple regression analysis (124). Some wavelengths of interest in interferometry are listed in Table I. The yellow sodium D light has been most commonly used in refractometry. Refractive index data for that light are often designated by a subscript D. The average wavelength between both lines of the doublet is listed in Table I.

TABLE I
Typical Wavelengths of Light Used in Interferometry

Source	Wavelength, nm
Hg arc	404.7
Hg arc	435.8
Ar laser	488.0
Ar laser	514.5
Tl arc	535.0
Hg arc	546.1
Kr laser	568.2
Na arc	589.3
HeNe laser	632.8
Kr laser	647.1
Ruby laser	694.3

2. *Molar Refractivity*

A useful concept for the estimation of the refractive index from density data is the molar refractivity P, with can be defined (12) in terms of refractive index n, molecular weight M, and density ρ as

$$P = \frac{M}{\rho}\frac{n^2 - 1}{n^2 + 2} \tag{6}$$

The molar refractivity is best determined from one set of refractive index and density measurements. Less reliably, it can also be obtained by the addition of atomic refractivities and contributions due to the bonding between the constituent elements in a compound (107,149). The molar refractivity of gases has been shown to be remarkably independent of density over a hundredfold pressure range (12). The values for water at different temperatures, shown in Table II, shows a slightly larger variation. As a result, the

TABLE II
Molar Refractivity of Water at Different Temperatures

Temperature °C	Density ρ g/cm³ (61)	Refractive index n_D (61)	Molar refractivity P (Eq. 6)
20	0.99823	1.33299	3.71228
30	0.99567	1.33192	3.71097
40	0.99224	1.33051	3.70942
50	0.98807	1.32894	3.70899
60	0.98324	1.32718	3.70907
70	0.97781	1.32511	3.70819
80	0.97183	1.32287	3.70759
90	0.96534	1.32050	3.70753

refractive indices at different temperatures, computed from the densities at those temperatures and the molar refractivity at 20°C, differ somewhat from the measured refractive indices (Table III). The relative error of the predicted refractive index change decreases from 10 to 5% with increasing temperature change.

Tabulated values of the quantity $(n^2 - 1)/(n^2 + 2)$ are provided in Table IV for the use of Eq. 6. They are well suited for linear interpolation.

For multicomponent systems, an average molar refractivity $\bar{P}$ can

TABLE III
Prediction of the Refractive Index of Water at Different Temperatures Based on the Densities Given in Table II and the Molar Refractivity at 20°C

Temperature °C	Refractive index, n_D Predicted (Eq. 6)	Measured (61)
20	1.3330	1.3330
30	1.3320	1.3319
40	1.3308	1.3305
50	1.3293	1.3289
60	1.3275	1.3272
70	1.3255	1.3251
80	1.3233	1.3229
90	1.3210	1.3205

TABLE IV
Values of the Quantity $(n^2 - 1)/(n^2 + 2)$ for Use in Calculating Molar Refractivities

n	$\frac{n^2 - 1}{n^2 + 2}$	n	$\frac{n^2 - 1}{n^2 + 2}$
1.330	0.204012	1.350	0.215173
1.331	0.204573	1.351	0.215727
1.332	0.205134	1.352	0.216281
1.333	0.205695	1.353	0.216835
1.334	0.206256	1.354	0.217388
1.335	0.206816	1.355	0.217940
1.336	0.207376	1.356	0.218493
1.337	0.207935	1.357	0.219045
1.338	0.208494	1.358	0.219596
1.339	0.209053	1.359	0.220147
1.340	0.209611	1.360	0.220698
1.341	0.210169	1.361	0.221249
1.342	0.210727	1.362	0.221799
1.343	0.211284	1.363	0.222348
1.344	0.211840	1.364	0.222898
1.345	0.212397	1.365	0.223447
1.346	0.212953	1.366	0.223995
1.347	0.213509	1.367	0.224543
1.348	0.214064	1.368	0.225091
1.349	0.214619	1.369	0.225639

TABLE V

Prediction of the Refractive Index of Sulfuric Acid–Water Mixtures from the Molar Refractivities of the Pure Components Listed in Table VI and the Density of the Mixtures

wt. % H_2SO_4	M H_2SO_4	f H_2SO_4 (Eq. 8)	f H_2O (Eq. 8)	$\bar{P}$ (Eq. 7)	$\bar{M}$ (Eq. 9)	ρ (115)	n_D^{20} Predicted (Eq. 10)	n_D^{20} Literature (116)
2	0.2064	0.0037	0.9963	3.7489	18.3151	1.0118	1.3355	1.3355
5	0.5260	0.0096	0.9904	3.8058	18.7827	1.0317	1.3390	1.3385
9	0.9719	0.0178	0.9822	3.8863	19.4446	1.0591	1.3437	1.3435
17	1.9362	0.0363	0.9637	4.0657	20.9198	1.1168	1.3534	1.3525
35	4.4963	0.0900	0.9100	4.5893	25.2230	1.2599	1.3756	1.3745
70	11.4906	0.2998	0.7002	6.6325	42.0190	1.6105	1.4222	1.4240
90	16.6497	0.6231	0.3769	9.7820	67.9080	1.8144	1.4358	1.4325

be based on the contributions from each constituent of molar refractivity P_i

$$\bar{P} = \sum f_i P_i \tag{7}$$

where f_i is the mole fraction of component i and is defined in terms of the molarities c_i (number of gram moles of solute i per liter of solution) of the constituents

$$f_i = \frac{c_i}{\sum c_i} \tag{8}$$

If an average molecular weight $\bar{M}$ is similarly defined as

$$\bar{M} = \sum f_i M_i \tag{9}$$

Eq. 6 then becomes for a mixture

$$\sum f_i P_i = \frac{\sum f_i M_i}{\rho} \frac{n^2 - 1}{n^2 + 2} \tag{10}$$

where ρ and n are the density and refractive index of the mixture. The refractive index of mixtures (and solutions) can be estimated by use of Eq. 10 from the molar refractivity of the pure components and the density of the mixture. Results of such an estimate for sulfuric acid-water mixtures are shown in Table V. The predicted change in refractive index over that of pure water seems to be within 5% of the measurements over the entire concentration range (except for a 9% deviation at 5 wt.%). Because no refractive index data of pure sulfuric acid could be found in the literature, an extrapolated value, derived from the average molar refractivity of the 80% mixture, was used for determining the molar refractivity of sulfuric acid listed in Table VI.

The same procedure has been applied to predict the refractive index of aqueous copper sulfate solutions. The molar refractivity of the solute (Table VI) has been based on the density of the hydrated

TABLE VI

Determination of the Molar Refractivity P of Pure Components from Molecular Weight M, Density ρ, and Refractive Index n. Data used in the Computations in Tables V, VII, and VIII

Component	M	ρ	n	P
H_2O	18.016	0.99821	1.3330	3.7125
H_2SO_4	98.08	1.8305	1.4162	13.4525
$CuSO_4 \cdot 5H_2O$	249.69	2.284	1.532	33.8757

TABLE VII
Prediction of the Refractive Index of Aqueous Copper Sulfate Solutions from the Molar Refractivities of the Pure Components Listed in Table VI and the Density of the Solutions

		n_D^{20}	
M $CuSO_4$	ρ^{20} (61)	Predicted (Eq. 10)	Literature (116)
0.05	1.007	1.3347	1.3343
0.1	1.015	1.3360	1.3357
0.4	1.062	1.3440	1.3441
0.7	1.108	1.3515	1.3520
1.0	1.153	1.3588	1.3596
1.3	1.198	1.3661	1.3670

crystal and the average of its three principal refractive indices (61). The refractive index of solutions, thus derived, is compared in Table VII with measurements from the literature. As had been observed in the case of sulfuric acid, the relative error in the predicted refractive index change (over that of pure water) is largest at low concentrations. The error decreases to about 3% at concentrations above 0.4 *M*.

The use of the molar refractivities of pure components employed above results in surprisingly good predictions of the refractive

TABLE VIII
Prediction of the Refractive Index of Copper Sulfate–Sulfuric Acid–Water Mixtures from the Molar Refractivities of the Pure Components Listed in Table VI and the Density of the Mixtures

			n_D^{25}	
M $CuSO_4$	M H_2SO_4	ρ^{25} (124)	Predicted (Eq. 10)	Measured (124)
0.0500	0.492	1.0358	1.3393	1.3391
0.0992	0.496	1.0435	1.3406	1.3404
0.2998	0.494	1.0737	1.3454	1.3455
0.0494	0.983	1.0659	1.3445	1.3442
0.1017	1.008	1.0744	1.3457	1.3457
0.2970	0.979	1.1034	1.3508	1.3506
0.5001	0.997	1.1320	1.3546	1.3550
0.0502	2.013	1.1261	1.3543	1.3543
0.2990	2.020	1.1625	1.3599	1.3599

index of copper sulfate–sulfuric acid–water mixtures, considering that no adjustable parameters were used in these predictions (Table VIII). Over the range of compositions for which reliable measurements are available, the relative error in the predicted refractive index change is at most 3%.

3. *Spectral Dependence of Refractive Index*

Refractive indices usually decrease with increasing wavelength (Fig. 3*b*). Any change in sensitivity due to this effect is, however, not significant. The use of shorter wavelengths in interferometry still results in greater sensitivity because of the increased phase difference obtained for a given optical path difference (Eqs. 4 and 5).

Measuring the refractive index of the same solution at different wavelengths, in principle, provides the opportunity to determine the concentration of more than one solute. The optical dispersion has been used for the interferometric determination of temperature and composition of different gas mixtures (118) and plasma diagnostics (63). For example, to determine the concentration of two solutes by this method, a refractive index dependence on composition and wavelength, as illustrated schematically in Fig. 4*a*, is necessary. Unfortunately, the wavelength-dependence of the refractive index of most electrolytes over the visible spectral range is too similar to provide independent measurements. Estimates of the spectral

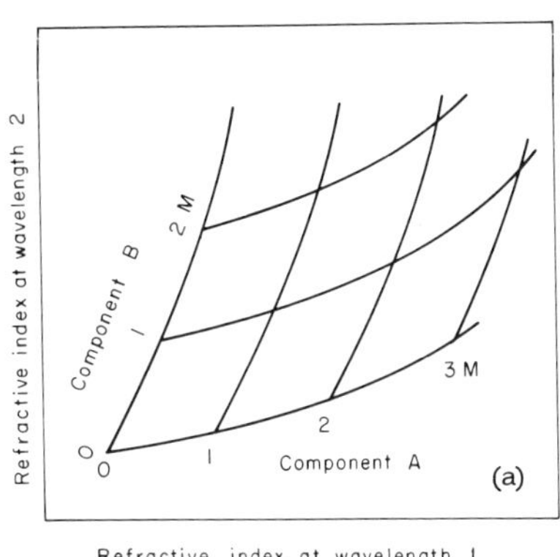

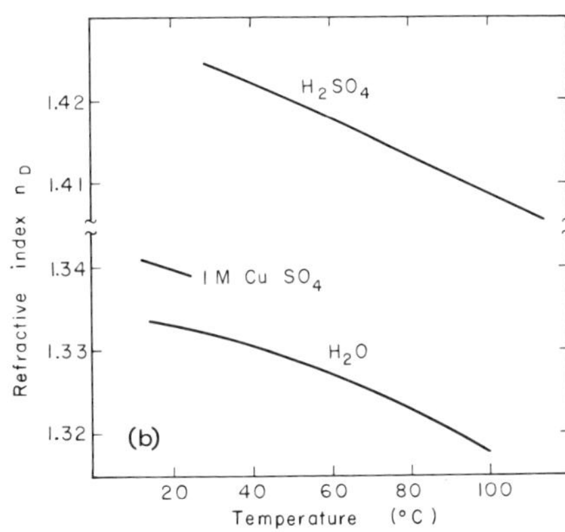

Fig. 4. (*a*) Determination of the concentration of two separate components of a solution from the refractive index at two different wavelengths (schematic). (*b*) Temperature dependence of the refractive index of H_2SO_4, 95–96% (110); 1 *M* KCl (116); 1 *M* $CuSO_4$ (116); H_2O (61).

dependence of the refractive index of aqueous cupric sulfate-sulfuric acid solutions, based on molar refractivities given in Table IX, support this conclusion. The optical dispersion of several gases has similarly been found too small to be useful (149).

TABLE IX
Molar Refractivities at Different Wavelengths Computed from Refractive Indices Shown in Fig. 3*b* and Densities Given in Table VI

Wavelength, λ nm	Molar refractivity, P		
	H_2O	H_2SO_4	$CuSO_4 \cdot 5H_2O$
434	3.7871	14.0375	34.5004
486	3.7539	13.9536	34.2099
589	3.7125	13.8132	33.8438
656	3.6942	13.7569	33.6840

4. *Temperature Dependence of Refractive Index*

Since the effect of temperature on refractive index cannot easily be separated from that of solute content, it is important to know what temperature variations can be tolerated in the analysis of concentration fields.

The temperature coefficient of the refractive index, in general, increases with increasing concentration and temperature. Some examples of refractive indices at different temperatures are given in Fig. 4*b*. For a 1 *M* $CuSO_4$ solution at 18°C, the change in refractive index per °C is 1.88×10^{-4}. A comparison with Fig. 2*a* shows that, at constant temperature the same change in refractive index corresponds to a change in concentration of 7.8×10^{-3} *M*. For most interferometric concentration determinations it is therefore necessary to reduce local temperature variations to well below 1°C.

III. Interferometry of Electrochemical Refractive Index Fields

1. *Instrumentation*

A. Introduction. Double beam interferometers of greatly varying designs have been described in the literature. Some have been used in gas dynamic (105,117,151,155) or biological studies (52) and can equally be used for electrochemical purposes. A large number of designs have been developed for other purposes, such as length measurements (22) or the testing of optical components and instruments (18,21,59,146). Some of these interferometers can be adapted for purposes of interest here. Typical representatives of different kinds of interferometers and their characteristic features will be discussed in this section.

For the mapping of electrochemical refractive index fields, particularly mass transfer boundary layers on electrodes, an interferometer should satisfy the following requirements:

1. Provide for arbitrarily spacing and orienting interference fringes, when observing a homogeneous medium.
2. Provide for placing the virtual origin (localization) of interference fringes in the object.
3. Possess high geometrical resolution.
4. Have high image intensity.
5. Provide large separation of sample and reference beams.
6. Have sample beam cross specimen only once.
7. Be immune to vibration.

The flexibility provided by the feature of arbitrary orientation and spacing of interference fringes allows one to choose optimum conditions for the evaluation of interferograms. It will usually be desirable to orient the fringes parallel to the expected refractive index gradient. A large fringe spacing may be preferred for the observation of small fringe displacements. With infinite fringe spacing in a homogeneous medium, the condition of "interference contrast" is realized. Under this condition, interference fringes in an inhomogeneous medium follow contour lines of equal optical path. With the interference phenomenon located in the object, a true image of the object can be obtained. Thus two-dimensional refractive index fields may be observed. If the interference fringes originate at infinity, rather than at the finite object distance, aspheric optical elements have to be employed to focus both planes in one direction each. One-dimensional concentration fields resulting from diffusion (51), sedimentation (109), and electrophoresis (94) can be satisfactorily observed this way.

Most electrochemical applications of interferometry require the observation of small areas at high magnification (112). Good optical resolution is therefore necessary. This requirement is similar to that in interference microscopy (52,79) but runs contrary to the commonly encountered observation of large fields at low magnification in

TABLE X
Classification and Characteristics of Some Double Beam Interferometers

Classification	Example	Observable refractive index fields	Fringe orientation	Fringe spacing	Fringe localization	Geometric resolution	Intensity	Vibration sensitivity	Separation of reference beam
Division of wave front	Rayleigh	One-dimensional	Fixed	Fixed	Fixed	Low	Low	High	Medium
Division of amplitude by:	Jamin	One-dimensional	Fixed	Variable	Fixed	Medium	High	Low	Medium
Reflection	Mach-Zehnder	Two-dimensional	Variable	Variable	Variable	Medium	High	High	Large
Polarization	Lebedev	Two-dimensional	None	None	None	High	Low	Low	Small
Diffraction	Zernike phase contrast	Two-dimensional	None	None	None	High	High	Low	Small

aerodynamic and optical testing applications. Short exposure times are desirable for the observation of transient phenomena. The high image intensity required for that purpose is usually associated with a compact construction, which is also favorable for immunity from vibration. A large separation between sample and reference beam facilitates the use of a well-defined reference medium. Purposefully small separations are used for *differential* measurements.

One of the advantages of double beam interferometry over multiple beam interferometry is the simple optical configuration of single passage of the sample beam through the specimen. The double pass feature, accomplished in some instruments by use of a mirror behind the specimen, although it doubles the sensitivity of the instrument, considerably complicates the quantitative analysis of optical artifacts. Due to light deflection in the object or misalignment of the mirror, the reflected beam does not necessarily retrace the path of the incoming beam. In the schematic instrument diagrams which follow, the simple lenses shown usually represent compound lenses in practice.

Due to the statistical nature of light emission from classical monochromatic sources, time-independent interference can be observed only between two light beams, which are said to be mutually *coherent*. Such beams possess a stationary point-by-point phase relationship and can be visualized as replica of each other. Theoretical treatments of coherence and double beam interference have been given by Steel (131). The way by which the two coherent beams are produced provides a frequently used classification of double beam interferometers (14,30,39,79,130) and is used in Table X for some specific examples. The alternatives basically are *division of wave front*, where laterally separated parts of a wave are selected, and *division of amplitude*, where a wave is normally split by partial

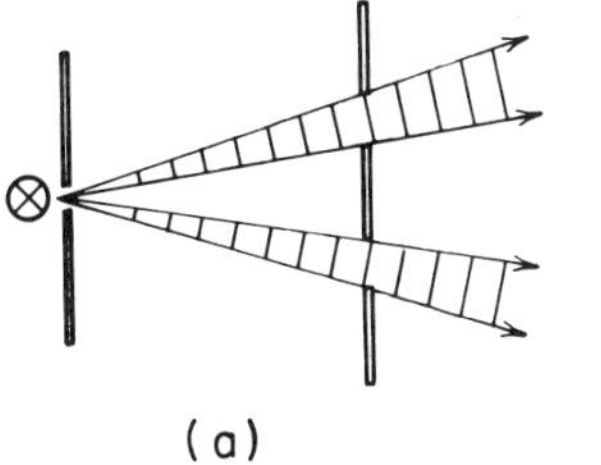

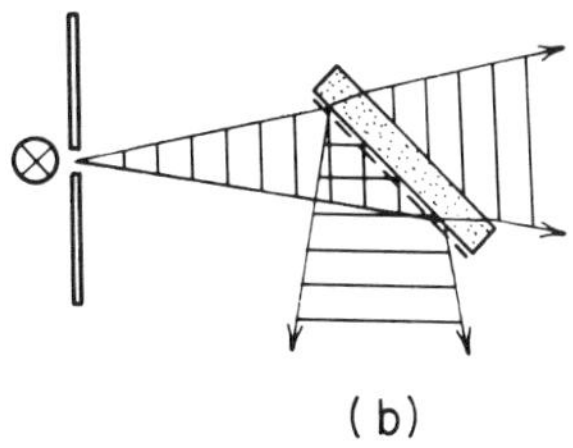

Fig. 5. Production of two coherent light beams: (*a*) division of wave-front by a double slit, (*b*) division of amplitude by partial reflection.

reflection and transmission (Fig. 5). Diffraction and polarization can be considered special cases of amplitude division (129). Strictly speaking, however, the different classes can not be completely separated, e.g., partial polarization is associated with amplitude division by reflection. The use of laser light greatly relaxes the need to produce exact duplicate beams for use in interferometry, because different parts of a single mode laser beam separated longitudinally or laterally possess a high degree of phase correlation at macroscopic distances. Spurious interference patterns due to diffraction are, however, often encountered with the use of laser light.

B. Rayleigh Interferometers. The Rayleigh interferometer is the principal instrument of present interest which employs division of wave front. Associated with this principle of operation are usually a large instrument size and low image intensity, because the division of wave front has to occur at a large distance from a small source in order to result in two coherent waves. According to the Van Cittert-Zernike theorem (16), the diameter D of the region illuminated almost coherently by a uniform, quasi-monochromatic circular source of radius r and mean wavelength $\bar{\lambda}$, located at a distance R is

$$D = \frac{R}{r} 0.16\bar{\lambda} \tag{11}$$

According to Eq. 11 a double slit of 5-mm spacing would have to be located at least 286 cm from a source of 1-mm diameter, illuminated by green mercury light.

In its classical form (14), illustrated in Fig. 6*a*, the Rayleigh interferometer can be used only for refractive index determinations of homogeneous specimens and is therefore more properly called an interference refractometer (23,151). A double slit C, illuminated by a small source A, produces two coherent beams which are made parallel by lens B. Superposition of the two mutually inclined beams by lens G results in interference fringes parallel to the intersection of wave fronts in the image H of the source. The spacing of interference fringes is determined by the angle between the interfering beams, and infinite fringe spacing cannot be achieved. Any change in the optical path length of the specimen beam with respect to the reference beam results in a lateral displacement of the interference fringes. This shift is either measured by a filar eyepiece I or is restored by the rotation of two mutually inclined glass plates in a Jamin compensator E (54,141). Apart from its early application in refractive index measurements of gases, the Rayleigh interference refractometer

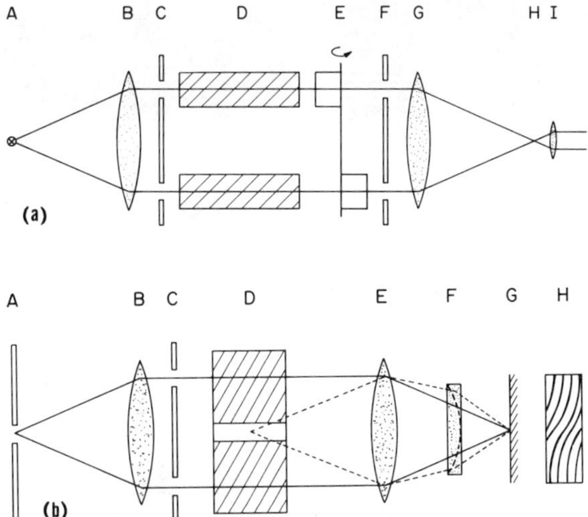

Fig. 6. (*a*) Rayleigh interference refractometer: *A*—point source, *B,G*—spherical lenses, *C,F*—double slits, *D*—specimen and reference chambers, *E*—compensator, *H*—plane of interference fringe formation, *I*—eyepiece. (*b*) Rayleigh interferometer for the observation of vertical refractive index variations, plan view diagram: *A*—point or slit source, *B*—spherical lens, *C*—double slit, *D*—specimen and reference cells (refractive index variations in vertical direction), *E*—spherical lens for imaging slit *A* (with interference fringes) in horizontal direction, *F*—cylindrical lens with horizontal axis for imaging cell *D* in vertical direction, *G*—film plane, *H*—fringe pattern in film plane for a diffuse liquid junction.

has been used for differential measurements between selected narrow regions in a concentration field resulting from diffusion (113).

By the addition of a cylindrical lens (109), the Rayleigh interferometer can be used for the observation of one-dimensional refractive index fields (51,94). The resulting arrangement is shown in Fig. 6*b*. The purpose of the cylindrical lens *F* is to focus the specimen in combination with the spherical lens *E* in the direction in which refractive index variations occur (e.g., vertical) without disturbing the formation of Rayleigh interference fringes. The slit source remains focused in the direction normal to the first one (e.g., horizontal) by lenses *B* and *E* with the cylindrical lens acting as a plano-parallel plate. Refractive index variations along the vertical direction of the cell result in a localized horizontal displacement of the vertical interference fringes (*H*).

An example of the resulting interferograms is shown on the left

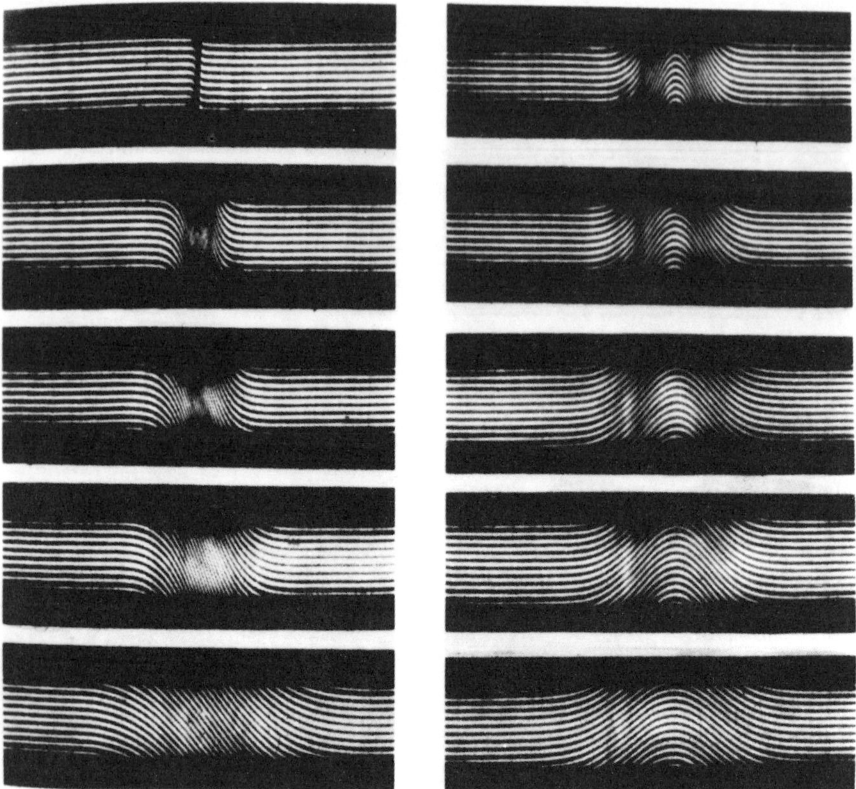

Fig. 7. Rayleigh interferograms, observation of the diffusion of a 0.2% solution of sucrose against water, Svensson (134). Left: refractive index profiles (normal interferogram). Right: refractive index derivative profiles, obtained with two identical diffusion boundaries, slightly shifted with respect to each other.

side of Fig. 7. The interference fringes on the right side represent the refractive index derivative. Such fringes have been obtained by mechanical (134) or purely optical (135) differentiation.

The basic Rayleigh interferometer can be modified in many ways. A simplification consists in omitting the second spherical lens E in Fig. 6*b*. Source A is then focused on the image plane G by lens B alone. As a result, the object is traversed by a slightly converging light beam. Without the cylindrical lens, this arrangement is also called the *Young interferometer*. The use of a single spherical lens can be combined with that of a parallel beam by placing a plane mirror behind the cell (109). Characteristics of the resulting double-pass instrument have been discussed in Section A above. A greatly increased image intensity can be obtained by the use of multiple light sources (137). These multiple sources (slits or pinholes) are arranged in such a way that the interference pattern due to each source alone is in registry with that of the neighboring source. Thus in contrast to the small number of fringes with decreasing intensity from the center, obtained with a single source (23,154), a large number of fringes of equally high intensity can be obtained. A schematic of a *multiple source* instrument which also employs a double light pass through the cell is given in Fig. 8 (108). Lens *D*,

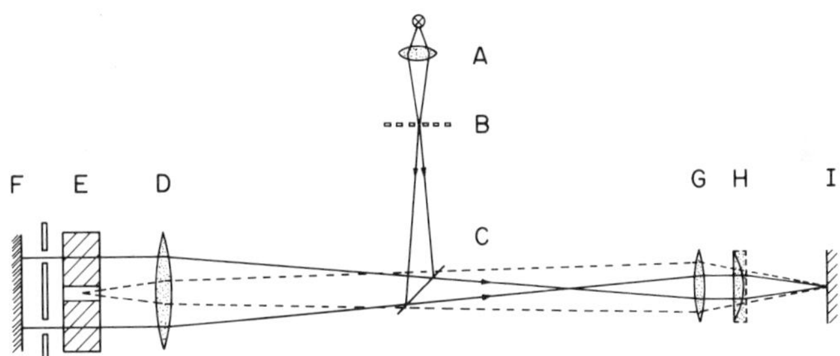

Fig. 8. Multiple source Rayleigh interferometer with double pass through the specimen. Perkin-Elmer (108), Chapman (24): *A*—monochromatic lamp, *B*—multipoint source, *C*—narrow mirror above reflected beams DG, *D*—spherical lens, *E*—specimen and reference cells, *F*—plane mirror, *G*—spherical lens for imaging cell in vertical direction, *H*—cylindrical lens with vertical axis for imaging interference fringes in horizontal direction (with lens G), *I*—photographic plate.

since it is traversed in both directions, serves the purpose of the two lenses *B* and *E* in Fig. 6*b*. Because the cylindrical lens is now part of a magnifying system of the primary interferogram, its axis is at right angles to the one shown in Fig. 6*b*. An example of an interferogram from an instrument with 162 point sources is given in Fig. 9 (62). A modified Rayleigh interference refractometer with three apertures has been reported for the resolution of very small optical path changes (0.001λ) by electro-optical means (76).

C. Jamin Interferometers. The Jamin interferometer employs division of amplitude by reflection from front and back of a plano-parallel glass plate. Like the Rayleigh interferometer, it has originally been used as a refractometer for uniform media (14). The same

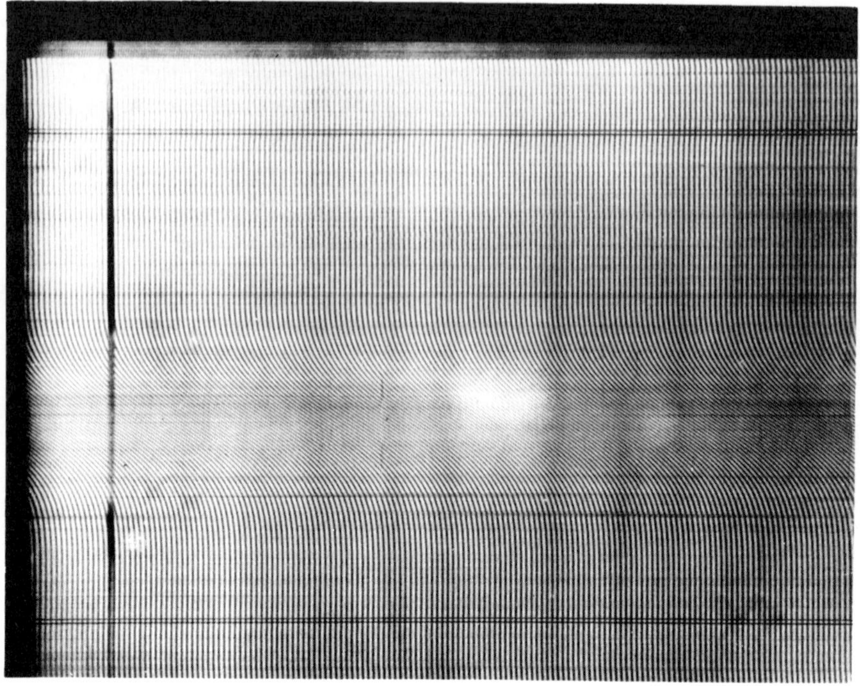

Fig. 9. Multiple source Rayleigh interferogram. Diffusion of 1.081 vs. 1.180 *M*-$CuSO_4$, 1.5 hr after forming boundary, Hsueh (62).

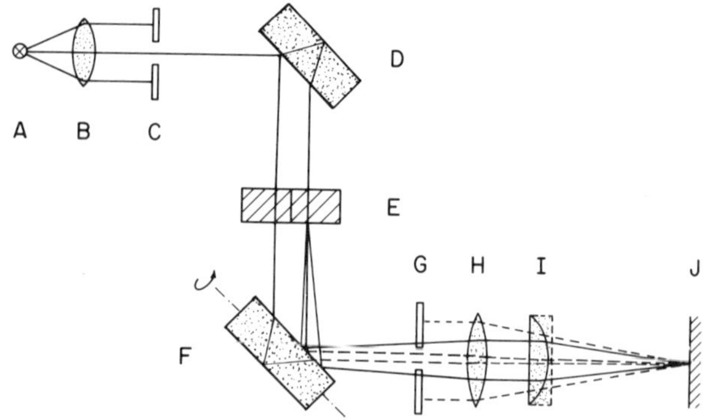

Fig. 10. Jamin interferometer arranged for the observation of horizontal refractive index variations, plan view diagram, Ibl, Barrada, and Truempler (64): *A*—mercury arc lamp with filter for 546.1 nm line, *B*—condenser, *C*—vertical slit, *D*—beam-splitting Jamin plate, *E*—object and reference cells, *F*—beam-uniting Jamin plate (tilted around horizontal axis for producing horizontal interference fringes), *G*—diaphragm for selection of reflected beams, *H*—spherical lens for imaging fringes from infinity in vertical direction, *I*—cylindrical lens with vertical axis for imaging object in horizontal direction, *J*—image plane.

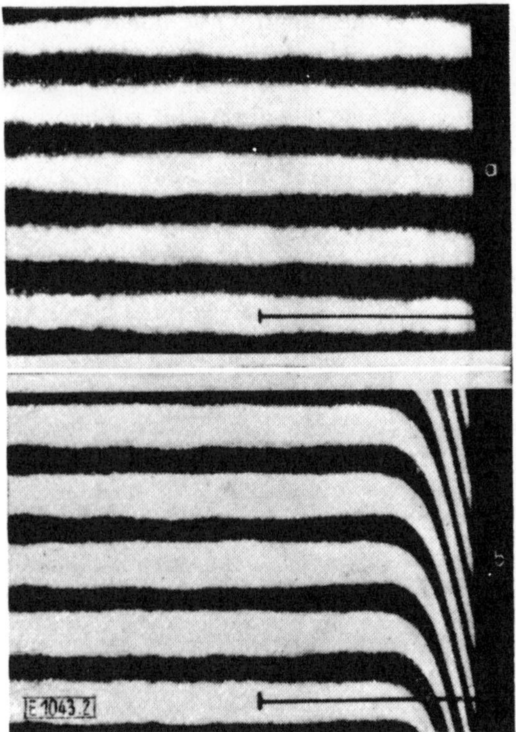

Fig. 11. Jamin interferogram of cathodic mass transfer boundary layer in 0.6 *M* $CuSO_4$ obtained with the interferometer described in Fig. 10. Scale 1 mm, shadow of electrode on the right. Ibl and Muller (65): (*a*) before electrolysis, (*b*) current density 2.2 mA/cm^2, natural convection, vertical cathode.

compensator can be used for both. Characteristics of the two refractometers have been compared by Kuhn (83). By the addition of a cylindrical lens, as with the Rayleigh interferometer, one-dimensional refractive index fields can be observed. The schematic of such an instrument is given in Fig. 10 (64). A collimated beam from a vertical slit *C* is reflected from the front and back sides of a plano-parallel plate *D*. The resulting two beams traverse object and reference cells *E* and are reunited by plate *F*, identical to *D* in thickness. Diaphragm *G* selects the desired reflections. The spherical lens *H* images the source (and therefore the interference fringes) in the vertical direction; the cylindrical lens *I* with vertical axis images the object in the horizontal direction without disturbing the formation of interference fringes which originate at infinity. Interferograms obtained with this instrument are given in Fig. 11.

A detailed theory of the Jamin refractometer (without cylindrical lens) has been given by Schoenrock (123). He shows that horizontal interference fringes which are straight, evenly spaced, and of low interference order can be produced by tilting the Jamin plates around a horizontal axis with respect to each other. Rotation of a plate around a vertical axis is found to displace horizontal fringes in a proportionate way and has been used in a compensating Jamin interferometer (2). Usable vertical interference fringes cannot, however, be produced this way. The introduction of two tilted plates as additional optical elements has been described by Antweiler (3) for producing vertical interference fringes with a Jamin interferom-

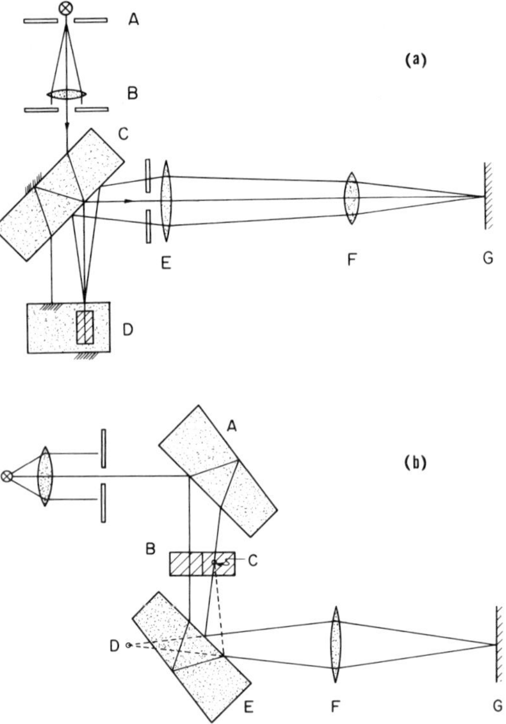

Fig. 12. Modified Jamin interferometers. (*a*) Lotmar interferometer (97): *A*—monochromatic light source, *B*—collimator lens, *C*—Jamin plate, *D*—cell and reference mirror, *E*—field lens, *F*—objective lens, *G*—image plane. (*b*) Sirks-Pringsheim interferometer: *A,E*—wedge-shaped glass plates, *B*—object and reference, *C,D*—virtual origin of interference fringes, *F*—objective lens, *G*—image plane.

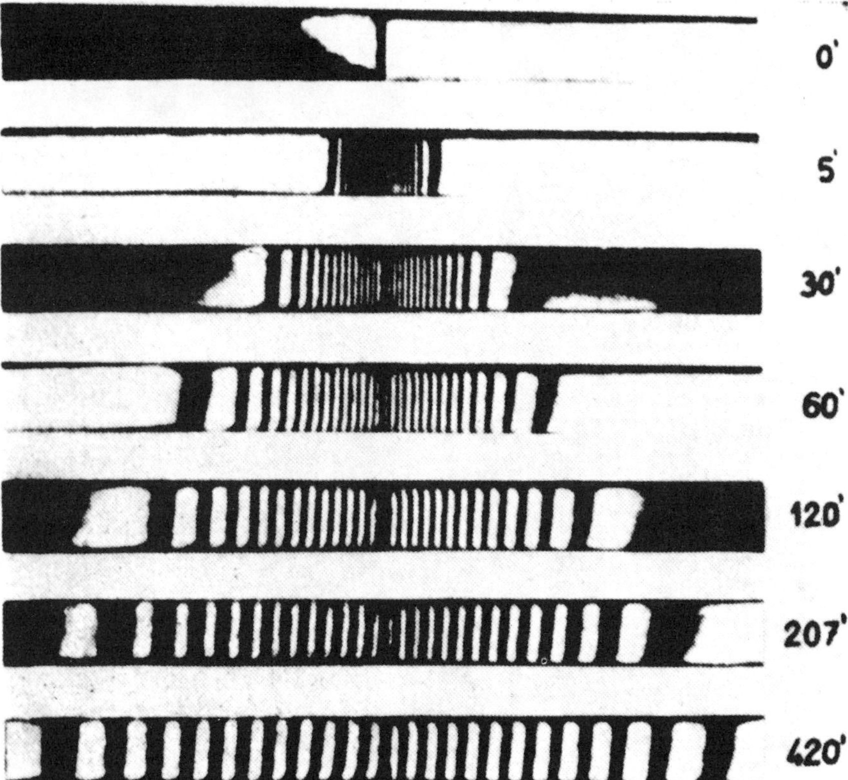

Fig. 13. Lotmar interferograms of the diffusion of two cadmium sulfate solutions (9% vs. 19%) observed with increasing time (97).

eter. The Jamin plates are aligned parallel to each other in this arrangement and a cylindrical lens provides an image of the cell in the vertical direction.

Several modifications of the Jamin interferometer have been proposed by *Lotmar* (96), who also described the construction of a compact single plate, double pass instrument (97), schematically shown in Fig. 12*a*. Interferograms of a diffusion experiment observed with this equipment are shown in Fig. 13. Operation with Jamin plates in parallel position has been reported by Antweiler (1). A spherical lens with a large depth of field (small relative aperature) is employed to bring simultaneously cell and interference phenomenon into focus. In this mode of operation, illustrated in Fig. 14, a homogeneous specimen appears of uniform intensity and interference

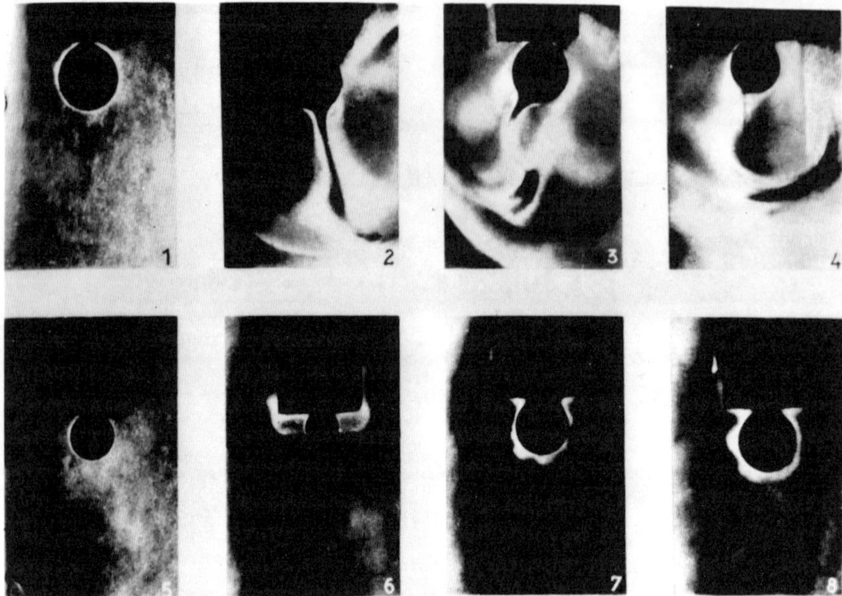

Fig. 14. Jamin interference contrast pictures of the anomalous convection on a mercury drop electrode, Antweiler (1): (*1*) undisturbed diffusion layer during the reduction of a 0.02 *N* $FeCl_3$ solution; (*2–4*) downward convection in the above solution, at lower potential; (*5*) undisturbed diffusion layer during the reduction of a 0.02 *N* $NiCl_2$ solution; (*9*) bell-shaped upward convection in the above solution, at lower potential; (*7–8*) eddy convection during the reduction of a 0.1 *N* $NiCl_2$ solution.

fringes form only in regions of varying refractive index. Variations smaller than those necessary for a change in optical path by a full wavelength manifest themselves in gradual intensity changes. The resulting image resembles one obtained by Schlieren techniques, which has led to the term "interferometric Schlieren system" for this mode of operation (which can be achieved with other interferometer arrangements too). The term "interference contrast" seems more appropriate, because the local image intensity is still a measure of refractive index, not refractive index derivative, as in Schlieren pictures.

The plano-parallel Jamin plates are responsible for the fact that the interference fringes originate at infinity, because interfering rays leave the plate assembly parallel to each other. Replacing the Jamin plates by wedge-shaped plates provides a means of moving the virtual origin of the interference phenomenon to a finite distance. This possibility is realized in the *Sirks-Pringsheim interferometer* (14,80), which employs plates with vertical apex (Fig. 12*b*) that result in vertical interference fringes. Their virtual origin *C* can be placed inside the object *B* by slight rotation of one of the plates around a vertical axis. Possibilities for application of this instrument remain largely unexplored.

Wedge-shaped Jamin plates are also employed in the *Dyson interference microscope* shown in Fig. 15 (8,37,80,128). The beam

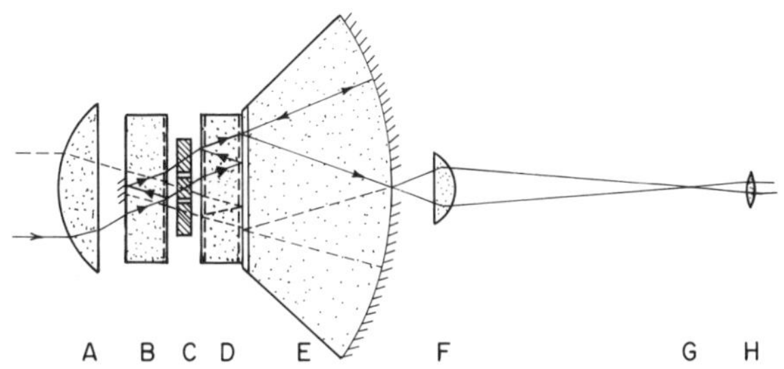

Fig. 15. Dyson interference microscope: *A*—condenser, *B*—wedge-shaped beam splitting plate with partially reflecting right face and totally reflecting center on left face, *C*—object with surrounding reference, *D*—wedge-shaped beam uniting plate with partially reflecting faces, *E*—portion of glass sphere with central opening on totally reflecting surface, where image of object is formed, *F*—microscope objective, *G*—image plane, *H*—eyepiece.

splitting plate *B* provides a reference for each ray passing through the object by reflection on a totally reflecting center section. Reference rays pass through the periphery of the specimen and are united with the corresponding specimen rays by plate *D*. The purpose of the element *E* with spherical reflecting surface is to produce an image of the object in the nonreflecting center part of the surface. This image is accessible for examination by a high-power (short working distance) microscope objective *F*. Translation, rotation, and tilting of plates *B* and *D* provide for the manipulation of interference fringes, including observation with interference contrast (infinite fringe spacing). Although the Dyson interference microscope offers some attractive features for purposes of interest here, no applications in this area seem to have been reported.

Although light intensities achieved with Jamin interferometers are inherently greater than those of a Rayleigh interferometer, it may be desirable to increase intensity. This could be necessary for observing transient phenomena or to compensate for restrictions in the observed area in order to maintain one-dimensionality of concentration profiles. A simple means to increase the light output, which does not seem to have received much attention, consists in altering the reflectivities of the Jamin plate surfaces. The plates normally employed have no coating on their front which results in a reflectivity of about 4% (89). The solid mirror coating on the back can be assumed to have 95% reflectivity. As illustrated in Fig. 16*a* the reflection of a beam from such a plate results in highly

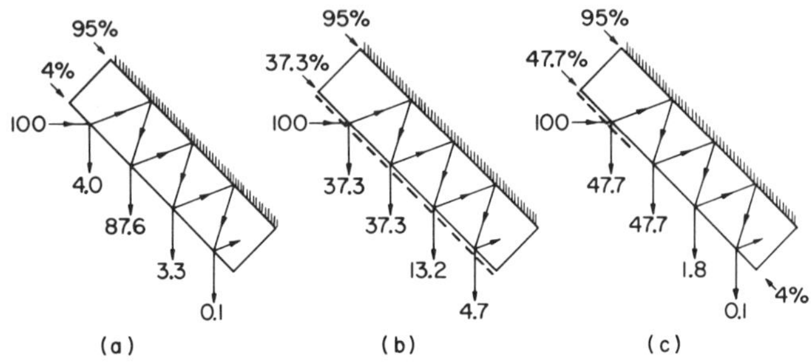

Fig. 16. Intensities of beams reflected from a Jamin plate with opaque reflecting coating on back side (95% reflectivity) and different front surfaces: (*a*) bare (4% reflectivity), (*b*) uniform (loss-free) coating of 37.3% reflectivity, (*c*) partial coating of 47.7% reflectivity.

uneven intensities of the first two reflected beams which are used in an interferometer. Equal intensity in these two beams can be achieved by a uniform coating on the front side of the plate (Fig. 16*b*). However, the total optical power contained in the two is now lower, with the balance being wasted in higher order reflections. These losses can be reduced by applying a coating over only part of the front surface in such a way that the incident beam is divided by the coating while the first reflection from the back of the plate exists through the uncoated section (Fig. 16*c*). Application of an antireflection coating to this section would result in a further gain of about 3%. The use of two Jamin plates of the kind illustrated in Fig. 16*b* for beam splitting and uniting results in an almost fourfold increase in light intensity of an interferometer, as compared to the use of uncoated plates. Two plates of the type indicated in Fig. 16*c* would produce an eightfold improvement.

The principal advantage of the Jamin interferometer is its simple, compact construction. The instrument is surprisingly insensitive to vibration, easy to operate, and provides good light intensity. These advantages primarily derive from the stiff connection between two mirror-pairs (front and back side of Jamin plates) which is also the cause of the main shortcomings of the design: the separation of the two coherent beams is limited by the thickness of available plano-parallel glass plates and the interference fringes are localized at infinity. The aspherical lens system bridges the gap between the finite object distance and the infinite fringe origin by sacrificing one coordinate direction of the object geometry. Only one-dimensional concentration fields (as in the Rayleigh interferometer) can therefore be investigated. Any significant variation in the refractive index profile over the area observed would result in a blurring of all interference fringes (which are replicas of each other). In many electrochemical applications, one-dimensional refractive index fields can be produced (at the expense of light intensity) by restricting the observed area to a sufficiently narrow region with an optical slit. In order for refractive index variations to result in lateral interference fringe displacements, the refractive index gradient should be parallel to the direction of the fringes.

Care must be taken in orienting the axis of the cylindrical lens normal to the interference fringes. Its rotation around the optical axis can result in distortions of the interference fringes which have not been adequately investigated.

D. Mach-Zehnder Interferometers. The four-plate or Mach-Zehnder configuration is probably the most versatile interferometer for the observation of refractive index fields. This instrument has been extensively used in heat transfer (32) and wind-tunnel measurements where usually large areas have to be observed (46,125). The primary characteristic of the Mach-Zehnder interferometer is that it allows one to arbitrarily orient, space, and localize the interference fringes. Fringes and object can simultaneously be brought to focus with ordinary spherical lenses, and thus two-dimensional refractive index fields can be observed.

The classical arrangement of components is shown in Fig. 17*a*. Collimated light from a point or extended source A is divided by a partially reflecting mirror E (division of amplitude). The resulting mutually coherent beams pass through object and reference chambers I and F and, after reflection by the completely reflecting mirrors G and H, are united by another partially reflecting plate J. An

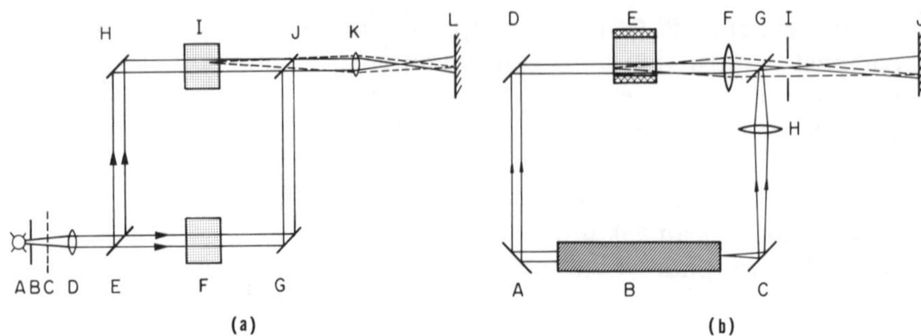

Fig. 17. Mach-Zehnder interferometers. (*a*) classical form: *A*—source, *B*—pinhole, *C*—color filters, *D*—collimating lens, *E*—beam splitting plate, *F*—reference cell, *G*, *H*—deflection mirrors, *I*—specimen cell, *J*—beam uniter plate, *K*—objective lens, *L*—image plane. (*b*) Modified form with laser light source: *A*, *C*, *D*—first surface deflection mirrors, *B*—HeNe gas laser with dual beam output, *E*—specimen cell, *F*—objective lens, *G*—beam uniter plate with dielectric coating, *H*—reference lens, *I*—spatial filter (slit), *J*—image plane. After Beach, Muller, and Tobias (6).

objective lens *K* forms an image of the object *I*, together with the interference pattern, on the film plane *L*. The light throughput is optimized by use of multilayer dielectric coatings with negligible absorption on the beam splitting and beam uniting plates. The position of the objective lens *K* after the beam uniter plate *J* introduces a large separation between object and objective, a configuration that is necessary for the observation of large objects but is not favorable for high optical resolution.

When used with conventional light sources, alignment of the instrument requires the use of elaborate procedures (57,156) in order to insure that light emitted at a given instant from any point of the source *A*, after traversing one arm *EGJ* of the interferometer, arrives at the same point at the same time in the image plane *L* as the light which had traversed the other arm *EHJ*. Otherwise, due to the time and space dependence of the phase of light emitted from an incoherent source, no interference phenomena can be observed. One can imagine that the adjustment of the four mirror plates *E*, *G*, *H*, and *J*, with their many translatory and rotatory degrees of freedom, to satisfy the above requirements, is no small task (47,86, 155). The reference cell *F* serves to equalize the optical path lengths of the two interfering beams, which should be the same within a fraction of a wavelength, if white light is to be used, and can be different by as many as a hundred wavelengths with typical monochromatic sources. The colorless zero order interference fringe, resulting from the use of white light, is used to establish fringe displacements across unresolved steep gradients or discontinuous boundaries (121). The large separation between specimen and reference beams, another characteristic of the Mach-Zehnder interferometer, results in large-sized instruments, which often have to be placed on massive foundations to reduce the effect of vibrations on the stability of optical components.

Detailed analyses of the Mach-Zehnder interferometer can be found in the literature (11,55,75,121,156). Qualitatively, the formation and localization of interference fringes, characteristic of extended light sources, can be explained with the help of Fig. 19*a* (14,55). A coherent pair of sample and reference beams can be arranged to enter the objective lens K slightly displaced and mutually inclined with respect to each other. Their common virtual origin P determines the localization of the interference phenomenon, which can be made to coincide with the object. Orientation and spacing of the interference fringes are determined by the angle between the two beams. Let W_1 and W_2 represent mutually coherent wave fronts in specimen and reference beams originating from the same wave front before division by the beam splitter. For the formation of low-order interference fringes in the object, the interferometer plates are adjusted in such a way that the virtual location W_1' and W_2' of the two corresponding wave fronts intercept in the virtual origin P of the interference phenomenon. The image formed in the film plane L then shows a bright interference fringe at the image P' of P oriented parallel to the line of intersection of W_1' and W_2' with neighboring dark and bright fringes determined by the local separation between the two wave fronts, similar to the interference in a tapered film. Infinite fringe spacing (interference contrast), is obtained with W_1' and W_2' parallel to each other.

The use of laser light sources greatly reduces the alignment requirements of double beam interferometers. Any part of the cross section of a single mode laser beam interferes with any other part of that cross section, because of the complete lateral (also called spacial) coherence of such a beam. At the same time, the great longitudinal (also called temporal) coherence of laser light makes it possible that large path differences between the two arms of an interferometer (meters for a HeNe laser) can be tolerated without noticeable decrease in the contrast of interference fringes.

A modified Mach-Zehnder interferometer, which takes advantage of the flexibilities offered by the use of laser light (6), is illustrated in Fig. 17*b*. In contrast to the conventional version of the instrument, the beam-splitting plate has been omitted and the two mutually coherent light beams are derived from the opposite ends of a helium-neon gas laser *B*, placed in one arm of the interferometer and adjusted for single-mode output. This arrangement results in a compact construction which is not sensitive to vibration. The location of the objective *F* between specimen *E* and beam uniter plate *G* allows the use of lenses with high numerical aperture (and associated resolution) close to the object, but necessitates the use of a reference lens *H* to provide a reference beam of a convergence equal to that of the object beam. Since high-contrast interference fringes are obtained with unequal path between the two coherent beams, the reference cell could be omitted. As a result, the interferometer can be used in a traveling mode, e.g., for scanning boundary layers in a flow channel with its inadvertently varying dimensions, without continuous adjustment of a reference cell.

A common problem in the use of laser light is the formation of spurious diffraction patterns due to light scattering from optical imperfections, such as dust particles, anywhere in the light path. A diaphragm *I* (Fig. 17*b*), which acts as a spacial filter and is located in the convergence plane of sample and reference beams, serves to attenuate such scattered waves, since they do not come to a focus in this plane. If, in order to eliminate undesired interference patterns, the diaphragm has to be closed sufficiently to impair the geometric resolution of the objective, a slit can be used with its long dimension in the direction in which the full resolution of the lens (presently 5×10^{-4} cm) is desired.

The fringe pattern is photographed by a camera back *J* with focal plane shutter. Due to the high light intensity available from even a low-powered laser, exposure times of 10^{-3} sec have been found sufficient for a film of intermediate sensitivity (Kodak Plus X). Beam expanders have later been added between mirrors *A*,*D* and *C*,*G* in order to enlarge the area of uniform intensity in the center of the beam (7).

Interferograms obtained with the laser interferometer sketched in Fig. 17*b* are shown in Fig. 18. They illustrate the development of an electrochemical mass transfer boundary layer under laminar forced convection conditions, with copper being deposited from a 0.1 *M* copper sulfate solution at 66 mA/cm^2. It can be seen that,

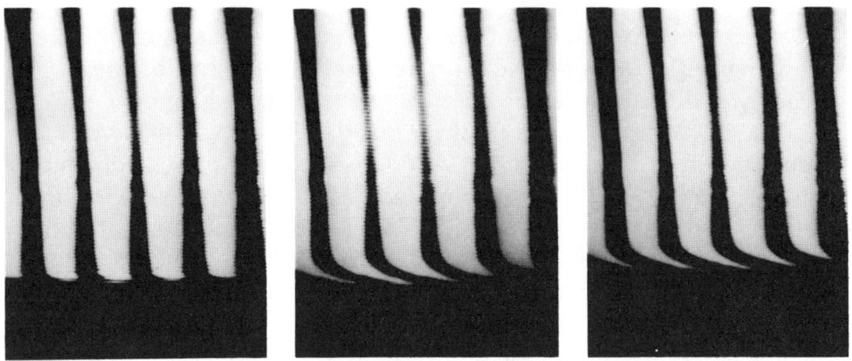

Fig. 18. Boundary layer development observed with interferometer of Fig. 17 (*b*). Cathode, 4 cm from leading edge, 66 mA/cm^2, 0.1 *M* $CuSO_4$, Re = 1100. Before electrolysis, and 2 and 20 sec after start of electrolysis.

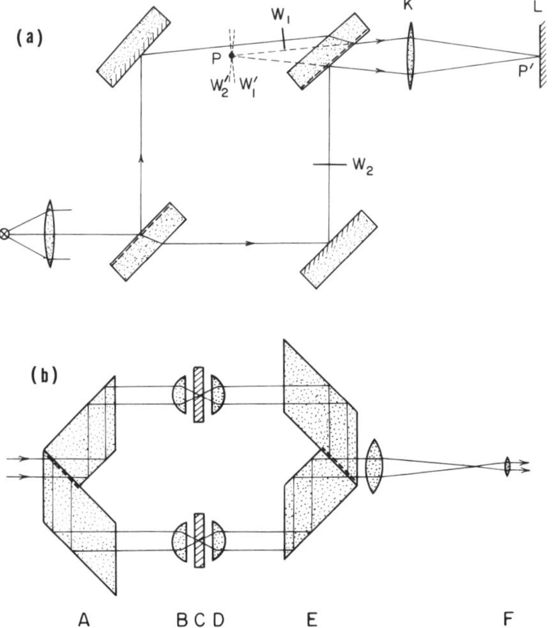

Fig. 19. Mach-Zehnder interferometers. (*a*) Localization and orientation of interference fringes: W_1, W_2—coherent wavefronts in specimen and reference beams respectively; W_1', W_2'—virtual location of W_1 and W_2 as imaged by the objective lens *K* on film plane *L*. (*b*) Leitz-Horn interference microscope: *A*—beam splitting prism assembly, *B*—condensers, *C*—object and reference, *D*—matched objectives, *E*—beam uniting prism assembly, *F*—eyepiece.

2 sec after the beginning of electrolysis, the boundary layer is still growing. After about 8 sec a steady state concentration profile is reached, which is the same as that shown after 20 sec.

Ease of operation and high geometrical resolution have been achieved in a special way in another variation of the four-plate interferometer principle, the *Leitz-Horn interference microscope*, schematically shown in Fig. 19*b*. The mirror surfaces are provided by two sets of prisms *A* and *E* which accomplish a superposition of the images of two identical microscopes, one looking at the object, the other at a reference (8,43,80). Matched microscope objectives of numerical aperture up to 1.36 can be used in this equipment resulting in a diffraction-limited resolution of about 0.25 μ. Specimen and reference beams are 6.2 cm apart. Compensators, located in both arms of the interferometer between prism *A* and condenser *B* as well as between objective *D* and prism *E* (133) provide for equalizing the optical path in both arms of the interferometer for use of white light and allow the user to arbitrarily space and orient interference fringes.

E. Michelson Interferometers. The Michelson interferometer, schematically shown in Fig. 20, employs the same partially reflecting plate *B* for beam splitting and uniting. By its nature, the instrument is of a double-pass type with large separation between sample and reference beams. Different interference phenomena can be observed (5,130) which have been analyzed quantitatively (132,142). Of interest here are fringes of equal thickness or Fizeau fringes, originating near the virtual locations of the mutually inclined sample and reference mirrors and running parallel to the line of intersection

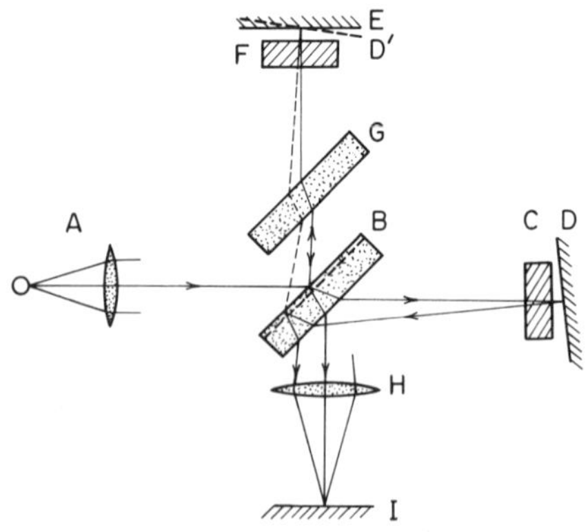

Fig. 20. Michelson interferometer: *A*—collimated light source, *B*—beam splitting and uniting plate, *C*, *F*—object and reference cells, *D*, *E*—object and reference mirrors, *D′*—virtual position of mirror D, *G*—compensating plate, *H*—objective lens, *I*—image plane.

between them. These fringes are similar to those from thin film interference in an air wedge formed by the two mirror surfaces E and D', provided that the difference in phase change between internal and external reflection on the beam splitter B is properly considered. However, the fringe profile shows a strictly sinusoidal intensity distribution, characteristic of double beam interference.

In order to equalize the optical path through glass in both arms of the interferometer, necessary for use of white light, a compensating plate G of equal thickness and dispersion as the beam splitter B can be added (41). With identical path length in both arms a colorless, zero order fringe is then observed with white light in the center of the field of view (or over the entire field with parallel plates). In contrast to Michelson's original arrangement, a collimated light source is commonly used now (132). With a sufficiently small source the interference phenomenon can be de-localized to appear in focus together with the object F.

Many modifications of the Michelson interferometer have been reported (5,143). Among them, the *Twyman-Green interferometer* is extensively used for the testing of optical components. The *Zeiss-Linnik interference microscope* for examining reflecting surfaces is discussed in a later section. Despite its popularity in many applications, the Michelson principle has been rarely used for the observation of refractive index fields in fluids.

F. Polarizing Interferometers. The birefringence (double refraction) of optically anisotropic media can be used to produce two coherent light beams. Devices of principal interest for this purpose (35,133) are the Savart plate (147), which produces two parallel beams, and the Wollaston prism (71), which produces two divergent beams as shown in Fig. 21. The two resulting beams are polarized

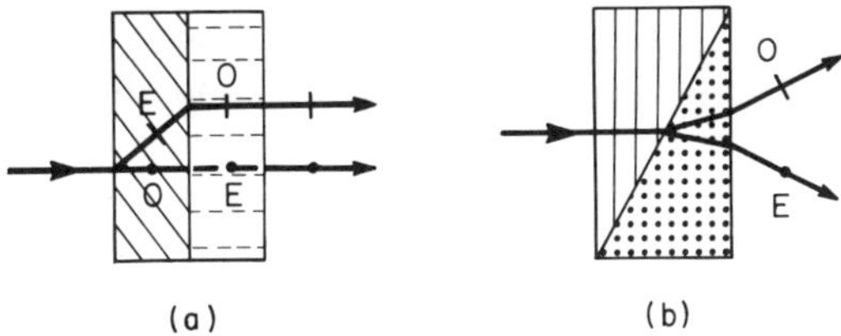

Fig. 21. Beam splitting devices for polarizing interferometers; ordinary (O) and extraordinary (E) rays with polarization indicated by direction of electric vector. (*a*) Savart plate, producing linear shear, (*b*) Wollaston prism, producing angular shear.

at right angles to each other. In order to bring two cross-polarized beams to interference, mutually parallel components of the electric vector have to be selected for superposition. Polarizing elements (filters or prisms) are used for this purpose as polarizers and analyzers on both sides of the beam splitters.

A large number of interferometer designs based on the above beam splitters and their modifications have been developed in recent years (37,42). In many of these instruments, the phase in the image of the object is compared with that of another image of the object rather than with that of a reference object. Interferometers of this type are also called *shearing interferometers* (18,139). They are widely used in optical testing (59,133,146) in order to avoid the use of a precise reference component. The two images may show an angular or a linear displacement (shear). If the lateral displacement is larger than the object, a blank area in the object plane may serve as a reference. If the displacement of the two images is smaller than the geometrical resolution of the optical system, a differential interferogram, indicative of the optical path gradients, results. Linear shear in the axial direction results in the superposition of a sharp image of the object with one which is out of focus. The local phase of the latter represents an average of an extended area of the specimen and may therefore be uniform (80,133). An inadvertent axial shear, due to an optical path difference between ordinary and extraordinary rays in the beam splitter, is often associated with lateral shear. Radial shear (58,59,129) employs two concentric images of different magnification. Rotational and folding shear are also used (131). Different kinds of shear are associated with different ambiguities in the interpretation of results (80).

The *Lebedev interference microscope* (also called Jamin polarizing interferometer) employs a beam splitter with lateral shear. It is available in several commerial forms and is shown schematically in Fig. 22*a* (8,30). A birefringent crystal plate C provides a lateral shear (like a Savart plate), which depends on the thickness of the plate and is chosen depending on the power of the objective lens (typically 50–500 μ). A half-wave retardation plate D rotates the plane of polarization of both beams by 90° so that they are reunited in a symmetrical way by a second identical crystal plate F. In an early electrochemical boundary layer study, the lateral beam separation has been large enough to have the object illuminated entirely by the extraordinary beam. Straight interference fringes have been introduced with a quartz-wedge compensator (119).

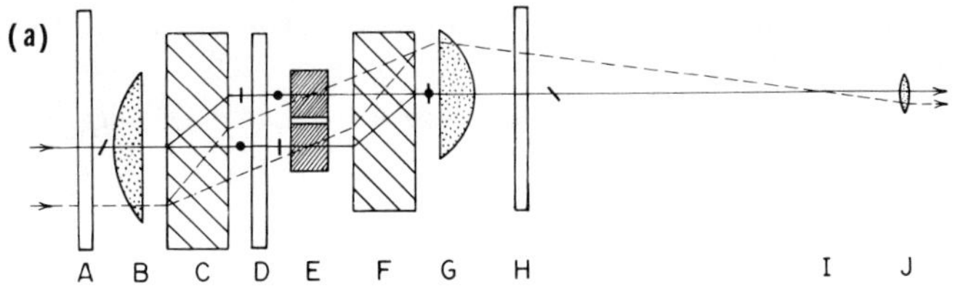

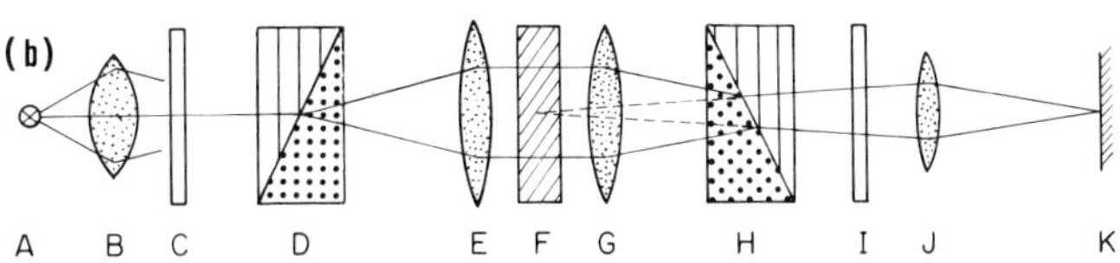

Fig. 22. Polarizing interferometers. (*a*) Lebedev interference microscope: *A*—polarizer, *B*—condenser, *C*, *F*—birefringent crystal plates, *D*—half-wave plate, *E*—object and reference, *G*—objective lens, *H*—analyzer, *I*—image plane, *J*—eyepiece. (*b*) Smith interference microscope: *A*—light source, *B*—condenser, *C*, *I*—polarizer, *D*, *H*—Wollaston prisms, *E*, *G*—objective lenses, *F*—object, *J*—eyepiece, *K*—image plane.

The *Smith interference microscopc* (Fig. 22*b*) is based on the use of Wollaston prisms as beam splitters. The *Nomarski interferometer* is a variation of this principle for use with high power objectives (44,103). It produces a differential interferogram and can also be adapted for use with reflected light. Another modification for the simultaneous recording of concentration and concentration gradient profiles has been reported by Weinstein (153). Wallaston prisms can also be used for the observation of macroscopic objects (42), larger than the size of the prisms. Equipment of this type for the use in gas dynamic studies has been described by Oertel (104).

Polarizing beam splitters can also be used for producing a double image in the eyepiece of a microscope (8,37,80). In another form of polarizing interferometers, beam splitting is achieved by the use of double focus lenses, made from birefringent materials (30,44,52, 129).

G. Gouy Interferometer. The Gouy interferometer probably requires the smallest instrumental effort of any interferometer (48). Interference is due to the superposition of different parts of a beam which have been deflected by a refractive index field. The technique could be classified as an interferometric one, employing division of amplitude, but, in part, it is also a Schlieren technique, because it employs light deflection.

The schematic of an instrument employing parallel light through the specimen is shown in Fig. 23. Refractive index variations in the object *D*, typically due to a diffusing boundary, cause locally varying light deflection (50,51). Equally deflected beams, from symmetrical parts of the boundary, are superimposed in the focal plane *F* of the objective *E*. Depending on their phase difference, a dark or bright interference fringe results. If lens *E* is omitted, the object can be made to be traversed by converging light.

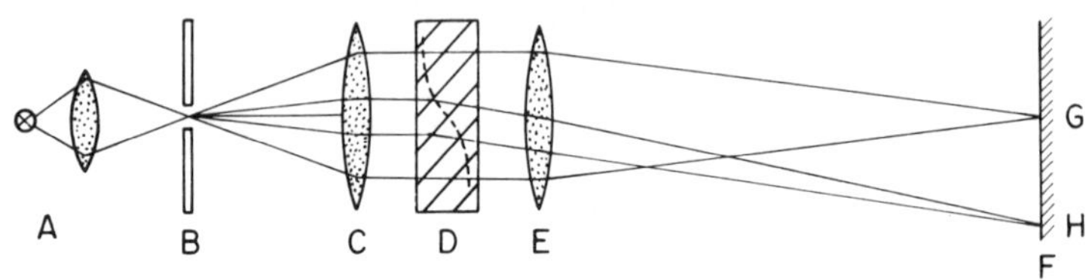

Fig. 23. Gouy interferometer (side view): *A*—illuminator, *B*—horizontal slit, *C*—collimating lens, *D*—cell (with refractive index profile), *E*—objective lens, *F*—image plane, *G*—undeflected slit image, *H*—deflected slit image.

A detailed analysis of fringe formation has been given by Gosting and Kegeles (73), and experimental tests of the technique have been performed by Longsworth (92). The general use of the Gouy interferometer is restricted by the requirement that the refractive index profile in the object be symmetrical and its general shape be known for the interpretation of results.

H. Diffraction Interferometers. Different orders of diffracted light coming off a diffraction grating (Fig. 24) are mutually coherent and can therefore be used for interferometry (91). Gratings of the

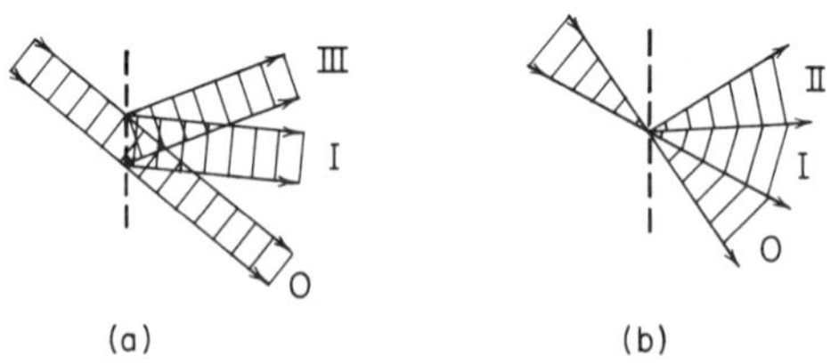

Fig. 24. Beam splitting by diffraction gratings. Diffraction orders indicated by Roman numerals. (*a*) Parallel beam, (*b*) convergent beam.

reflection or transmission type may be used. They should be blazed in such a way as to concentrate the light flux in the desired diffraction orders. A diffraction interferometer employing four transmission gratings in a collimated monochromatic light beam is sketched in Fig. 25*a* (151). The gratings are analogous to the four plates of a

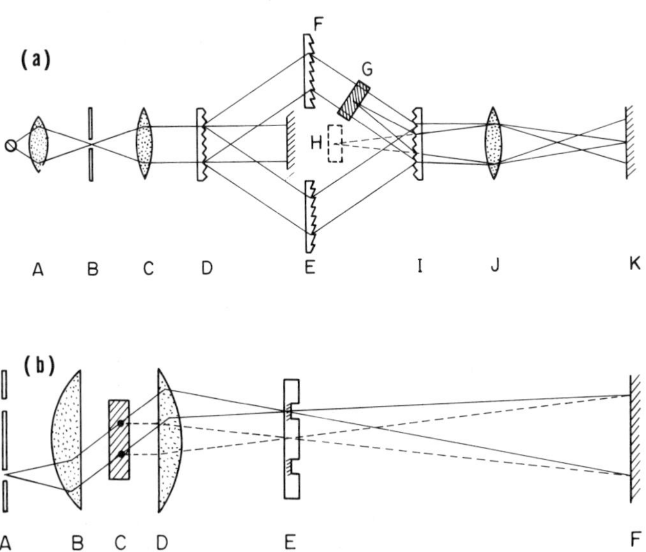

Fig. 25. Diffraction interferometers. (*a*) Four-grating interferometer, Weinberg (151): *A*—monochromatic light source with condenser, *B*—pinhole, *C*—collimating lens, *D*, *I*—diffraction gratings operating in plus and minus first order, *E*, *F*—diffraction gratings operating in second order, *G*—object, *H*—virtual location of object, *J*—objective lens, *K*—image plane. (*b*) Zernike phase contrast microscope: *A*—diaphragm with annular opening, *B*—condenser lens, *C*—object, *D*—objective lens, *E*—annular phase plate in rear focal plane of objective, *F*—image plane.

Mach-Zehnder interferometer. The beam splitting and uniting gratings *D* and *I* work in plus and minus first order, the beam deflection gratings *F* and *E* in second order. Two gratings in a converging beam have been used in the original *Kraushaar diffraction interferometer* (77) and a later modification of it (106). For complete separation of adjacent diffraction orders it becomes necessary under these conditions to superimpose zero and first order beams resulting from the extreme angles of incidence (151) (Fig. 24*b*).

The principal advantage of diffraction grating interferometers, apart from low cost, seems to be that the alignment of gratings is much less critical than that of corresponding mirrors. Weinberg (151) has shown that the two types of beam splitters differ by a factor of at least 368 in sensitivity to angular misalignment. Low optical speed, due to the poor efficiency of gratings, is one of the shortcomings of grating interferometers; another is the appearance of spurious interference fringes due to grating imperfections (151). Although fringe spacing can be controlled by a tilting of gratings, the condition of interference contrast can, therefore, usually not be achieved. The combination of a diffraction grating with a double beam interferometer and its use for adjusting zero path difference has been analyzed by Kahl and Sleator (72).

Diffraction in the object is the beam-splitting process employed in the *Zernike phase contrast microscope*. Phase and amplitude of diffracted and undiffracted light are altered in such a way that interference of the two in the image results in intensity variations which are sensitive to small local phase variations in the object. Thus the diffracted light from a phase object is converted to simulate that of an amplitude object. Several detailed analyses of phase contrast principles (10,157,158) as well as more qualitative discussions (9,36,127) are available in the literature. The schematic of a commonly used form of phase contrast microscope is given in Fig. 25*b* (8). An annular diaphragm A in the focal plane of the condenser B provides a hollow cone of illumination on the object C. Undiffracted light, originating from a given point in diaphragm A (solid lines), passes as a parallel beam through the object, while diffracted light (dashed lines) is deflected in the object. In the rear focal plane of the objective lens, the undiffracted light forms an image of the annular diaphragm while most of the diffracted light passes through other parts of this plane. The spacial separation of diffracted and undiffracted light makes it possible to adjust relative amplitude and phase of both by means of an annular phase plate. Typically, the undiffracted light is attenuated by partial absorption and advanced in phase by a quarter wavelength. Superposition of diffracted and undiffracted light in the image plane F then results in an interference contrast picture. Variations in intensity repeat themselves with every increment of one wavelength optical path difference in the object. Since interference fringes cannot be introduced at will in the picture to resolve ambiguities of interference order, and estimating fractional orders requires local photometry, the phase contrast microscope is not particularly suited for quantitative analysis (150), although optical path differences of about 5×10^{-4} wavelengths can be recognized (36). Nevertheless, the phase contrast principle has been used as a refractometer for the precise determination of D_2O in H_2O (69). A phase contrast telescope (120) can be used for the observation of large objects.

Division of amplitude by diffraction is also employed in *holographic interferometry*, which is discussed in the chapter by Srinivasan.

2. *Interpretation of Interferograms*

A. Conventional Interpretation. In the conventional interpretation of interferograms, the changes in optical path length (or phase) of the light through a specimen of fixed dimension are taken as a direct measure of local refractive index variations. Corrections to this first-order approximation, due to light deflection and diffraction, are neglected in this approach and will be considered later.

As pointed out in discussing the capabilities of different interferometers, interferograms of most instruments can be classified as being obtained with either "finite" or "infinite" fringe spacing. Kinder (75) has compared interferograms of several objects with finite and infinite fringe spacing. The interpretation of both is different (53). In the first case, a field of continuously varying phase, generated by the instrument, is superimposed on the object and results in an interference pattern, typically straight fringes, with an object of uniform thickness and refractive index (47). A sketch of such a field is given in Fig. 26*a*. Any change in local phase due to the object is added to the predetermined instrumental phase variation and results in a displacement of interference fringes (Fig. 26*b*), because a different phase (interference order) is now associated with a given point in the image, and interference fringes represent lines of constant total phase. Since fringe displacement is a measure of optical path change due to the object, originally straight, equidistant fringes, deformed by a one-dimensional concentration field, can be interpreted as plots of concentration versus distance (67). Why this interpretation is valid only for equidistant fringes and one-dimensional refractive index fields can be seen with the aid of Fig. 26*b*, where the heavy lines represent the center of interference fringes, with the associated interference order noted in wavelengths. A fringe of zero order is arbitrarily chosen here; with white light, it would be the colorless high contrast fringe. Along a line AD, following

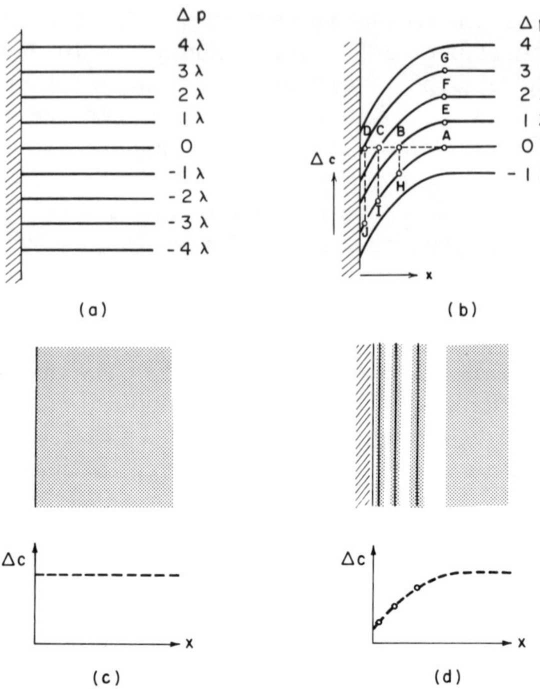

Fig. 26. Conventional interpretation of interferograms, (*a*) and (*b*) with finite fringe spacing, (*c*) and (*d*) with infinite fringe spacing: (*a*) instrumentally generated interference fringes with local phase given in wavelengths; (*b*) displaced fringes due to a refractive index field; (*c*) uniform object; (*d*) concentration field like the one used in (*b*).

the original straight fringe of zero order, the local interference order increases by one wavelength each, as successive fringes are intersected at points B, C, and D and is the same as in points E, F, and G, respectively. With the distance DJ equal to AG, etc., the local displacement of the zero order fringe is a measure of the local phase change. If all fringe spacings AE, EF, etc., are equal, the curved fringe AJ represents a linear plot of interference order or phase, and to a good approximation of concentration c versus distance x. It must be remembered, however, that the curved fringe AJ represents the phase at the location of the fringe, and, only for a one-dimensional refractive index field is the local fringe displacement, e.g., distance DJ, a measure of the local phase along the original, straight fringe, e.g., in point D.

Fig. 27. Two-dimensional interferogram with finite fringe spacing; heated horizontal tube in air under natural convection, Hansen (56).

The conventional evaluation of a two-dimensional interferogram, such as the one shown in Fig. 27, requires that one establish the local phase variation due to the object along a desired cross section, preferably parallel to the original fringe pattern. Graphical interpolation techniques for this purpose have been discussed by Schardin (121).

With infinite fringe spacing (interference contrast), the interferogram of an object of uniform optical thickness is of uniform intensity throughout and can assume any value between constructive and destructive interference (Fig. 26*c*). Phase variations due to the object results in local intensity variations which repeat themselves with each wavelength of optical path increase. As a result, interference fringes, which represent lines of constant phase change due to the object, appear. The interferogram of a concentration field, like the one shown in Fig. 26*b* for finite fringe spacing, is sketched in Fig. 26*d* for infinite fringe spacing. The concentration profile along a selected cross section is derived in the familiar way from the interference fringes, which now represent concentration contours (Fig. 26*d*). The

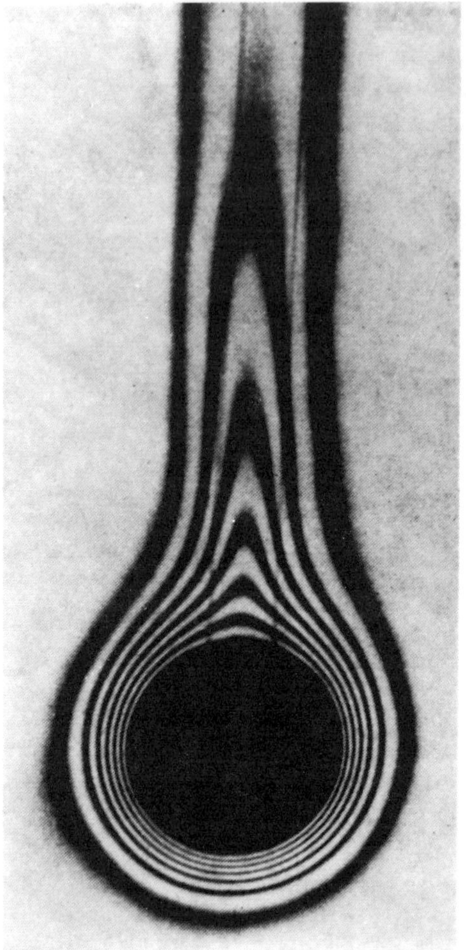

Fig. 28. Two-dimensional interferogram with infinite fringe spacing, heated horizontal tube in air under natural convection. Eckert and Drake (32), used with permission of McGraw-Hill Book Company.

same procedure can be applied to two-dimensional refractive index fields, like the one shown in Fig. 28.

The reconstruction of three-dimensional refractive index fields from (two-dimensional) interferograms presents special problems. A theoretical discussion has been given by Iwata and Nagata (70). Some fields of rotational symmetry have been analyzed by decomposing them into concentric layers of finite thickness and numerically integrating the optical path length (155,121). In an effort to provide additional data of a three-dimensional object, focusing refractometry has been employed to obtain refractive index profiles at a cell wall (7). More generally applicable methods, such as the use of simultaneous interferometric views from different directions, possibly by holographic techniques, need to be developed.

The resolution of interferometric refractive index determinations depends on what minimum change in interference order can be detected. With finite fringe spacing, a fringe displacement of 0.1 fringe separations can usually be detected visually. The unaided estimation of fractional orders with infinite fringe spacing is more problematical. In either case, interference colors, resulting from the use of white light, provide increased visual resolution of small phase variations. Tables of interference colors (40) show that in the more sensitive regions of the color series, changes in optical path of 10 nm (approx. 0.02λ) result in distinct color changes. However, the dispersive properties of the object will have to be considered under these circumstances (23,26). Apart from noise due to optical imperfections, photographic grain, etc., the gradual intensity variation across a double beam interference fringe is often responsible for some uncertainty in defining fringe position. Photographic processes are available to introduce steep flanks or equidensity lines in the fringe profile (78,82,88). Photoelectric scanning devices, which resolve a fringe pattern into finite density increments, have recently become available to serve a similar purpose more quantitatively. A resolution of 0.01λ should be expected from such instruments. The direct photometry of interference fringes has been shown to resolve optical path differences of 0.001λ (76). Microdensitometry of photographed interference patterns can yield a similar resolution.

According to Eq. 3,

$$\Delta n = \frac{\Delta p}{d} \tag{12}$$

the minimum resolvable refractive index variation Δn for a given resolution limit for the change in optical path length Δp is inversely proportional to the extent d of the object in the light propagation direction (cell thickness). Refractive index and concentration changes resulting in an optical path change of a full wavelength are given in Fig. 29 as a function of cell thickness. The minimum resolvable refractive index or concentration variation would be 0.1

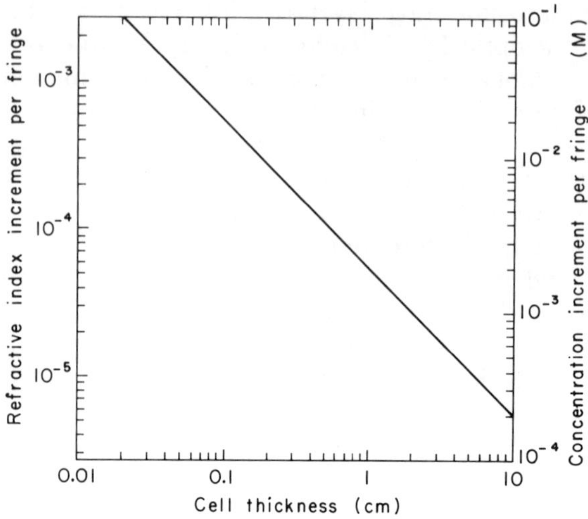

Fig. 29. Refractive index increment Δn, and corresponding concentration increment of a 0.5 *M* $CuSO_4$ solution ($n = 1.3468$), necessary to shift interference fringes by one fringe spacing ($\Delta p = \lambda$) in a cell of thickness *d*. (Eq. 12) Single pass interferometer, green Hg light (546.1 nm). Practical resolution limits of interferometry are 0.1 to 0.001 times the values indicated.

to 0.001 times the values indicated, depending on the resolution of fringe displacement.

B. Consideration of Light Deflection

(a) Introduction. Valid refractive index profiles often cannot be derived from interferograms by the conventional techniques of interpretation. The reason for this difficulty is that light does not propagate along a straight line in the presence of refractive index variations normal to the propagation direction. Errors due to this effect can be particularly large in condensed media, because large refractive index gradients are frequently encountered there. Errors in interferograms due to light deflection have been considered by several authors and found negligible. In part, the validity of such conclusions is limited to gaseous media (47,155) and refractive index variations of small extent at a large distance from the imaging optics (20).

The physical reason for light deflection (or refraction) in refractive fields lies in the dependence of propagation velocity on refractive index (Eq. 1). Thus different elements of a wave front advance to different degrees in a time interval, resulting in a tilt of the wave front. The direction of the wave normal, which indicates the propagation direction of the wave or ray, is therefore changed. The angle of deflection is proportional to the component of the refractive index gradient normal to the propagation direction and the distance over which the light encounters this gradient

$$d\phi = \frac{1}{n}\frac{\partial n}{\partial y}\,dx \qquad (13)$$

Equation 13 is also the basis of Schlieren optics (122).

Light deflection results in two kinds of errors in an interferogram. The first is a geometrical distortion due to the lateral displacement and change in direction of transmitted light. The second error is a distortion of the measured refractive index, because the deflected light travels on a longer path and passes through regions of different refractive index.

(b) Constant Refractive Index Gradient. For a one-dimensional, linear refractive index profile with planes of constant refractive index parallel to the incident beam, the geometrical light path, responsible for geometrical distortions, as well as the optical path length, responsible for refractive index distortions, can be derived in closed form. Since linear refractive index approximations can be used to gauge the magnitude of possible errors in more complex fields, some essential results are derived below.

Geometrical Light Path. One approach to calculate the light path in an isotropic inhomogeneous refractive index field of constant gradient is based on the principle of Fermat (13,60) which states that the medium is traversed by the light in such a way that the optical path length or the time necessary for traversal is minimum.

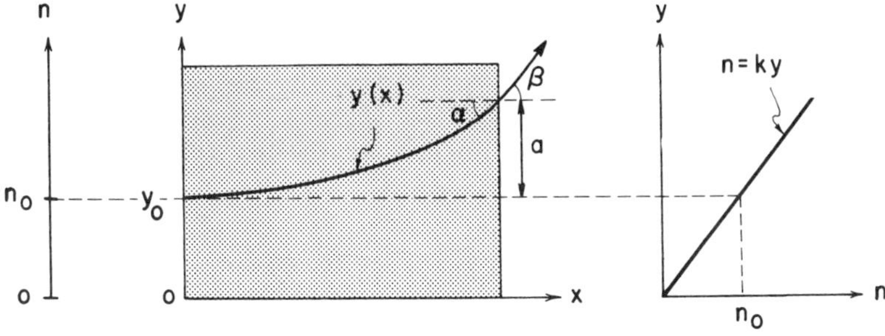

Fig. 30. Light deflection in uniform refractive index gradient field: definition of coordinates.

The light path can then be found by use of the techniques of variational calculus. Geometrical and refractive index coordinates to be used are defined in Fig. 30. The optical path length is

$$\int n(y)\, ds \equiv \int n(y) \sqrt{\left(\frac{dx}{dy}\right)^2 + 1}\, dy = \min \tag{14}$$

or, with Eq. 1 ($n = c/v$)

$$\int \frac{ds}{v(y)} \equiv \int \frac{\sqrt{(dx/dy)^2 + 1}}{v(y)}\, dy = \min \tag{15}$$

Any extremum has to satisfy Euler's equation, i.e., for the integrand of Eq. 14

$$F(y) \equiv n(y) \sqrt{\left(\frac{dx}{dy}\right)^2 + 1} \tag{16}$$

the condition for minimum is

$$\frac{\partial F}{\partial x} - \frac{d}{dy}\left(\frac{\partial F}{\partial (dx/dy)}\right) = 0 \tag{17}$$

Inserting Eq. 16 in Eq. 17 results in

$$0 - \frac{d}{dy}\left(\frac{n(y)(dx/dy)}{\sqrt{(dx/dy)^2 + 1}}\right) = 0 \tag{18}$$

Integration of Eq. 18 gives

$$\frac{n(y)(dx/dy)}{\sqrt{(dx/dy)^2 + 1}} = C_1$$

or

$$\frac{dx}{dy} = \frac{1}{\sqrt{(n/C_1)^2 - 1}}$$

therefore

$$x = \int \frac{dy}{\sqrt{(n(y)/C_1)^2 - 1}} + C_2 \tag{19}$$

For the special case of a linearly increasing refractive index, depicted in Fig. 30

$$n(y) = ky \tag{20}$$

the light path is determined as follows: insert Eq. 20 into Eq. 19 and change variables

$$\frac{ky}{C_1} = z, \qquad \frac{dy}{dz} = \frac{C_1}{k}$$

so that

$$x = \frac{C_1}{k} \int \frac{dz}{\sqrt{z^2 - 1}} + C_2$$

which can be integrated (29) to give

$$x = \frac{C_1}{k} \cosh^{-1} \left(\frac{ky}{C_1}\right) + C_2 \tag{21}$$

Since a general expression of the form

$$X = B \cosh^{-1} Y + C$$

is equivalent to

$$Y = \cosh \frac{X - C}{B}$$

one obtains from Eq. 21

$$y = \frac{C_1}{k} \cosh \frac{k(x - C_2)}{C_1} \tag{22}$$

The constants of integration C_1 and C_2 are determined with the initial conditions, which are

$$\text{for } x = 0, \qquad \begin{aligned} y &= y_0 \\ n &= ky_0 \equiv n_0 \\ \frac{dy}{dx} &= 0 \end{aligned} \tag{23}$$

Insert Eq. 23 into Eq. 22

$$\frac{y_0 k}{C_1} = \cosh \frac{-kC_2}{C_1} \tag{24}$$

Equation 24 must be true for all values of k, for $k \to 0$

$$\frac{n_0}{C_1} = \lim_{k \to 0} \cosh \frac{-kC_2}{C_1} = 1$$

therefore

$$C_1 = n_0 \tag{25}$$

With Eq. 23 and Eq. 25 it follows from Eq. 24 that

$$\cosh \frac{-kC_2}{n_0} = 1$$

and therefore, because $k \not\equiv 0$ and n_0 finite

$$C_2 = 0 \tag{26}$$

Thus Eq. 22 for the curved light path becomes

$$y = \frac{n_0}{k} \cosh \frac{kx}{n_0} \tag{27}$$

The chain line Eq. 27 may be developed in a series

$$y = \frac{n_0}{k} + \frac{kx^2}{n_0 2!} + \frac{k^3x^4}{n_0{}^3 4!} + \frac{k^5x^6}{n_0{}^5 6!} + \cdots \tag{28}$$

or, with $y_0 = n_0/k$

$$y = y_0 + \frac{x^2}{y_0 2!} + \frac{x^4}{y_0{}^3 4!} + \cdots$$

A hyperbolic cosine function for the light path (60) and an equivalent formulation (47) can be found in the literature. An approximate form of Eq. 28 has been given by several authors (46,51,148). The angular light deflection α, due to a refractive index gradient k normal to the propagation direction experienced over a distance x, is

$$\tan \alpha = \frac{dy}{dx} = \sinh \frac{kx}{n_0} \tag{29}$$

$$\tan \alpha = \frac{kx}{n_0} + \frac{k^3x^3}{n_0{}^3 3!} + \frac{k^5x^5}{n_0{}^5 5!} + \cdots \tag{30}$$

Due to refraction at the cell-air interface the inclination β of a deflected beam, after entering into air (Fig. 30) is larger than the angle α given by Eq. 29. Application of Snell's law for the angle of refraction then results in

$$\sin \beta = (n_0 + ka) \sin \left[\tan^{-1} \left(\sinh \frac{kx}{n_0}\right)\right] \tag{31}$$

where a is the lateral beam displacement at the cell exit given by Eq. 33 and n_0 is the refractive index where the beam enters the cell. Numerical data for α and β are given in Fig. 31*a*. It can be seen that deflection angles of several degrees can be commonly expected. The current density, added in a separate scale to this figure and included on some of the following ones, results in interfacial concentration gradients of the magnitude indicated, under consideration of diffu-

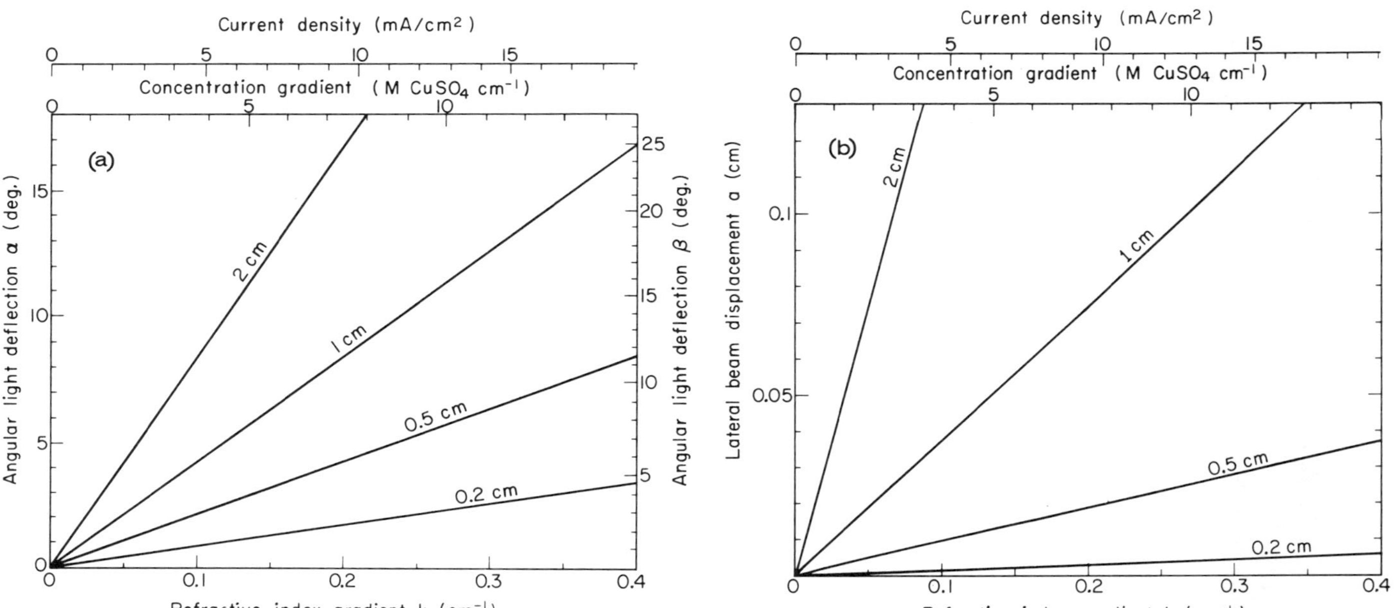

Fig. 31. Light deflection in a constant refractive index gradient. (*a*) Angular deflection β and α with and without consideration of refraction at the cell exit, respectively (Eqs. 31 and 29). Cell thickness $x = 0.2, 0.5, 1$, and 2 cm. Concentration gradient associated with refractive index gradient valid for 0.5 M $CuSO_4$ ($n = 1.3468$, $\delta n/\delta c = 0.027/M$). Current densities indicated result in the associated concentration gradient at the surface of a cathode in 0.5 M $CuSO_5$ in the absence of convection. (*b*) Lateral beam displacement at the cell exit (Eq. 33). Cell thickness $x = 0.2, 0.5, 1$, and 2 cm. Refraction inside cell wall not considered. Concentration gradient and current density scales as in (*a*).

ion and migration in a 0.5 *M* $CuSO_4$ solution ($D = 4.62 \times 10^{-6}$, $t_{cat} = 0.304$ (33,45).

The local lateral displacement of a deflected beam

$$a = y - y_0 \tag{32}$$

is indicated in Fig. 30. From Eq. 27 it is found to be

$$a = -\frac{n_0}{k} + \frac{n_0}{k} \cosh \frac{kx}{n_0} \tag{33}$$

$$a = \frac{kx^2}{n_0 2!} + \frac{k^3 x^4}{n_0{}^3 4!} + \frac{k^5 x^6}{n_0{}^5 6!} + \cdots \tag{34}$$

These equations can be found in the literature (67,87). Some computed data are given in Fig. 31*b*. They show that for all but the thinnest cells, the lateral beam displacement *a* can easily exceed 10^{-2} cm, the order of magnitude of typical diffusion layers. Angular light deflection and lateral displacement are used as input parameters in subsequent computations.

The curved light path in a linear refractive index field can also be derived by integation of Snell's law of refraction (66,87). For this purpose, the refractive index field is divided into layers of constant refractive index with light refraction occurring at each boundary. Equation 27 also results from this approach.

Optical Path Lengths. The optical path length p observed in the interferogram, is found for a deflected beam by integrating the geometrical path length s over the local refractive index n

$$p_2 = \int_0^x n \, ds \tag{35}$$

With Eq. 20 and Eq. 27

$$n = n_0 \cosh \left(\frac{kx}{n_0}\right)$$

and

$$ds = \sqrt{1 + \sinh^2 \left(\frac{kx}{n_0}\right)} \, dx$$

the optical path length becomes

$$p_2 = n_0 \int_0^x \cosh^2 \left(\frac{kx}{n_0}\right) dx = \frac{n_0{}^2}{k} \left[\frac{1}{4} \sinh \frac{2kx}{n_0} + \frac{kx}{2n_0}\right]_0^x + C \tag{36}$$

Since for $x = 0, p = 0$, the constant of integration is $C = 0$. Therefore

$$p_2 = \frac{n_0 x}{2} + \frac{n_0^2}{4k} \sinh \frac{2kx}{n_0} \tag{37}$$

Expansion in a series results in

$$p_2 = n_0 x + \frac{1}{3!} \frac{2}{n_0} k^2 x^3 + \frac{1}{5!} \left(\frac{2}{n_0}\right)^3 k^4 x^5 + \cdots \tag{38}$$

These expressions have also been given by Ibl (67), and the first two terms of Eq. 38 have been derived by Caldwell (19).

The error in optical path length, due to light deflection, is responsible for the error in concentration derived from an interferogram. The error can be defined most simply as the difference in optical path length p between the deflected beam 2 (Fig. 32) and a hypothetical undeflected beam either entering (beam 1) or leaving

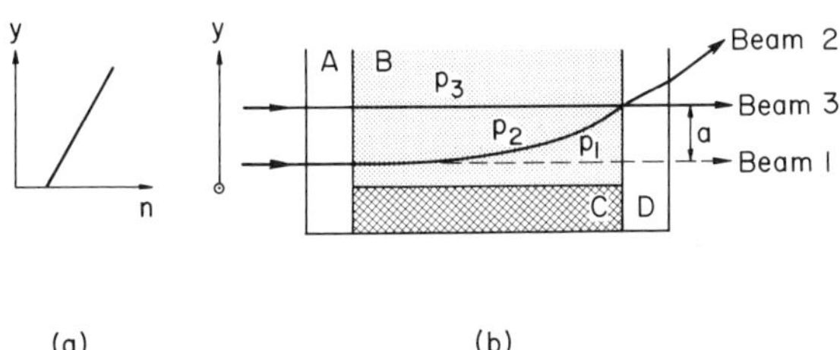

Fig. 32. Optical path length errors due to light deflection near a working electrode: (*a*) refractive index profile, (*b*) cross section of electrolysis cell. Deflected Beam 2 and undeflected reference Beams 1 and 3 that enter and leave, respectively, the refractive index field with the deflected beam: p_1, p_2, p_3—optical path lengths of the three beams inside the cell; *A*, *D*—cell walls; *B*—electrolyte; *C*—electrode.

(beam 3) the cell at the same point as the deflected one. The resulting errors are

$$p_2 - p_1$$

and

$$p_2 - p_3$$

They can be expressed in number of fringe shifts by dividing the optical path length error by the wavelength of light. For one-dimensional, constant refractive index gradient fields, the error can be computed in closed form. With Eq. 37

$$p_2 = \frac{n_0 x}{2} + \frac{n_0^2}{4k} \sinh \frac{2kx}{n_0}$$

and, according to Fig. 30

$$p_1 = n_0 x \tag{39}$$

the first error is

$$p_2 - p_1 = -\frac{n_0 x}{2} + \frac{n_0^2}{4k} \sinh \frac{2kx}{n_0} \tag{40}$$

or

$$p_2 - p_1 = \frac{1}{3!} \frac{2}{n_0} k^2 x^3 + \frac{1}{5!} \left(\frac{2}{n_0}\right)^3 k^4 x^5 + \cdots \tag{41}$$

The other reference optical path length is

$$p_3 = (n_0 + ka)x$$

With Eq. 33

$$p_3 = n_0 x \cosh \frac{kx}{n_0} \tag{42}$$

The second error therefore is

$$p_2 - p_3 = \frac{n_0 x}{2} + \frac{n_0^2}{4k} \sinh \frac{2kx}{n_0} - n_0 x \cosh \frac{kx}{n_0} \tag{43}$$

or

$$p_2 - p_3 = -\frac{k^2 x^3}{3!\, n_0} + \frac{3k^4 x^5}{5!\, n_0^3} + \cdots \tag{44}$$

Optical path length errors derived in the literature (34,114,138) cannot easily be compared with the above results. Numerical results of Eq. 40 and Eq. 43 are illustrated in Fig. 33. The fringe shifts given there can be related to refractive index (or concentration) errors by use of Fig. 29. In a 1-cm thick cell, for instance, an error of 100 fringes corresponds to an error in concentration of 0.2 *M*. An important result of this analysis is that the two errors are of opposite sign. In order to place reliable limits on possible interferometric errors, however, the optical path length analysis will have to be further refined, as indicated below.

(c) Effect of Focusing. The above analysis does not consider the effect of the imaging optics in an interferometer. The purpose

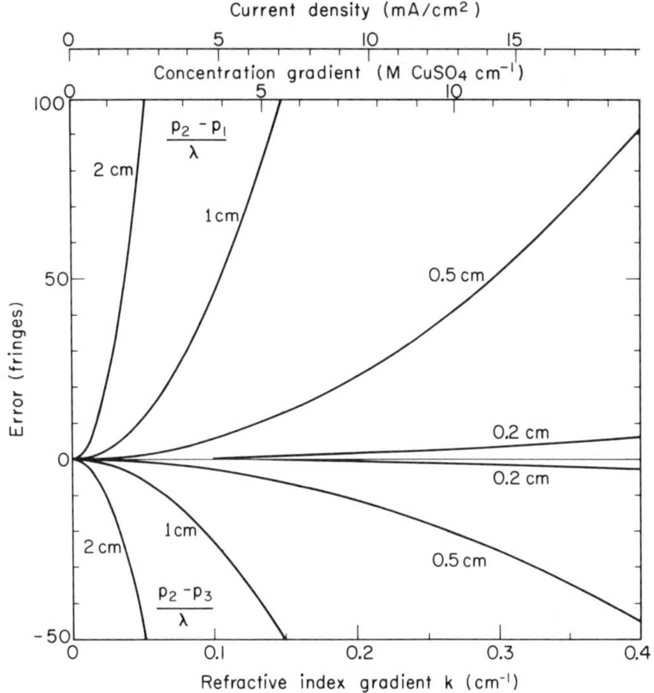

Fig. 33. Errors in optical path length due to light deflection in constant refractive index gradient for cell thickness x of 0.2, 0.5, 1, and 2 cm. Wavelength 546.1 nm, refractive index $n = 1.3468$ (0.5 M $CuSO_4$), effects of cell wall and focusing not considered. Concentration gradient and current density scales as in Fig. 31.

of this optics is to bring a selected plane in the object to focus in the image. This plane is optically conjugate to the image plane. It is referred to in the following as "plane of focus," but should not be confused with the planes that contain the primary and secondary focal points of the objective lens. If light deflection occurs in the object, the image is determined by the virtual origin of the deflected light in the plane of focus. Under these circumstances the shape of the image can be expected to depend on the choice of the plane of focus. This effect is illustrated in Fig. 34 for a cathodic boundary layer with the electrode shadow as the object. A light beam C, entering the cell at the surface of electrode B and parallel to it, is deflected toward the bulk of the solution A and leaves the cell at point D. With the objective lens G focused on the cell wall facing

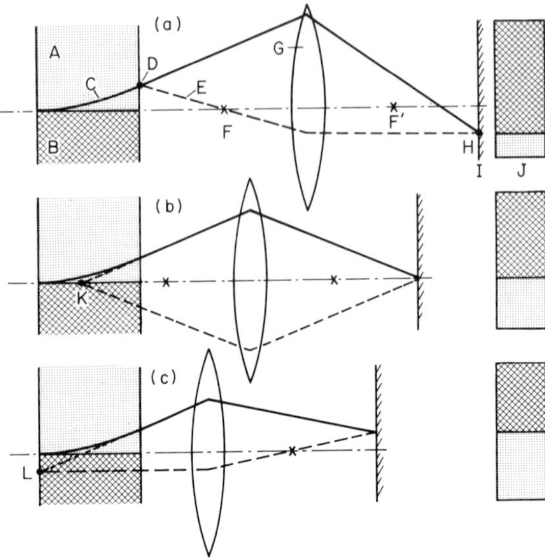

Fig. 34. Effect of choice of focal plane on geometrical distortion due to light deflection: (*a*) exit side of cell focused, (*b*) optimum focus, (*c*) entrance side of cell focused (refraction at cell-air interface and in cell wall not shown). *A*—electrolyte, *B*—electrode, *C*—deflected beam, *D*, *K*, *L*—virtual locations of electrode surface, *E*—auxiliary ray, *F*, *F'*—primary and secondary focal points of objective lens, *G*—objective lens, *H*—image of electrode surface, *I*—image plane, *J*—picture of electrode shadow.

the camera (Fig. 34*a*), the shadow of the cathode surface in this plane appears at *D* and its image in film plane *I* is *H*. Thus the electrode shadow appears advanced into the solution side of the interface due to the presence of the refractive index gradient. Focusing on the cell wall facing the light source (Fig. 34*c*) results in a virtual origin *L* of the same deflected beam. The electrode shadow now appears receded into the electrode. For an intermediate focusing position (Fig. 34*b*), the virtual origin *K* of the deflected beam coincides with the electrode surface. The electrode shadow is, therefore, not displaced in the image.

In the schematic of Fig. 34, refraction at the cell exit as well as inside the cell wall has been neglected. Even in the *absence* of refractive index gradients, these effects result in an axial displacement of the virtual location *E'* of an immersed object *E* (Fig. 35*a*). Since the light is assumed here to be incident normal to the first cell wall,

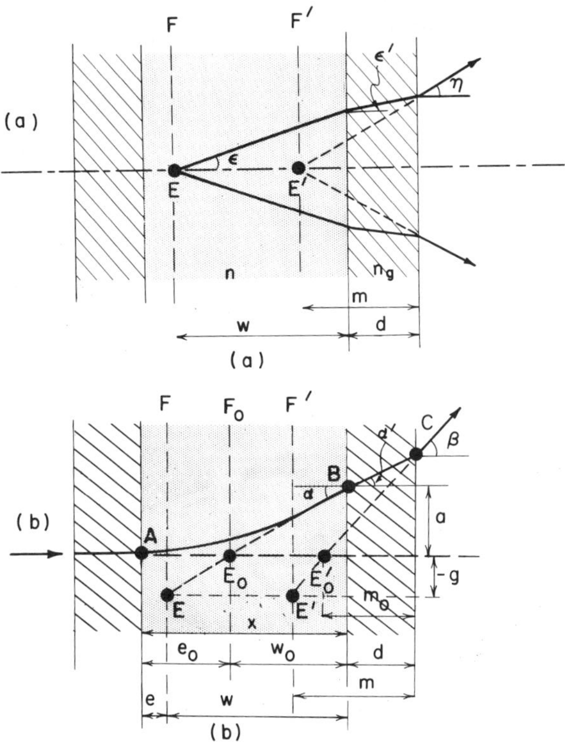

Fig. 35. (*a*) Effect of focusing in the absence of light deflection: *E*—real, immersed object, *E'*—virtual location of *E*, *F*—real plane of focus in cell, *F'*—virtual plane of focus (focus in air). (*b*) Effect of focusing in the presence of light deflection. *ABC*—deflected light beam, *F*—real plane of focus, *F'*—virtual plane of focus, *e*—location of plane of focus, *g*—lateral displacement of virtual beam origin. Location e_0 of the plane of focus F_0 for which the virtual origin E_0' of a deflected light beam (e.g., electrode shadow) is not displaced laterally.

refraction in the second one (facing the camera) will only be considered.

The application of Snell's law (Eq. 45)

$$n \sin \varepsilon = n_g \sin \varepsilon' = \sin \eta \tag{45}$$

results in Eq. 46 for the virtual location m of an immersed object (Fig. 35*a*).

$$m = \cot \eta (w \tan \varepsilon + d \tan \varepsilon') \tag{46}$$

For small angles ε from the optical axis, this equation reduces to

$$m = \frac{w}{n} + \frac{d}{n_g} \tag{46a}$$

where w is the thickness of the medium of refractive index n and d is the thickness of the glass wall of refractive index n_g. Thus a real plane of focus F inside the cell is transformed into a virtual plane of focus F' by refraction effects. (F' would be the real plane of focus in the absence of the cell, Fig. 35*a*.) If the imaging optics of an interferometer are focused on an immersed target in plane F, a preferred experimental procedure, the image is determined by the virtual location of the target in plane F'.

The effect of focusing in the *presence* of light deflection is illustrated in Fig. 35*b* for an arbitrary plane of focus F in the cell (with associated virtual plane of focus F'); a deflected light beam ABC appears to originate from point E' with a lateral displacement g from its true origin A. This lateral displacement can be formulated as

$$g = a - (x - e) \tan \alpha \tag{47}$$

Values for the variables a and α are obtained from Eqs. 33 and 29 respectively.

As Eq. 47 and Fig. 35*b* show, the lateral beam displacement g is independent of the thickness d of the cell wall for a given choice of the plane of focus F. Some typical results are given in Fig. 36*a*. A positive displacement signifies a movement toward the solution side of a cathodic boundary layer; a negative displacement, a movement toward the electrode side. The results illustrate that the lateral displacement of the virtual beam origin (and therefore the geometrical distortion of an image) strongly depends on the location of the plane of focus and that it can assume values larger than the dimension of typical mass transfer boundary layers.

The location e_0 of the plane of focus F_0 (inside the cell) which results in no lateral beam displacement (e.g., no movement of the electrode shadow) in a medium of constant refractive index gradient can be calculated similarly (Fig. 35*b*).

$$e_0 = x - a \cot \alpha \tag{48}$$

The location of this special plane of focus is independent of the glass wall thickness. Typical results of Eq. 48, given in Fig. 36*b*, show that its position, very close to halfway across the cell, is almost unaffected by changes in refractive index gradient. Although some

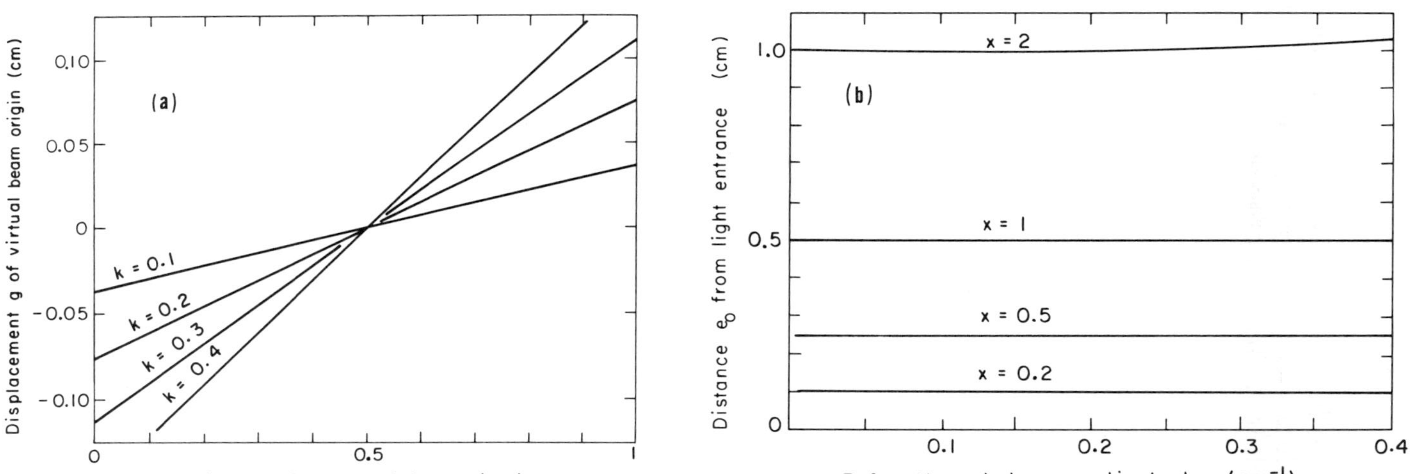

Fig. 36. Light deflection in a medium of uniform refractive index gradient k. (*a*) Lateral displacement g of virtual beam origin due to focusing. Independent of cell wall thickness. Cell thickness $x = 1$ cm. (*b*) Location e_0 of plane of focus for zero displacement of electrode shadow due to light deflection. Independent of cell wall thickness. Cell thickness $x = 0.2$, 0.5, 1, and 2 cm.

geometrical distortions in boundary layers can be avoided by this choice of plane of focus, its use is not of great practical interest because it is difficult to establish and does not minimize errors of other parameters in an interferogram (e.g., local concentration) (7). Most of the analysis to follow is restricted to focusing on an immersed object on the inside face of the cell wall facing the light source. (This is the preferred mode of operation for the observation of cathodic boundary layers (7) because it avoids the superposition of beams with different phase.)

The error in phase due to light deflection can be represented by the difference in optical path length between a deflected beam AC of length p_4 (Fig. 37) and a hypothetical undeflected beam EI of

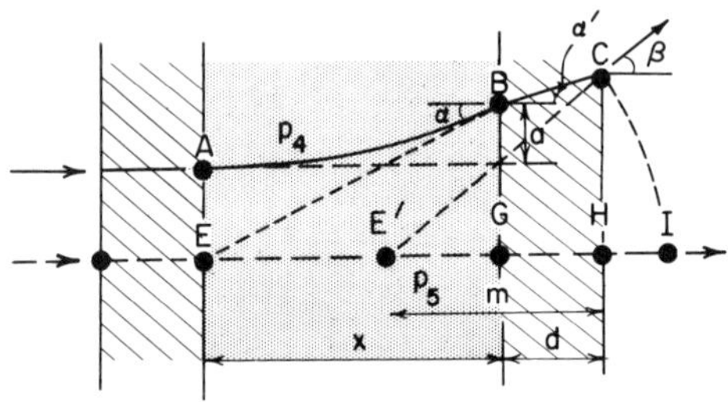

Fig. 37. Determination of the optical path error due to light deflection. Plane of focus at entrance of light into the medium of variable refractive index (plane AE). AB—optical path length p_2, AC—optical path length p_4, EI—optical path length p_5.

length p_5 passing through the same virtual origin E' in the virtual plane of focus. Points C and I lie on a circle centered in E'. Beyond points C and I, the imaging optics introduces no phase difference between both beams. The optical path length error for a point in the interferogram which corresponds to the image of E' is

$$\begin{aligned} p_4 - p_5 &= (p_2 + \overline{BC}n_g) - [x(n_0 - k\overline{AE}) + dn_g + \overline{HI}] \\ &= \left(p_2 + \frac{dn_g}{\cos\,[\sin^{-1}(\sin\beta/n_g)]}\right) \\ &\quad - \Bigg[n_0 x - kx(x\tan\alpha - a) + dn_g \\ &\qquad - \cot\beta\left(x\tan\alpha + d\tan\left(\sin^{-1}\left(\frac{\sin\beta}{n_g}\right)\right)\right)\left(1 - \frac{1}{\cos\beta}\right)\Bigg] \end{aligned} \tag{49}$$

As defined before, x and d are the thickness of cell and glass wall, respectively, n_0 and n_g are the refractive indices at the light entrance into the liquid and inside the glass wall, respectively, k is the refractive index gradient normal to the direction of the entering light, and p_2 is the optical path length of the deflected beam between points A and B. The values of p_2, a, α, and β have to be inserted from Eqs. 37, 33, 29, and 31, respectively.

With an increase in glass wall thickness, the optical path $\overline{BC}$ of the deflected beam inside the glass wall is increased more than the path $\overline{GH}$ of the undeflected comparison beam. However, the component $\overline{HI}$ of the latter is increased, so that the path difference between deflected and undeflected beams remains very closely the same. In effect, the independence of g, e_0, and $(p_4 - p_5)$ on wall thickness is achieved by a displacement of the imaging optics, necessary to keep the real plane of focus F in a fixed location while the thickness d is changed.

Some results of Eq. 49, expressed in fringe shifts, are given in Fig. 38. They illustrate that, for focusing on an immersed object at the entrance side of the light into the cell, the glass wall thickness is not an important source of optical path (concentration) error.

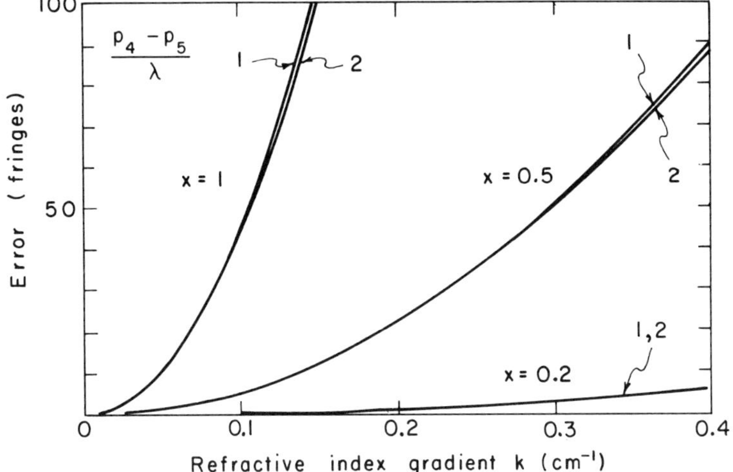

Fig. 38. Error in optical path length $(p_4 - p_5)/\lambda$ due to light deflection in a refractive index field of constant gradient k. Plane of focus on side of cell at which light enters the liquid medium ($e = 0$). Cell thickness $x = 0.2$, 0.5, and 1 cm, $n_0 = 1.3468$, $\lambda = 5.461 \times 10^{-5}$ cm; curves 1, 2—cell wall thickness $d = 0$ and 1 cm, respectively, cell wall refractive index $n_g = 1.51$.

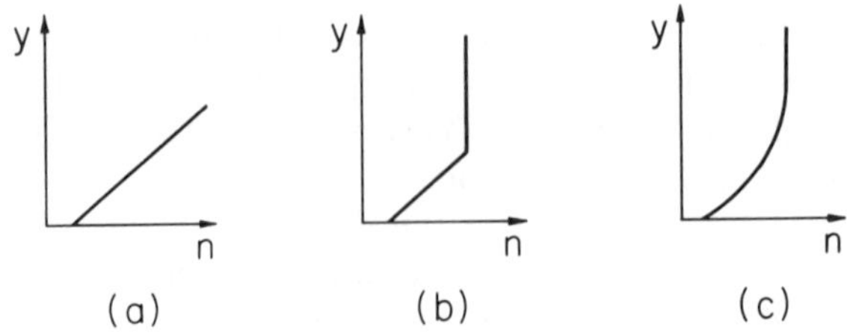

Fig. 39. Models of mass transfer boundary layers for light deflection: (*a*) constant concentration gradient, (*b*) Nernst model, (*c*) realistic concentration profile.

(d) Nernst Boundary Layer Model. All the considerations of light deflection discussed so far have been based on an unlimited linear refractive index profile (Fig. 39*a*). For the consideration of light deflection errors in mass transfer boundary layers (Fig. 39*c*), the Nernst model (Fig. 39*b*) can be used as a first approximation. This model has the advantage in that analytical solutions of the kind outlined above can be employed. For deflected beams that remain entirely inside the boundary layer, the previous solutions are directly applicable. The discontinuous change of refractive index gradient at the edge of the boundary layer results in optical artifacts which, although generally more severe than with more realistic refractive index profiles, serve to illustrate some problems of light deflection in boundary layers.

Near the outer edge (solution side) of a Nernst-type boundary layer, a light beam propagates only over part of the cell depth l within the boundary layer, as indicated in Fig. 40*a*. The curved part AB can be described by the previously derived equation, while the straight part BC continues with the slope of the curve at point B. The lateral displacement b of a beam beyond the boundary layer can be derived by use of Eq. 30. Disregarding third order terms, it is

$$b = \frac{kx}{n_0}(l - x) \tag{50}$$

where x is the abscissa of point B (Fig. 40*a*). The displacement b reaches a maximum value b_m for a beam which leaves the boundary layer in the middle of the cell ($x = l/2$)

$$b_m = \frac{kl^2}{4n_0} \tag{50a}$$

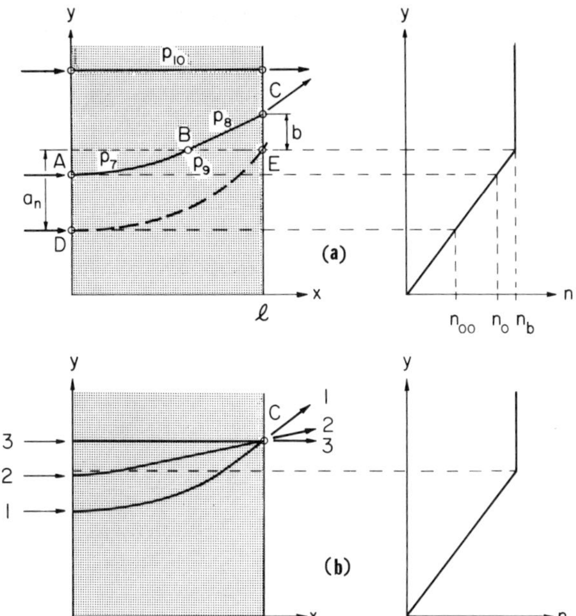

Fig. 40. Light deflection near the edge of a Nernst-type boundary layer. (*a*) Optical path lengths: p_7—deflected curved beam AB inside boundary layer; p_8—deflected straight beam BC outside boundary layer; $p_9 = p_7 + p_8$; p_{10}—undeflected beam in bulk refractive index. (*b*) Beam crossover. Two deflected (*1*, *2*) and one undeflected beam (*3*) exit at every point C.

For a beam *DE*, which travels entirely inside the diffusion layer, the lateral beam displacement a_n is, according to Eq. 34, approximately

$$a_n = \frac{kl^2}{2n_{00}} \tag{51}$$

It can be seen from Eq. 50 that a pair of deflected beams 1 and 2 of different optical path exit at any point located within a distance b_m outside the Nernst diffusion layer (Fig. 40*b*). In addition, an undeflected beam 3 also passes through the same points.

The optical path length p_9 of a deflected beam near the edge of the boundary layer can be determined as the sum of the path lengths p_7 and p_8 of the curved and straight parts (Fig. 40*a*). For focusing on the light exit side of the cell, the optical path difference between boundary layer and bulk, a measure of the local fringe shift, is

$$p_9 - p_{10} = p_7 + p_8 - l(n_b) \tag{52}$$

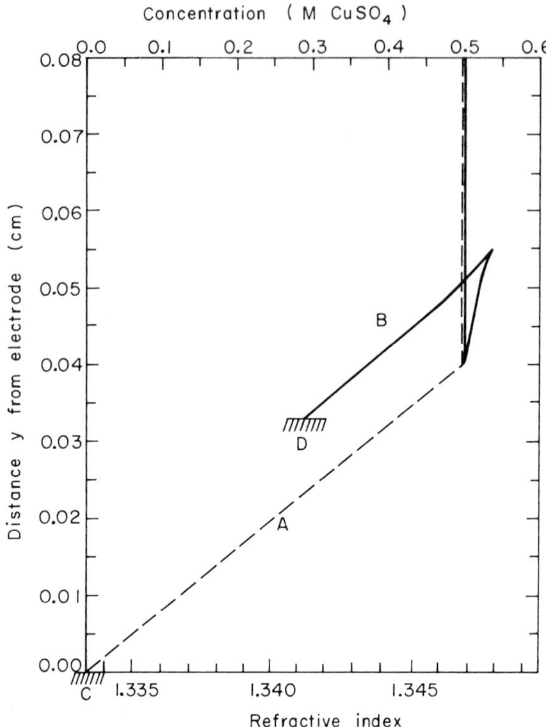

Fig. 41. Computed interference fringe B for a Nernst-type boundary layer of refractive index profile A (Eq. 52). Focus on cell exit, effect of cell wall thickness neglected. C—location of electrode surface, D—image of electrode surface. Bulk refractive index $n_b = 1.3468$, refractive index gradient $k = 0.33\ \text{cm}^{-1}$, cell width $l = 0.5$ cm. wavelength $\lambda = 5.461 \times 10^{-5}$ cm. Beach (7), and McLarnon (99).

An example of this computed optical path profile is given in Fig. 41, curve B. The superposition of three different fringe shifts, due to rays 1 to 3 shown in Fig. 40*b*, immediately outside the boundary layer, can be expected to result in a superposition of different fringe systems in this region. To a good approximation, the linear part of the fringe runs parallel to the true refractive index profile (curve A).

(e) Realistic Boundary Layers. Mass tranfer boundary layers in convective diffusion show a refractive index profile of the shape illustrated in Fig. 39*c*. The refractive index gradient and, therefore, the degree of light deflection are largest at the electrode and decrease continuously toward the bulk solution. The deflected light successively traverses regions of varying refractive index gradient and may, as seen in the Nernst model, enter the uniform bulk solution. Numerical techniques are best suited to analyze light propagation in such media (7).

In order to compute an interference fringe pattern based on an assumed refractive index profile, the optical paths of a small number (10–20) of deflected beams, which enter the cell parallel to each other and usually parallel to the electrode surface, but at different distances from it, are determined. Each beam is divided into a large number (100–200) of straight segments, as schematically shown in Fig. 42*b*. The slope of a typical segment *HJ* is determined

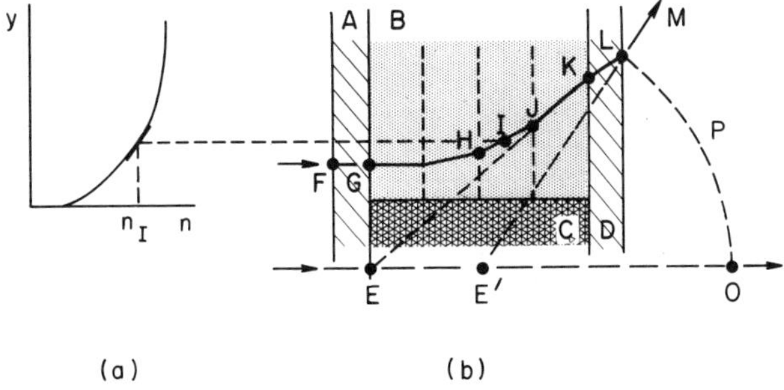

Fig. 42. Numerical computation of a deflected light beam in a boundary layer. (*a*) Refractive index profile. (*b*) Cross section of cell: *A*, *D*—cell walls, *B*—electrolyte, *C*—electrode, *GL*—deflected light beam, *P*—equiphase circle centered in virtual beam origin *E*′.

from the slope of the previous segment, the length of the segment, the refractive index n_I at its center (Fig. 42*a*) and the component of the refractive index gradient in the direction normal to the beam at the center of the segment.

The optical path length of a deflected beam, including refraction *KL* in the cell wall, is obtained by summing the product of local refractive index n_I and geometrical path length *HJ* of each beam element. From position and slope of the beam *M* leaving the cell, the virtual origin *E*′ of the deflected beam in the virtual plane of focus is determined. The fringe shift in the interferogram relative to the bulk solution is due to the difference in optical path length between deflected beam *GL* and a hypothetical undeflected beam

EO of the same virtual origin *E'*, passing through a cell filled with bulk solution. Thus phase and position of one point in the interferogram have been established. Repetition of this procedure with 10–20 beams yields a computed interference fringe.

Figure 43 illustrates how the above procedure can provide computed interference fringes (*b*) to (*e*) for an assumed refractive

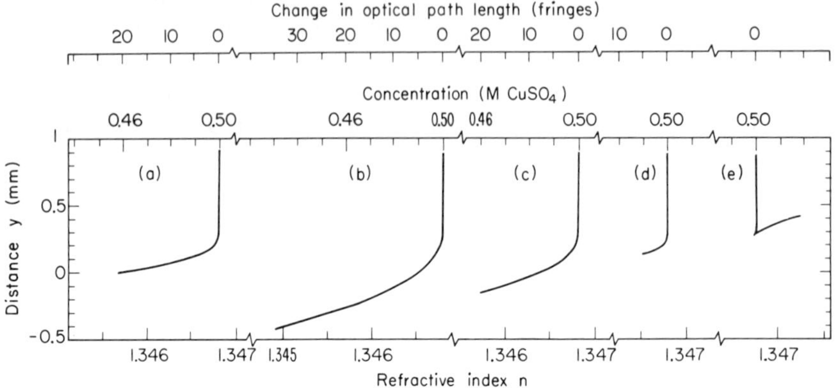

Fig. 43. Interference fringes (*b*) to (*e*) computed for refractive index profile (*a*) for different choices of the plane of focus (inside cell). Cell width 1 cm, glass wall thickness 1.27 cm, current density 4.5 mA/cm^2, wavelength 5.461 × 10^{-5} cm; (*b*) focus 3.4 mm inside glass wall farthest from the camera; (*c*) focus on inside face of glass wall farthest from the camera; (*d*) focus inside cell, 3.75 mm from the inside face of the glass wall farthest from the camera; (*e*) focus inside cell, 7.5 mm as above. Beach (7).

index profile (*a*) with different choices of plane of focus. It can be seen that both boundary layer thickness and interfacial concentration are seriously falsified by the light deflection, and the error strongly depends on the choice of the plane of focus. A common observation with such computed interference fringes is that the location of the outside edge of the boundary layer is not affected by light deflection (except with very unfavorable focus as in Fig. 43*e*).

Observed interference fringes similar to the computed ones of Fig. 43 are shown in Fig. 44. The dependence of apparent boundary layer thickness and interfacial concentration on the choice of the plane of focus can be seen. The double value of phase, found in Fig. 43*e* seems to be responsible for the extraneous interference fringe system seen in Fig. 44*c* and *d*. This effect, due to the crossing

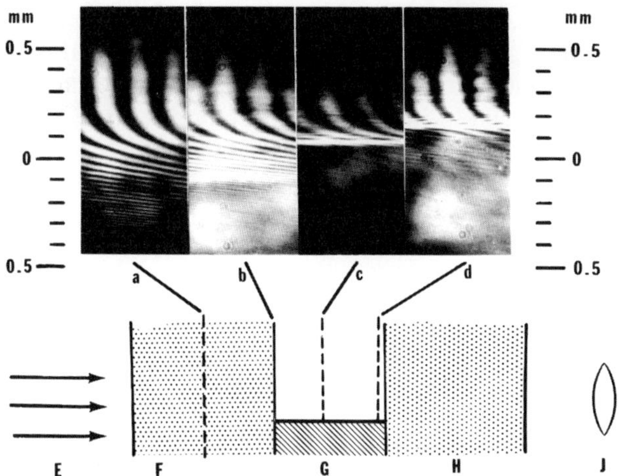

Fig. 44. Effect of the change of focal plane (inside cell) on the appearance of interference fringes for a cathodic deposition boundary layer. Planes of focus indicated in cross section of cell. (*a*) Interferogram with the focal plane inside the glass wall farthest from the camera, (*b*) interferogram with the focal plane on inside face of the glass wall farthest from the camera, (*c*) interferogram with the focal plane approximately half way across the cell, (*d*) interferogram with the focal plane near the inside of the glass wall near the camera. *E*—incoming collimated light, *F*—glass wall farthest from the camera, *G*—electrolyte and electrode, *H*—glass wall near the camera, *J*—camera lens; current density = 5.0 mA/cm^2; cell width 1.9 cm; glass width = 1.27 cm. The zero of the vertical scale defines the electrode shadow before the boundary layer was formed. Beach (7).

of differently deflected rays, is best avoided by focusing on the side of the cell where the light enters (farthest from the camera).

The numerical analysis of light deflection in refractive index fields can easily be adapted to investigate the effect of other parameters on the interferogram. For instance, the effect of misalignment of the cell, with the light not incident parallel to the electrode surface has been described (7).

Due to the large number of variables involved in the numerical analysis of light deflection in refractive index fields, it is difficult to generalize results. An attempt has been made in Figs. 46–47 to correlate data that should allow an experimenter to estimate interferometric light deflection errors under a wide range of experimental conditions for three quantities: interfacial concentration, interfacial concentration gradient, and boundary layer thickness. The computations have been based on a refractive index profile of the algebraic form given in Eq. 53 for a cathodic boundary layer of constant thickness δ (0.4 mm), constant bulk refractive index n_b (1.3468), and variable interfacial refractive index n_i (1.333–1.345). For all computations $\partial_n/\partial_c = 0.027\ (M\ CuSO_4)^{-1}$.

$$\frac{n - n_i}{n_b - n_i} = 1 - e^{-Y^2}\left[\frac{A}{1 + BY} - \frac{C}{(1 + BY)^2} + \frac{D}{(1 + BY)^3}\right] \tag{53}$$

This profile (an approximation to an error function complement) has been found to approximate a boundary layer under convective diffusion (7). In Eq. 53, n is the refractive index at a distance y from the electrode surface. The dimensionless distance Y from the electrode is

$$Y = \frac{y}{\delta} \tag{54}$$

where δ is a mathematical boundary layer thickness. Physical boundary layer thicknesses of different definition can be determined from Fig. 45, which shows concentration profiles employed in the computations. The quantities used for the other parameters in Eq. 53 are

$$A = 0.34802 \qquad B = 0.47047$$

$$C = 0.09588 \qquad D = 0.74786$$

All the results presented in Figs. 46–47 are for focusing on an object located at the inside of the cell wall farthest from the camera. This procedure largely compensates for the effects of variable cell wall thickness. The scatter of points is due to the graphical evaluation of computed interference fringes, made up of a limited number of points.

The error in concentration at the interface is shown in Fig. 46*a*. For a 2-mm thick cell, the error is unexpectedly negative and amounts to at most −8% of the bulk concentration. This error is two-thirds of the largest one computed for an unlimited linear concentration profile (Fig. 38). For the 1-cm cell, the error changes sign and increases up to 16% with increasing current density, which is much less than the error shown in Fig. 38.

The error in concentration gradient at the interface, shown in Fig. 46*b*, is at most 1% for the 2-mm cell and up to 50% for the

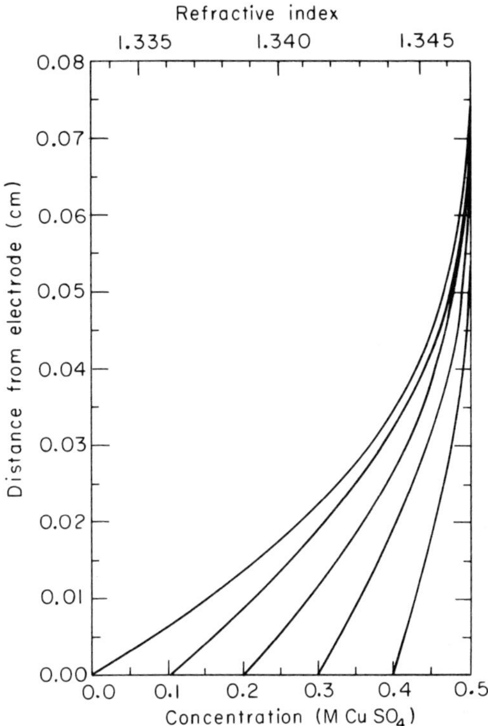

Fig. 45. Concentration profiles of boundary layer models employed for computations in Figs. 46–47. McLarnon (99).

1-cm cell. Because of the parallel displacement of the interferogram near the interface due to refraction in the glass wall, the influence of the wall thickness is expected to be very small.

Since, with the present choice of focal plane, the outside of a boundary layer (solution side) is normally not affected by light deflection, the error in boundary layer thickness (Fig. 47*a*) is entirely due to the displacement of the electrode shadow, as demonstrated in Figs. 43 and 44. For a 1-cm cell, the error is comparable to the one calculated for a constant refractive index gradient (Fig. 36*a*) at interfacial refractive index gradient 0.1, but only $\frac{1}{5}$ at $k = 0.4$. In agreement with the previous results, the effect of cell wall thickness is not noticeable.

Similar to the analysis for a constant refractive index gradient, the position e_0 of the plane of focus F_0 inside the cell (Fig. 35*b*) can

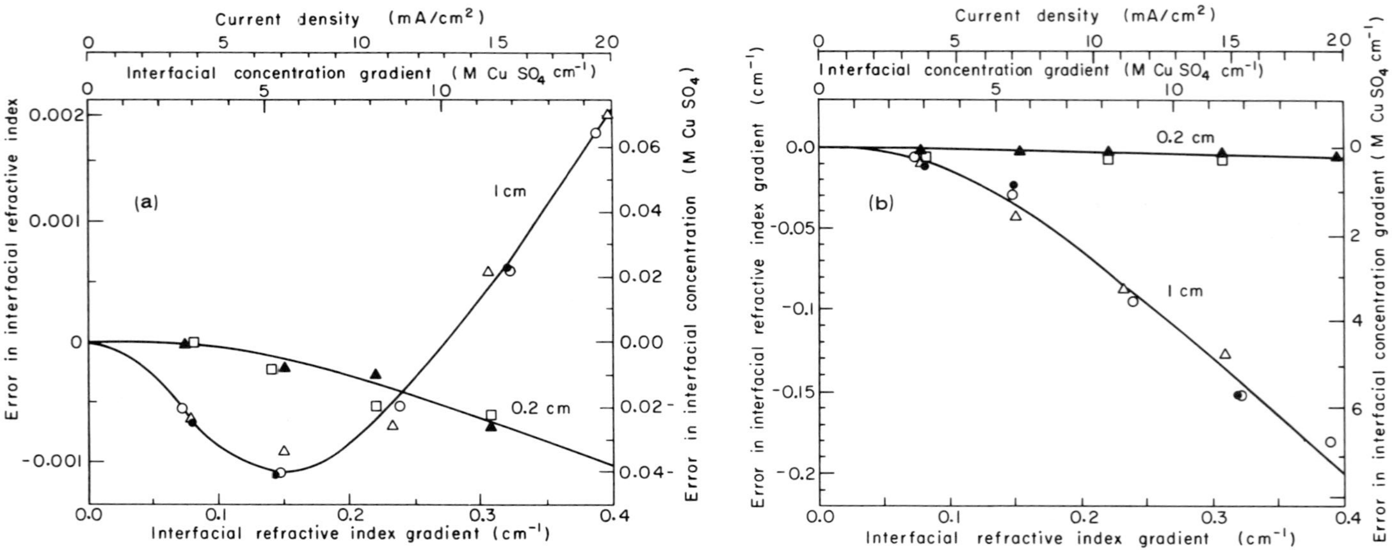

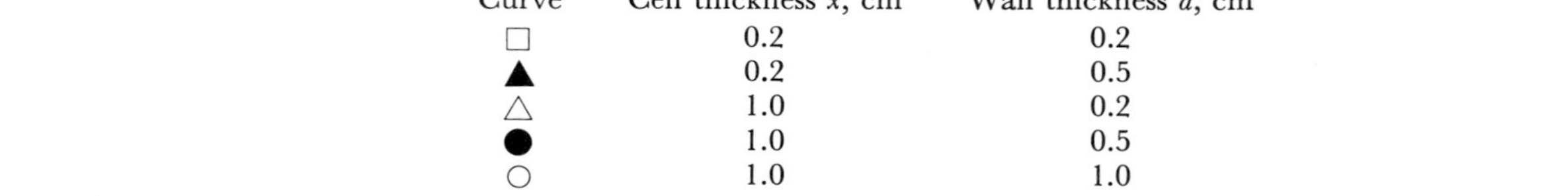

Fig. 46. Interferometric errors due to light deflection in a boundary layer model for convective diffusion (Fig. 45). Cell thickness 0.2 and 1 cm, wall thicknesses as indicated below. Boundary layer thickness 0.4 mm, bulk concentration 0.5 M $CuSO_4$, focusing on immersed target on inside surface of cell wall where light enters the solution. McLarnon (99).

Curve	Cell thickness x, cm	Wall thickness d, cm
□	0.2	0.2
▲	0.2	0.5
△	1.0	0.2
●	1.0	0.5
○	1.0	1.0

(a) Error in apparent interfacial concentration. (b) Error in apparent interfacial concentration gradient.

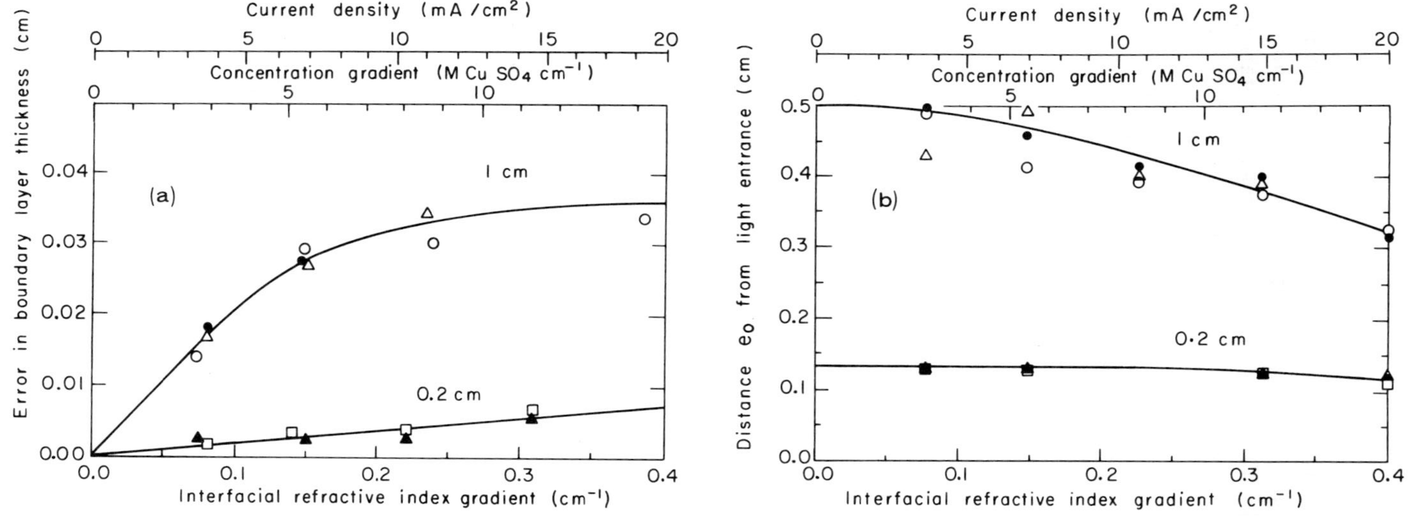

Fig. 47. Interferometric errors due to light deflection. Data as in Fig. 46, McLarnon (99). (*a*) Error in apparent boundary layer thickness. (*b*) Location e_0 of plane of focus in cell for zero displacement of electrode shadow.

be calculated, for which the image of the electrode (electrode shadow in interferogram) is not displaced by light deflection. Figure 47*b* gives some results. Compared to the previous computation (Fig. 36*b*), the position of the plane of focus in the cell for zero electrode displacement is slightly closer to the entrance side of a 1 cm thick cell and slightly farther from it for the 2-mm thick cell.

It can be concluded that computations for a constant refractive index gradient often provide a reasonable estimate of light deflection errors in boundary layers.

For the precise interpretation of measured interference fringes as refractive index profiles, a second step has to be added to the above computations. Based on a comparison of observed and computed interference fringes, the assumed refractive index profile has to be modified to improve agreement between the two, and the computation is repeated. Several cycles are usually necessary to produce satisfactory agreement (7).

(f) Anodic Boundary Layers. The considerations of the two preceding sections have been restricted to boundary layers with refractive index decreasing toward the electrode surface. Boundary layers with refractive index increasing toward the electrode surface, such as those typically encountered in anodic metal dissolution, cause the light to be deflected toward the electrode. If all the light which strikes the electrode surface is absorbed, rays entering the cell closer to the electrode than the lateral beam displacement do not contribute to the picture (Fig. 48*b*). In contrast to the cathodic boundary layers, an overlapping of differently deflected beams is now best avoided by focusing the cell on the plane where the light exits. If the electrode is sufficiently smooth, light which strikes the

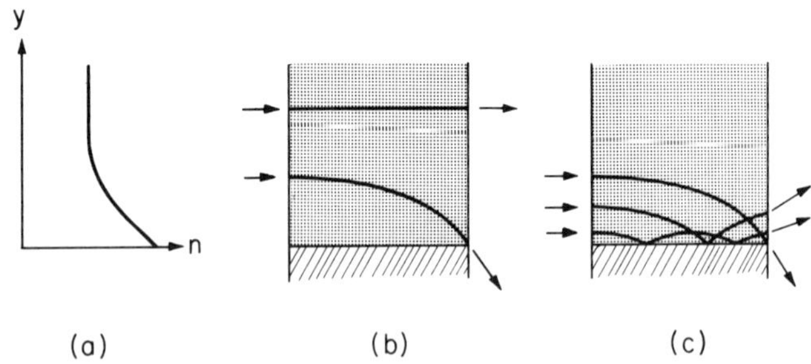

Fig. 48. Light deflection in anodic boundary layer: (*a*) refractive index profile, (*b*) undeflected beam in the bulk and deflected beam which leaves cell at the electrode surface, (*c*) reflection of deflected beams at the electrode surface.

surface is reflected (Fig. 48*c*), and a much more complex optical situation exists, particularly if a reflected beam is deflected back to the electrode surface and reflected again.

C. Diffraction Effects. The closest distance to an electrode, to which the composition of a solution can be determined by interferometry, is basically limited by light diffraction. For a two-dimensional (infinitely thin) object in focus, the diffraction-limited resolution of a microscope is usually formulated as

$$r = 0.61 \frac{\lambda_0}{\text{N.A.}} \tag{55}$$

where λ_0 is the light wavelength in vacuum and N.A. the numerical aperture of the objective lens defined as

$$\text{N.A.} = n \sin \theta \tag{56}$$

Here n is the refractive index in the object space; θ is the angle between the optic axis and the marginal ray entering the objective at its periphery. For coherent illumination, the numerical factor in Eq. 55 is 0.77 (15).

Even the simplest flat electrode is, however, of finite thickness, and diffraction effects can be expected to be more complex. The microscopic resolution limit is, therefore, usually not attained in electrochemical interferometry. Since a theory of diffraction from a planar object does not seem to be readily available, a discussion of diffraction problems will have to remain largely qualitative.

Only a selected cross section of an extended electrode can be in focus, with other parts being out of focus to various degrees. It is these defocused regions that are probably the most serious cause for degraded resolution due to diffraction in an interferometer. These regions can be expected to produce diffraction fringes similar to those observed behind a straight edge. Their real or virtual origin is projected in the plane of focus. The interaction of these diffraction fringes with the interferogram can result in local changes of geometry and apparent phase.

Amplitude and phase of a Fresnel diffraction pattern behind a straight edge can be derived from Cornu's spiral (71). An example is given in Fig. 49. It is important to keep in mind that across the diffraction pattern, the phase varies continuously. Superposition with the continuously varying phase of an interferogram can, therefore, lead to quite unexpected results (Fig. 50). The distance of diffraction maxima and minima from the geometrical shadow

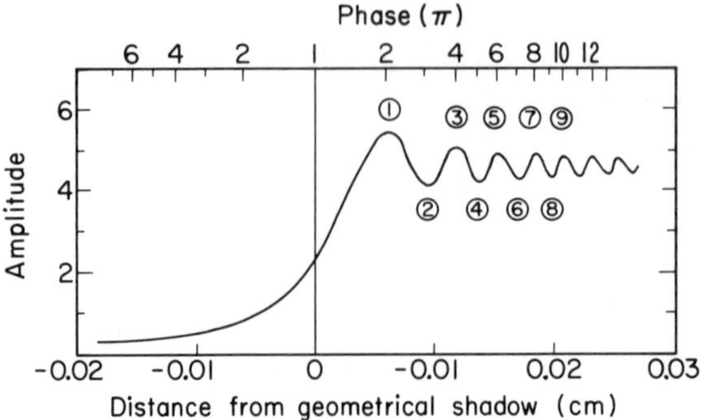

Fig. 49. Amplitude and phase in a diffraction image 9.2 mm behind a straight-edge located at abscissa zero. Wavelength 546.1 nm.

depends on the distance behind the straight edge (degree of defocusing) as shown in Fig. 51. It can be seen that for a 1-mm defocusing, the separation between the first two intensity maxima is 0.02 mm. Such a diffraction pattern will be resolved by interferometers of interest here. If half the distance between the first diffraction maximum and the true location of the edge is taken as the uncertainty of the location of the edge, diffraction from one end of a 1-cm wide electrode, focused on the other end, introduces an uncertainty of electrode position of 0.003 cm. This distance is much smaller than the light deflection in the same cell under most conditions.

A more realistic appraisal of diffraction phenomena in interferometry, in addition to being based on a better theory of diffraction, will also have to include the effect of light deflection.

3. *Applications*

A. Diffusion and Electrophoresis. The determination of diffusion coefficients by interferometric techniques involves the observation of the concentration profile which develops between two solutions of different concentration as a function of time. Evaluations of results with different degrees of sophistication can be found in the literature (25,27,85,136). A technique which is based on the observation of regions far removed from the liquid junction, where concentration gradients are small, and light deflection errors are more likely to be negligible, has been employed by Chapman (24). This technique is also unaffected by the sharpness of the initial

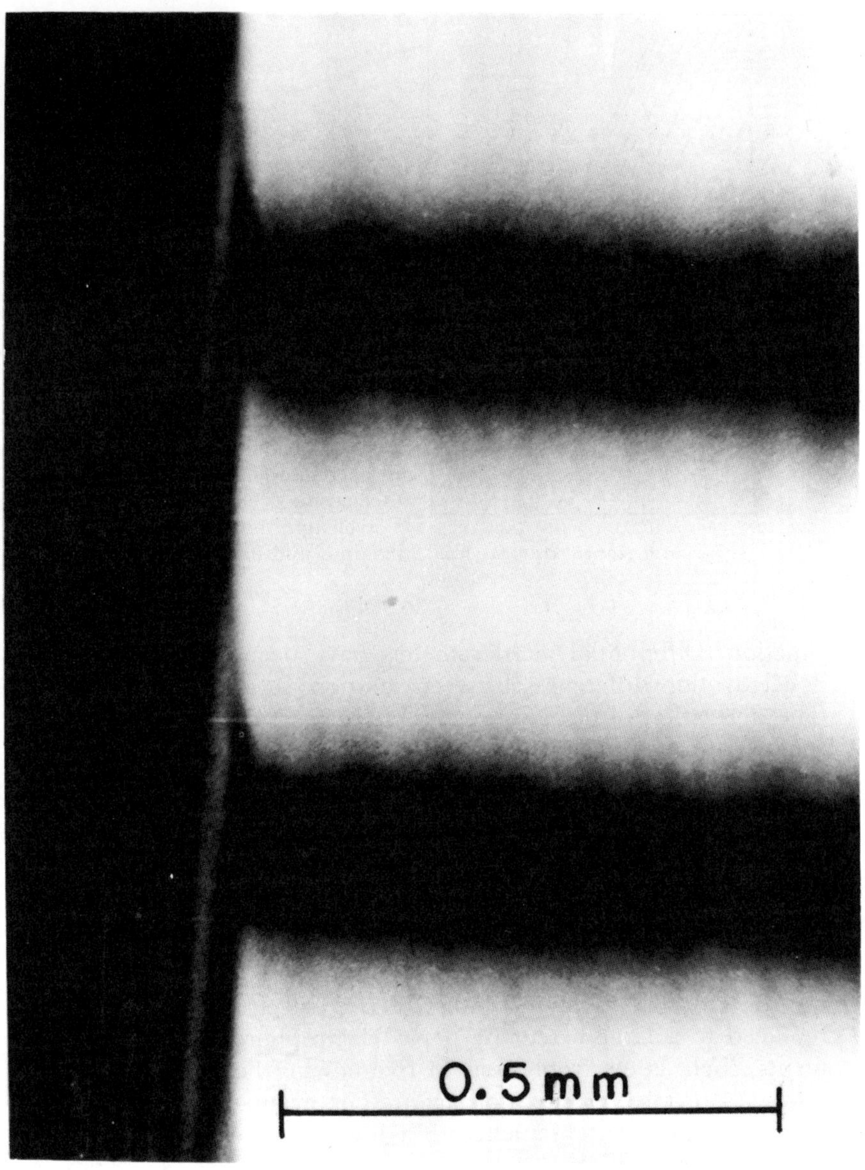

Fig. 50. Diffraction effects in an interferogram near an electrode surface.

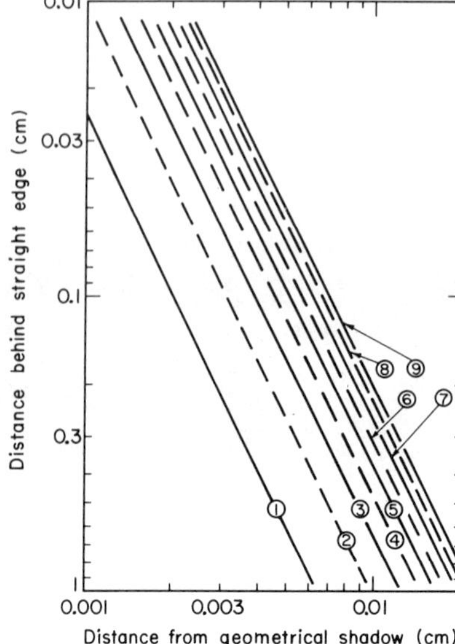

Fig. 51. Position of diffraction maxima (solid lines) and minima (broken lines) as a function of distance (defocusing) behind a straightedge. Wavelength 546.1 nm, numbering of maxima and minima as on Fig. 49.

junction. Differential interferometers have been used for very low concentration differences between solutions at the liquid junction. The derivation of diffusion coefficients from such measurements is somewhat different (17,139).

Diffusion coefficients have been determined with different types of interferometers, among them Jamin-types (25,85), Rayleigh-types (20,27,93), Mach-Zehnder (19), Polarizing (17,111,139), and Gouy (28,49) types. Interferograms of diffusing liquid junctions are shown in Figs. 7, 9, and 13. Optical techniques for diffusion studies have been reviewed by Gosting (51).

Electrophoresis has lost some of its importance for the application of interferometry because the use of zone electrophoresis has been replaced to a large extent by paper electrophoresis. New developments, such as electrophoresis in free-flowing liquids (100) could, however, reverse the situation again. The evaluation of an electrophoretic separation observed with a Jamin interferometer has been illustrated in a number of examples in the literature (47,48,98). A review of electrophoretic techniques has been given by Longsworth (95).

B. Mass Transport Boundary Layers. Boundary layers on working electrodes have been investigated by interferometry under

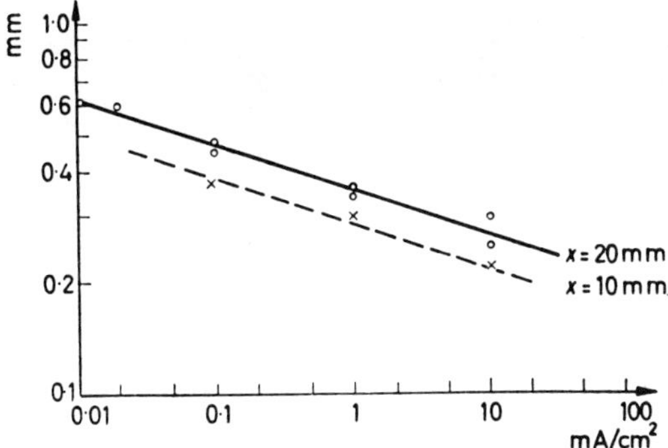

Fig. 52. Boundary layer thickness (defined as distance from the electrode at which concentration reduction is 10% of that at the interface) on a 3 cm high vertical cathode, observed 1 and 2 cm from the bottom, under natural convection. 0.6 *M* $CuSO_4$. Ibl (68).

forced (90) and natural (64,65) convection conditions, as well as in the absence of convection (62,144,145). Interferometers employed in these studies include Jamin (64,65), Mach-Zehnder (90,144,145), and Rayleigh (62) types. Interferograms of mass transfer boundary layers have been shown in Figs. 11, 14, and 18. Data on boundary layer thickness under natural convection, derived from interferograms, are given in Fig. 52. While concentration profiles derived by the conventional method at low current densities are close to those expected (Fig. 53), deviations found at higher current densities (90) could well be due to light deflection effects. A displacement

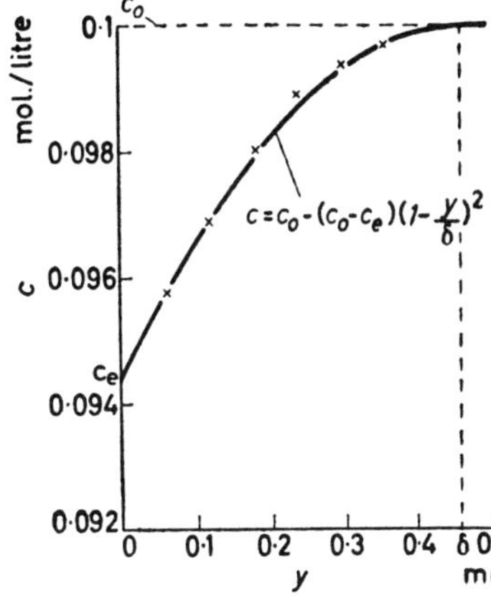

Fig. 53. Comparison of an interferometrically measured concentration profile (×) with a theoretical one (——) 0.1 *M* $CuSO_4$, 0.32 mA/cm². Ibl (68).

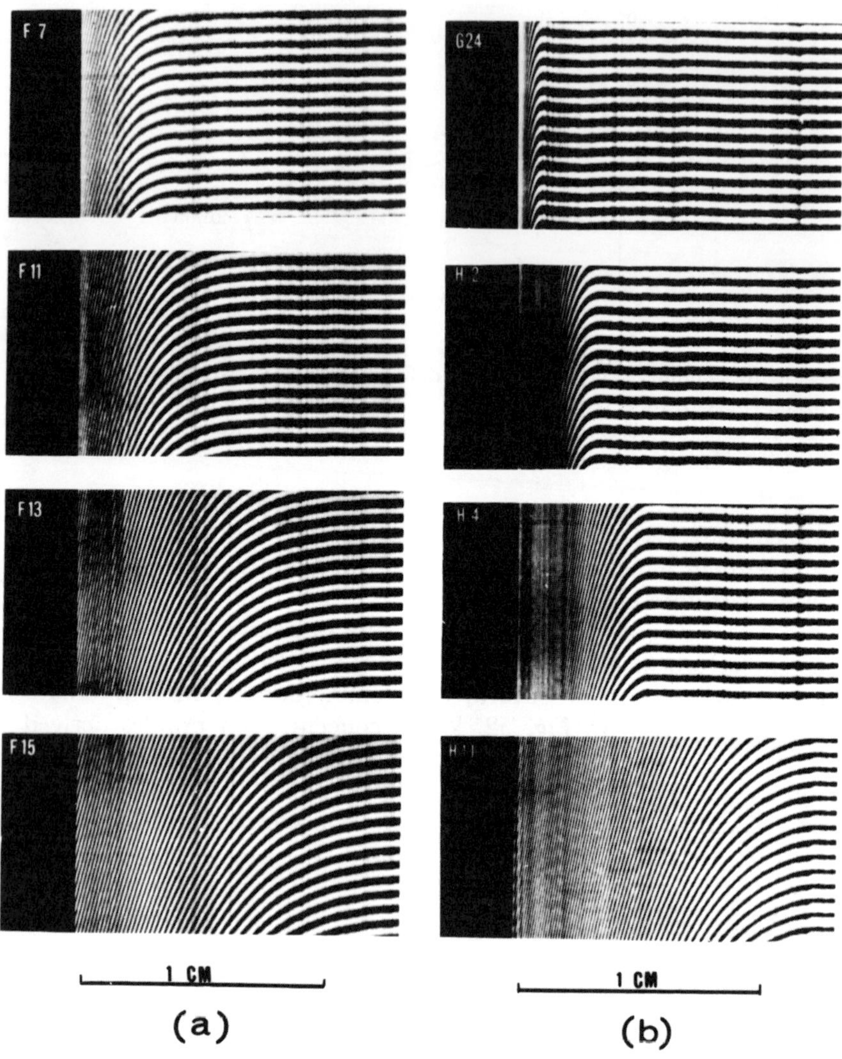

Fig. 54. Boundary layers in stagnant electrolytes at constant voltage, observed with a Rayleigh interferometer Hsueh (62). (*a*) 0.0385 *M* $CuSO_4$; initial current density 0.244 mA/cm^2; times: 40, 103, 263, and 475 min. (*b*) 0.1 *M* $CuSO_4$ + 1 *M* H_2SO_4; times: 9, 40, 106, and 450 min.

of the electrode shadow at high current densities has been demonstrated (145). In the presence of a supporting electrolyte, the interferogram represents a superposition of two boundary layers which possess opposite concentration gradients and are, in general, of different thickness. The result can be a maximum in the refractive index profile (Fig. 54*b*) or a more abrupt change from the bulk refractive index compared to the boundary layer with a single solute (Fig. 54*a*). Qualitatively, interferograms of electrochemical boundary layers are similar to those found in heat transfer and nonelectrochemical mass transfer (31,126).

C. Outlook. Recent advances in the quantitative understanding of electrochemical processes require a detailed knowledge of transport properties. To a large extent, interferometry is capable of providing data for use in models of transport processes. Together with the application of established interferometric techniques, the use of a number of instruments not previously considered for such purposes could result in important advances in the design and use of electrochemical devices.

The systematic consideration of light deflection errors, barely initiated at present, should make it possible to produce results of greatly improved reliability. The computational techniques developed for the consideration of light deflection can be adapted for the analysis of some 3-dimensional concentration fields. It is to be hoped that the present limitation of interferometry to one-component electrolytic solutions can be overcome by a combination with other techniques, such as spectroscopic ones.

IV. Interferometry of Electrode Surfaces

Among optical techniques used for surface finish evaluation (152), double beam interferometry provides some particularly attractive techniques for observing the microtopography of electrode surfaces (4). Compared to multiple beam interferometry, usually considered for this purpose (133,140), double beam interference microscopes are basically capable of using objectives of higher numerical apertures, because of the absence of a reference mirror between lens and object. Also, they avoid the sampling of an extended region to contribute to the interference phenomenon at one point of the image (82). Thus smaller surface features can be resolved. Another advantage is that the specimen surface does not have to be coated with a highly reflecting material (4,80), but can be inspected without alteration. The sinusoidal intensity distribution, typical of double beam interference fringes, is often considered a disadvantage for surface topographic studies (4). It will be shown, however, that this feature can be turned into an advantage by the use of microdensitometry.

Complete separation of sample and reference beams for the examination of opaque surfaces is realized in a modification of the Michelson interferometer, the Zeiss-Linnik interference microscope (8,43,143), schematically shown in Fig. 55. This instrument employs pairs of matched objective lenses in specimen and reference beams and a replaceable reference mirror, which can be chosen with a reflectivity close to that of the specimen for good fringe contrast. Auxiliary optical elements, not shown in Fig. 55, allow one to control spacing and orientation of interference fringes.

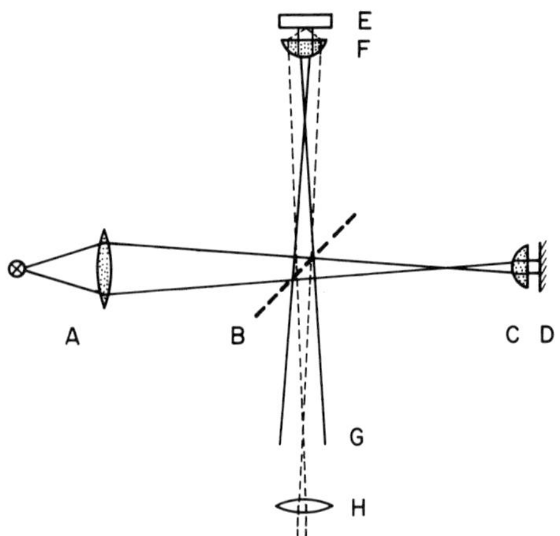

Fig. 55. Zeiss-Linnik interference microscope: *A*—illuminating system, *B*—beam splitter, *C*, *F*—matched objective lenses, *D*—reference mirror, *E*—object, *G*—image plane, *H*—eyepiece.

A sample interferogram of an optically polished metal surface is shown in Fig. 56. For the interpretation of such an interferogram, it must be remembered that optical path differences are doubled because of the reflection of light from the specimen. An interference fringe shift of one fringe spacing therefore represents a local displacement of the reflecting surface by only half a wavelength of the light

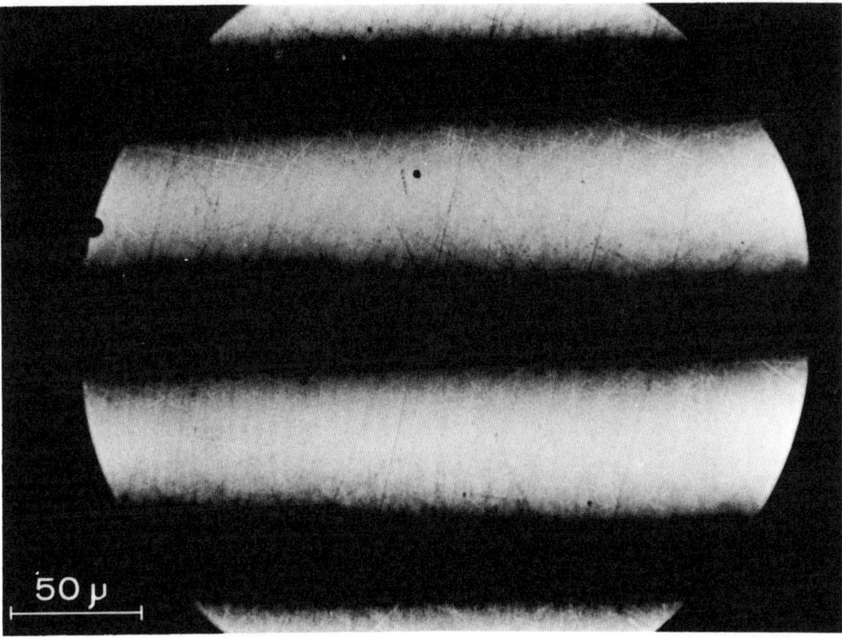

Fig. 56. Interferogram of an optically polished surface of type 316 stainless steel. Wavelength 535.0 nm, Zeiss-Linnik interference microscope. Objective 60 ×, N.A. 0.63.

employed. Thus the displacement of interference fringes of approximately 0.08 fringe spacings due to the few prominent scratches seen on Fig. 56, represent a surface deviation of about 200 Å.

The depth of the finer polishing marks is difficult to estimate by visual inspection of the interferogram. However, a microdensitometer scan of an enlarged transparency along the shoulder of an interference fringe (parallel to its main direction) provides a quantitative analysis of these details (Fig. 57, solid line). The local optical transmission can be correlated with the fringe displacement by a similar scan normal to the direction of the interference fringes (Fig. 57, broken line) and a contour scale can be derived from such a scan based on the linear change in phase between fringes. This scale automatically takes all photographic nonlinearities and process variations into account. Since the scale is compressed at both ends, the surface profile scan is preferably chosen in the center part of the transmission range, where variations in surface positions of 5 Å seem

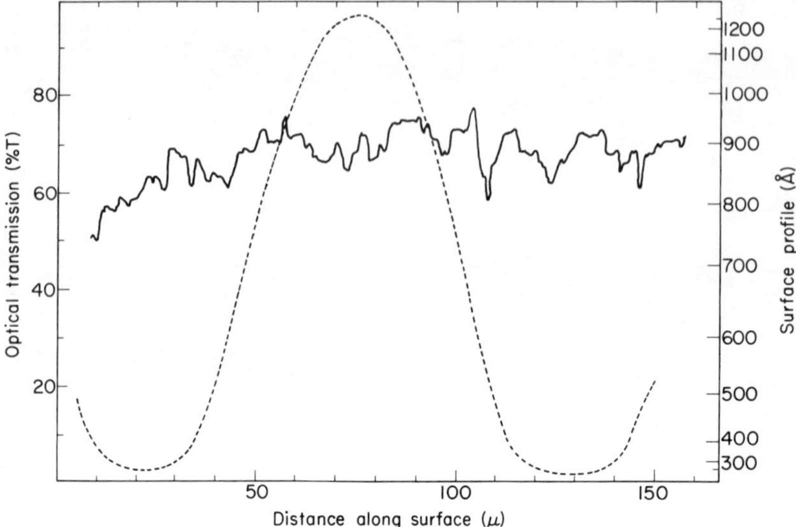

Fig. 57. Microdensitometry of interference micrograph shown in Fig. 56. Solid line: scan parallel to interference fringe, resulting in surface profile. Broken line: scan normal to interference fringes, resulting in (nonlinear) surface profile scale shown on the right.

to be resolvable. This compares with a resolution of 20–30 Å reported for the use of photographically generated equidensity lines (82). In good agreement with the visual estimate, the deepest local surface variation found in Fig. 57 is about 200 Å. This figure is further confirmed by the electron micrograph of a replica of the same surface, shadowed at a small angle (Fig. 58). The more shallow polishing marks, recorded in Fig. 57, are also visible in Fig. 58.

The detailed analysis of interferograms assumes all intensity changes to be due to interference. It should be asserted separately (e.g., by blocking the reference beam) that other factors affecting local image intensity, such as scattering or absorption in the object, or the effect of imperfections in the instrument, can be neglected. Since the phase of the object wave is measured relative to that of the reference wave in the Zeiss-Linnik interference microscope and many other interference microscopes for reflected light (80), smoothness and flatness of the reference mirror are essential for reliable results. Any effect of the reference mirror on the interferogram could be detected by slightly displacing the object in the field of view.

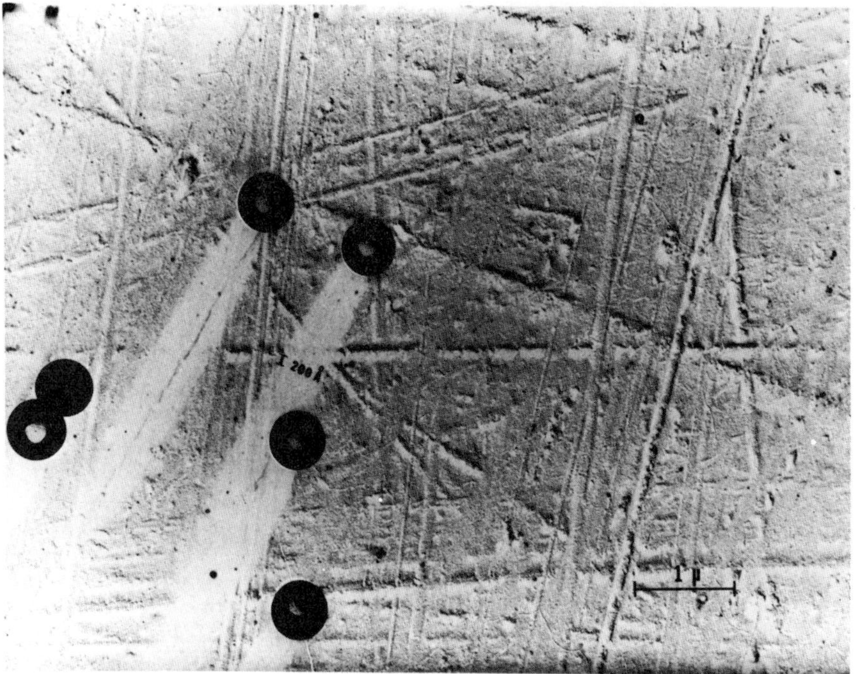

Fig. 58. Electron micrograph of a replica of the surface shown in Figs. 56 and 57. Diameter of latex spheres 0.53 micron. Depth of deepest scratches 200 Å.

Instrument designs which avoid the use of a reference plane have been described by Krug et al. (81).

Polarizing interference microscopes for reflected light (44) usually produce differential interferograms in which the image intensity is a measure of the surface slope in a given direction. This mode of operation produces a relief-like appearance of the object which can be very pleasing to the eye, but is less suited for quantitative evaluation. Other types of interference microscopes for reflected light have been described (38,80).

Interference microscopy can be expected to find increasing use for the characterization of electrode topographies. New applications dealing with crystallographic and metallurgical factors can be forseen as well as *in situ* observations of electro-crystallization and dissolution processes.

Acknowledgments

This work was conducted in part under the auspices of the U.S. Atomic Energy Commission. Some of the analyses given in Section III, part 2, Bb, Bd, and C were originally carried out at the Swiss Federal Institute of Technology in collaboration with Professor N. Ibl.

References

1. Antweiler, H. J., *Z. Elektrochem.*, **44**, 719 (1938).
2. Antweiler, H. J., *Kolloid Z.*, **115**, 130 (1949).
3. Antweiler, H. J., *Chem. Ing. Tech.*, **24**, 284 (1952).
4. Baird, K. M., in *Advanced Optical Techniques*, A. C. S. Van Heel, Ed., North-Holland, Amsterdam; Wiley, New York, 1967, Ch. 4.
5. Baird, K. M., and G. R. Hanes, in *Applied Optics and Optical Engineering*, Vol. IV, Part I, R. Kingslake, Ed., Academic Press, New York, 1967, Ch. 9.
6. Beach, K. W., R. H. Muller, and C. W. Tobias, *Rev. Sci. Instr.*, **40**, 1248 (1969).
7. Beach, K. W., Ph.D. Thesis, University of California, Berkeley, June 1971 (UCRL-20324), also *J. Opt. Soc. Am.*, **63** (1973), in press.
8. Benford, J. R., and H. E. Rosenberger, in *Applied Optics and Optical Engineering*, Vol. IV, Part I, R. Kingslake, Ed., Academic Press, New York, 1967, Ch. 2.
9. Bennett, A. H., H. Jupnik, H. Osterberg, and O. W. Richards, *Phase Microscopy*, Wiley, New York, 1951, Ch. II.
10. *Ibid.*, Ch. VII.
11. Bennett, F. D., and G. D. Kahl, *J. Opt. Soc. Am.*, **43**, 71 (1953).
12. Born, M., and E. Wolf, *Principles of Optics*, 2nd ed., Macmillan Co., New York, 1964, Ch. II.
13. *Ibid.*, Ch. III.
14. *Ibid.*, Ch. VII.
15. *Ibid.*, Ch. VIII.
16. *Ibid.*, Ch. X.
17. Bryngdahl, O., *Acta Chem. Scand.*, **11**, 1017 (1957).
18. Bryngdahl, O., in *Progress in Optics*, Vol. IV, E. Wolf, Ed., North-Holland, Amsterdam, 1965, Ch. II.
19. Caldwell, C. S., Ph.D. Thesis, University of Washington, 1955.
20. Calvert, E., and R. Chevalieras, *J. Chim. Phys.*, **43**, 37 (1946).
21. Candler, C., *Modern Interferometers*, Hilger and Watts, London, 1951, Ch. VI and VII.
22. *Ibid.*, Ch. VIII and IX.
23. *Ibid.*, Ch. XX.
24. Chapman, T. W., Ph.D. Thesis, University of California, Berkeley, November 1967 (URCL-17768).
25. Chatterjee, A., *J. Am. Chem. Soc.*, **86**, 793 (1964).
26. Craven, E. C., *Proc. Phys. Soc.*, **57**, 97 (1945).

27. Creeth, J. M., *J. Am. Chem. Soc.*, **77**, 6428 (1955).
28. Dunhop, P. J., and L. J. Gosting, *J. Am. Chem. Soc.*, **75**, 5073 (1953).
29. Dwight, H. B., *Tables of Integrals and other Mathematical Data*, Macmillan, New York, 1960, 3rd ed., p. 160.
30. Dyson, J., in *Concepts of Classical Optics*, by J. Strong, Freeman, San Francisco, 1958, Appendix B.
31. Eckert, E. R. G., and R. M. Drake, *Heat and Mass Transfer*, McGraw-Hill, New York, 1959, Ch. 9, 11.
32. *Ibid.*, Ch. 11.
33. Emanuel, A., and D. R. Olander, *J. Chem. Engr. Data*, **8**, 31 (1963).
34. Forsberg, R., and H. Svensson, *Opt. Acta*, **1**, 90 (1954).
35. Françon, M., in *Proc. Symposium Interferometry*, Natl. Phys. Lab., 1959, H. M. Stationery Office, London, 1960, p. 129.
36. Françon, M., *Progress in Microscopy*, Row, Peterson & Co, Evanston, Ill., 1961, Ch. II.
37. *Ibid.*, Ch. III.
38. *Ibid.*, Ch. IV.
39. Françon, M., *Optical Interferometry*, Academic Press, New York, 1966, Ch. II.
40. *Ibid.*, Ch. III.
41. *Ibid.*, Ch. IV.
42. *Ibid.*, Ch. VII.
43. *Ibid.*, Ch. XVI.
44. Françon, M., in *Advanced Optical Techniques*, A. C. S. Van Heel, Ed., North-Holland, Amsterdam; Wiley, New York, 1967, Ch. 2.
45. Fritz, J. J., and C. R. Fuget, *J. Phys. Chem.*, **62**, 303 (1958).
46. Goldstein, R. J., *Rev. Sci. Instr.*, **36**, 1408 (1965).
47. Gooderum, P. A., G. P. Wood, and M. J. Brevoort, *36th Ann. Rep. Natl. Adv. Comm. Aeronautics*, Report 963, p. 289, U. S. Gov. Print. Office, Washington, D.C., 1951.
48. Gosting, L. J., E. M. Hanson, G. Kegeles, and M. S. Morris, *Rev. Sci. Instr.*, **20**, 209 (1949).
49. Gosting, L. J., *J. Am. Chem. Soc.*, **72**, 4418 (1950).
50. Gosting, L. J., and L. Onsager, *J. Am. Chem. Soc.*, **74**, 6066 (1952).
51. Gosting, L. J., in *Advances in Protein Chemistry*, Vol. XI, M. L. Anson, K. Bailey, and J. T. Edsall, Eds., Academic Press, New York, 1956, p. 429.
52. Hale, A. J., *The Interference Microscope in Biological Research*, Livingstone, Edinburgh, 1958, Ch. II.
53. *Ibid.*, Ch. III, IV.
54. Hansen, G., *Z. Instrumentenkd.*, **50**, 460 (1930).
55. Hansen, G., *Z. Tech. Phys.*, **12**, 436 (1931).
56. Hansen, G., *Zeiss Nachr.*, **3**, 302 (1940).
57. Hansen, G., *Z. Instrumentenkd.*, **60**, 325 (1940).
58. Hariharan, P., and D. Sen, *J. Sci. Instr.*, **38**, 428 (1961).
59. Hariharan, P., and D. Sen, *Opt. Acta*, **9**, 159 (1962).
60. Heydebrand und der Lasa, E., von, *Ann. Phys.*, [5] **37**, 589 (1940).
61. Hodgman, C. D., R. C. Weast, and S. M. Selby, Eds., *Handbook of Chemistry and Physics*, 38th ed., Chemical Rubber Publishing Co., Cleveland, Ohio, 1956.

62. Hsueh, L., Ph.D. Thesis, University of California, Berkeley, December 1968 (URCL-18597).
63. Hugenschmidt, H., and K. Vollrath, in *Proc. Ninth Itnl. Congr. High-Speed Photography*, W. G. Hyzer and W. G. Chace, Eds., Society of Motion Picture and Television Engineers, New York, 1970, p. 86.
64. Ibl, N., Y. Barrada, and G. Truempler, *Helv. Chim. Acta*, **37**, 583 (1954).
65. Ibl, N., and R. Muller, *Z. Elektrochem.*, **59**, 671 (1955).
66. Ibl, N., Private communication, 1956.
67. Ibl, N., *Proc. Seventh Meet. Int. Comm. Electrochem. Thermodyn. Kinet., Lindau, 1955*, Butterworths, London, 1957, p. 112.
68. Ibl, N., *Proc. Eighth Meet. Int. Comm. Electrochem. Thermodyn. Kinet., Madrid, 1956*, Butterworths, London, 1958, p. 174.
69. Ingelstam, E., E. Djurle, and L. Johansson, *J. Opt. Soc. Am.*, **44**, 472 (1954).
70. Iwata, K., and R. Nagata, *J. Opt. Soc. Am.*, **60**, 133 (1970).
71. Jenkins, F. A., and H. E. White, *Fundamentals of Optics*, McGraw-Hill, New York, 1957, 3rd ed., Part II.
72. Kahl, G. D., and D. B. Sleator, *Rev. Sci. Instr.*, **36**, 993 (1965).
73. Kegeles, G., and L. J. Gosting, *J. Am. Chem. Soc.*, **69**, 2516 (1947).
74. Kern Optical Company, Aarau, Switzerland, Micro-Electrophoresis Apparatus LK 30, Pamphlet.
75. Kinder, W., *Optik*, **1**, 413 (1946).
76. Kinzly, R. E., *Appl. Opt.*, **6**, 137 (1967).
77. Kraushaar, R., *J. Opt. Soc. Am.*, **40**, 480 (1950).
78. Krug, W., and E. Lau, *Phys. Bl.*, **15**, 11 (1959).
79. Krug, W., J. Rienitz, and G. Schulz, *Contributions to Interference Microscopy*, Hilger and Watts, London, 1964, Ch. 3.
80. *Ibid.*, Ch. 4.
81. *Ibid.*, Ch. 8.
82. *Ibid.*, Ch. 9.
83. Kuhn, H., *Rep. Progr. Phys.*, **14**, 64 (1951).
84. Labhart, H., and H. Staub, *Helv. Chim. Acta*, **30**, 1954 (1947).
85. Labhart, H., W. Lotmar, and P. Schmid, *Helv. Chim. Acta*, **34**, 2449 (1951).
86. Lamla, E., *Z. Instrumentenkd.*, **62**, 337 (1942).
87. Lamm, O., *Z. Phys. Chem.*, **A 138**, 313 (1928).
88. Lau, E., and W. Krug, *Equidensitometry*, Focal Press, London, 1968.
89. Levi, L., *Applied Optics*, Wiley, New York, 1968, p. 332.
90. Lin, C. S., R. W. Moulton, and G. L. Putnam, *Ind. Eng. Chem.*, **45**, 640 (1953).
91. Lohmann, A., *Opt. Acta*, **9**, 1 (1962).
92. Longsworth, L. G., *J. Am. Chem. Soc.*, **69**, 2510 (1947).
93. Longsworth, L. G., *Rev. Sci. Instr.*, **21**, 524 (1950).
94. Longsworth, L. G., *Anal. Chem.*, **23**, 346 (1951).
95. Longswoth, L. G., in *Electrophoresis*, M. Bier, Ed., Academic Press, New York, 1959, Ch. 4.
96. Lotmar, W., *Helv. Chim. Acta*, **32**, 1847 (1949).
97. Lotmar, W., *Rev. Sci. Instr.*, **22**, 886 (1951).
98. Lotmar, W., *Plasma*, **1**, 209 (1953).
99. McLarnon, F. L., private communication, 1971.

100. Mel, H. C., *J. Chem. Phys.*, **31**, 559 (1959).
101. Merwin, H. E., in *Int. Crit. Tables*, Vol. 7, E. W. Washburn, C. J. West, N. E. Dorsey, and M. D. Ring, Eds., McGraw-Hill, New York, London, 1930, p. 21.
102. Nebe, W., *Analytische Interferometrie*, Akademische Verlagsges., Leipzig, 1970.
103. Nomarski, M. G., *J. Phys. Radium*, [8] **16**, 9S (1955).
104. Oertel, H., in *Proc. 5th Intl. Congress High-Speed Photography*, J. S. Courtney-Pratt, Ed., Society of Motion Picture and Television Engineers, New York, 1962, p. 525.
105. Oppenheim, A. K., P. A. Urtiew, and F. J. Weinberg, *Proc. Roy. Soc.*, **A 291**, 279 (1966).
106. Pandya, T. P., and F. J. Weinberg, *Proc. Roy. Soc.*, **A279**, 544 (1964).
107. Partington, J. R., *An Advanced Treatise on Physical Chemistry*, Vol. 4, *Physico-Chemical Optics*, Longman Green Co., New York, 1953, Sec. XA.
108. Perkin Elmer Corporation, Norwalk, Conn. Instructions, Model 238 Electrophoresis Apparatus, Sept. 1963.
109. Philpot, J. St. L., and G. H. Cook, *Res. Appl. Ind.*, **1**, 234 (1948).
110. Pickard, R. H., A. H. J. Houssa, and H. Hunter, in *Int. Crit. Tables*, Vol. 7, E. W. Washburn, C. J. West, N. E. Dorsey, and M. D. Ring, Eds., McGraw-Hill, New York, London, 1930, p. 65.
111. Podalinskii, A. V., *Russ. J. Phys. Chem.*, **37**, 636 (1963).
112. Rienitz, J., U. Minor, and E. Urbach, in *Optical Instruments and Techniques*, K. J. Habell, Ed., Chapman and Hall, London, 1962.
113. Roegener, H., *Z. Elektrochem.*, **47**, 164 (1941).
114. Roegener, H., *Kolloid Z.*, **105**, 110 (1943).
115. Roth, W. A., and K. Scheel, Eds., *Landolt-Boernstein, Physikalisch-Chemische Tabellen*, Vol. I, Springer-Verlag, Berlin, 1923, Table 87m.
116. Roth, W. A., and K. Scheel, Eds., *Landolt-Boernstein, Physikatisch-Chemische Tabellen*, Vol. II, Springer-Verlag, Berlin, 1923, Tables 174, 179, 185, 186.
117. Rowe, R. L., *Instrum. Soc. Am.* (*ISA*) *Trans.*, **5**, 44 (1966).
118. Sadovnikov, G. V., B. M. Smolskiy, and V. K. Shchitnikov, *Heat Transfer Sov. Res.*, **1**, 32 (1969).
119. Samarcev, A. G., *Z. Phys. Chem.*, **A 168**, 45 (1934).
120. Saunders, M. J., and A. G. Smith, *J. Appl. Phys.*, **27**, 115 (1956).
121. Schardin, H., *Z. Instrumentenkd.*, **53**, 396, 424 (1933).
122. Schardin, H., in *Engebnisse der exakten Naturwissenschaften*, Vol. 20, F. Hund and F. Trendelenburg, Eds., Springer-Verlag, Berlin, 1942, p. 303.
123. Schoenrock, O., *Z. Instrumentenkd.*, **62**, 209, 241 (1942).
124. Selman, J. R., Ph.D. Thesis, University of California, Berkeley, June 1971 (UCRL 20557); see also S. K. Arapskoske and J. R. Selman, March 1971 (UCRL 20510).
125. Shaftan, K., and D. Hawley, *Photographic Instrumentation*, Soc. Phot. Instrumentation Engrs., Washington, D.C., 1962, p. 19, 98
126. Simon, H. A., and E. R. G. Eckert, *Int. J. Heat Mass Transfer*, **6**, 681 (1963).
127. Slayter, E. M., *Optical Methods in Biology*, Wiley-Interscience, New York, 1970, Ch. 13.

128. *Ibid.*, Ch. 14.
129. Steel, W. H., in *Progress in Optics*, Vol. V, E. Wolf, Ed., North-Holland, Amsterdam, 1966, Ch. III.
130. Steel, W. H., *Interferometry*, Cambridge University Press, 1967, Ch. 1.
131. *Ibid.*, Ch. 4 and 5.
132. *Ibid.*, Ch. 6.
133. *Ibid.*, Ch. 9.
134. Svensson, H., *Acta Chem. Scand.*, **3**, 1170 (1949).
135. Svensson, H., *Acta Chem. Scand.*, **4**, 1329 (1950), **7**, 159 (1953).
136. Svensson, H., *Acta Chem. Scand.*, **5**, 72 (1951).
137. Svensson, H., *Acta Chem. Scand.*, **5**, 1301 (1951).
138. Svensson, H., *Opt. Acta*, **1**, 25 (1954).
139. Thomas, W. J., and E. McK. Nicholl, *Appl. Opt.*, **4**, 823 (1965).
140. Tolansky, S., *Surface Microtopography*, Interscience, New York, 1960.
141. Tolanski, S., *An Introduction to Interferometry*, Wiley, New York, 1962, 3rd ed., Ch. 5.
142. *Ibid.*, Ch. 8.
143. *Ibid.*, Ch. 9.
144. Tvarusko, A., and L. S. Watkins, *Electrochim. Acta*, **14**, 1109 (1969).
145. Tvarusko, A., and L. S. Watkins, *J. Electrochem. Soc.*, **118**, 248 (1971).
146. Twyman, F., *Prism and Lens Making*, Hilger and Watts, London, 1952, 2nd ed., Ch. 12.
147. Van Heel, A. C. S., in *Concepts of Classical Optics*, by J. Strong, Freeman, San Francisco, 1958, Appendix D.
148. Weinberg, F. J., *Optics of Flames*, Butterworths, London, 1963, Ch. 1.
149. *Ibid.*, Ch. 2.
150. *Ibid.*, Ch. 4.
151. *Ibid.*, Ch. 8.
152. Weingraber, H. von, in *Proc. Intl. Production Engr. Res. Conf.*, Carnegie Inst. Tech., Sept. 1963, M. C. Shaw, Ed., Am. Soc. Mech. Engrs., New York, 1963, p. 659.
153. Weinstein, W., *Nature*, **172**, 461 (1953).
154. Williams, W. E., *Applications of Interferometry*, Methuen, London; Wiley, New York, 1950, p. 10.
155. Winckler, J., *Rev. Sci. Instr.*, **19**, 307 (1948).
156. Winkler, E. H., Naval Ordnance Laboratory Report, NOLR 1077.
157. Wolter, H., in *Encyclopedia of Physics*, Vol. 24, S. Fluegge, Ed., Springer-Verlag, Berlin, 1956, p. 589.
158. Zernike, F., *Physica*, **9**, 686, 974 (1942).

Holography and Holographic Interferometry in Electrochemistry

VAKULA S. SRINIVASAN*

Biosciences & Electrochemistry Department
Systems Group of TRW Inc.
Redondo Beach, California

* Present Address: Department of Chemistry, Bowling Green State University of Ohio, Bowling Green, Ohio 43403.

Contents

I. Introduction

Holography, as the Greek roots of its name (holos, whole; graphein, to write) imply, is a technique of recording that captures the entire content of the light reflected or transmitted by an object. By the term "entire content of the light," we mean the spatial distribution of intensity, the wavelength, and the phase of the light. The first two properties are used in conventional photography, where the image is recorded by intensity distribution of the light reflected by the object on a photographic emulsion by some optical means.† In holography, all three properties are used; by making use of the phase content of light we are able to eliminate the use of lenses in imaging the object on the photographic emulsion. The phase of the light gives information regarding the mutual position of the parts which together constitute the object. Thus the hologram records all the information carried in the light waves emanating from the scene, the size and the shape of the object, the relative brightness of every point on the object, and the position of the object in space from all angles that can be "viewed" by the recording medium.

The concept of holography, also known as lensless photography or wave front reconstruction [as it was called by its inventor, Dennis Gabor (15,16)] has developed into a powerful new optical technique through advances in laser technology. The invention of the laser has facilitated this recording of phases of light, reflected or transmitted by different parts of the object, relative to each other.

The whole technique of holography comprises two steps: (1) the photographic recording of the scene in its entirety—the light waves emanating from the object are recorded on the light-sensitive medium, along with a "coherent" background to provide phase reference; (2) the reillumination of the hologram (developed photographic plate) with a coherent light source. During this process the original wave pattern is regenerated from the hologram. A first consequence of this property is the achievement of a true

† The response of the photographic emulsion is usually wavelength-dependent. The emulsion can be sensitized for any selected wavelength.

three-dimensional image, whose perspective changes with viewing angle. The reconstructed pattern, like the original wave pattern, can be examined with conventional optical instruments with the result that different regions of the recorded scene can be examined at leisure and in detail from a single holographic plate. This "post-exposure-focusability" property of a hologram eliminates the depth-of-field limitations of conventional imaging systems. A second important consequence of the hologram's property of recreating the original wave pattern both in amplitude and phase, is the ability to compare interferometrically one reconstructed wave pattern with another. Interference phenomena occur when two holographically re-created wave patterns are superposed. Likewise, one can create interference between a single holographically re-created wave pattern and a real wave pattern. The resulting interference fringes are quantitatively related to the changes in optical path lengths in a scene that occurs between the two holographic exposures or between the first holographic exposure and the real scene.

Mass transport near electrode-solution interfaces and the motion of liquids near dropping mercury electrode have been investigated by classical interferometric methods (2–4,21–24,33,36–45,48,49). The classical techniques demand the use of high quality optical elements and precision alignments. In holographic interferometry, the preparatory and alignment procedures are far less critical. The advantages of holographic interferometric methods over classical interferometric methods will become evident in the following sections. The following sections present a discussion of the concepts of holography and holographic interferometry; the experimental arrangement; and a critical evaluation of the technique as applied to electrochemistry.

II. Principles

1. *Interference*

According to classical electromagnetic theory of light, the instantaneous complex disturbance at any point, associated with the electric field, can be represented by a complex scalar function, given by an amplitude factor, a phase factor, and a time harmonic factor.

$$U = A \exp(i\phi) \exp(i\omega t)$$

In this function, A is the amplitude factor, this amplitude modulus being the actual physical quantity; the first exponent part represents the phase part. The time harmonic factor, exp $(i\omega t)$, where ω is the frequency, is usually omitted for brevity's sake. It is implicitly assumed that the secondary wave issuing from the point is of the same frequency as the incident wave (6). So, conventionally, the above function is written as

$$U = A \exp(i\phi) \tag{1}$$

The basic idea of holography is to record this amplitude and phase of a scattered electromagnetic wave front, together with a suitable coherent background, in such a way that it is possible to reproduce the electromagnetic field distribution of the original wave front at a later time. The light source illuminating an object should emit waves with well-defined phase relationship, so that the scattered light waves from the object may also have well defined phase content. This well-defined phase relationship between waves is known as coherence. In conventional light sources like tungsten lamps or gas discharge tubes, the light waves are emitted by various electrons or the excited gas molecules, without any mutual relation in time or space. The resulting light wave does not possess any specific phase relationship. This is called incoherent light. The fluctuations produced by any such physical source are too rapid to be followed by the eye or by any physical detector. If two different light beams originate from the same monochromatic point source (like a sodium lamp), the fluctuations in the two beams are in general correlated, and the beams are said to be completely or partially coherent, depending on whether the correlation is complete or partial. In beams from different sources, the fluctuations are completely uncorrelated, and the beams are said to be mutually incoherent. That is why, in holography, the source illuminating the object and the one providing the coherent background are the same. The laser (acronym for light amplification by stimulated emission radiation) is a light source with a defined phase content (32). The atoms in the laser are pumped to an excited state by means of electrical discharge or flash lamps. When the population of the excited atoms are in great preponderance, the system can be stimulated to produce a cascade of photons, in step, all in the same frequency, by triggering the emission of energy that drops the atom from the excited state to a ground state. The trigger is a photon of this quantum of energy, which on striking an excited atom, causes it to emit another photon at the same frequency. Thus the light wave produced is in step with the triggering one. The wave thus produced travels through the laser tube and in turn stimulates the emission of radiation in step, from the other excited atoms. This process builds up intense radiation, with highly correlated light waves traveling through the laser column. The laser is essentially a cavity resonator with two reflecting mirrors, one of which is partly reflecting at the ends. The waves that go to ends of the column are reflected back and forth along its axis. The result is the emergence of a highly correlated, intense, unidirectional light beam through the partly reflecting mirror. Thus the laser is a good light source of high coherence.

Coherence is a necessary factor in holography. Holography itself is based on the phenomenon of light interference. It is relevant here to discuss a few salient points about this optical phenomenon, which makes it possible to record the phase content of the light wave. When two waves of the same wavelength arrive at the same point, interference will occur. There is either an increase in the electric field amplitude at this point (constructive interference), or the amplitude of each wave cancels the other (destructive interference). The phase difference between these two waves arriving at this point depends upon the distance traveled by each with respect to the other. The intensity I of light can be defined as the time average energy flux perpendicular to the direction of energy flow, and it is proportional to the square of the electric field amplitudes. Thus when two coherent beams of equal intensity that have traveled different distances arrive at a point, all intermediate cases can occur—from maximum brightness to total darkness. The contrast in brightness is related to the degree of coherence between these two beams of light. If two beams of incoherent light are superposed, no such interference is observed because the total intensity of light is the sum of the intensities of individual beams.

The whole concept can be written in the following manner. The electric fields for two waves at a plane are given by

$$U_1 = A_1 \exp(i\phi_1) \tag{2}$$

and

$$U_2 = A_2 \exp(i\phi_2) \tag{3}$$

If the beams are coherent, the total complex disturbance at a point in the plane is

$$U = U_1 + U_2 = \exp(i\phi)[A_1 + A_2 \exp[i(\phi_2 - \phi_1)]] \tag{4}$$

and the resultant light intensity at the point

$$I_R = |A|^2 = UU^* = [A_1^2 + A_2^2 + A_1 A_2 \exp[i(\phi_2 - \phi_1)] + A_1 A_2 \exp[-i(\phi_2 - \phi_1)]] \tag{5}$$

where the superscript asterisk indicates a complex conjugate. Or the same equation can be written as

$$I_R = A_1^2 + A_2^2 + 2A_1A_2 \cos(\phi_2 - \phi_1) \tag{6}$$

Thus we see that in addition to constant intensity terms A_1^2 and A_2^2, there is a periodic term $2A_1A_2 \cos(\phi_2 - \phi_1)$, which changes sign depending upon $(\phi_2 - \phi_1)$. If the light waves under consideration are plane wave fronts, depending upon the position of a point on the plane, maxima or minima of light are seen. Loci of points fulfilling this condition appear as bright or dark fringes.

If the beams are incoherent, the time average for the last term of Eq. 6 is

$$\frac{1}{\tau}\int_0^{\tau} \cos(\phi_2 - \phi_1)\, dt = 0$$

for the response time of the detector. Then the total intensity is the sum of the intensities due to individual beams. Thus

$$I_R = I_1 + I_2 = (U_1U_1^* + U_2U_2^*) = A_1^2 + A_2^2 \tag{7}$$

The periodic part of the equation vanishes. Thus it is clear that the phase relation between two beams and the degree of correlation that exists between the fluctuations in these two beams determine and, conversely, are revealed by, the distinctness of the interference effects to which the beams give rise on superposition.

2. *Coherence*

The contrast in interference fringes is highly dependent upon the coherence of the light beams interfering. This gives rise to certain considerations of the source of light. The coherence effects are separated into two categories. The first, termed temporal or longitudinal coherence, results from considerations of finite spectral width of the radiation. The second, termed spatial or transverse coherence, results primarily from angular source size considerations.

A. Temporal Coherence. If the light from a monochromatic point source is divided into two beams that are made to interfere with each other, just as in a Michelson interferometer, it is found that the fringes are distinct if the optical paths of the two interfering beams are nearly equal; but as the difference in optical path is increased, the visibility of the fringes decreases and the fringes eventually disappear. This arises because no source is strictly monochromatic and the waves emitted by atoms are not infinite (continuous) waves but are discharged as wave trains. The coherence of the two light beams that may interfere is linked with the duration and consequently with the length of the wave trains. The length of the wave train determines the bandwidth of the radiation emitted by the atoms. A large number of wave trains pass at random time intervals during the time required to make an observation. At first, assuming that all the wave trains are identical, we divide the wave train into two trains of equal length as we divide the beam. When the optical paths of these beams are the same, the wave trains arrive at the point of observation P at the same time. But, if the difference between their optical paths is greater than the length of the wave train, one of these two wave trains has passed the point of observation before the other has arrived. Then there is no interference at P due to pairs of wave trains derived from the same wave train emanating from the source. The wave trains superposed at P at any instant are derived from different incident wave trains, and because they arrive at random, in rapid succession, their contribution to the interference terms average to zero over the relatively long time required to make an observation. A wave train that is weakly damped is almost a sinusoidal oscillation and therefore monochromatic. Therefore, a monochromatic source emits long wave trains. From the concepts mentioned here, it is possible to derive equations that represent the degree of interference that can be observed between two paths from a source, differing in transmit time by τ. An arbitrary criterion for the τ is chosen to provide enough contrast for the fringes; and the factor $c\tau$, having the dimension of length, is called coherence length where c is the velocity of light. This property is also referred to as temporal or longitudinal coherence. As a laser source is highly monochromatic, temporal coherence is usually high as compared to sodium light, etc. The light from the laser possesses such a high degree of temporal coherence that two beams of light coming from the laser, even though the path traveled by each differs by several centimeters, can interfere on the recording plane, like a photographic plate, producing a hologram.

B. Spatial Coherence. The spatial coherence in a light beam designates coherence between two points in the field illuminated by a light source. The degree of coherence between these points describes the contrast of the interference fringes that are obtained when these two points are taken as a secondary source. The important and generally unexpected conclusion is this: even an extended source, composed of millions of statistically incoherent oscillators, can produce a coherent field at these two points, provided only the angular diameter of the source, as seen from the plane of these two points, is small compared to the separation of the points. In other words, if two slits are placed near the extended source, no interference fringes will be seen. If, however, the same two slits are placed far enough away from this noncoherent source, interference fringes with good contrast can be obtained, as the angular diameter of the source decreases with distance. But it should be remembered that as one goes farther away from such a source, the intensity of light decreases.

Thus the limitation on temporal coherence can be overcome by adjusting the path traveled by the interfering beams to be nearly the same. The tolerance in adjustment depends upon the coherence length of the source, depending upon how "pure" or monochromatic the light is. But the chief requirement for interference to occur over a large area is that the interfering beams be spatially coherent. Every secondary source point must be capable of interfering with every other secondary source point. Because lasers provide an intense coherent source point, they easily fulfill the condition for spatial coherence. The earlier work on holography was done, in fact, with high-pressure gas lamps providing an intense source of light, but only microscopic objects could be investigated because of the small available angle. But again, high-pressure and high-temperature lamps, because of increased line broadening, have reduced temporal coherence. A weak low-pressure mercury lamp can be made to have long coherence length (20 cm), but the coherent power available in any usable angular field of view is much smaller than the weakest of He/Ne lasers.

C. Localization and Visibility of Fringes. A brief discussion of the fringes will facilitate understanding of the experimental arrangement presented later (7,14). All the arrangements used in interferometric experiments, including holographic interferometry, may be represented generally as devices by which light from a source is made to reach points in a region of space by two or more different paths.

Let us consider a monochromatic point source S, of wavelength λ_0, as shown in Fig. 1. For the sake of clarity consider two rays SA_1B_1P, SA_2B_2P, originating at source S, and after traveling through space the rays arrive at point P. The travel through space, are shown as discontinuities A_1B_1 and A_2B_2 in the figure. The rays might

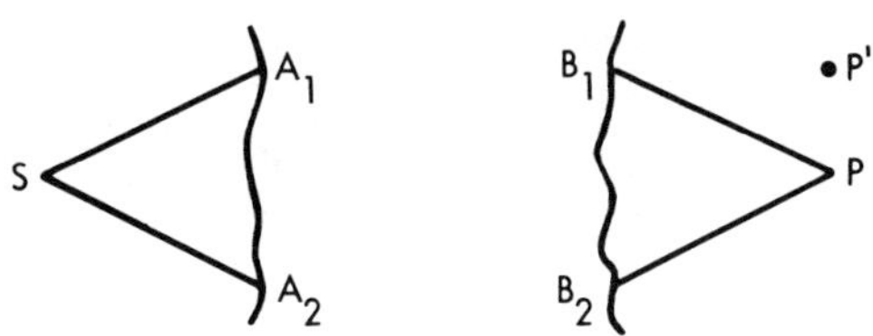

Fig. 1. Interference of rays from a point source. A_1B_1 and A_2B_2, shown as discontinuities, are the two different light paths. The fringes produced by these rays are nonlocalized.

have traveled through different distances, so that the path difference between these two rays at point P is

$$\delta_0 = SA_1B_1P - SA_2B_2P \tag{8}$$

The value of δ_0 depends upon the position of P. Likewise, one can find another point P' which has a different path difference δ_0'. The state of interference at points P and P' is determined by the corresponding path differences. But there could be points like P for which δ_0 has the same value. These points exist on any plane in the region common to the two paths from the point source S. Likewise, there could be another set of points on another plane for point P', and so on. Thus in any region where the two beams are superposed, if we hold a screen, we obtain at all points on the screen an interference pattern that is determined from one point to another. Even if we move the screen through the region of overlap, we can still see the same interference pattern, because in each new position there are corresponding points fulfilling the conditions discussed above. These interference fringes are said to be nonlocalized. Their visibility (in terms of contrast) is dependent upon only the relative intensities of the two waves. Such fringes are always obtained with a point source.

However, if the source is an extended source, like a sodium lamp used in earlier experiments, or an object illuminated with diffused laser light, one can assume that such a source is made up of incoherent point sources, each of which gives rise to a nonlocalized interference pattern; the total intensity at each point then is the sum of the intensities in these elementary patterns. If the phase difference at P is not the same for all points of the extended source, the elementary patterns are mutually displaced in the vicinity of P, and the visibility of the fringes at P is less than with a point source. In general, this mutual displacement, hence, the reduction in the visibility of the fringes, increases with the extension of the source, at a rate depending on the position of P. As the source extends about S, the visibility at some positions of P may remain at or near its value with a point source, when elsewhere it might have fallen to zero. Then the fringes are said to be "localized." Thus if a source has a certain size, the phenomenon of interference is localized on a surface that is the focus of the points of intersection of the rays which originate from incident rays coming from S. When the fringes are localized, this presents a small problem from the experimental point of view. The phenomenon causing the shift in the fringes, as we see later, might be in a place different from the location of P. Under those circumstances, taking the photographic picture of the system and the fringes together poses some difficulty. Nonlocalized fringes from these beams (not the image of fringes in space) emerging from a point source can be focused on the focal plane of a lens of the imaging system. Under these circumstances it will appear that fringes originate from the source at infinity. Ibl and Muller (23) used this approach to photograph the fringes at infinity, using another cylindrical lens to focus the plane of the electrode in the horizontal axis.

We discussed the high spatial coherence of a laser source. The wave front from an appropriately adjusted laser source has the same phase across the entire front. It is true that two plane wave fronts (by focal plane illumination) traveling in the same direction from a single laser source interfere and the fringes produced are nonlocalized. In holographic interferometry we can create this condition with a parallel reference beam and a scene beam. The two beams interfere, producing grating-like fringes in space. In fact, we used this approach in studies of concentration gradients by holographic interferometric technique, using transmitted light near the electrode.

When a scatter diffuser is used to illuminate a scene, wave fronts are generated in all directions. Under these circumstances in holographic interferometry, the location of fringes becomes important. The concept of localization of fringes is discussed in a later section. While recording the hologram, the localization of fringes is not a serious problem. We record the interference of a separate reference beam (usually planar) and the wavefronts from each point of an object scattering the incident light. Interference fringes occur on any plane, where these wave fronts and the reference wave front are superposed.

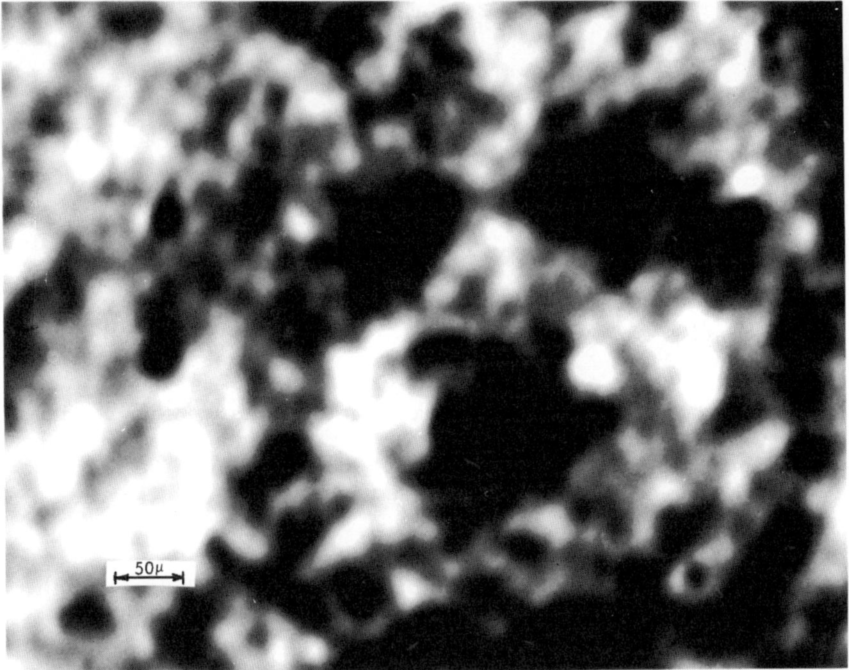

Fig. 2. Photomicrograph of a hologram due to a spherical reference wavefront and the wavefronts from a scattering object. The wavefronts are coming in all directions. The mean angle between the reference source and the object is about 45°.

3. Holography

A. Elementary Theory. Basically a hologram is a record of the interference fringes produced by the light wave front reflected from or transmitted through a scene and a reference wave front. A photomicrograph of a hologram illuminated by ordinary light is shown in Fig. 2. Literature related to interferometry in general is applicable to holography as well. We shall briefly review the elementary mathematical principles behind the factors affecting the formation of a hologram, the characteristics of reconstructed images and the optical systems for forming a hologram (28,50).

For simplicity, Fig. 3 is taken for analysis. An incident beam of plane radiation is partly deflected by a triangular prism providing the reference beam. The object or the scene intercepts the other part of the original beam. The object diffracts the incident radiation to

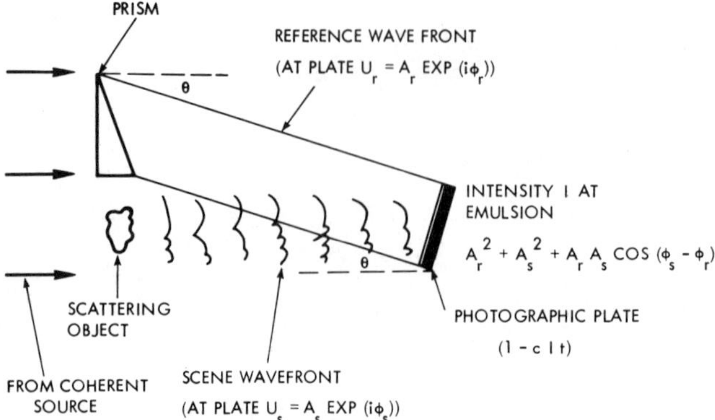

Fig. 3. Schematic of the recording of a hologram of a scattering object, with a plane reference wave. Note that the reference wavefront is normal to the hologram plate. The intensity variation at the photographic plate and the amplitude transmission factor are shown in the figure.

generate a field with magnitude $A_s(x, y)$ and phase $\phi_s(x, y)$ at the photographic plate. Here x and y are taken as the coordinates on the photographic plate, considering it as two-dimensional. The corresponding field generated by the reference beam on the photographic plate has an amplitude $A_r(x, y)$ and phase $\phi_r(x, y)$.

Thus we have on the photographic plate a field U_r,

$$U_r = U_r(x, y) = A_r(x, y) \exp(i\phi_r(x, y)) \tag{9}$$

due to reference beam and

$$U_s = U_s(x, y) = A_s(x, y) \exp(i\phi_s(x, y)) \tag{10}$$

due to the scattered beam from the scene.

The resultant amplitude at the plate is the complex addition of the two intermixing traveling waves, being equal to $(U_s + U_r)$.

The intensity at the plate, which determines the total exposure at each point (x, y), is given by the square of the resultant amplitude of the total field at each point; namely

$$(U_s + U_r)(U_s + U_r)^* \tag{11}$$

where the asterisk denotes the complex conjugate.

The light transmitted by the exposed and processed photographic plates is characterized by the so-called *H-D* photographic curve (Hurter-Driffield Curve). This curve relates the amount of light transmitted by the plate and E, the exposure, which is defined as the product of total incident intensity I and the exposure time t. For small changes in exposure on the linear portion of the curve, the amplitude transmission factor α defined as the ratio of the complex amplitude of the wave emerging from the plate to that of the wave incident on the plate (6), can be expressed as

$$\alpha = (1 - cE) \tag{12}$$

where c is a positive constant and E the exposure. Thus the amplitude transmission factor of the hologram after the exposure and processing is

$$\alpha = 1 - k(U_r + U_s)(U_r + U_s)^* \tag{13}$$

where $k = ct$. One can recollect that the second term contains a periodic term.

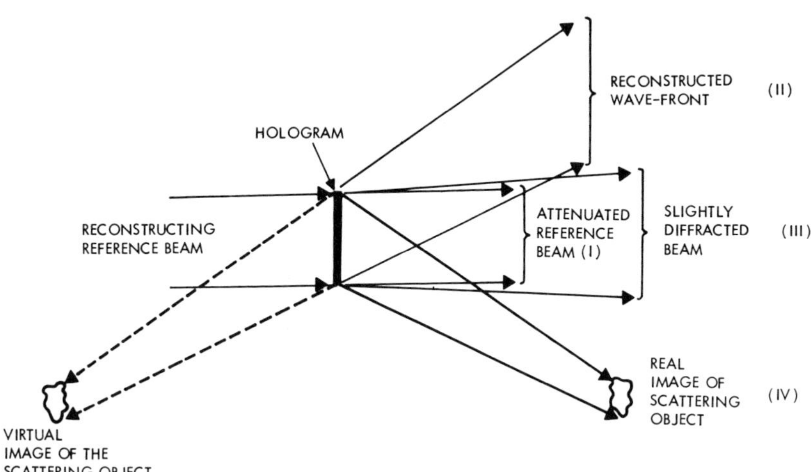

Fig. 4. Holographic reconstruction and image formation from a hologram. Note that the reconstructing wavefront is normal to the hologram plate. The numerals correspond to the terms in Eq. 15.

When the hologram is reilluminated (Fig. 4) with the original reference beam U_r the complex light pattern emerging from the photographic plate will be $U_t = \alpha U_r$,

$$U_t = U_r[1 - k(U_r + U_s)(U_r + U_s)^*] \tag{14}$$

Multiplying the term gives

$$U_t = \underset{\text{I}}{[1 - k(A_r)^2]\, U_r} - \underset{\text{II}}{k(A_r)^2 U_s} - \underset{\text{III}}{k(A_s)^2 U_r} - \underset{\text{IV}}{k U_r U_r U_s^*} \quad (15)$$

Thus the transmitted wave front has four components.

The meaning of each term from the physical understanding is as follows:

1. The first term (I) is the reference beam attenuated by its transmission through the plate.
2. The second term (II) contains U_s the function for the original wave front. This function is multiplied by the square of the amplitude (intensity) of the reference beam and a constant.
3. The third term (III) contains a constant multiplied by the intensity factor from the diffraction of the scene alone and the amplitude function of the light reilluminating the hologram. The scattered light from the object has created a varying pattern of darkness on the photographic granules, like an ordinary photograph. Thus the third term represents the reference wave as seen through the photographic plate exposed to the granular pattern of the laser-illuminated scene alone.
4. The last term (IV) represents the "conjugate" or real image. By definition, real image is the one that can be caught on a screen or located in space. The location and identification of the real image depend upon how the hologram is taken and how it is reconstructed. In the special case where the reference beam is a plane wave of uniform intensity, normally incident on the plate, the identification is easy. In this special case, the product $U_r U_r$ is a constant and the effect of the conjugate on U_s is to represent a wave which converges on the downstream side of the plate to form a real, but pseudoscopically inverted image of the subject.

Thus when we identify all the terms of the equation, we come to the conclusion that the light coming from a reilluminated hologram is the same as the superposition of light from sources represented by the individual terms in the equation. One term of this equation represents the original scene both in relative amplitude and phase. Strictly speaking, the phase is reversed by exactly 180° relative to the reference beam, because of the negative sign in the second term (II). In most cases this phase reversal is not important because one is only concerned with phase relationship in the image itself and not between the image and the reference beam. The theory predicts that, when observed with conventional optical instrumentation, one cannot tell the original wave from its reconstruction.

One should note that the reconstructed image is positive in the above example. The phase reversal is due to the use of a photographic negative. If the contact print of the above hologram is used, there will be another 180° phase reversal, but the image intensity will be the same. The hologram behaves like a grating, always yielding a positive image. If, however, one uses the reference beam in the same axis as the scattering object (Gabor hologram) one gets a negative image with the original hologram. The reader should refer to the reference cited for further discussions.

B. Recording of a Hologram. To obtain a hologram, a source with sufficient spatial and temporal coherence is needed. The holocamera system must be mechanically and thermally stable and the recording must be done on a medium that has high resolution to record the fine interference pattern. We will briefly look into the theoretical requirement of each (28,29,30).

Coherence limitations have already been discussed. Temporal coherence limits the size of the object that can be holographed. The effect of poor spatial coherence is that, from any point on the hologram, only a portion of the object will reconstruct. Pulsed lasers have poor spatial coherence. The lasers should operate in a single mode (TEM_{00} mode), to have a high degree of spatial coherence, as discussed in the experimental section.

Once the light beam is divided into two parts, the path lengths should remain stationary within $\frac{1}{4}$ of the wavelength of light, during exposure. Otherwise the fringe pattern moves, washing out the hologram. The optical path length is a function of the refractive index of the air. Reasonable stability of air temperature and pressure should be maintained during the experiment. If one meter path at normal atmospheric pressure undergoes a pressure change of 0.05 in. of mercury or temperature change of 0.31°F, the effective path changes by λ.

If the maximum angle between the reference beam and the object beam is θ, then the maximum number of interference fringes per unit length, known as spatial frequency $f_{\max}$, is given by

$$f_{\max} = \frac{2 \sin (\theta/2)}{\lambda} \tag{16}$$

where λ is the wavelength of light from the laser. The photographic medium should have sufficient resolution to record this frequency.

If f_{max} is small because of small θ, the demand on the film is less.

If all the above conditions are met, a hologram will be recorded.

C. Types of Holograms. The diffraction pattern due to an object, depending upon the distance of the recording medium from the object, is known as either near-field or Fresnel diffraction, or far-field or Fraunhofer diffraction. The holograms taken under these conditions are known as Fresnel holograms or Fraunhofer holograms (11). There is a type of hologram known as Fourier-transform hologram, which is a photographic record of the interference pattern between a Fourier-transform of the object amplitude distribution and coherent background arising from a point reference in the plane of the object. These various types of holograms arise because of mathematical operations. Though holography has been viewed from this angle, it is sufficient for us to know that the basic arrangements to take a hologram depend upon how the coherent background is provided to record the interferogram on the photographic plate. Various names have been given to these arrangements; however, they can be grouped into two broad categories: (1) in-line or one-beam holograms and (2) off-axis or two-beam holograms.

D. In-Line (Gabor) Hologram. A schematic of such an arrangement is shown in Fig. 5, for a microscopic object, with large unobstructed space around it, illuminated by a collimated laser beam (16). The scattered light from the object is the scene beam and the portion not obstructed by the object acts as a reference. This arrangement provides spatial and temporal matching at the same time, with a weakly coherent source. With increasing scattering angle from the object, however, there is increasing mismatch and, consequently, the contrast of the fringes decreases. This, in turn, limits the solid angle of the recorded cone of rays (i.e., numerical aperture) and, hence, the resolution of the reconstructed image. Further, during reconstruction, the real and the virtual images (see Fig. 5) lie on the same axis so that viewing is made difficult. However, one can focus a microscope on the real image so that the virtual image is beyond the depth of focus. Hence this one-beam arrangement is useful for microscopic studies.

Fourier-transform holography can be considered an in-line arrangement. Unlike the Gabor arrangement, where the reference source is far from the photographic plate as compared to the object, the reference source is a point located in the same plane as the object. The advantage is that the spherical waves originating from this point source and from each point on the object, have approximately

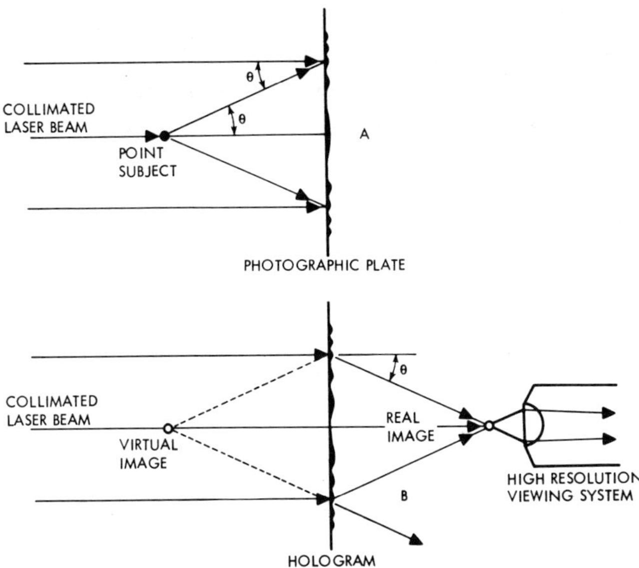

Fig. 5. Basic Gabor in-line arrangement. *A*. Hologram formation of a point source. *B*. Hologram reconstruction of the images. The blackening of the lines corresponds to the intensity distribution of incident light and that of the transmitted light respectively. As the scattering angle increases the spatial frequency increases. The microscope aids in separating the two images.

the same curvature traveling in the same direction. Hence the demand on the coherence of the source and the resolution of the film are less. During reconstruction, unlike the Gabor hologram, the images of Fourier-transform hologram are separated in space. A Fourier-transform hologram has been used for both transmitted light and reflected light. It is easy to take and has a high resolution for plane objects. Reconstruction is easier. Stroke and co-workers (50) have studied this arrangement exhaustively. This type of arrangement will work only for limited cases. The off-axis arrangement is convenient for electrochemical studies.

E. Off-Axis Hologram. Carrier frequency or off-axis reference beam holograms are most commonly used. The simple arrangement is shown in Fig. 3. This arrangement, introduced by Leith and Upatnieks (30), is superior in many respects. The scene and the reference beams fall upon the hologram recording plate at an angle to one another and the subject does not intercept the reference beam.

Unfortunately, this arrangement does not automatically provide spatial and temporal matching and does not always lend itself to use with sources with limited coherence. Because the angle is larger, high resolution film for recording is needed. The real and the virtual images are separated. Both reflecting and transmitting objects can be holographed.

The off-axis technique is the practical arrangement for electrochemical studies, because the reference beam can be folded in from any angle desired. One might be interested either in the transmitted light near the electrode for the study of concentration gradients near an electrode by holographic interferometry, or in the reflected light from an electrode for the study of the surface. Microscopic study of the projected real image is equally possible with this arrangement. Usually, the available commercial lasers have sufficient coherence that optical path matching within the coherence length is no problem. Further, we are interested only in a small area near, or on, an electrode for a given view and so the demand on spatial coherence is not too high.

F. Reconstruction of Images. During reconstruction, two images are recovered. The image formed behind the hologram plate is called the virtual image, and the one formed in front of the hologram plate is called the real image. Ordinarily, only one of the images is visible at one time due to emulsion thickness. Each image has its practical use.

The virtual image is produced by the exact replica of the original reference beam. The reconstructing beam should be identical with the original reference beam in wavelength, divergence, and the angle of incidence. Under these circumstances the image is free of aberrations and is identical in size and location to the hologram, as was the original object. This image has all the characteristics of the three-dimensional object, and is used for holographic stored beam interferometry.

The real image is formed when the hologram is illuminated with the exact conjugate of the reference beam. This implies that the light beam is of the same wavelength, the angle of incidence is exactly 180° from that of the reference beam, and the beam has the same degree of convergence as the reference beam had divergence. It means that now the reference beam illuminates the hologram from the opposite side. Now the image is formed exactly in the space where the original object had been and it is on the observer side of the plate. This real image is not suited for visual observation and it is said to be pseudoscopic. It is difficult to explain how the image looks unless one observes it. When a convex surface is holographed and reconstructed, this surface looks convex in the virtual image, because it is on the same side of the reconstructing beam. But the same surface appears concave in the real image, because it is on the opposite side of the reconstructing beam. When one's head moves to change the viewing angle, the real image moves in the opposite direction and therefore it is not suitable for three-dimensional study. But it is suitable for microscopic study, for recording two-dimensional images or magnification without lenses. The latter is achieved by varying the degree of convergence.

Aberrations are introduced when the hologram is reconstructed with a different orientation of the plate or different wavelength. The image may be magnified or distorted (34). The reconstructing light may have less coherence than the one used for making the hologram. Filtered arc lamps have been used for this purpose. While looking at a reconstructed image, we are looking only at a small portion of the image through the hologram; whereas in making the hologram, the wave fronts from each point of the object have to be recorded on the entire face of the hologram plate.

The reconstructed image appears granular, due to the spatial coherence of the light illuminating the hologram. Each scattering point in the system creates its own miniscule interference pattern in space, which results in a speckled appearance. If one attempts to reduce this granularity by increasing the incoherence with the use of ground glass in the reconstructing beam, while viewing, there is reduction in the resolution (31).

We can briefly summarize the characteristics of the holographic recording and reconstruction (35). Each point on the hologram plate records information from every point on the object, from its angle. This gives rise to the perspective in holographic imaging process. Since each portion of the hologram contains the total information, even if the hologram is broken to pieces, each piece can reconstruct the whole scene from its "view." The loss is in image resolution, because the aperture is decreased. An ordinary photograph records an image by density variations of the darkened silver from the halide; but in holography, the recording is done in a grating pattern. The darkened silver acts more as a filter for producing lines with contrast. These lines can exist even as refractive index or thickness changes in the supporting gelatinous materials. Therefore, reconstruction is quite possible with a bleached hologram, from which only the darkened silver has been removed. This is called a phase hologram. In principle, any monochromatic point source can be used for reconstruction. A laser is not necessary. Because our viewing aperture is small, point sources have sufficient spatial coherence for reconstructing that much angular view. When we change the wavelength during reconstruction, the lateral and longitudinal magnification change, not always in the same proportion. However, the concept of reconstructing in a different wavelength is used in acoustic and microwave holography (10,17).

4. Holographic Interferometry

The technique of holography makes available a new type of interferometer. In all the two-beam interferometric techniques, one beam serves as reference for comparison with the test beam. The fringes so produced shift if there is any perturbance in the test beam. In these methods, the test beam and the reference beam travel through two different paths before they are incident on the plane of observation. The reconstructed wave front of a scene from a hologram can serve as the comparison beam. However, this comparison beam, as we discussed earlier, traveled the same path as the test scene wave front is traveling now; that is, the comparison beam and the test beam have a common path. Thus holographic interferometry is a common-path interferometry. Because both the test beam and the comparison beam are traveling through the same optical arrangements, whatever optical imperfection exists for one beam exists for the other as well, and the effects due to imperfections cancel out. This is unique for holographic interferometry and differential interferometry becomes easy. Two types of holographic interferometry will be described (8,9,20).

A. Stored-Beam Holographic Interferometry. In this technique, one first makes an ordinary hologram of the subject to be studied (9). This wave front from the subject serves as the comparison beam. After development, the single-exposed hologram is replaced in the exact position in which it was exposed and reilluminated with the exposing beams. The reference beam will be diffracted by the hologram producing the virtual image of the subject in the same position that the subject occupied during exposure. If the subject has not moved since the exposure, a portion of the light that is reflected by or transmitted through the subject will now pass directly through the developed photographic plate, so that the true subject is also viewed simultaneously with the virtual image through the photographic plate. Thus any optical device, including the human eye, intercepts the superposition of two light wave patterns, one from the holographic image where the object was during exposure and another from the object itself, from wherever the object is, while being viewed. Since both the wave patterns are coherent, they can interfere. Wherever the light waves from these two secondary sources (object and the image) are 180° out of phase, dark bands appear. Strictly speaking, the holographic image itself has 180° phase reversal; thus if no movement of the subject occurred during this process and the photographic plate was without any stress, the entire scene would be dark. This does not happen in practice, for the intensity of light from the image or the object is not always perfectly matched, so there is no completely dark field.

If some portions of the scene have moved or been subjected to conditions that produce refractive index gradients, so that the optical path produced is different for the light waves from different parts of the object, the scene will appear to be covered with dark bands where the two images are 180° out of phase. These fringes can be quantitatively interpreted to account for the displacements on the object. If there is an immovable fiduciary line on the object it will stay dark throughout the viewing. This could be the solidly held circumference of a vibrating diaphragm. The immediate next dark line on the image during viewing will indicate that the optical path of light from that region has changed by one wavelength or the object in that region has moved through one half wavelength and so forth.

If we are to look at the fringes with transmitted light, the gradients causing the changes in optical path produce alternate dark and bright bands, where the integrated optical path change corresponds to one half wavelength (between adjacent dark and bright fringes). However, the shape of these bands depends upon the geometry of the refractive index field in the scene. These fringes are known as infinite fringes. The whole sequence of stored beam interferometry is shown in Fig. 6. The main advantage of this technique is that it permits real-time viewing. The great disadvantage of this arrangement is the rigidity required of the setup, and the accuracy with which the plate must be replaced on the setup is very high. Even the stress introduced in the emulsion by the development process is capable of sufficiently bowing the 0.04 in. glass photographic plate so that perfect realignment cannot be obtained for wide-angle subjects.

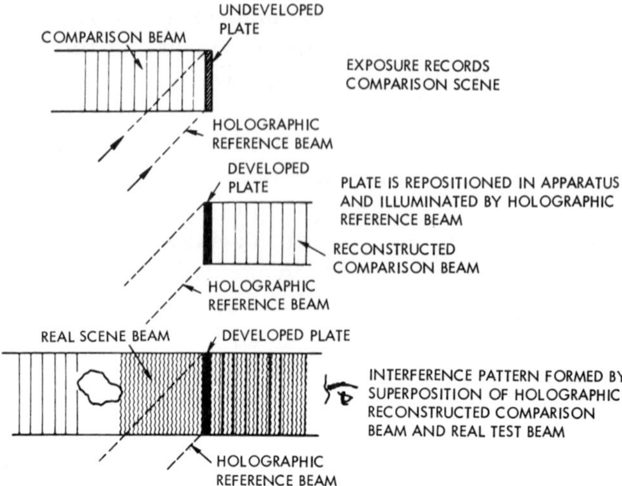

Fig. 6. Schematic of the operation of stored-beam holographic interferometric technique.

For the stored beam technique, a two-beam arrangement is used with a very narrow angle between the reference beam and the scene beam. The components should be rigidly mounted. The holder of the hologram plate should be so designed that the plate may be removed for development and later reinserted in its original position.

In this arrangement, one more interesting experimental control can be achieved. The preceding paragraphs discussed the study of the movement of the object by rigidly mounting the optical components and placing the hologram exactly in the same position. If we know that the object is rigidly mounted, the superposed, reconstructed holographic image can be slightly tilted about any desired axis by either tilting the hologram about that axis or the optical component that folds in the reconstructing reference beam. Both these deliberate misalignments change the angle of incident reconstructing reference beam and, as described earlier, this affects the location of the reconstructed virtual image. Consequently, fringes appear on the images, but the shape and angle of the fringes can be manipulated. As the photographic plate is planar, the fringes appear linear, but the direction of fringes is controlled by the direction of the tilt. This control is true whether the hologram is of a transparent object like liquids or gases in a glass cell (transmission hologram) or of an opaque body like an electrode. These fringes are similar to the fringes produced in the Mach-Zehnder interferometer, and are commonly known as "finite fringes." This deliberate misalignment to produce finite fringes compensates for the defects in the processed photographic plate, but the use of finite fringes in holography is specific.

A ground glass may be placed behind the transparent subject under investigation, so that the light traverses the subject from a variety of angles. This ground glass remains behind the subject, when making the hologram and reconstructing as well. A reflectance hologram, too, is made with a diffusedly illuminated subject, but the relocation requires greater precision. During the experiment the fringes are seen easily, making the three-dimensional nature of the subject and the fringes apparent. However, the ground glass now acts as a set of secondary source points for illuminating the object. This decreases the spatial coherence. During the interferometric experiment, both the real object and the virtual image act as diffusedly lit tertiary sources, emitting wave fronts in all directions. As established previously, these fringes are localized. Whether looking at the finite fringes or infinite fringes, when working with diffused illumination, the aperture of a copy camera during reconstruction should be adjusted for a proper f number to capture both the fringes and the object. Because of the ability of human eyes to accommodate, when one directly looks at the reconstructed object it appears that the fringes and the object are in the same plane. Direct viewing of the subject without the ground glass is like viewing a transparency with a point source like a penlight, held at a distance. Holographic interferometry being a common-path interferometry, the beams (both the comparison beam and the test beam) are traveling in the same direction, producing nonlocalized fringes everywhere, including at infinity. To view these fringes, it is convenient to place a lens on the output side of the hologram and view with one's eye on the focal point of the lens where all the rays cross. With such a telecentric system the entire system can be viewed at once; in the absence of a diffuser one can see the fringes, like coarse grating, extending in space and can project them on a white paper, just by intercepting the interfering beams (20).

B. Double-Exposure Holographic Interferometry. A second type of holographic interferometry is called double-exposure holographic interferometry (20). Any standard holographic arrangement can be used. A first exposure is made with the subject or scene in its initial state. This is the comparison exposure. The change to be studied is introduced in the subject or scene, after which a second exposure of duration equal to the first is made. Then the photographic plate is developed and reconstructed. For reconstruction, as in conventional holography, only the reference beam is used. If the same setup is used, the scene beam is blocked. In essence, we are simultaneously reconstructing two holograms, taken at two different times. The light wave patterns recorded from the object emerge from the hologram, superposed on each other. We are then able to observe the interference between two waves that were originally separated in time: these sequences are shown in Fig. 7.

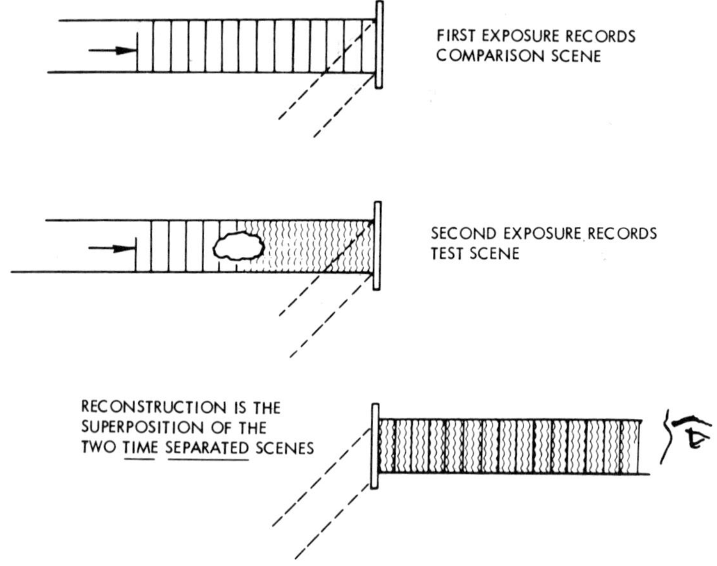

Fig. 7. Schematic of the operation of a double-exposure interferogram.

The reconstruction geometry is not at all critical. The precision needed for relocation of the plate in the case of stored beam interferometry is not needed. Both the images are locked in together, and therefore any variation in the reconstruction process affects both of them essentially the same way, keeping the interference patterns the same. Though the relocation problem is made easier, the advantage of real-time viewing is lost. We have information about the difference between two instants. The interference fringes appear whenever there is 180° phase difference between these two images. By using a conjugate reference beam, as described earlier in the section on holography, we can reconstruct the real image in space, with the fringes in space. Then the fringes can be looked at with a microscope.

The mathematical theory is similar to the one described earlier. We have recorded two individual holograms with the same reference beam. Reconstructing with a wave similar to the original reference wave regenerates the components as described earlier.

Other variations of holographic interferometry have been reported. Double-exposure holographic interferometry on two individual plates, aligned during exposure, has been reported. More than two exposures on a single hologram plate, from a scene disturbed through a sequence, is also suggested. A technique known as time-averaged holographic interferometry can be considered as an infinite exposure technique. A periodically vibrating object, when holographed, on reconstruction shows fringes related to the periodic displacements of different parts of the object. It is sufficient for us to confine our arrangements to stored-beam interferometry and double-exposure holographic interferometry for the electrochemical studies.

C. Interpretation of Interference Fringes. Holographic interferograms made with transmitted light through the test medium using a collimated beam are interpreted in the same fashion as conventional interferograms (Mach-Zehnder). In the case of infinite fringe arrangements, whether real-time or double-exposure, each fringe line is a locus of points of equal optical path length. The optical path length change

$$\delta = \int (n_2 - n_1)\, ds \tag{17}$$

where the integral is taken along the viewing direction and n_1 and n_2 are the refractive indices during first and second exposures. The change in the optical path for two adjacent fringes is one wavelength. These fringes appear as lines, the distances between them determined by the value in the integral. But a more dramatic way of obtaining the interferograms is through the finite fringes. This arrangement is used in conventional interferometry also. The finite fringe interferogram has already been discussed. By deliberately changing the angle of one of the beams between two exposures or adjusting the plates as described earlier for stored-beam interferometry, a background of fringes will be obtained. The object in this case displaces the fringes from their normally straight position. In investigating a transparent field, where the refractive index changes at different points during the experiment, the fringes are likewise shifted. The optical path change at each point of the object or scene may be obtained by measuring the transverse displacement of the fringes at various object or scene points, relative to unperturbed fringes outside this field where the change is occurring. Thus the finite fringe technique lends itself easily to data reduction. The second advantage is that the bending of the fringe in the transverse direction eliminates the ambiguity of direction which accompanies infinite fringe arrangement, where we measure the distance between fringes for gradient.

In electrochemical work we prefer to use parallel beam illumination, with the concentration gradient being perpendicular to the direction of the light beam. The following analysis is given for a finite fringe arrangement during stored-beam interferometric studies. In Fig. 8, the finite fringe pattern is given when no current is flowing

Fig. 8. Finite fringe pattern when no current is flowing through the cell. The fringes have been deliberately adjusted to be normal to the electrode.

through the cell. Figure 9 shows the interferogram of the diffusion layer with the direct recording of the concentration profile, in terms of the fringe shift, at a particular time. The distance between the fringes is given by the tilt of the hologram plate. Each point on a given fringe corresponds to the same phase for the plane wave front.

In Fig. 9, during concentration polarization, an additional path difference at B is caused by changes in the refractive index. The

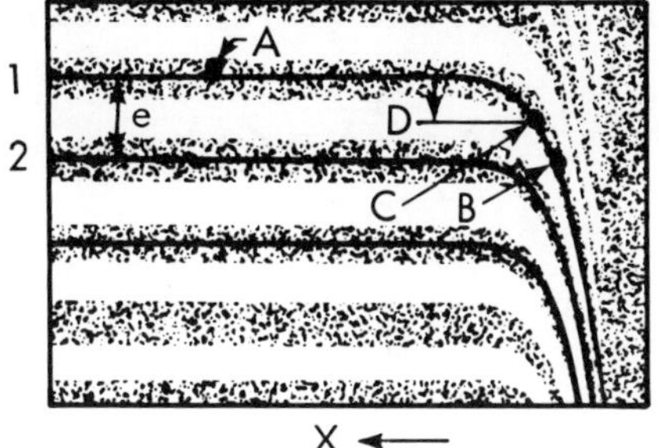

Fig. 9. Interferogram of the diffusion layer with direct recording of the concentration profile. The concentration change at any point with respect to the bulk can be calculated, in terms of fringe shifts and refractive index.

change in the optical path difference is $\Delta\delta = (n_0 - n_1)d$, where d is the thickness of the cell through which the light is passing, and n_0 and n_1 are the refractive indices of the solution in the bulk and at the point under consideration in the diffusion layer. When the concentration change is small, $\Delta n = K(C^0 - C)$, where K is the proportionality constant relating to the index of refraction and concentration; C^0 and C are the concentration of the solution in the bulk of the solution and in the plane parallel to the electrode, at which the point B is situated; and Δn is the difference in refractive indices. So one gets

$$\Delta\delta = Kd\,\Delta C \tag{18}$$

The optical path difference for the interfering rays changes by one wavelength in passing from one dark fringe to another dark fringe (or one bright fringe to another bright fringe). With no externally applied current to the electrochemical cell, the concentration at the point B will be the same as in the bulk and the phase is represented by fringe 2. When the diffusion layer builds up, fringe 1 shifts and passes through B at a point in time. The concentration difference ΔC, between this plane and the bulk has caused this shift, causing an optical path difference of $\Delta\delta = \lambda$. Fringe 1 is displaced at B from its undisturbed position (Fig. 8) by a distance equal to the distance between two adjacent fringes corresponding to a $\Delta\delta$ value of λ. In the position C where the fringe displacement is equal to $e/2$, $\Delta\delta$ is equal to $\lambda/2$; and, in general, $\Delta\delta$ is proportional to the fringe displacement b and the equation can be written as

$$\Delta\delta = \frac{b}{e}\lambda = Kd\,\Delta C \tag{19}$$

or

$$\Delta C = \frac{b}{e}\frac{\lambda}{Kd} \tag{20}$$

where λ, e, K, and d are constants for a given set of experimental conditions and ΔC can be directly plotted on the ordinate of the interferogram. Thus the shift of any one particular fringe gives the concentration profile. The intersection of the disturbed line of that fringe and the electrode surface is to be taken as the origin for the concentration profile. Since refractive index does not allow separate determination of concentration profiles of the components in an electrode process, the method is limited to binary salts.

It should be emphasized that this interpretation of fringes is the same as that of conventional interferometry.

D. Three-Dimensional Fringes. The fascination of holography and holographic interferometry is its three-dimensionality. The preceding section did not deal with this property associated with holography, but rather with a plane wave front traveling through the transparent medium and the fringe shifts associated with optical path changes, as viewed from the plane of observation. It was previously mentioned that the introduction of ground glass for diffuse illumination of front-lighted surfaces and rear-illuminated transparent subjects gives rise to localized fringes. These fringes, on reconstruction, appear three-dimensional and can be examined over a range of angles limited by the boundaries of the hologram plate. The positions of these fringes in space depend upon the nature of the changes that give rise to the interference patterns and often have a location different from the object. Yet, in some cases, visibility of the fringes depends on the direction and angular aperture of the lens used to view them.

Because of the complexity of the interpretations of the fringes, no detailed analyses are available in literature, although excellent interferograms are obtained which may be viewed over a wide range of angles, yet one must find the index of refraction as a function of position in space. The general solution to this problem for transparent subject has not been found. This solution will be of interest to electrochemists working on mass-transport problems associated with hydrodynamics near electrodes.

For a general appreciation of the complexity of the problem a simple case is presented.

Consider that a double-exposure holographic interferometric study of a solution surrounding a point electrode is made with diffused rear-illumination. Before the experiment, the refractive index of the solution is n_1. The reference hologram is taken. Assume that after a brief electrolysis a uniform sphere of refractive index n_2 is formed around the point and the second exposure is taken. Thus we are comparing light through two identical (imaginary) spheres whose refractive index changed from n_1 to n_2 between these two exposures. This configuration is shown in Fig. 10.

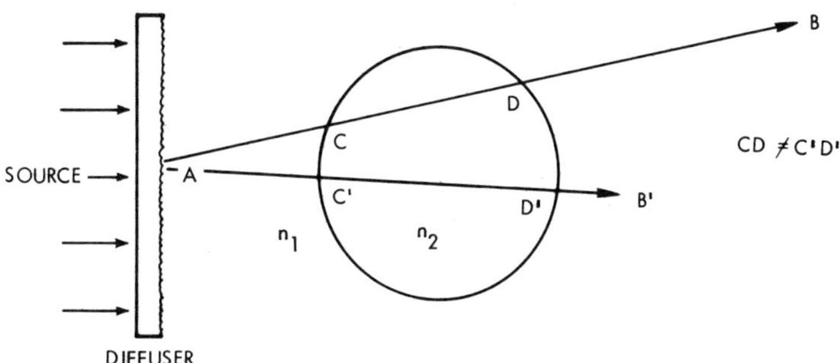

Fig. 10. Optical path change through a sphere of uniform refractive index. During an event the refractive index of the sphere changes from n_1 to n_2. Two different wavefronts with paths *AB* and *AB'* are considered.

To determine the location of the fringes, first consider any arbitrary ray in the scene such as *AB*. The relative phase of the two light waves along this ray *AB* is found from the difference in optical paths during the two exposures. The only portion of the path contributing to this difference is the chord *CD* interior to the circle. The number of waves along the path *CD* is $n(1/\lambda_0)$ where 1 is the length of the chord *CD*, λ_0 is the wavelength in vacuum, and n is the index of refraction. Thus the optical path difference, in number of waves, from the two exposures is $(n_2 - n_1)(1/\lambda_0)$. If this number is an integer, then constructive interference from the two exposures will occur. If this number is an integer plus $\frac{1}{2}$, then destructive interference will occur.

In this example, the important point to note next is that this relative phase depends on the length of the path 1. Thus all chords of the same length will give the same phase difference between the two exposures.

Figure 11 shows a number of paths for a given viewing angle, all tangent to the same circle. Therefore, the chords and, hence, path differences are all the same for all of these paths. It is apparent that

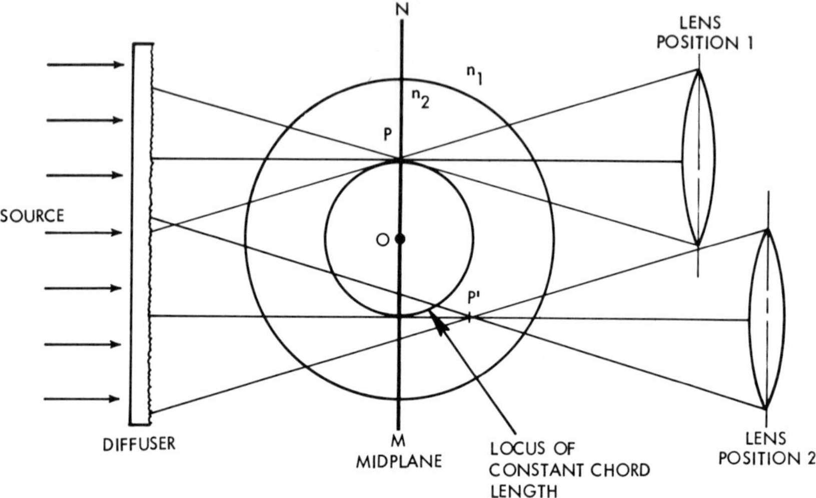

Fig. 11. Location of infinite fringes in a spherically symmetrical system. During an event the refractive index changes from n_1 to n_2 inside the uniform sphere. The inner circle represents the locus of chords of equal length, representing constant optical path. The interference fringes appear to be localized in the midplane NOM.

these paths all cross at approximately the same point *P*. Thus the relative phase difference between exposures is essentially the same for all rays of the cone intercepting the lens focused on *P*. As the point *P* is moved in the midplane *MN*, thc phase difference will change because of changes in the chord length. Thus, in the focal plane of the camera or eye, a set of concentric interference fringes will appear, even though a fairly wide aperture is used. In contrast, consider a point *P′* which is not on the midplane. A wide aperture lens focused on this point receives light from many different chord lengths. Consequently, the interference is not the same for all rays entering the aperture and the observed fringes will be much reduced in contrast. The conclusion for this example is that the fringes appear to be located at the midplane of the sphere, apparently going out of focus as the focus of the viewing system is moved away from the midplane. As with real objects, the depth of focus is increased by using a smaller aperture.

The foregoing example can be generalized to the case of a sphere with an index which is an arbitrary function of radius only. The same midplane property will occur. Spherical shocks from electric sparks and explosions, and spherical index variations about point electrodes in solution, can thus be handled quantitatively. For shocks from high velocity models which give conical or cylindrical geometry, an approximation to the above occurs where viewing is done perpendicular to the rotational axis. In these cases, the rotational symmetry dominates, giving an approximation to the focusing behavior described above. From whatever angle one looks, these fringes, for symmetric systems, appear around the midplane. If the system is not symmetric, when view angles change, the optical path changes and, consequently, the apparent location of the fringes changes. In dealing with such systems we encounter two variables, the change in refractive index as we move away from the reference point and the variation of the physical path traversed by the rays gathered by the imaging optical system.

In looking at the fringes created by reflected light (front-lighted with parallel beam or diffused light), similar considerations arise. But here the fringes are investigated not in terms of the change in the refractive index, keeping the total physical path constant, but in terms of the change in the optical path due to physical movement of the object under investigation. Because of the movement of the surface, the point which originally reflected light in one direction now either changes the direction of reflection or changes the distance through which the light beam is traveling by a parallel displacement. So the translational movement or the rotational movement of the point makes the difference in fringe location. In general, with parallel illumination, pure translational movements give fringes at infinity and pure rotational movements give fringes near the surface. If the surface is illuminated with diffused light, the location of the fringes is more complicated (1,18,19,47).

5. Holographic Microscopy

It is interesting to note that holography was invented to augment the field of electron microscopy. Usually two techniques are used:

(1) lensless holographic microscopy, and (2) lens-assisted holographic microscopy (13,25).

It has been established that the in-line (Gabor) holographic arrangement is suitable for microscopy. The reconstructed real image is magnified by a conventional microscope; the great advantage is the enormous depth of focus one can obtain. Conventional microscopes, depending upon the fidelity of recording, have depths ranging up to several millimeters. Thus a transient phenomenon occurring in a depth of solution can be holographically recorded with a pulsed laser, and the phenomenon can later be investigated leisurely at different depths. The chief problem is that a site under investigation should not be obstructed by the one preceding or following it. A two-beam arrangement may be used as well, but in that case the conjugate reference beam should be used for reconstruction.

In the conventional microscope, the magnification and the depth-of-focus both depend on the objective and the eyepiece. The distance between the objective and the eyepiece is always kept at a standard value. During focusing the whole system moves together. In lens-assisted holography or holomicrography, the objective and the eyepiece are separated. The magnified image from the objective is holographed with an off-axis arrangement. The image from this reconstruction is used for further magnification. The reconstructed image can also be used for dark-field view, phase contrast, etc. Though the depth of field is not as high as the first method, the resolution is higher. In this technique a resolution of 1 μ, with a depth of variable field of 150 μ, has been reported (51).

For electrochemical investigations, the lensless holography is suitable because of the dimensions of cells encountered and the depth of interest in the cell. The questions to be answered are whether or not there is an advantage in this arrangement and whether or not the resolution is adequate.

A. Resolution in Holography. The resolution or resolving power of an optical system is its capacity for imaging fine detail. This subject has been a controversial one, due largely to the subjective nature of human judgment in image quality (11,46). However, there is an optical constant, which largely determines the resolving power of microscope objective, known as numerical aperture.

$$\text{n.a.} = n \sin \theta \tag{21}$$

where θ is the angle of the marginal ray with respect to the optical axis of the lens or hologram plate and n is the refractive index of the medium between the object and the lens. Thus the larger the size of the hologram, the greater the numerical aperture, hence, the resolution. In principle, the resolution of the holographically reconstructed image is limited by the size of the hologram, the subject-to-hologram plate distance, and the wavelength. The first two determine the numerical aperture. However, for the extremely large angles of incidence on the photographic emulsion, a small change in the emulsion in the developing process of the plate will cause appreciable aberrations in the reconstruction. When the reconstruction requires that the image-carrying beam travels the substrate, the substrate should be uniform in optical path to allow controlled aberration. This imposes a serious restriction because it requires that the substrate consist of good quality optical glass with surfaces that are precisely figured and polished. In practice, the resolution is further restricted by nonideal properties of the emulsion, other physical distortions, and grain noise of the silver halide. If λ is the wavelength of the laser and n is the refractive index of the surrounding medium, the fringe spacing as a function of the scattering angle is given as $\lambda/n \sin \theta$. Thus if the film is not capable of resolving greater than $K_{\max}$, the maximum angular radius of the pattern on the hologram plate is

$$\theta_{\max} = \sin^{-1} \frac{\lambda K_{\max}}{n} \tag{22}$$

The maximum useful numerical aperture of the observing system is $n \sin \theta_{\max} = \lambda K_{\max}$. The microscope resolution for coherent illumination is given by $\lambda/n \sin \theta_{\max}$ from which is derived an image resolution equal to that of the film $1/K_{\max}$. In addition to the limiting factors of film resolution, there are the other factors that determine the image quality as described above. Scattering of the light by film grains and submicroscopic particles causes a background "speckle" pattern. In the case of Gabor in-line technique, the out-of-focus images give rise to diffraction patterns, both of which sometimes severely mask the details of the image in focus. Thus the resolution in holography can be generally improved by using very fine grain emulsions, keeping all optical surfaces free of dust, etc. A resolution to about 10 μ has been obtained in our laboratory in the lensless arrangement.

III. Apparatus and Technique

1. Components

The simple two-beam arrangement is preferred for our electrochemical studies, because of its versatility. The laser beam can be divided into two parts, one for scene illumination and the other for reference. In whatever manner these two beams are brought together, the net path difference between the two should not exceed the coherence length of the laser.

For constructing a hologram, a source, a stable optical configuration, and a recording system are needed. For reconstruction, in addition, a copy camera is needed. The object under investigation is an electrochemical cell or an electrode, in which the refractive index changes are produced. First, we will look individually into each component of the holographic system (or holocamera, as it is called commercially) and then describe an arrangement.

A. Source. A continuous wave (CW) laser of 15 mW or more power is needed. If a pulsed or Q-switched laser is used, the output should be at least 0.01 J. The tolerance for pulse duration is dependent upon how fast a moving object should be holographed. For example, with a laser of pulse duration of 20 nsec, of wavelength 700 nm, an object moving at a velocity of 6.2 m/sec can be holographically recorded. However, in electrochemistry, the events are often much slower than that. Table I lists the essential characteristics

TABLE I
Lasers

Active medium	Pumping method	Wavelengths
Helium neon	Glow discharge (CW)	632.8 nm
Argon II	Hot or cold cathode discharge (pulsed or CW)	488.0 nm and 514.5 nm (principal lines); 457.9 nm, 476.5 nm, and 501.6 nm useful
Ruby (Cr^{+3} in Al_2O_3)	Optical absorption of blue, green light from Xenon flash lamps (pulsed)	694.3 nm

of some important commercially available lasers. While using the Ar laser, only one wavelength should be selected for recording purposes, by using filters. Pulsed lasers have short coherence length.

If the hologram is recorded with a pulsed laser, a CW laser is needed for reconstruction. For our purposes, any CW laser with the above power requirement will do. One can expect a greater resolution with shorter wavelength, as seen in Eq. 22.

The laser provides a highly directional beam with very little divergence, but being an optical cavity, it can operate in many modes. Though the commercially available lasers are supposed to operate predominantly in one mode only, there is the presence of higher orders. Lasers which generate wave fronts having single phase across the wave front possess complete spatial coherence. The presence of other modes suggests the existence of more than one phase relationship. To produce a uniphase wave front from the laser beam, spatial filters are used.

B. Spatial Filter. The principle of the spatial filter is as follows. The beam of coherent light can be focused down to a diffraction-limited spot site only if the wave front has the same phase across the entire surface with no nodes, and the edges of the wave front have to pass through a mechanical aperture. To produce a uniphase wave front we can use the converse of this requirement (5). The beam of light from the laser is focused at a point in space, just at the entrance of a small mechanical aperture like a pinhole. Only that portion of light that converges at this aperture will emerge from the pinhole with a uniphase wave front. The focusing arrangement is usually a $60\times$ corrected microscope objective. The aperture is a fine pinhole of about 5–20 μ in diameter, placed after the lens, on a movable mechanism. The whole arrangement, the microscope and the pinhole, is known as spatial filter. The emergent beam contains a bright circular pattern in the center with the intensity slowly falling toward the periphery. This spot is known as the Airy disc.

C. Working Table. A stable flat-surfaced granite slab with shock isolation is needed for mounting optical components. The shock isolation is not needed for pulsed techniques. These types of tables are commercially available. The previously discussed tolerance for vibrations during experiments is an actual experimental problem. The quality of the hologram is affected by the floor of the building in which the laboratory is situated, and by how much can be spent for the isolation mount and the slab. This is a common problem with all interferometric methods. It is preferable to have a plastic or wooden hood over the entire assembly to minimize drafts of air causing pressure and temperature changes during experiments. A commercially available lathe-bed optical bench with proper vibra-

Fig. 12. Precision-adjustable holographic plate holder. After processing, the plate can be precisely aligned in any axis.

tion isolation can be used. In communications to technical journals, "coffee-table" holograms, and "hand-held" polaroid holograms have been reported. These holograms are made with very narrow angle beams (Fourier type) and are not of high enough quality for interferometric experiments described here.

D. Optical Components. Front-surface mirrors, $1/4$ wavelength flat, on mounts with preferably *x-y* rotational adjustments, antireflection coated beam splitters with rotational adjustments, lenses (both positive and negative) of convenient focal lengths, a shutter, ground glass plate, coated prisms, and neutral density filters are standard requirements. These materials need not be of good quality. Quite a few of them can be purchased in surplus stores. The stability of the components is achieved by waxing down these elements on solid supports.

E. Photographic Plate Holder. This too is available commercially. Figure 12 shows a view of a homemade holder. This holder has adjustable rotation about three axes. These adjustments are made by micrometer thread screws, which are particularly useful for finite fringe adjustment and in real-time interferometry when compensation is needed for emulsion shrinkage and subtle changes in the location of the plate. The plate holder is capable of holding 4×5 in. hologram plates up to $\frac{1}{4}$ in. thick; or an adapter can be used to hold 1×3 in. plates. The hologram plate-holding mechanism seats the plate against the positioning mechanism with reproducible spring compression. In all cases, the plate may be removed for development and then replaced into its original position with the precision required for real-time interferometry. The plate-holder is mounted on a cast iron base, which can be fixed either onto the surface of the granite table or on an optical bench. The high quality bearings used have no backlash. Thus, one has a conveniently adjustable and extremely stable plate holder.

F. Photographic Plates. Two important characteristics of the photographic materials used in holography are (1) color sensitivity and (2) resolution.

As laser light is monochromatic, it is always advantageous to have the photographic material sensitized for this region. The recording then can be done with minimum light. Since panchromatic films reach their peak sensitivity before the red region, special films are preferred. Agfa-Gevaert plates and films are available sensitized in the wavelengths corresponding to ruby laser, He/Ne laser, and Ar II laser regions. These films are available at different resolutions.

In an ideal case, the resolution is limited by the resolution of photographic material. The spatial frequency for two parallel beams of wavelength 0.8 μ with an angle of 45° between them when incident on a photographic plate is about 1250 lines/mm. For a smaller angle the number will be less. Only a photographic emulsion that has this resolution can form a good hologram. We always assumed a linear behavior of a photographic emulsion. That is, the intensity of the interference pattern as recorded on the photographic plate varies sinusoidally, as given in the intensity equation (Eq. 6). In fact, the recording of the intensity is nonsinusoidal. The greater the deviation, the poorer will be the reconstruction. Higher order images are observed with nonlinear emulsions. The emulsions should be thin (a few wavelengths thick) for the hologram to be considered two-dimensional. The darkened developed plates should be about 0.25 density unit to stay within this linear region. The optimum exposure can be achieved with short experience.

It is well known that high sensitivity and high resolution are conflicting demands on a photographic material. Less-sensitive materials have less grain noise too. A compromise has to be reached among the available intensity from the laser, the exposure time (determined by coherence duration of the wave and variation of the setup), and quality of the image, before selecting the photographic material. Some photographic emulsions useful for holography are given in Table II, with their approximate resolution limits. These numbers are only guide lines. Each type of emulsion is processed according to manufacturer's recommendations.

TABLE II
Emulsion Characteristics

Type	Resolution (lines/mm)	Sensitivity (light)
Kodak 649F	>3000	Panchromatic
Kodak SO243	500	Panchromatic
Agfa Scientia 8E75	3000	For ruby laser
10E75	2800	
Agfa Scientia 8E70	3000	For He/Ne laser
10E70	2800	
14C70	1500	
Agfa Scientia 10E56	2800	Ar laser
Recordak AHU	~ 250	Panchromatic (better at green)

G. Copy Camera. A variety of photographic techniques and equipment can be used for recording the reconstructed holographic image. A view camera with extension bellows can be used with a polaroid back. For studying fine fringes during real-time interferometric studies, it is better to use a film camera with a view-finder that shows the fringes magnified, so that one can monitor the scene as it is being photographed.

In copying the holographic reconstruction, the image (real or virtual depending upon the choice) is focused upon just as the actual object would be. The exposure time is determined by trial and error. As the laser light is monochromatic, the readings on the light meters used for intensity measurements should be weighted. We found that for Panchromatic films, for He/Ne laser light, the assumption of an ASA rating of twice that of the film gave satisfactory results. The camera is mounted on a proper axis to provide the view.

2. *Holography Technique*

A basic schematic of the two-beam arrangement is shown in Fig. 13. Any convenient arrangement in which the distances traveled by

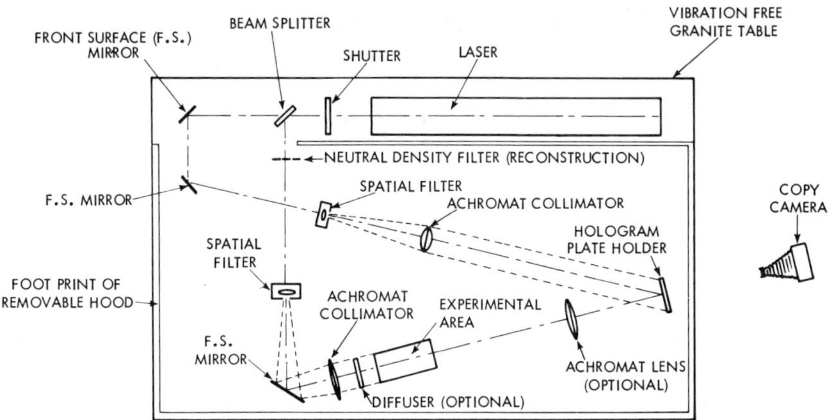

Fig. 13. Basic off-axis arrangement for holography and holographic interferometry.

the beams to the photographic plate are approximately the same will be satisfactory.

The straight beam from the laser, passing through a shutter device, is split into two beams by the beam splitter. After the convenient division of these beams, each has to be passed through a spatial filter to improve its spatial coherence. The beam passes through a corrected microscopic objective and concentrates at a point in space, before diverging in the form of a cone. The divergence depends on the numerical aperture of the objective. Just in front of this spot a pinhole of about 7 μ in diameter is mounted. The hole is usually punctured in a thin disc of metal mounted on a movable mechanism which can be manipulated in the x-y plane, normal to the beam. The pinhole is carefully aligned so that it is in the same axis as the bright spot. Only then can one see faint light emerging from the other side of the pinhole. This is observed by holding a white sheet on the other side. Otherwise, there will be complete blackout, the light cone being stopped by the metal disc. Once this is achieved, one moves the pinhole toward the spot, always watching that light is emerging from the other side of the disc. Just when the pinhole is located in the focal point of the objective a uniform, bright, circular disc is seen on the white screen. Minor adjustments are then needed to improve on the Airy disc. The light intensity is usually less than it would have been in the absence of the filter, but we are getting light in only one mode, the lowest TEM_{00}. The other modes existing as lobes around the center spot are effectively filtered out. The "spatially" filtered light emerges as a bright cone from the pinhole in both the beams.

Collimating lenses are mounted on the optic axis of each beam. The sizes and focal lengths of the lenses depend upon how big a scene is to be recorded and what the intensity ratio between the reference beam and the scene beam should be. Smaller diameter lenses with shorter focal lengths produce brighter beams than the larger diameter lenses with longer focal lengths of the same f-number. But the area covered by the beams is correspondingly less. We may recollect that the transmission function involves the amplitude of the reference wavefront. To produce a good hologram the intensity of the reference beam should be 2–5 times that of the scene beam. The collimation is done by moving the lens along the optic axis until the pinhole is in the focal point of the lens. Under these circumstances the size of the beam emerging from the lens has the same diameter everywhere in space. Collimation is checked by holding a white sheet normal to the direction of propagation at two distances and seeing that the diameter of the beam does not change appreciably. Though this method is crude, it is sufficient for our experiments. If desired, a diffuser may be introduced behind the scene for pronounced three-dimensional effects. The diffuser produces rays in all directions from a previously collimated beam. These rays pass through the transparent subject, like the electrochemical cell, through an imaging lens and to the hologram. The function of the imaging lens is to produce a real image of the cell in reconstruction, so that real-time interferometry may be observed on a screen at the focal plane of the lens. The lens may be located as close to the hologram as possible without intercepting the reference beam. The experimental subject must be located at least one focal length from the lens. The hologram is not at the focus of the lens. Rather, the lens causes the rays incident on the hologram to be convergent. As far as the hologram is concerned, it is part of the scene. Hence, upon reconstruction of the holographic image by the reference beam, these converging rays are reproduced, creating a real image at the focal plane of the lens (even though the reconstructed rays do not pass through the lens during reconstruction). In the presence of the subject beam, the experimental subject is also imaged at that plane. Changes in the subject cause interference of the two images, and the interference fringes are visible on the viewing screen. The viewing screen may be either an opaque screen or a ground-glass

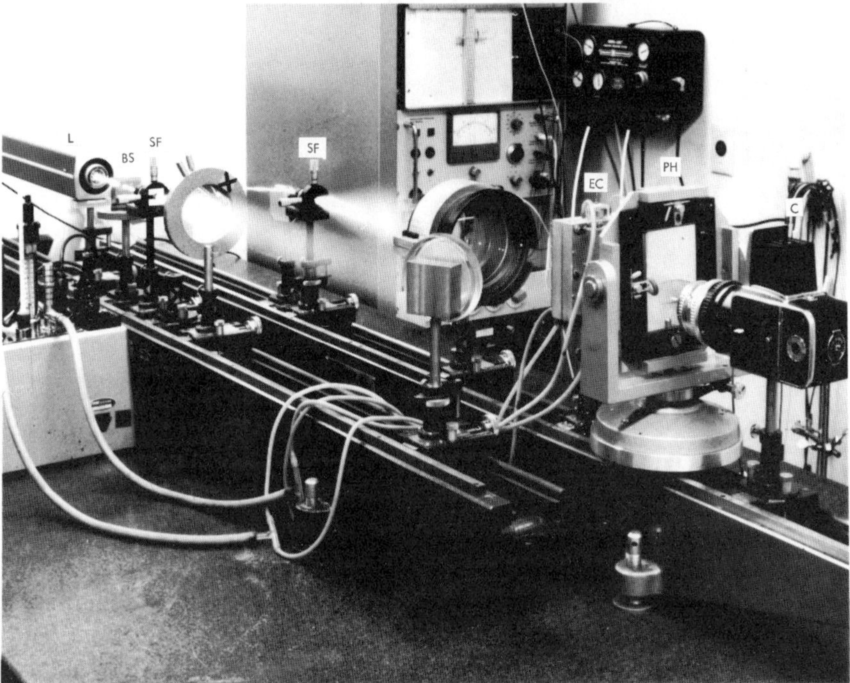

Fig. 14. A total view of the arrangement. *L*, Laser; *BS*, beam splitter; *SF*, spatial filters; *EC*, electrochemical cell; *PH*, photographic plate holder; *C*, copy camera. The thermostat and the electrochemical polarizing unit are also in view.

screen. If desired, the imaging lens can be omitted from the setup; then the virtual image of the subject would be reconstructed. In that case, the subject would normally be located as close as possible to the hologram to make the virtual image close to the copy camera. The third lens can also be used to create a real image from the virtual image by locating the lens so that it operates on the rays coming from the hologram during reconstruction. Either the virtual or the real image can be photographed. In the case of the focused real image, substituting photographic film for the viewing screen will produce a photograph without using the copy camera. For our back-illuminated, transparent, electrochemical cell we avoided the use of the diffuser and the imaging lenses. The imaging of the virtual image and the object was done with the copy camera during reconstruction. With the copy camera, both the object and the fringes

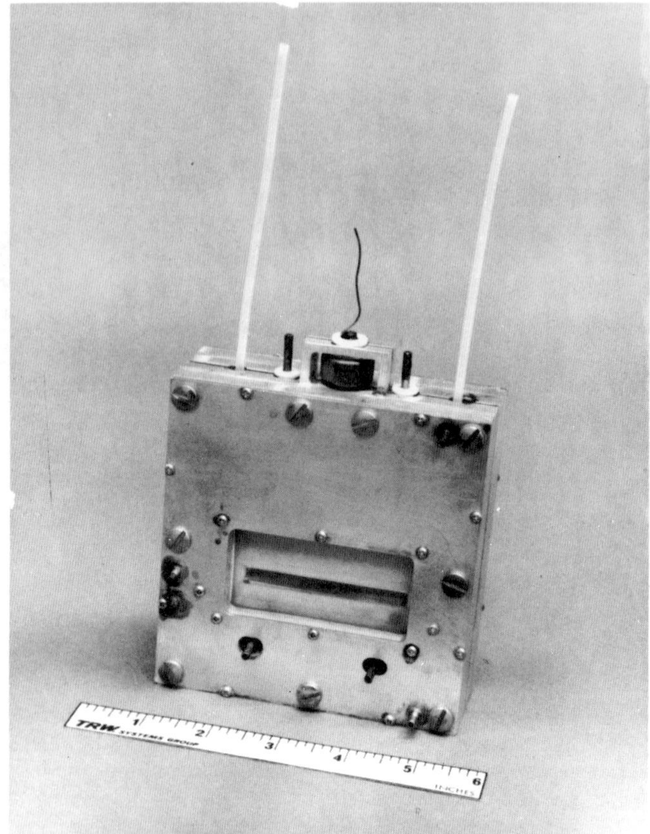

Fig. 15. A closer view of the electrochemical cell. The central slit is the region where the electrochemical process occurs. The jacket is for thermostating liquid.

could be photographed simultaneously. A photograph of our actual arrangement, mounted on two lathe beds, on a vibration-free bench, is shown in Fig. 14. Figure 15 shows a close view of the electrochemical cell used for mass-transport studies constituting the scene. The electrodes are two thin, flat, rectangular strips of suitable metal under investigation, mounted on suitable supports. The gap between the electrodes is filled with a solution of the binary electrolyte. Two flat glass pieces hold the solution in the gap. The collimated beam, parallel to the electrodes, passes through the glass plates and is incident on the photographic plate. The design of the cell is such that the electrodes can be thermostated by a water jacket and the distance between the electrodes can be varied.

After the hologram of the quiescent subject has been made and re-placed in the plate holder, real-time interferometry is done by reconstructing the holographic image simultaneously with the illuminating of the subject, as small changes are allowed to occur. In order to maximize the contrast of the interference fringes, the brightness of both the reconstructed image and the subject should be equal. A reconstructed image will often be much less bright than the illuminated subject itself, unless the illumination of the subject is reduced below the level at which the hologram was made. The main purpose of the neutral filter is to allow convenient attenuation of the subject beam to maximize the interference-fringe contrast.

Finite fringe interferometry can be done with the arrangement described above either in the double-exposure or the real-time techniques. The finite fringes are produced by a small rotation of one of the mirrors or the hologram after the holographic exposure of the experiment in its undisturbed state has been made. In the case of double exposure interferometry, a second exposure is made with the subject in its disturbed state and with a mirror or the hologram slightly rotated. The hologram will reconstruct both images, and these will interfere and cause a modulation of the finite fringes. In the real-time interferometry case, the spacing of the finite fringes may be determined after the hologram is developed, while the holographic image is reconstructed and the mirror or hologram is given small rotary displacements. A rotation of about 1 minute of arc will produce approximately 1 finite fringe per millimeter. The fine-thread adjustable rotation of the plate-holder makes finite-fringe interferometry particularly easy.

IV. Applications

In this chapter the applications of holography and holographic interferometry in electrochemical studies are discussed (26). This is a technique, and like all techniques it has its specificity and limitations. One or more of the following variables are under the control of an electrochemist: the potential of the electrodes, the current passing through the electrochemical system, and the concentration of the electrolyte. The optical techniques in general are utilized to monitor the third variable near the electrode. Further, any physical change of the electrode associated with any of these variables is also of interest and only optical techniques can provide this information. Holography and holographic interferometry fill in the requirements for the last two approaches of an optical technique in its own way.

1. Holographic Microscopy

The conventional microscope has a short depth of focus. Consequently, the bubble formation or a nucleus of growth occurring over a larger area and larger depth cannot be photographed at the same time. These time-dependent phenomena can be frozen in time by pulse holography. The reconstructed real image can be investigated later. But as we saw earlier, granularity imposes limitations. It becomes practically difficult to distinguish the image from the granular structure below 10 μ. If one were to use a lens-assisted imaging process the resolution could be improved to 1 μ but the depth of focus is determined by the imaging lens, while viewing the virtual image.

2. Holographic Interferometry

The most practical application is holographic interferometry. We saw earlier how complicated the interpretation of the three-dimensional fringes are. Figure 16 shows the magnified, reconstructed real image of a hanging mercury drop in a dilute solution of copper sulfate. The hologram was taken with the double exposure technique, with diffused laser light illuminating the mercury drop. The first exposure was done before the electrolysis began. The second exposure was done 1 min after electrolysis. The processed hologram plate was illuminated with a conjugate reference beam. The projected three-dimensional interferogram of the mercury drop was photographed through a microscope. A similar photograph of a cylindrical copper electrode in copper sulfate solution is shown in Fig. 17. In these pictures we have combined the holographic microscopy with double-exposure holographic interferometry. The electrochemical reactions occur very close to the electrode and the diffusion layer does not extend beyond a millimeter before natural convection sets in. Yet one can see clearly the mass transport process near the spherical electrode. Compared to the size of the electrode, this diffuse region is quite small. This is quite a contrast to the study of three-dimensional fringes in the wake of a bullet in the air, where the shock waves compress the air, producing changes in refractive index at much farther distances compared to the size of the bullet. The point electrode approximation cannot be valid for practical reasons;

Fig. 16. Reconstructed, magnified, real image of a hanging mercury drop. The hologram was taken by the double exposure technique, with He/Ne, 25 mW laser, with a very fast film (Agfa 10E70). The infinite fringes appear in the midplane, as discussed in the text.

however, it should be possible some day, with increased sophistication in numerical methods, to investigate quantitatively the complicated phenomenon of natural convection near such electrodes by three-dimensional interferometry.

For the time being, we are restricted to finite fringe interferometry to study mass-transport phenomena near the electrode. Concentration gradients can be observed by holographic interferometry in real time in the vicinity of the electrode during the electrochemical process. The time dependence of the fringe shift, which in turn depends upon the rate of mass transport of the electrolytes, can be theoretically analyzed. The overall mass transport is controlled by diffusion, electromigration, and convection. The fringe shift depends on the change in the refractive index of the solution near

Fig. 17. Reconstructed, magnified real image of the cylindrical copper electrode. Double-exposed hologram was used for reconstruction. The infinite fringes appear very close to the electrode.

the electrode and therefore depends upon the mass transport of all the ions. The current measured depends upon the mass transport of the electroactive species only. In a binary electrolyte system, the fringe shift can be correlated to the current under steady-state and nonsteady-state conditions considering both electromigration and diffusion. In systems with supporting electrolytes, the migration current of electroactive species is minimized and the current is due to diffusion only. However, one optically monitors all the ions in the solution. Until steady state is reached, the fringe shifts are related to migration of supporting electrolyte components, also.

Thus the mass transport properties of electrolytes can be optically monitored in pure solutions, as well as in the presence of supporting electrolytes. A theoretical treatment of systems with supporting electrolytes is complicated. The analysis associated with such fringes is no different from those obtained from setups with Jamin plates or the Mach-Zehnder interferometer.

As we saw earlier, the fringe shift is associated with a change in the optical path produced by a change in the refractive index in the plane under investigation with respect to a reference plane, where no change has occurred. It is assumed that the physical path traversed by the light has not changed. If we look at Eq. 20, again we notice that for fringe shift, with $K = 3 \times 10^{-2}$ refractive index unit/mode, depth $d = 1$ cm, and wavelength = 6328 Å, ΔC comes to about 2 millimoles. Conversely, a change of this concentration on the surface of the electrode with respect to bulk, whether due to electrolysis or surface adsorption, assuming a gradient existing in the solution, produces one fringe shift. A fractional fringe shift can be interpreted depending upon the quality of the fringe and the ingenuity of the experimenter.

However, the change in the refractive index can also be brought about by temperature difference. Though the temperature measurement by holographic interferometry is not in the field of electrochemistry, we would like to mention that a difference of 0.05°C in solutions, with 1-cm path for the beam produces an observable fringe shift. One should remember that this is a differential measurement, and there is no need for alarm if, during reference exposure and real-time test studies, the existing thermal gradients do not change. Yet, this effect exists, a proper thermostating method is essential, and no large current should be passed through the cell upsetting the existing thermal equilibrium.

These interferometric shifts are due to integrated changes occurring over the entire path of the light, as seen from the viewing angle. For quantitative analysis, uniformity of the medium through the entire light path in any particular medium through the entire light plane is assumed. Yet, this may not be true in practice. To keep errors in the uniformity at a minimum the fringes are very closely spaced (10–20 lines/mm). The fringe shift is calculated as a dimensionless quantity. It is the ratio of the fringe displacement to the distance between the fringes. Consequently, one can attain higher precision in the determination of concentration gradients. In the case of rotating electrode (Fig. 18) at low rotational speed in finite fringe experiments, one has to consider the variation of optical path from the center of the disc. We can keep fringes close together to permit calculation of the mean optical path along a radius from the center. As we progress away from the center to the periphery, the

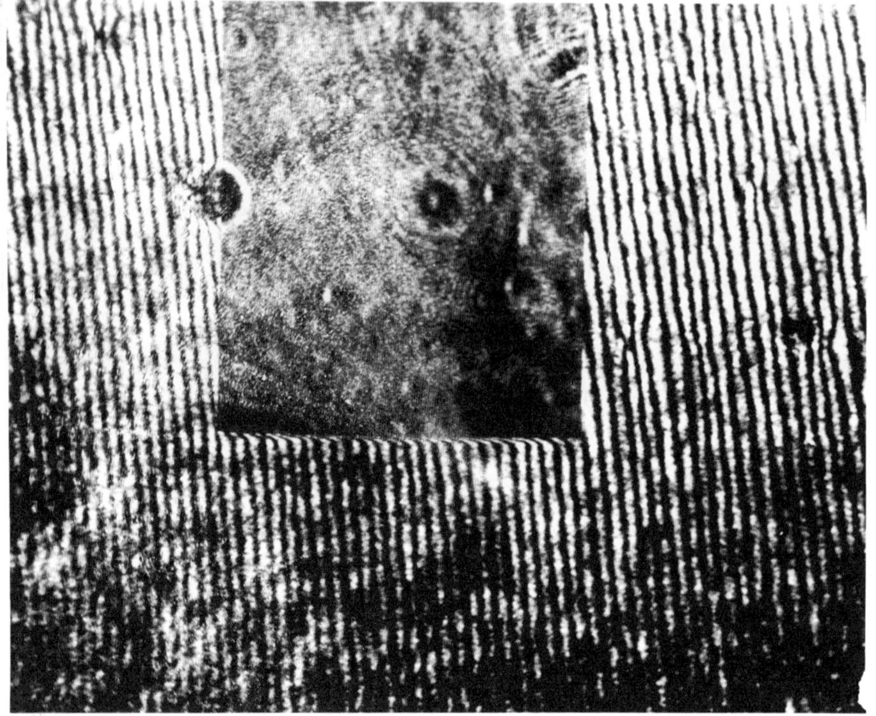

Fig. 18. Finite fringes with a rotating disc electrode, at very low speed. Optical path changes as a function of radial distance.

mean optical path for the light in the boundary layer changes slowly to zero. As a consequence the fringe shift decreases, for the same concentration gradient, as we proceed away from the center. For calculations, these fringes should be converted to concentration gradients near every point at the electrode-solution interface. The fringes shown in the figure are for very low rotational speed. These are particularly interesting because no electrochemical mass transport theory is so far available to explain them at low rotational speed. At higher speeds the diffusion layer is too small to be seen interferometrically.

Discussions have come up relative to adsorption of gases in solutions on electrode surfaces. In all the above discussions, a refractive index change is assumed as a result of an electrochemical process. If the system involved is an oxidation-reduction process, with both the species existing in solution, there will be no fringe shift unless there is an appreciable change in the refractive index properties. The same thing applies to the study of crystal growth around nuclei on the surface of an electrode. There will be a concentration gradient between the nucleus and surrounding medium, as the nucleus grows as a dentrite. In principle, if one studies the system by reflection stored-beam interferometry with plain electrodes, prior to experiment as the reference scene, one can follow the growth of the dentrite. The question is whether the dimension of the dentrite is sufficient to produce the observable fringes. It is very doubtful, considering the dimensions and the speckled pattern from the reconstruction.

It has been suggested that by using two wavelengths from a laser source (Ar II), one can produce two sets of finite fringes and from the periodicity of the fringes between two colors one can improve on the resolution of the fringe shift. It is not practical in holographic interferometry. In our discussions we have assumed a two-dimensional hologram. Discussions on three-dimensional holograms are beyond the scope of the present report. Each wavelength will produce its own hologram on the same photographic plate, as each has its own spatial frequency. Thus we have two holograms on the same plate. During the reconstruction with the same two wavelengths, we have two sets of holograms reconstructed by these two wavelengths leading to holographic cross-talk and confusion. In all these interferometric methods, going to higher resolution is meaningless, because, then one has to take the bending of light in a concentration gradient into consideration. This has been treated by Ibl (21).

A promising application of reflectance holograms is the study of stress corrosion. One can subject the specimen to stress and displacement of the surface due to stress can be seen by holographic interferometry. The stress pattern can be deduced from these fringes. The specimen can be exposed to a hostile environment and the effect of stress on corrosion can be optically followed by time lapse studies. Besides, the stress patterns occurring on the surface of electrode materials as a result of plating in or plating out of materials can be looked into. The actual displacement as a result of stress release or stress formation can be seen by reflectance holographic interferometry. Similar observations have been made for the substrates when the deposited metal was evaporated.

The actual loss of material due to corrosion can be followed by holographic interferometry using surface reflectance holograms. A loss depth of one quarter wavelength can produce a change of one half wavelength in the path of reflected light. This will amount to the appearance of a dark or bright band on the holographic image, on locations where the attack has taken place. The corrosion pattern can be followed. However, before each measurement the corrosive environment should be removed. Further, one should be sure that there is no stress release during this process, causing actual physical movement of the surface.

One can qualitatively look at the effect of corrosion-inhibiting coatings on materials. The results are visual and dramatic. The transmission hologram of a cell with flat wall was taken to provide the comparison scene. The reconstructed image was adjusted for finite fringes. The corrosive liquid was poured into the cell. The fringe pattern would not change because the solution was of uniform refractive index. The material under investigation was introduced into the liquid. The photograph of one such study is shown in Fig. 19. The arrow in the figure shows the region of separation between inhibitor-coated area and the unprotected area. The attacked regions show fringe shifts. Here, we are concerned about visual monitoring of the system without any serious consideration to the geometry or how the optical path changes.

Transparent film formation on electrodes, like creeping of the solution on a platinum electrode, or, film formation on the surface at some applied potential can be studied using the clean electrode prior to film formation as the comparison scene. The very thin film can produce phase shift because of the change in the reflectivity equations according to Fresnel equations. So far, this author has not been successful, probably because of poor intensity matching between wave fronts from the real electrode system and the holographically re-created comparison scene. It could also be possible that the phase difference between these two wave fronts was not sufficient to produce fringes of good contrast.

To summarize, holographic interferometry offers the following advantages over classical methods of interferometry. The alignment and preparatory procedures are far less critical than for many conventional forms of interferometry. The common path nature of the holographic interferometer provides flexibility, permitting the use of inferior quality optical elements. By introducing a diffuser behind the scene, one can produce wave fronts in all directions, thereby enabling one to record three-dimensional information. During reconstruction, the image and the fringes can be examined from

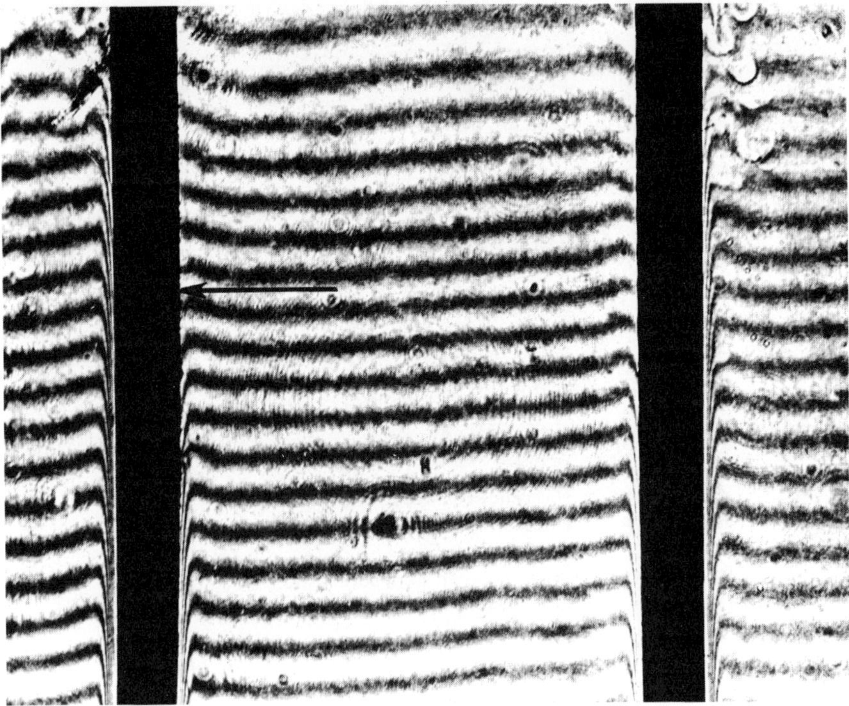

Fig. 19. Visual display of the corrosion inhibition due to surface coating. The material on the left has been coated with the inhibitor. The arrow demarcates the exposed and coated areas.

different directions. This procedure opens an approach to look into localized electrochemical activity, nature of flows which do not have rotational symmetry, and performance of asymmetric electrodes. Further, it allows one to observe changes which occur in the subject as a function of time, producing a differential interferogram.

Acknowledgment

The above contribution was made possible through a research contract from ERC, NASA, Cambridge. The author wishes to express his thanks to his colleagues at the Physical Research Center, TRW Systems, Redondo Beach, California, for many helpful discussions.

List of Symbols

A	Amplitude modulus for the electric field
A_r	Amplitude modulus for the reference beam
A_s	Amplitude modulus for the scene beam
α	Amplitude transmission factor
b	Fringe displacement
b/e	Fringe shift
c	Constant; velocity of light
C	Concentration at a point under consideration
C°	Bulk concentration
d	Depth of the optical cell
δ_0	Path difference
$\Delta\delta$	Change in the optical path difference
e	Distance between the adjacent fringes
E	Exposure
$f_{\max}$	Spatial frequency
ϕ	Phase angle
ϕ_r	Phase angle for the reference wave front
ϕ_s	Phase angle for the scene wave front
I	Intensity
k	Reduced constant for a given exposure time
K	Proportionality constant for the change in refractive index for one mole of change in concentration
$K_{\max}$	Resolution of the recording medium, in cycles per mm
λ	Wavelength in the medium
λ_0	Wavelength in vacuum
n	Refractive index of the medium
n_0	Refractive index of the bulk solution
n_1	Refractive index of the solution at the point under consideration
n.a.	Numerical aperture
ω	Frequency
s	Distance
t	Time; exposure time
τ	Detector response time; coherence time
U	Scalar function for the electric field
U_r	Electric field due to reference wave front
U_s	Electric field due to scene wave front

References

1. Alexandrov, E. B., and A. M. Bonch-Bruvich, *Sov. Phys.-Tech. Phys.*, **12,** 258 (1967).
2. Antweiler, H. J., *Z. Elektrochem.*, **43**, 596 (1937).
3. Antweiler, H. J., *Z. Elektrochem.*, **44**, 719 (1938).
4. Antweiler, H. J., *Z. Elektrochem.*, **44**, 831 (1938).
5. Bloom, A. L., *Spectra-Physics Laser Technical Bulletin, No. 2,* Spectra-Physics, Inc., Mountain View, Calif. (1963).

6. Born, M., and E. Wolf, *Principles of Optics*, Pergamon Press, New York, 1965, p. 453.
7. *Ibid.*, pp. 291–300.
8. Brooks, R. E., L. O. Heflinger, and R. F. Wuerker, *I.E.E.E. J. Quantum Electronics*, *QE2*, p. 275 (1966).
9. Brooks, R. E., L. O. Heflinger, and R. F. Wuerker, *Appl. Phys. Lett.*, 248 (1966).
10. Cutrona, L. J., E. N. Leith, L. J. Porcello, and W. E. Vivian, *Proc. I.E.E.E.*, **54**, 1026 (1966).
11. DeVelis, T. B., and G. O. Reynolds, *Theory and Applications of Holography*, Addison-Wesley, Reading, Mass., 1967.
12. El Sum, H. M. A., *Advanced Holography*, Plenum Press, New York, 1969.
13. Ellis, G. W., *Science*, **154**, 1195 (1966).
14. Francon, M., *Optical Interferometry*, Academic Press, New York, 1966.
15. Gabor, D., *Nature*, **161**, 777 (1948).
16. Gabor, D., *Proc. Roy. Soc.*, **A 197**, 454 (1949).
17. Goldfischer, L. I., *J. Opt. Soc. Am.*, **55**, 247 (1965).
18. Haines, K. A., and B. P. Hidebrand, *I.E.E.E. Trans. on Inst. and Measurement*, **IM-15**, 149 (1966).
19. Haines, K. A., and B. P. Hildebrand, *Appl. Opt.*, **5**, 595 (1966).
20. Heflinger, L. O., R. F. Wuerker, and R. E. Brooks, *J. Appl. Phys.*, **37**, 642 (1966).
21. Ibl, N., *Proceedings of the 7th Meeting C.I.T.C.E.*, Butterworth, London, 1957, p. 112.
22. Ibl, N., I. Barrada, and G. Truempler, *Helv. Chim. Acta*, **37**(**1**), 583 (1954).
23. Ibl, N., and R. Muller, *Z. Elektrochem.*, **59**, 671 (1955).
24. Ibl, N., and R. Muller, *J. Electrochem. Soc.*, **105**, 346 (1956).
25. Knox, C., *Science*, **153**, 989 (1966).
26. Knox, C., R. R. Sayano, E. T. Seo, and H. P. Silverman, *J. Phys. Chem.*, **71**, 3102 (1967).
27. Leith, E. N., and J. Upatnieks, in *Progress in Optics*, Vol. VII, North-Holland, Amsterdam, 1967.
28. Leith, E. N., and J. Upatnieks, *J. Opt. Soc. Am.*, **52**, 1123 (1962).
29. Leith, E. N., and J. Upatnieks, *J. Opt. Soc. Am.*, **53**, 1377 (1963).
30. Leith, E. N., and J. Upatnieks, *J. Opt. Soc. Am.*, **54**, 1295 (1964).
31. Leith, E. N., and J. Upatnieks, *S.P.I.E. Journal*, **4**, 3 (1965).
32. Lengyel, B. A., *Introduction to Laser Physics*, Wiley, New York, 1966.
33. Lin, L. S., R. M. Moulton, and G. L. Putnam, *Ind. Eng. Chem.*, **45**, 640 (1953).
34. Meir, R. W., *J. Opt. Soc. Am.*, **55**, 987 (1965).
35. Metherell, A. F., H. M. A. El-Sum, and J. J. Dreker, *J. Acoust. Soc. Am.*, **42**, 733 (1967).
36. O'Brien, R. N., *Rev. Sci. Instr.*, **35**, 803 (1964).
37. O'Brien, R. N., *Nature*, **201**, 74 (1964).
38. O'Brien, R. N., *J. Electrochem. Soc.*, **111**, 1300 (1964).
39. O'Brien, R. N., *J. Electrochem. Soc.*, **112**, 951 (1965).
40. O'Brien, R. N., *J. Electrochem. Soc.*, **113**, 389 (1966).
41. O'Brien, R. N., *Can. J. Chem.*, **35**, 932 (1957).
42. O'Brien, R. N., and C. Rosenfield, *Nature*, **187**, 935 (1960).

43. O'Brien, R. N., and C. A. Rosenfield, *J. Phys. Chem.*, **67**, 643 (1963).
44. O'Brien, R. N., C. A. Rosenfield, K. Kinoshita, W. F. Yakymyshyn, and J. Leja, *Can. J. Chem.*, **43**, 3304 (1965).
45. O'Brien, R. N., F. W. Yakymyshyn, and J. Leja, *J. Electrochem. Soc.*, **110**, 820 (1963).
46. Parrent, G. B., Jr., and G. O. Reynolds, *S.P.I.E. J.*, **3**, 219 (1965).
47. Powell, R. L., and K. A. Stetson, *J. Opt. Soc. Am.*, **55**, 1593 (1965).
48. Samarcev, A. G., *Z. Phys. Chem.*, **A 168**, 45 (1934).
49. Samarcev, A. G., and L. Y. Kurtz, *J. Phys. Chem.* (*U.S.S.R.*), **5**, 1424 (1934); *CA*, **29**, 5722 (4) (1934).
50. Stroke, G. W., *An Introduction to Coherent Optics and Holography*, Academic Press, New York, 1966.
51. Van Ligten, R. F., *J. Opt. Soc. Am.*, **57**, 564A (1967).

Optical Microscopy in Electrochemistry

A. C. SIMON
Electrochemistry Branch
Naval Research Laboratory
Washington, D.C.

Contents

I. Introduction

A survey of the literature pertaining to electrochemistry reveals that, in the decade from 1960 to 1970, there was a marked increase in the number of papers wherein optical microscopy was used in the solution of electrochemical problems. There is no question that this increase reflects the many recent improvements in microscopes and their accessory equipment that have been made by the various microscope manufacturers. It is also true, however, that the increased use of microscopy in the field of electrochemistry is the direct result of the ingenuity of various investigators in the field, who have pioneered in applying the microscope to modern problems of electrochemistry.

In the subsequent discussion, the work of some of these investigators will be discussed. One of the problems connected with an article of this kind is in determining what work should be included. The field of electrochemistry is broad, and, in many papers pertaining to both electrochemistry and microscopy, there is a tendency to emphasize phases of the work that might more properly be considered as pertaining to other fields, such as those of metallography or analytical chemistry.

To avoid both the necessity for decision and the possibility of criticism, it was decided to leave this decision to others. The criterion for determining whether an article was to be considered as pertaining to electrochemistry was usually its source. If the paper has been accepted by any of the journals devoted primarily to electrochemistry, it was considered to be pertinent. Such journals include the *Journal of the Electrochemical Society*, *Electrochemical Technology*, *Electrochimica Acta*, *Soviet Electrochemistry* (*Elektrokhimiya*), *Berichte der Bunsengessellschaft für physikalische Chemie* (*former Zeitschrift für Elektrochemie*), *Transactions of the Faraday Society*, *Journal of the Electrochemical Society of Japan*, etc. It will be found that many of the authors cited, whose articles appeared in an electrochemical journal, also have had articles on the same subject accepted in journals unrelated to electrochemistry. In some instances some of these are referenced, particularly when the papers form a series pertaining to the same subject. In most instances, the papers cited have sufficiently complete reference lists of their own to enable an interested reader to pursue the subject in greater detail. Occasionally, examples of electrochemical uses could not be found and it was necessary to use papers from other areas of science as illustrations.

Since many excellent books have been written upon the principles of microscopy and the use and care of the equipment, it will be assumed that the reader is here concerned primarily with the applications that have been made or could be made of the optical microscope in electrochemistry.

At the present time, there are relatively few places in the United States, or indeed in the world, where a mastery of microscope techniques and the principles of microscopy are taught. Certain areas of microscopy, such as mineralogy or metallography, are taught in some detail, particularly abroad. In general, however, it is to be expected that the electrochemist who discovers a need for optical microscopy will be a novice in the use of the instrument, and with practically no background in techniques or methods.

The following review is written with the hope that it will offer a convenient source of information for both those with casual interest and those who intend to use microscopy in their investigations. The applications that are presented do not necessarily represent unique techniques, or instrumentation which apply only to the field of electrochemistry. In many cases, these ideas have already been applied in other areas of chemistry, or in other fields of science, such as metallography, petrography, mineralogy, bacteriology, zoology, etc.

The idea that microscopy is only an investigatory technique, to be self-acquired, has been rejected by Haselmann (126). However, many electrochemists will find it necessary to self-acquire this knowledge, and it may happen that such novices, equipped with a dedicated urge to solve their own problem, may succeed with a technique or application that the well-informed, trained microscopist would at once dismiss as impossible. Such success, however, does not mean that these novices have automatically become microscopists.

It is true, as Haselmann suggests, that the trained microscopist is capable of much more versatility than any self-taught novice and he is better able to apply his specialized knowledge to any of the various fields. At the same time, such a trained microscopist will have a much better overall grasp of the possibilities of the instrument.

Such versatile training often transfers the experience gained in one field to another, although sometimes, unfortunately, only after a considerable period of time. For example, microphotometry was developed a half century ago for use in mineralogy. Not until a much later date did metallographers take over this instrument and use it to physically characterize the optical constants of alloy constituents. In a similar manner, the inverted microscope, that was originally developed for metallography, at a much later date, was introduced as a new technique in the field of biological microscopy.

These two, by no means isolated, examples indicate how lack of communication may have retarded progress in individual fields because of ignorance of techniques well known in other areas. A repetition of this situation can be avoided in electrochemistry and should not be tolerated with the recent progress that has been made in methods for the search and retrieval of information.

II. Functions of Microscopy

1. Obtainment of Ultimate Magnification

There are many excellent books and journal reviews of the principles and functions of microscopy, only a few of which are cited here (4,57,61,64,275). These functions are, therefore, discussed only briefly in the following sections. This discussion is more to acquaint the uninitiated with the possibilities of the microscope and to introduce examples of its use in electrochemistry, than to undertake an explanation of theoretical principle.

The principal function of the microscope is magnification. How this is accomplished is shown, greatly simplified, in Fig. 1. A convex lens of very short focus is brought above the object to be magnified and just beyond its focal point. This lens, called the objective, forms a considerably magnified real image at some point above the objective. A second lens is placed so that it forms an enlarged virtual image of the real image produced by the objective. This double magnification is the basis of the compound microscope. The primary magnification of the object depends upon the relative distance between object and objective and between the real image and objective. The ordinary microscope has a tube length of about 150 mm, having the objective at the lower end and the eyepiece at the upper. If, for example, the distance from object to objective should be 5 mm, the initial magnification becomes $\frac{150}{5} = 30$ times. A common eyepiece magnification is 10 times so that the total magnification would then be $30 \times 10 = 300$ times.

It is usual to define optical magnification as diameter magnification rather than area magnification. Magnification is usually designated by some value of magnification such as $500\times$, which would

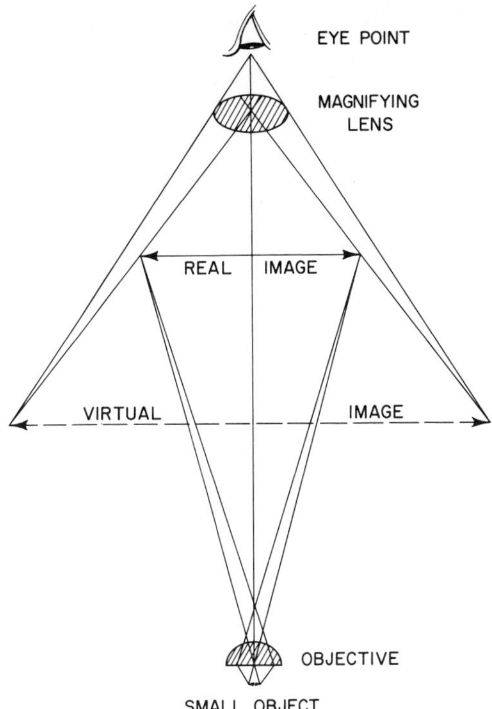

Fig. 1. The principle of the compound microscope.

be translated as a diameter magnification of 500 times the original object diameter. In photomicrographs it is also customary to employ a scale superimposed on the photograph, which is equal to a designated number of microns at the pictured magnification.

The National Bureau of Standards has recently decreed that the official term for the commonly accepted micron shall henceforth be micrometer, designated by μm instead of μ. In the same manner, nanometers (nm) will be used instead of millimicrons (mμ). This new terminology is used in this article and presumably will soon be used on all photomicrographs. In examining such photomicrographs it may serve as an aid in estimating magnifications to remember that the smallest particle visible to the unaided eye is about 40 μm.

Naturally, neither the eyepiece nor the objective need be as simple as pictured. Eyepieces usually consist of two or more lenses, which may or may not be corrected for astigmatism, spherical and chro-

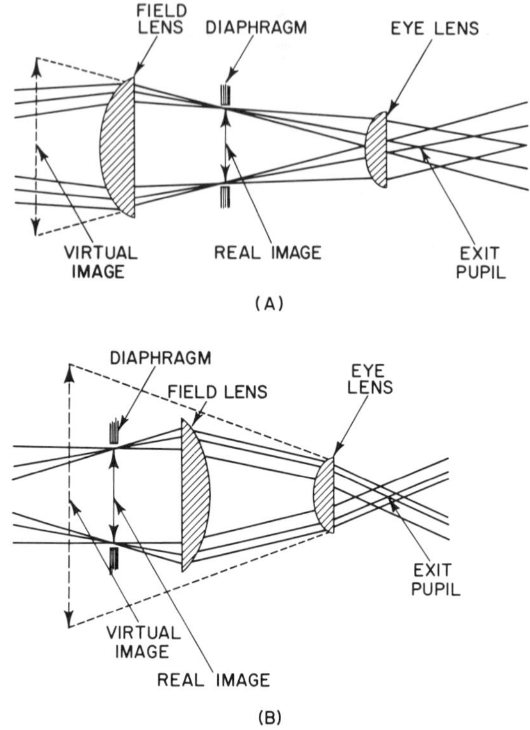

Fig. 2. The construction of the Huygens (*a*) and the Ramsden (*b*) oculars.

matic aberration, etc. Most modern lenses are also antireflection coated to reduce internal flare and thus improve contrast. There are two general types of eyepieces which are used for actual viewing. The Huygens, or negative eyepiece (Fig. 2*a*), consists of two convergent lenses, a so-called field lens and an eye lens. These two lenses have focal lengths in the ratio of 3:1, respectively. They are arranged so that the distance between them is twice the focal length of the eye lens. It is to be particularly noted that in this type of eyepiece, the rays from the objective must pass through the field lens before forming the real image. The real image is then enlarged as a virtual image by the eye lens.

The Huygens eyepiece, or ocular, gives a smaller image than the positive ocular but it has the advantage of a large field of view, very bright and sharp at the center. It is considerably less sharp at the outer part of the field. The lenses may be corrected to form an achromatic doublet or triplet but more commonly are uncorrected, because the aberrations in one lens tend to neutralize those in the other. Such lenses are usually of 10× or less.

The other type of eyepiece is the Ramsden, or positive, eyepiece (Fig. 2*b*). In this eyepiece the two convergent lenses are of equal focal length, and they are spaced at two-thirds of the distance of the individual focal lengths. In this case, the objective forms a real image before the rays pass through the field lens, as shown in Fig. 2*b*. Such an eyepiece acts like an ordinary single lens magnifying glass to enlarge the image of an object placed close in front of it. It is, therefore, called a positive eyepiece. In contrast, the Huygens eyepiece forms no image of a real object placed close in front of it.

The Ramsden ocular is capable of higher magnification than the Huygens eyepiece; the lenses are usually corrected for better color separation and a flatter field. This eyepiece also has the advantage that a micrometer scale placed at the diaphragm is simultaneously imaged at the exit pupil, along with the real image from the objective, and thus gives a direct measure of the primary image.

Various other types of oculars are made. Compensating eyepieces are those which have been corrected so as to compensate for the residual errors left in the objective. Needless to say, these must be used with the correct objectives. Wide-field eyepieces have been made for search work but these need special, large draw tubes. They may also have large distortions in the outer portion of the field, nullifying their possible advantage. In photographic work a properly corrected, concave, diverging lens may be used to form an eyepiece which is only useful for this photographic work, since no image appears at the eyepoint. These eyepieces do have the advantage of superior flatness of field. Such oculars are usually called amplifying eyepieces. Eyepieces are also available with very high exit pupils, so that they can be conveniently used while wearing spectacles.

Eyepieces are usually furnished in a choice of several magnifications. Usual numbers are 5×, 7.5×, 10×, 15×, and 20×. Increasing magnification narrows the field of view and somewhat reduces contrast. It will be found that, for most microscopes, the 10× and 15× eyepieces will be the most used. The higher magnification obtainable with 20× eyepieces, for most purposes, is offset by a number of disadvantages.

Objectives are also much more complicated than the drawing would indicate. Various names are given to objectives depending on the number of aberrations that have been corrected. Achromatic lenses are corrected for spherical aberration in one color only, but in addition have correction for chromatic aberration by the superimposition of two spectral color regions at the same focus point.

Apochromatic objectives use fluorite lenses to correct for spherical aberrations in two regions of the spectrum and these excellent lenses also have three colors superimposed to correct for chromatic aberrations.

The performance and correction of the so-called Fluorite or semiapochromatic objectives are intermediate between the two above types.

Objectives vary more in the magnifications offered than do eyepieces. Seldom, however, does the primary magnification exceed 100 ×. As the magnification increases, the working distance between objective and object decreases. The working distance has been plotted in Fig. 3 for a number of different manufacturers and for lenses of various magnification. It will be noticed that the variation in working distance of lenses of equal magnification decreases as the magnification increases, and, that as the magnification exceeds about 40 × the working distance between the front of the objective lens

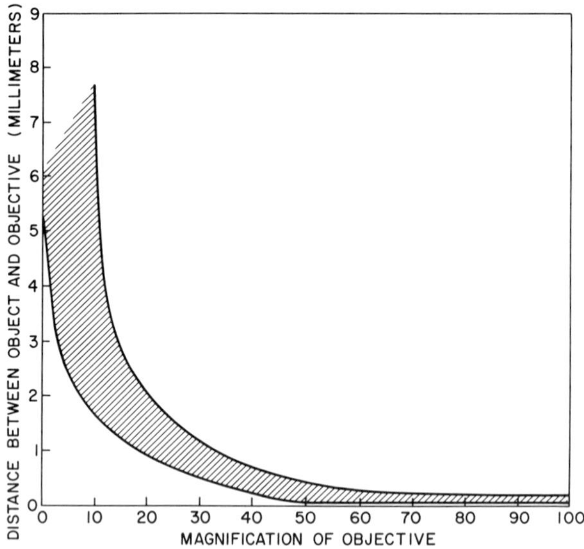

Fig. 3. The relationship between the working distance of the objective (distance from front lens of the objective to the surface of the object) and its magnifying power.

and object becomes less than one millimeter. Magnifications above 60 × usually require oil immersion objectives.

The depth of field for the objective also decreases with the magnification. The depth of field refers to depth of image that appears to be simultaneously in sharp focus for one position of the objective. For a given magnification, this will appear to be greater in visual work than in photomicrography. In visual work the observer habitually changes the focus, thus seeing successive layers of the specimen in sharp focus. The eye can also accommodate to slightly out of focus positions of the real image. In photomicrography, however, there is no such successive shifting of focus or accommodation possible, so that the limited depth of field becomes more quickly noted. Figure 4 gives an indication of the very limited depth of field available and of its decrease with increasing magnification.

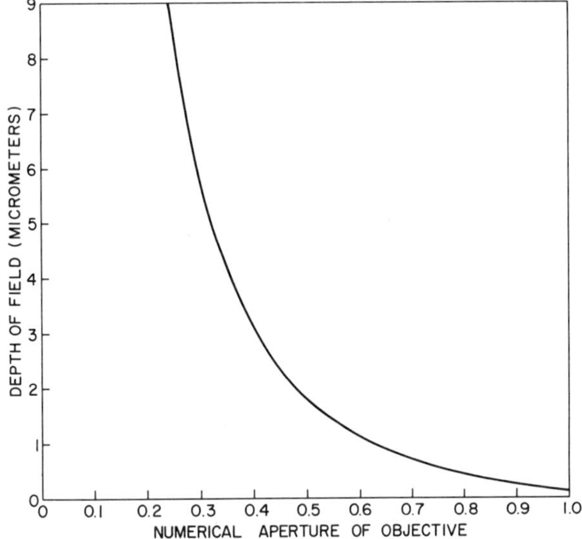

Fig. 4. The effect of variation of the objective numerical aperture on the depth of the field that is in focus.

Depth of field and depth of focus are used interchangeably by many authors but it has been correctly pointed out (275) that depth of focus is the depth of sharp focus found in the photographic image plane, which can be measured by movement of the camera back and not by movement of the objective. Shillaber (275) has given the formula for camera depth of field as related to the numerical aperture (N.A.)

$$\text{Depth of field} = \frac{\lambda\sqrt{n^2 - (\text{N.A.})^2}}{(\text{N.A.})^2} \tag{1}$$

where λ = wavelength, n = index of refraction of the medium. For visual work the formula given below may be used.

$$\text{Depth of field} = \frac{250 \text{ mm}}{M^2} \tag{2}$$

where M = the magnification.

$$\text{The depth of focus} = \text{depth of field (Eq. 1)} \times M^2 \tag{3}$$

It is obvious from this that the depth of focus is much greater than the depth of field. Practically, this means that a very small movement of the objective may throw the desired object completely out of focus, so that a very precise fine focus adjustment is required for medium and high magnifications. On the other hand, once the objective is in focus, the camera film plane can be varied over a fairly long distance without having an effect on the image quality. Its movement will, however, vary the magnification.

Contrary to what is popularly believed, the ultimate magnification possible with the microscope is not as important as is the resolving power. The resolving power of a microscope objective determines its ability to reveal closely adjacent structural details as actually separate and distinct. A true measure of resolving power, shown both by experimentation and various calculations, is the numerical aperture of the lens and this is universally employed to rate objectives. Numerical aperture (N.A.) may be defined as

$$\text{N.A.} = \eta \times \sin \tfrac{1}{2}\text{A.A.}$$

where: η = the lowest refractive index of the system from the object through to first lens of the objective; A.A. = the angle between the most divergent rays which can pass through the objective to form an image.

The resolution is increased with radiation of short wavelength, and by filling the space between object and front lens of the objective with a material of high refractive index. Oblique illumination also increases the resolution, and by means of a condenser the light can be made to converge upon the object from all azimuths, as a cone of rays. Theoretically, there is no limit to the magnifying power of the microscope. Practically, cumulative aberrations, loss of light, and resolution restrictions limit magnification to 1000 times the numerical aperture. Above this value "empty" magnification occurs, with larger size but no additional detail.

If light of short wavelength is employed, so that the effective resolving power is increased, magnifications to 3000–4000 × are possible. The limit of resolution is usually given as

$$\text{Limit of resolution} = \lambda/2\ \text{N.A.} \tag{4}$$

where λ = the wavelength of the light.

Practically, this means that about 0.0002 mm (0.2 μm) may be considered the limit for visual work (Fig. 5) and 0.0001 mm (0.1 μm) the limit for ultraviolet work. However, structures far below the limit of resolution may be detected by diffraction patterns obtained with special types of illumination.

Resolution ultimately depends not only upon the numerical aperture but also upon how well the manufacturer has eliminated distortion, aberration, and glare, and upon how well the various lenses are aligned with one another. Cost and quality increase

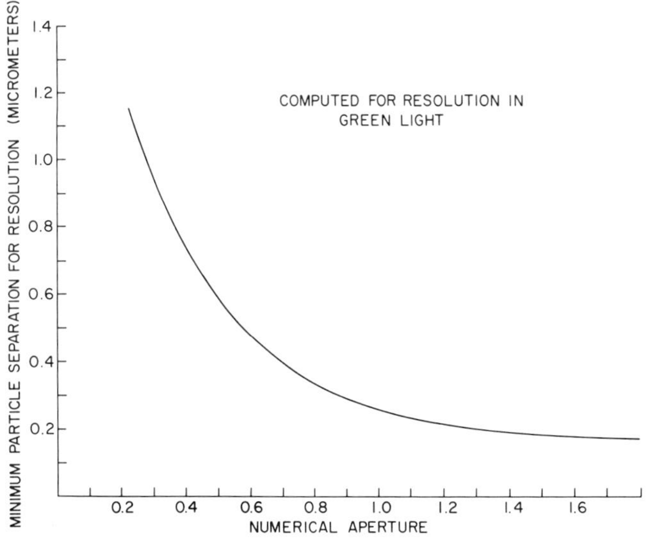

Fig. 5. An illustration of the variation in resolution with objectives of different numerical aperture.

together. Despite this truism, microscopes of different manufacture may possess special features that make one variety most suitable for a definite specialized application. Given the opportunity for selection, a potential user would do well to familiarize himself with the different instruments. Lists of the manufacturers of various types of microscopes, both general and for special applications, are available (5,298).

A. Suitable Samples and Their Preparations. There are two principal types of microscopes. One type employs only a single objective and, although it may also use binocular eyepieces, is therefore a monocular instrument. A second type employs two objectives and two eyepieces in completely independent viewing systems. This results in true binocular or stereo vision. Instruments of this latter type have the advantages of long working distance under the objectives, an apparent greater depth of field, and of possessing the property that movements as viewed under the microscope are not reversed. Such instruments are much used in the assembly of small instruments, dissection and other technical functions, as well as being valuable in research. They are suitable for the study of both transparent and opaque objects, and in many cases the necessary information can be obtained without special preparation of the specimen. This type of instrument has the disadvantage that the construction limits the possible magnification, which rarely exceeds 200 ×.

The monocular objective microscope, on the other hand, is capable of much higher magnification. Because of its high magnification, the samples to be viewed require very special tratment to insure that resolution will be suitable, and that the field of view will be flat enough to be viewed in its entirety with the limited depth of field that is available. In addition, if the object is to be viewed in transmitted light, the section must be made very thin and the two surfaces of this section should be parallel.

In biology and medicine, the section for transmitted light is obtained in this condition by microtoming thin sections of tissue from a specimen that has been mounted and suitably impregnated, frozen, or otherwise made sufficiently rigid, so that the knife cuts without excessive distortion of the specimen, or of the microtomed section. It would not seem that there would be many applications of this technique in other fields. However, surfaces suitable for microscopic examination have been obtained by microtoming across suitably mounted soft metals, such as lead and its alloys (193).

Another example wherein microtoming was employed was in obtaining thin sections for microscopy from dental wax-impregnated battery plates (182). Microtoming has also been used to obtain sections of ion-exchange membrane fuel cells after these had been suitably impregnated with a polyester resin (173).

For most electrochemical studies with the microscope, it would seem that greater use has been made of opaque materials. Opaque materials are best studied by reflected light, and for this purpose a smooth, flat, and highly polished surface is required. If the material is porous, it should be impregnated with some material which is capable of filling the voids and then hardening. Some waxes, epoxies, and polyesters have been employed for this purpose (7,215). If the material is brittle or inclined to crumble, it should be mounted in an external mount so that it may be properly held for polishing. Bakelite, epoxies, plaster of paris, Wood's metal, and many other substances have been used to embed and hold the specimen (10,48, 61,214,307). Recently, an investigation has been made of materials suitable for mounting media (262).

In mounting products of electrochemical reaction, one should be sure that no constituent is soluble in the impregnating or mounting medium and that no reaction takes place during the procedure. It has been found, for example, that certain epoxy formulations react with explosive violence with such oxides as the PbO_2 found in the plates of the lead-acid battery (278). Reaction was found to occur to a lesser extent when a polyester was used as the impregnant for sectioning and polishing the $Ag/Ag_2O/AgO$ electrode (310).

Regardless of whether an impregnant or mount is required, the sample must be sectioned and the sections then ground on successively finer grades of abrasive. Usually, mounted abrasives are used for this purpose, in the form of adhesive-backed paper discs which can be mounted on comparatively slow speed turntables. These discs can easily be replaced when the abrasive is dulled, or when changing to a finer grade. Water or oil is usually used as the lubricant and coolant. About 600 mesh abrasive (17 μm) is the finest that can be conveniently mounted on discs. Beyond this point, resort must be had to lapping on lead discs, or to using various cloth laps, impregnated with the finer abrasives. The American Society for Testing Metals has published the results of a symposium on the preparation of specimen surfaces (15) and similar instructions have been given elsewhere (11,21,120,137,154,244,248,259,321). Since the above methods employ wet grinding and the use of slurries for the final polishing, trouble may be experienced if water- or oil-soluble constituents occur in the specimen surface. Benedicks and Metts (30) used dry lead laps impregnated with a succession of classified abrasives. They also substituted a lacquer film as a base to retain the abrasive particles, instead of the lead surface. The method worked well for the harder materials but not for soft ones. A simpler dry method was developed especially for the dry polishing of sections from lead-acid battery plates (279). This became necessary because a corrosive electrochemical couple was developed between the lead of the grid and the PbO_2 of the positive active material during conventional wet methods of polishing. This method was later found to be applicable to the dry polishing of sections of the $Ag/Ag_2O/AgO$ electrode (310). It has more recently been used to determine the phase equilibria in the $AgSO_4/AgCl$ salt system (Fig. 6) and in

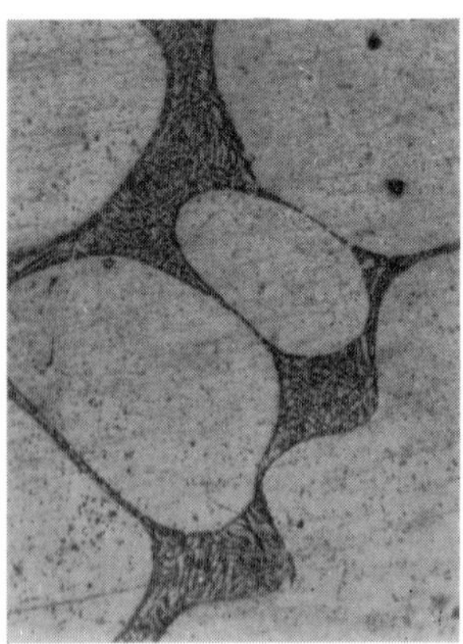

Fig. 6. The microstructure developed in a 95% AgCl–5% $AgSO_4$ system through the use of a brief water etch.

polishing various water-soluble crystals. It has also been found to give practically undistorted polishing of metals up to a hardness comparable to brass, but it does not appear to polish iron and steel (279). In this method, a Syntron vibratory polisher is used, which contains a sponge rubber pad as the lapping surface. Into this surface is rubbed 0.3 μm alumina powder, which polishes the specimen as it is moved around on the lap by the vibratory motion of the latter.

In mineralogy, use is made of thin slices of specimen, which are polished on both surfaces, and which are thin enough so that transmitted light can be used to study the crystal structure under the microscope. While these thin sections have usually been used only for mineralogy, their use would be advisable in the study of electrode changes during oxidation and reduction, and in the passivation and oxidation layers formed at electrode surfaces. A study has been made of electrodeposits of α and βPbO_2, using such thin sections (1). Mineralogical texts (7,49,61,124,157,312) give descriptions of how these thin sections can be prepared.

Many types of surfaces can be polished in a nearly distortion-free condition by electrolytic polishing or a combination of electrolytic and mechanical polishing. These procedures are usually quite specific for a particular material and details of this type of surface preparation must be sought elsewhere (59,100,150,297).

In many cases, the polishing technique will leave the surface in a highly polished condition but without distinguishable features. Just as specific stains have been used in biology and bacteriology to bring out certain structures, so must various etches be applied to metals or other materials in order to distinguish the various phases. This etch may function by dissolution of certain constituents, by the formation of thin films or by selective precipitation upon one phase. The etch may be simply water (Fig. 6) or it may be a mixture of several chemicals (16,28,105,154). Etching may also be accomplished in some cases by thermal evaporation (58), cathodic bombardment (224,247), or electrolytic procedures (150,189).

In many cases no etching is required, since the constituents of the section may vary in color, optical reflectivity, or hardness so that identification of the various phases present may be made on the basis of color, reflectivity, or variation in polishing relief. It is also of interest to note that vacuum evaporation of films upon the surface has been used both to increase resolution and to identify phases (165).

2. *Obtainment of Optimum Contrast*

A. Transmitted and Reflected Light. Light is used in the microscope optical system to conduct an enlarged image of the specimen to the eye, either as this specimen appears in transmitted light, or as it appears in the light reflected from its surface. The optical principles are quite similar in the two systems. When transmitted light is used, the illuminating beam is a solid cone of light concentrated on the specimen by a condenser lens mounted beneath the specimen. This condenser is usually equipped with an adjustable diaphragm to control the angular aperture of the cone of rays. In this manner, a compromise can be made between contrast and resolving power. The system is designed to produce even illumination across the field of view, with the minimum introduction of color.

There are several systems that may be used to obtain uniform illumination. The first method is shown in Fig. 7. The light from the source is imaged directly at the specimen plane. The lamp condenser is sometimes omitted but its presence will generally provide a source

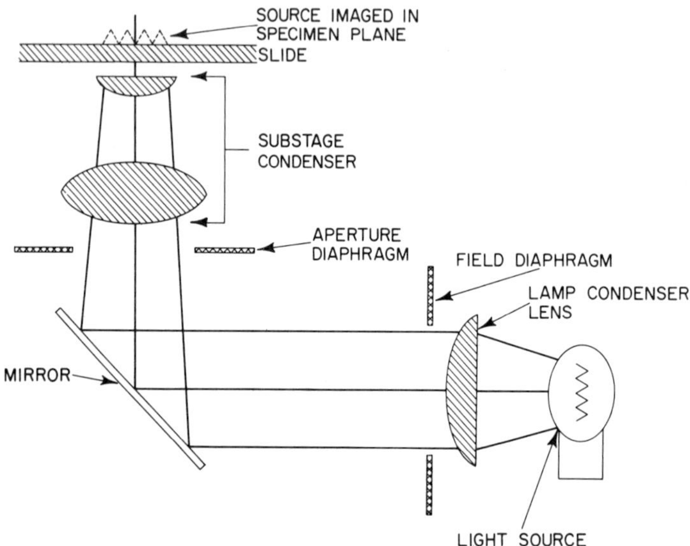

Fig. 7. An illustration of a method of specimen illumination in which the image of the light source is focused in the object plane. This method requires a large, uniform incandescent source in order to supply uniform illumination over the entire image field.

image large enough to fill the objective with light, even at low magnification. The field diaphragm at the lamp condenser regulates the size of the spot of light while the aperture diaphragm regulates the intensity. The method has been frequently used because there is a minimum of light loss. However, the incandescent portion of the filament must be relatively large in order to fill the objective with light at all magnifications, and the source must also have a very uniform intensity in order to provide uniform lighting of the specimen.

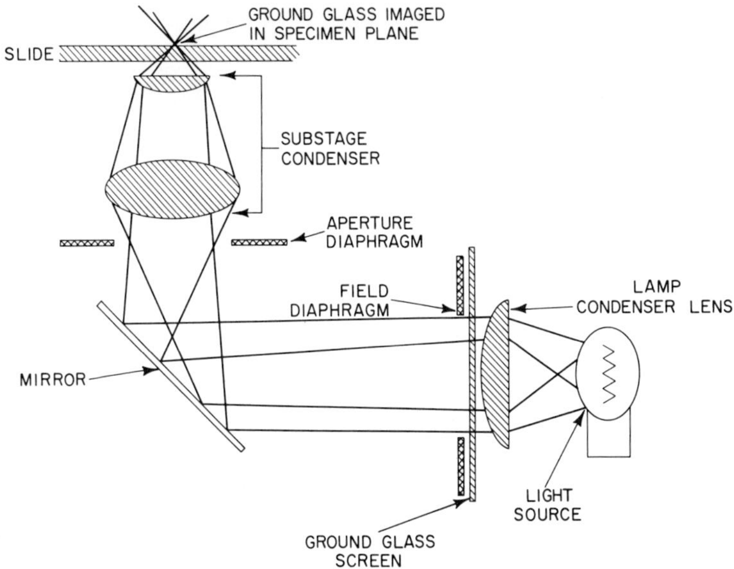

Fig. 8. An alternate method of specimen illumination, similar in principle to that shown in Fig. 7, but using a ground glass diffuser in front of the light source to increase uniformity. Image of the ground glass, rather than that of the source, is here focused in the specimen plane. This method is effective but imposes considerable reduction in the light intensity.

In order to eliminate these disadvantages, a modification was introduced as in Fig. 8. In this case, a ground glass diffuser is placed in front of the lamp condenser. This causes the diffuser to act as a uniform secondary light source and the condenser lens on the microscope is now uniformly filled with light. The image of the ground glass is focused on the object plane with uniform illumination. It may be necessary to move the image of the ground glass slightly out of focus to prevent the grain of the glass from appearing in the illumination. The disadvantage of this method is the very considerable loss of light at the ground glass screen.

The method of illumination now most often used and the one with the least disadvantages is that advocated by Köhler (166). In this system, the image of the light source is projected into the plane of the diaphragm of the substage condenser (Fig. 9). This causes the substage condenser to be uniformly filled with light, as could be verified if the eye were placed at the object position. This secondary,

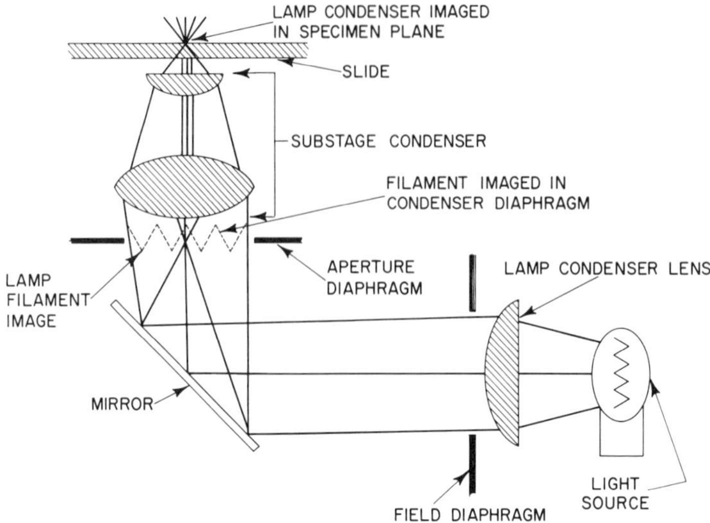

Fig. 9. The Köhler method of illumination. The light from the source is focused just below the substage condenser so that the latter appears uniformly filled with light. This secondary light source, the condenser, is then focused at the specimen plane. This method is generally considered to be the most effective.

very evenly illuminated, light source is then projected into the plane of the object.

In reflected light, the same systems may be used, except that the rays from the source are directed to a transparent reflector that reflects the beam of light downward into the objective and from there to the specimen surface. The illuminated image of the specimen surface is then reflected back through the objective and through the transparent reflector to the eyepiece (Fig. 10).

The diaphragms in the illustrations are an important part of the illumination system. The aperture diaphragm is placed so as to limit the cone of rays transmitted by the instrument and thus regulate the intensity of the light. When the aperture diaphragm opening is reduced, the illumination over the entire field is uniformly reduced, in much the same manner as the iris diaphragm of a camera determines the illumination upon the film. Although a lesser number of rays are present, they converge on each point of the image so that there is no reduction in size of the image.

Field diaphragms are placed so that they are simultaneously imaged with the field. Their opening and closing directly affects

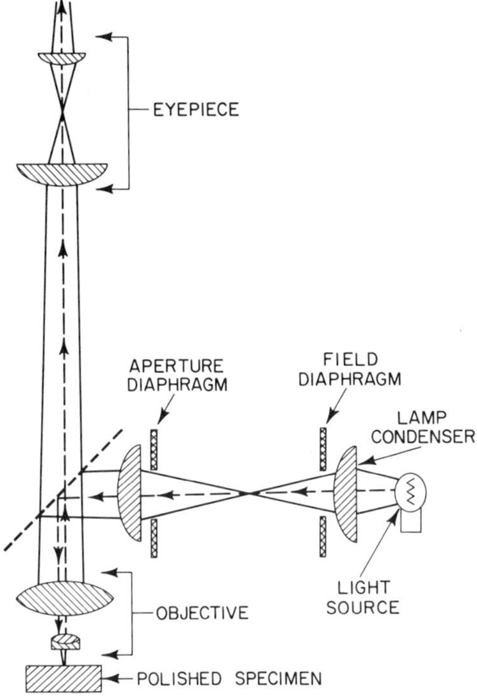

Fig. 10. The method of obtaining illumination of opaque objects. This method is used with metallurgical and petrographic microscopes (vertically illuminated or reflected-light microscopes), and is usually available as an accessory for transmitted-light microscopes. In inverted microscopes the opaque object is placed at the top so that the above diagram would be inverted. In such microscopes the rays to the oculars are also deflected so that a more convenient viewing position is obtained.

the field area but does not affect the brightness of the remaining field. Figure 11*a* shows the positions at which field diaphragms can be placed to limit the field area. The eyepiece field diaphragm is used to cut off that portion of the field in which the aberrations are the worst. The adjustable field diaphragm is usually effective with the objectives of low magnification. The area of the light source, imaged in this plane, can also be controlled by the lamp diaphragm in front of the lamp condenser lens (Fig. 9). Consequently, this diaphragm can also be considered as a field lens.

The positions at which aperture diaphragms may be placed are shown in Fig. 11*b*. Each of these can be regulated independently

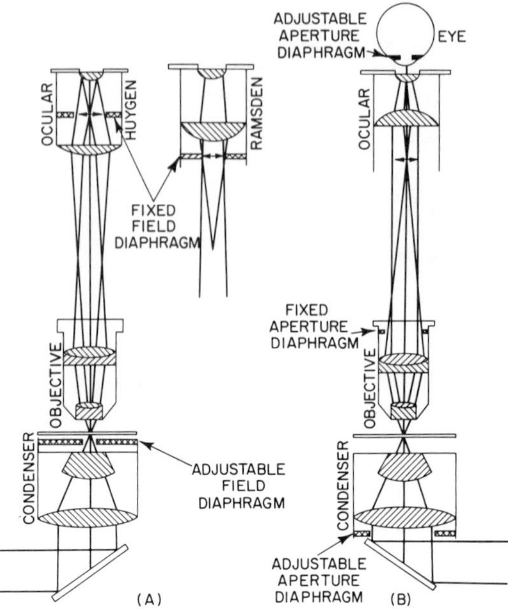

Fig. 11. An illustration arranged to show the approximate positions of the field and aperture diaphragms used with the compound microscope.

but their images are superimposed. Thus, the aperture diaphragm of the condenser is imaged in the focal plane of the objective, and this, together with that of the objective, are both imaged in the focal plane of the eyepiece. The aperture diaphragm of the objective determines the resolving power; that of the eye, the retinal image quality; while the condenser aperture diaphragm determines contrast.

In addition to illuminating the specimen area, a very important function of the employed lighting arrangement is to provide contrast in different parts of the imaged structure.

If convergence is decreased by reducing the aperture diaphragm, the illumination is rendered more nearly directional and contrast is increased. If overdone, however, the image outlines are surrounded by spurious haloes and resolution is reduced. If one side of the condenser lens is closed by an opaque screen and the rest left open, oblique illumination will be obtained which causes unequal lighting of the subject and creates the impression of shadow detail and increased depth. Contrast can also be obtained in colored subjects by introducing the proper complementary filters to change the overall color of the light.

In addition to these methods, various types of annuli or phase changing discs can be used to obtain additional contrast. These methods of obtaining contrast can be used in most cases with either transmitted or reflected light.

B. Darkfield Illumination. In this illumination system, the cone of light concentrated on the specimen has a dark central core (Fig. 12*a*) obtained by introducing an annular stop with an opaque central portion *A*, into the light path. An object at the apex of the illuminating cone will appear self-luminous against a dark background, provided that the surfaces of the object can reflect,

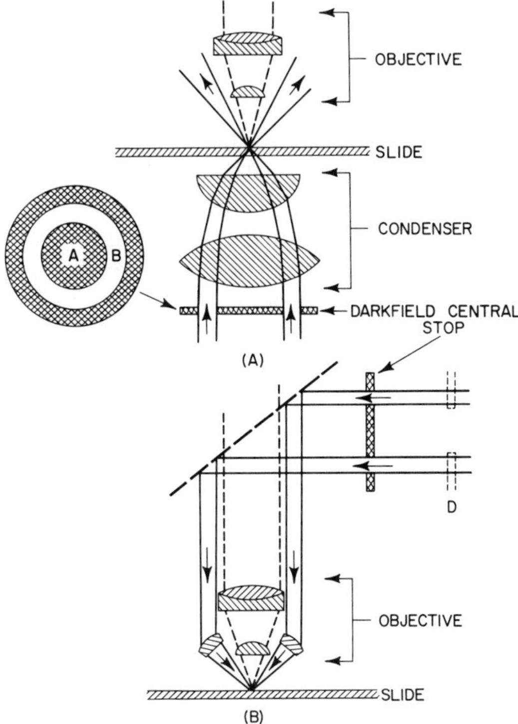

Fig. 12. The method used for obtaining dark field and color contrast illumination in transmitted and reflected light microscopes.

diffract, or refract any of the light rays so that they can be included by the objective. Such refracted or otherwise diverted rays are indicated by the dotted lines in the figure.

When reflected light is used, a different arrangement is necessary and for high magnifications the system is part of the objective itself. The central portion of the cone or rays is darkened as before by a central stop placed in the vertical illuminator. The cylinder of rays is then directed down the outside of the objective tube, either through air or within an enclosing lucite cylinder. At the lower end of the objective, the light is directed obliquely upon the object, as shown in Fig. 12*b*. This is accomplished either by a lens system or a reflecting surface. The rays so reflected are again at such an angle to the surface that no direct reflection into the objective takes place. This same objective can be used for bright field work by simply removing the darkfield stop and closing down the diaphragm *D*, so that the external cone of rays is cut off and light is reflected to the surface only through the central portion of the lens, as in ordinary objective usage.

In transmitted light, this form of illumination has proven useful for studying small living objects, suspended particles, or the motions of particles in an electric or magnetic field. In an extreme form, this type of illumination is used in an instrument known as the ultra-microscope. In this case, an intense beam of light is directed on the object from approximately right angles to the angular cone of rays entering the objective. By this arrangement, the presence and motions of particles considerably below the limit of resolution can be detected. This microscope cannot actually resolve the particles but detects the light scattering and diffraction occurring at their surfaces.

In reflected light, the darkfield image will show up fissures, grain boundaries, scratches, and other surface discontinuities. These surface irregularities will appear bright against a dark background. A disadvantage of this type of illumination is that very little structural detail is usually evident. Furthermore, out-of-focus detail will scatter light and cause background fogging, so that contrast is reduced. For this reason, very thin, transparent sections or relatively flat reflecting surfaces are required.

Using the same principle, it is possible to show the object and background in different contrasting colors. The phenomenon depends upon using a two color disc inserted into the optical path. The smaller central portion of the disc, *A* of Fig. 12*a*, is made just large enough to fill the full aperture of the objective. This central disc *A* is centered in the somewhat larger ring *B*, of a contrasting color. This latter ring is the full size of the substage condenser. Under proper conditions, this results in a central "dark" core for the cone of rays, just as in the usual darkfield. However, in this case, the background will be colored to correspond with *A* instead of being black. Any objects within the field will appear to be illuminated in the color shown by *B*. The main advantage of this system is the excellent contrast made possible for both viewing and photomicrography.

In general, not much use appears to have been made of darkfield illumination in electrochemistry. Indeed, its special characteristics are not often called for in most other fields of chemistry. However, in a study of CdS crystals containing the acceptor elements Cu, Ag, or Au, wherein two types of precipitates occurred, darkfield illumination was of assistance in distinguishing the two types (86). Powers (237) also used darkfield illumination to distinguish a phenomenon in which entities described as resembling cobwebs appeared in a dissolving zinc oxide film.

C. Polarized Light. The principles of polarized light microscopy could not be fully explained in the space available here. The general principles and applications can be found in numerous texts, a few of which are referenced (61,69,124,125,136). The essential parts of this microscope are diagramed in Fig. 13. The polarizer and analyzer may be calcite prisms or sheets of Polaroid. If calcite, the beam of light is separated into two directions of vibration by the polarizer and one of these is deflected and eliminated by sidewise total reflection at an inclined surface. If Polaroid, the microscopic acicular crystals of herapathite or other pleochroic substance strongly adsorb one direction of vibration. In either case, the light passing through the polarizer loses one vibrational direction and becomes strongly polarized. This polarized light is directed to the eye, still polarized. If a second piece of the same polarizing material is placed so that it intercepts the light from the object, and is turned so that it transmits in a plane at right angles to the polarizer, practically all of the light coming to the eye will be extinguished and the arrangement is known as "crossed polarizers." If a crystalline object with anisotropic properties is centered and rotated upon the stage, it will alternately brighten and darken at 90° intervals. This is because the crystal itself is adding a different polarization to the light, which at certain angles sufficiently changes the plane of polarization to, in effect, "uncross" the polarizer and analyzer. To determine the extinction

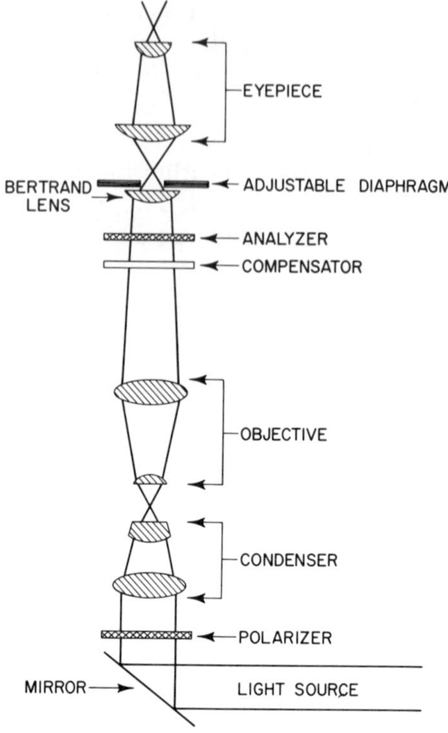

Fig. 13. The principle of the polarized light microscope.

angles and other quantitative information, the polarizing microscope is equipped with a circular revolving stage with an edge vernier that permits readings of 0.1°. Usually, the polarizer and analyzer are also provided with scales so that their relative orientations can be determined.

Two types of observation are made with the polarizing microscope, orthoscopic (parallel light) and conoscopic (convergent light). The orthoscopic examination of a crystal between crossed polars reveals only those optical characteristics parallel to the microscope axis. In this type of examination, the substage condenser lens is kept nearly closed and the Bertrand lens is removed from the optical path. With this arrangement, it can be determined whether the crystal is isotropic or anisotropic and polarization colors, extinction angles, the "fast" and "slow" vibration directions, and the birefringence can also be determined. These are factors which help in identification and classification of the crystal.

When a strongly convergent beam of light is passed through the crystal (conoscopic observation), it is possible to examine its optical characteristics in many directions at once. The objective, at its focal plane, forms a small real image of the object, and this image at each point is a focal point for the light which has passed through the crystal in definite directions. This causes interference such that curved interference bands are seen, which are colored in white light and dark in monochromatic light. These interference bands are symmetrically arranged around the optic axis and accompanied by dark "isogyres" which have shapes determined by the positions on the interference figure where rays, vibrating in directions parallel to the vibrations of the polars, are brought to a focus. The Bertrand lens is used to enlarge these figures to be seen in the eyepiece. Because only one crystal may be studied at a time, it is necessary that an adjustable iris be provided in the Bertrand lens cell, or in the illuminator, so that the isolation of a single crystal can be accomplished in the central portion of the field.

A study of these interference figures makes it possible to determine whether anisotropic substances belong to the uniaxial or biaxial system. It is also possible to differentiate isotropic substances from sections of crystals that may be merely normal to an optic axis. The figures may also be used to determine the positive or negative sign of the substance and to make qualitative estimate of the birefringence values. If the crystal is biaxial, it is then possible to determine the direction of the optic axial plane and vibration directions, and the size of the optic axial angle. The dispersion of the optic axes can be determined and with them the crystal system. The methods for making these determinations are given in the texts that were referenced above.

The polarizing microscope is found to be of assistance in the examination of many kinds of material. Fragments as small as 0.01 mm can be examined and much larger material can be fragmented to find crystals or fragments suitable for study. The amount of material needed is usually quite small. It is also possible to use polarizing microscopes to determine whether a material is homogeneous or heterogeneous and to distinguish between isomers. The studies with polarized light also include the stress patterns that are shown by materials under strain or stress.

The above discussion has applied in particular to the examination of transparent or translucent materials. Much the same kind of information can be obtained from reflecting surfaces (Fig. 14), but here the preparation of the sample becomes more critical and the

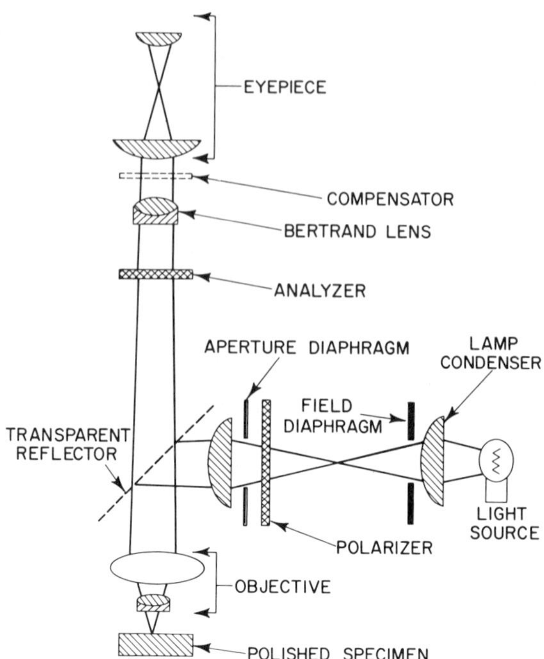

Fig. 14. An illustration of the slightly different arrangement of polarizer and analyzer required for use with a metallurgical or petrographic microscope.

chance for erroneous data greater. Care must be taken to avoid directional polishing; cleavage flaws and other effects introduced during polishing must be avoided; and distorted surface layers, resulting from grinding, must be removed. Transparent grains in the surface layer, which reflect not only from the polished surface but from under surfaces or cracks, are also a source of error.

Polarization from reflecting surfaces has been applied to the study of grain structure. Any anisotropic material when viewed under crossed polars will show varying intensity of the light reflected from each grain, and better contrast between grains is thus obtained. Isotropic (cubic) metals do not give any such effect under polarized light unless deeply etched or having had a film deposited on their surfaces.

One important effect of polarized light is in reducing surface reflections or glare. Therefore, a polished surface may be examined under polarized light to obtain a more realistic idea of its real appearance, and the various reflectivities obtained by bright field vertical illumination can be compared with the same field as it appears without the vertical reflection. Another property of polarized light is of particular interest to electrochemists. Using polarized light, the microscope objective may be focused upon an opaque surface within an electrolytic cell without the troublesome glare from the intervening glass surfaces, that would occur if unpolarized light were used.

Examples of the use of polarized light in electrochemistry are given by Rosenberg (252), who studied the strain patterns characteristic of $GaAs_{(1-x)}P_x$ alloy overgrowths on GaAs single crystal wafers, and by Prener (239), who studied the crystallographic properties of calcium fluor- and chlorapatite crystals (a phosphor constituent of fluorescent lamps).

D. Phase Microscopy. By the insertion of a ring diaphragm (annulus) under the condenser and the placement of a corresponding phase plate in the rear focal plane of the objective, the microscope becomes equipped for phase contrast (Fig. 15). The diameter of the central opaque portion of the condenser annulus diaphragm must be less than the cone of light accepted by the objective, otherwise all direct light would be cut off and the result would be darkfield illumination. Thus, compared with darkfield, the interior structure of small objects remains visible. Basically, phase contrast makes use of interference between the axial and oblique rays so as to improve the visibility of objects of low contrast, principally by emphasizing the boundaries between regions of different contrast. Hollow cone illumination is furnished by the condenser annulus. Rays from this annulus that are not diffracted by the object will pass through area B of the phase plate. Here, a transparent film retards the rays by $\frac{1}{4}$ wavelength (alternatively, this area is clear and area A is coated with the transparent film). Rays diffracted by the object do not pass through area B but are diffracted inside or outside of this region to area A. The result is that the two rays differ in wavelength and interfere with one another to increase the intensity variations so that the specimen appears to have density variations.

In transmitted light, the interference improves the visibility of objects of low contrast, principally at the boundary between regions with different refractive index. Small particles, in particular, are made more visible. One of the greatest advantages of phase contrast is that it makes possible the study of living organisms that might otherwise require devitalizing staining to make them visible.

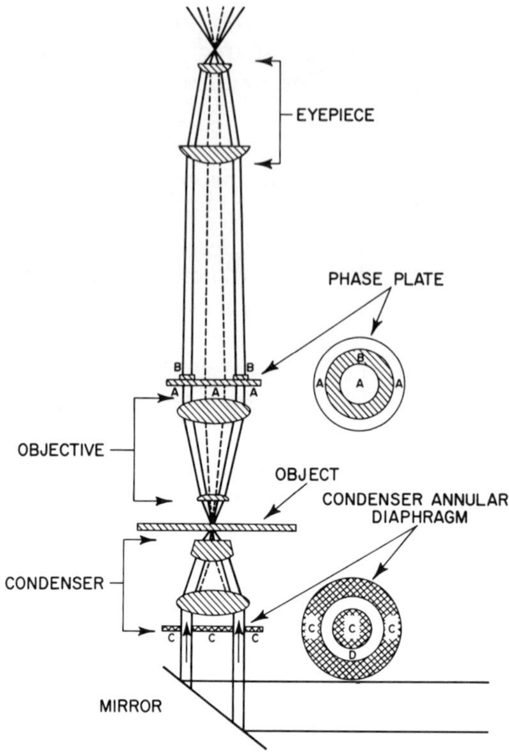

Fig. 15. Illustration of the modification required to adapt the compound microscope to work involving phase contrast.

In the case of reflected light microscopes, the phase differences emphasize the contrast between the boundaries of regions at different heights above the surface, even when these differences are small. This type of illumination is also insensitive to differences of surface reflectivity, so that height differences are not confused with other effects.

Phase contrast in reflected light has been used with success in metallographic studies where surface level differences existed between phases that were otherwise not distinctive. These differences in level might be introduced by either etching or polishing. Since increasing brightness occurs with increasing surface elevation, the difference between depressions and elevations can be determined as well as relative heights. Measurements have shown that phase contrast systems are sensitive to difference in level of 50 Å or less

(204). However, when the difference in level becomes too great (above 200 Å), a reversal sets in and increasing brightness then means decreasing surface elevation. This reversal makes interpretation of surfaces difficult unless it can be ascertained whether the phenomena being observed are above or below the limit of reversal. Another problem with phase contrast is that system-induced haloes may appear around each light-reflecting detail, so that these haloes might be confused with actual structure by the inexperienced. Wilska (318) claims to have overcome this problem with coatings of soot on the annuli instead of metallic coatings. The present author's experience with phase contrast has been limited to reflected images. In this application it has been found that, in most cases, phase contrast yielded no more information than could be obtained by relief polishing. This may have been because the subjects selected were not suitable, or, more likely, that the differences in level were too great to be suitable for this type of illumination. Other authors (60), however, caution against drawing conclusions from phase contrast observations until a thorough study has been made of the method. Ideally, different phase plates, with different degrees of contrast, should be tried on simple, familiar subjects before unknown subjects are observed. In various studies of electrochemistry closely associated with metallography, it would appear that phase contrast might have been profitably employed. The author has found no indication, however, that any investigator has employed it. The foregoing discussion has served, however, to introduce the subject of interferometry which has been used to a considerable extent in electrochemical investigations. For a fuller discussion of phase contrast the reader is directed to other sources (4,31,61,318,319).

E. Interference Microscopy. Interference microscopy is, from the standpoint of the electrochemist, the most important method of obtaining contrast, as well as a valuable method of measurement. There are many different ways in which this interference can be achieved and many different types of microscopes have been developed for the use of the different methods. The subject is too complex to be discussed here in the necessary detail. The book by Krug, Rienitz, and Schulz (171) is recommended for further study. Their book gives a highly informative discussion of the principles and problems of interference microscopy and discusses the different types of instruments that have been developed for its use. There is, in addition, a very comprehensive bibliography that includes most work done prior to 1964.

In some respects, phase contrast could be considered as interference microscopy but there are several significant differences. First, there is the single ray path of the phase microscope as compared with the two or more completely separate beams of the interference microscope. A second distinction is that there is no essential alteration of the image in the case of the interference microscope, whereas phase contrast images are so altered as, in some cases, to defy interpretation. Third, in interference microscopy, the measurement of path difference can be made precisely, and can be altered to obtain the best conditions. Fourth, in some styles of interference microscopes it is not necessary to restrict the aperture of illumination. Despite these points of difference, there are some types of interference microscopes that are difficult to separate in principle from phase contrast, and the opposite is also true.

While over a hundred types of interference microscopes have been described in the literature, all depend upon some form of separation of the light into two paths which are subsequently made to interfere with one another while superimposed upon the microscope image. This division may be at semitransparent surfaces so that one part of the beam is transmitted and the other is reflected. The double refraction properties of certain crystals can also be used to separate the beam into two portions. A diffraction grating or other method of diffraction can be used to split the beam into several orders, some of which can then be made to interfere. Beam division can also be made by adjacent stops in diaphragms or other beam splitting optical device.

The two or more beams that interfere may do so in different ways. In one case, the reference beam has no contact with the object but is independently produced and superimposed on the image arriving from the object. In a second case, the reference beam is from the object but is out of focus. In a third case, the reference image is sheared laterally with respect to the first. Without discussing in detail the advantages and disadvantages of these various methods, it would seem that the first method allows use of unrestricted apertures, there is an excellent image and good contrast of fringes, but its sensitivity to vibration makes necessary heavy and expensive construction. In method two, the highest magnifications of the objective may be used, the aperture is unrestricted, and the effect of vibration is reduced, but these instruments are difficult to adjust and the image formed may resemble that obtained in phase contrast, although this effect can be removed by proper instrumentation.

Instruments of the third type have some deterioration of image quality, but at low magnification have the advantage of simplicity and stability, and have been judged the best instruments for the measurement of thin films.

While the principles are similar, a further division of instruments is required for transparent and opaque objects, because here the transition from one type of illumination to the other can not be made as simply as was the case in the previously described methods of illumination.

The ability of phase contrast to disclose very small phase differences has been mentioned. However, the haloes associated with the image, the phenomenon of inversion, and the fact that satisfactory phase contrast is only obtained with small objects make interpretation of the phase contrast image difficult. These facts probably contributed to the rapid acceptance of interference microscopy. In interference microscopy minute phase shifts are not made directly visible, but quantitative measurements are easier to obtain and can be obtained more reliably. Another advantage is the absence of halo effects, and that all areas, large and small, having equal phase shift look the same.

New applications are constantly being found for interference microscopy. It has been used for qualitative and quantitative observations in biology and medicine because of its freedom from optical artifacts and the lack of necessity for killing the organism under study. Many of the small variations in optical thickness that occur within the cell or other organism would not be detected when viewed with phase contrast, but become easily detectable when interference microscopy is used. Similarly the method has been used to measure the index of refraction of small amounts of liquid (285), and the dry weight of biological objects (20,76).

In transmission microscopy, the interference microscope has been used to investigate and to measure the thickness of film coatings, thin film layers, crystal-step heights, polished thin sections, glass, foils, lacquers, and film replicas.

In reflected light microscopy, replication techniques and reflection coatings make it possible to use interferometry methods on almost any type of surface. Liquid lacquer replicas are ideally suited for this type of study. Due to their ease of distortion, however, it is better to use block replicas. Replicas, or the reflective surface itself, can be used to investigate and measure surface topology, scratch and etch depths, microhardness indentations, the surface finish of substances such as glass, wood, paper, crystals, metal, ceramic glasses, anodized coatings, abrasive papers, and surface roughness caused by machining, grinding, and electropolishing.

It would seem to be of particular interest to electrochemists that considerable application has been made of this technique in the measurement of transparent surface films such as those occurring in passivation or anodizing. Opaque films can be measured as well, after proper preparation.

Interference phenomena have been used to estimate the molecular diffusivities in liquids (220). In studies of the etching of germanium with and without ion bombardment the interference microscrope was used to estimate the surface roughness (240). The instrument was similarly used in a study of the chemical polishing of CdS (291), and in the electropolishing of Zn specimens (238). In a study of anion effects on copper deposition (122), the interference microscope was used with only limited success. More successful was the direct microscopic observation of the movement of growth steps during chemical vapor deposition of epitaxial layers of germanium on germanium (268). Here the topography of growth hillocks was determined and their heights measured.

F. Fluorescence Microscopy. Certain substances, when exposed to ultraviolet light, are excited to emit secondary radiation, usually of longer wavelength than that of the ultraviolet source and generally in the visible region. These secondary radiations cause the substance emitting them to appear luminous. This phenomenon is called fluorescence when the visible emission ceases quickly after the ultraviolet excitation is discontinued, and is called phosphorescence when the secondary emission continues for a considerable time without the necessity of continued ultraviolet excitation.

The method of fluorescence microscopy differs from those previously described, which depend upon observing the specimen in the light transmitted or reflected by it, in that the fluorescence specimen is self-luminous. The radiation source, usually, but not necessarily, ultraviolet radiation, does not contribute to the visible image. The parts of the object that fluoresce, therefore, appear bright against a more or less dark background. However, the techniques of brightfield, darkfield, and phase contrast, as described for visible light, can all be used effectively with ultraviolet light.

Only a limited number of substances show fluorescence, the most brilliant examples being found among minerals and certain chemicals. Much of this fluorescence is based on the presence of small quantities of the rare earth elements as impurities in the lattice structure of the mineral being examined. This makes possible deductions as to the conditions under which the mineral was deposited and concerning its geochemical structure. Fluorspar, apatite, felspar, and scheelite (and some uranium minerals) have fluorescent structures. Occasionally, mineral fluorescence is also the result of organic inclusions. Fluorescence offers an additional means for providing contrast between various areas of the specimen and in identifying some of its constituents.

The possibilities of fluorescence microscopy were greatly increased by the development of selective dyes with marked fluorescent colors (fluorochromes), which may be used for selective staining of certain organic and inorganic materials. Such dyes can also be used to detect flaws and cracks in various materials when left residual on the surfaces, or forced under pressure through the structure.

For best results with fluorescence microscopy, special equipment must be used. A high intensity ultraviolet source is needed and, to avoid loss of intensity, lamp bulb, condensers, and microscope optical lenses must be of fluorite or fused quartz. Interpretation of microscope fluorescence phenomena must be done with caution. The fluorescence results are often the result of unsuspected impurities. Other factors that cause variations are the concentration of the fluorescing material, the nature of the material associated with it, thickness of the layer being observed, the nature of the mounting medium, temperature, and the characteristics of the radiation used for excitation. Before the method can be used for any particular problem, these variables should be checked in a preliminary research. Despite these restrictions, the fluorescence microscopy has shown spectacular results in certain areas. It has been used to differentiate materials otherwise similar in structure and properties, to detect impurities in mixtures, and to detect variation in chemical reaction among seemingly similar materials.

Its use in electrochemical investigations has not been indicated (but see section on electroluminescence) but study of the book by Radley and Grant (242) and similar sources of information on fluorescence should certainly suggest possibilities.

G. Ultraviolet Light. The functions of ultraviolet and fluorescence microscopy are different. Ultraviolet microscopy is a function of resolution, while fluorescence microscopy is a function of contrast. In the latter case, a fairly wide band of wavelengths is employed.

In ultraviolet microscopy, a single wavelength is employed and the optics are designed to give nearly ideal correction at this specific wavelength. Early instruments of this type were very difficult to use because there was no provision for obtaining focus of the image. In later instruments of this type, the image was picked up on an ultraviolet sensitive image-orthicon and was viewed on a television screen. Later still, RCA developed an ultraviolet image converter tube which employed a photoemissive cathode, a single stage electron imaging system, and a fluorescent screen to convert the ultraviolet image into a visible image.

3. Measurement

A. Linear Measurement. In this and other methods to be discussed, the accuracy of microscopic measurement can be made very high, perhaps higher than can be obtained by any other method. However, as in any method, frequent checks against known values are essential. If the magnification is known accurately, then photomicrographs can be used for direct measurement, taking into account any possible dimensional change in the film or during processing. In traversing microscopes it is possible to move the object or the microscope over the necessary distance and this motion is read on a micrometer scale. Accurate positioning, in this case, is by a cross-hair eyepiece. Many microscopes have accurately graduated mechanical stages that can be used to move the object in this manner. An eyepiece with a scale mounted at its focal plane renders object and scale visible simultaneously. Since this scale gives a value for the image only, such a scale must be compared with a standard stage micrometer, while maintaining a given objective and tube length, to obtain definite values for the micrometer eyepiece scale. For very accurate measurements, filar micrometer eyepieces are used. In this instrument, a screw with a micrometer thread draws a cross-hair across the field containing the eyepiece scale. Estimation of partial divisions is then made from the amount of revolution of a graduated drum attached to the screw head. As in the previous case, this scale must be compared with a standard stage micrometer to obtain actual values at a given magnification.

B. Area Measurement. The area of simple shapes can be determined from linear measurement. Areas of irregular shape can be estimated by comparison with eyepiece micrometers which have concentric squares or circular rulings. More accurate measurements can be made with an eyepiece that has squares ruled in a grid. By counting the number of squares included in the object, its area can be found (129). In photomicrographs a planimeter may be used. For a long time, the method of cutting out the area from the photograph and weighing it, and then calculating from the weight per unit area, has been used with considerable success (77,145).

C. Volume Measurement. Volume measurement is discussed later in relation to quantitative microscopy, but, for particle determination, a reasonably accurate determination is possible with simple equipment if average values are sufficient. If all particles in a field of view are measured crosswise to the field and on a line bisecting the projected area of the particle (201), experimental comparisons have shown that this result gives values in satisfactory agreement with those of three actual dimensions of the individual particles. Its validity decreases as the particles become more elongated. The subject of actually determining the volume and size of particles has been widely discussed and a considerable bibliography is given by Chamot and Mason (61).

D. Depth and Thickness. The depth of field is so small with objectives of high magnification that simple yet accurate measurement can be made by focusing first at the lowest and then at the highest point of the object, and then noting the difference in the readings of the graduations on the fine focusing knob. A perhaps more accurate method employs the same focusing technique but uses a supplementary accurate micrometer to read the actual movement of the microscope tube.

If sections can be made, the thickness of a film or other desired material can be obtained from its cross section. By the use of very oblique sectioning (87,128,149,326), the apparent thickness of a film is considerably magnified so that more accurate measurement can be made of very thin material provided that the angle of sectioning is known. The limit of sensitivity is about 3000 Å.

In addition to the above direct methods of measurement for thin films the use of polarizing and interference microscopy offers even higher accuracy in some cases, and in many cases the method is nondestructive. The following description of some of these methods has been abstracted from the book by Schwartz and Schwartz (263). When a monochromatic light beam is directed upon a reflecting surface that possesses a step height (i.e., height equal to thickness of film), the beam is split into fringes which are spread $\frac{1}{2}$ wavelength apart. The deviation of the fringes can then be measured relative to the known interfringe spacing. This method is nondestructive (exclusive of the step height preparation) but requires a highly reflective surface and one that is flat. With single beam interferometry, the limit of sensitivity is 200 Å; with multiple beam, about 25 Å (171).

A microscope that has not been previously discussed is the light section or profile microscope. In this instrument, a line of light is focused on the surface by means of an illuminating unit inclined at an angle of 45° to the surface to be inspected. This line of intersection is then viewed by another microscope also inclined at 45° to the surface and 90° to the illuminator. Both units are in the same vertical plane. The surface profile appears as a band of light running across a black background and this light forms a magnified contour of the surface irregularities. Measurements of these irregularities are made with a micrometer eyepiece. The instrument is most suitable for surface finish measurements in which the scratches all run in one direction. When a transparent coating, above a low diffusing substrate, is viewed under the light sectioning microscope a multiple bright band is observed because of the multiple reflection between the coating-to-air and the coating-to-substrate surfaces. By measuring the separation between the proper light beams and applying a corrective factor, based on the optics of the system and the index of refraction of the film, it is possible to obtain the film thickness. This method has been applied to anodic coatings on aluminum (210) and to measure the glaze thickness on high density alumina insulation (41). The limit of sensitivity is about 0.1 μm in the 1 μm to 400 μm thickness range.

Another technique involves using a cylindrical cutter to cut a matching groove through the film (or a diffusion layer) and into the substrate. After etching and staining, two linear variables are measured under the microscope and substituted into a suitable equation, in order to calculate the thickness. Since the dimensions of the groove cutter are accurately known, it is found that this method is more reproducible than the oblique sectioning method. Sensitivity is claimed to be about 10 μm (203).

If the material can be cleaved along a crystal plane perpendicular to the surface film, a clean surface break results and after this is etched and stained, the cross section of the film can be measured under the microscope. Sensitivity of this method is limited to about 10 μm.

At the present time, there are two ways in which optical density measurements can be used to measure thickness. The first (87) is with a photograph taken of the interferometer pattern in the usual manner. A densitometer is then used to scan this photograph. This produces a density plot from which it is possible to accurately locate the fringe lines and the deviations (± 350 Å). In the second method, the film is mounted on a glass or silicon substrate. If on glass, visible light can be used, but if on silicon, it is necessary to use infrared light. Any attenuation is exponential ($e^{-\lambda t}$). This averaging technique necessitates that the light can be transmitted through the sample (29).

Another technique making use of infrared, but in reflection, requires that a scan be made of the reflectivity versus wave number of an infrared beam directed onto the layer surface. Constructive and destructive interference occurs at $\frac{1}{2}$ wavelength points. The minimum measurable thickness by this method is about 5000 Å.

An ordinary microscope equipped with a monochromatic filter and having a stage capable of rotation on an axis normal to the optical axis of the microscope has also been used to measure thickness (232). The film is rotated around this axis and the reflected light is thus observed at various angles. By observation of fringe minima, the film thickness can be determined.

E. Angular Measurement. Various means exist for angular measurements under the microscope. Some microscopes are equipped with accurately graduated rotatable stages, readable by vernier, so that very accurate angular measurements can be made, using a crosshair eyepiece. Less accurate but effective are eyepieces with graduations marking rotation (goniometer eyepieces). Various reticules, designed to fit into the eyepiece and to appear superimposed on the field of view, can also be used for fairly accurate angular measurements. The graduated stage and goniometer eyepieces are useful for measuring the extent of optical rotation, the apparent angles of crystals, and similar studies. It is, of course, feasible to make accurate angular measurements upon photomicrographs, using means of angular measurement not related to the microscope.

In crystal measurement, it should be borne in mind that the angles measured between edges do not necessarily correspond to angles measured between surfaces which these edges intersect, so that edge measurements do not necessarily equal the interfacial angles. When a sufficiently perfect crystal of large enough size is available (>1 mm in diameter), its interfacial angles are usually measured on a reflecting goniometer. In such instruments, the crystal is mounted on a needle attached to a rotatable drum, equipped with accurate graduations. Tilting and lateral adjustments are provided so that any zone of the crystal can be accurately located parallel to the axis of rotation of the drum. Light from a slit collimator then falls on the crystal and, as the drum is turned, the light from each face in that zone is reflected into a telescope equipped with crosshairs. A reading is taken when the slit image coincides with the crosshairs. This represents the simplest form of reflection goniometer. There are also two- and three-circle goniometers available, which measure the angles in all zones with one setting of the crystal. So-called universal stages are available for the microscope which convert it into an accurate three-circle goniometer.

In mineral or metallurgical specimens where one deals with flat surfaces, the measurement of crystal orientation can be accomplished by suitable etching to form etch pits in the surface of the crystals or grains and then measuring the reflections from such etch pits (183). Instruments have been developed that make this measurement relatively simple (294,306). These instruments are capable of determining normals to plane surfaces with a diameter as small as a few μm. Thus, the orientations of a variety of grains or crystals may be determined easily and accurately by measuring the directions of the normals to facets composing the etch pits.

4. Sampling and Testing Surfaces

A. Spectrography, Photometry, and Refractometry. Qualitative and quantitative study of materials having absorption spectra are possible with the microscope. Jelley (139) contributed the first of these types of microspectrographs. With this instrument, the image of any colored object, in natural or polarized light, could be focused on a slit of a spectrograph. This instrument is suitable for both visual and photographic observation and is useful for the microscopical study of various optical properties as determined in polarized light. It has been used to study the pleochroism of crystals, to determine the birefringence at a given orientation and the dispersion of that birefringence and the dispersion of the optic axis (140). A bibliography of early microspectroscopes and microspectrographs is given by Chamot and Mason (61). Early attempts were also made to study materials by emission spectroscopy (43,59,

234) using a spark source. In general, however, a spark source was too rapid and was insufficiently luminous to obtain a good spectrum. The commercial development of the ruby laser microprobe has made it possible to focus a minute beam, yet one with a tremendous energy for vaporization, upon the sample. The area to be analyzed is brought into focus under the microscope and centered under the eyepiece cross hairs. The laser beam is then flashed through the same optical system, at the same settings, so that the selected area is vaporized. At the same moment, a spark is discharged between two carbon electrodes just above the vaporized area. These electrodes have a potential across them (1–2 kV) high enough to raise the vaporized gases to spectral emission and this emitted light is passed through a spectrograph. The laser beam is so small that an analysis can be made of the fingertip without experiencing pain.

Systems of this type have been used to study inclusions in metals (257) and to analyze molten stainless steel while it is contained in a furnace (255). It has been used in histochemistry and the detection limits of some metals (Ca, Mg, and Cu) were found to range from 10^{-14} to 10^{-19} g (114). A study was made of the practicability of laser probe excitation in spectrochemical analysis (153,245), and it was concluded that the method was practical for microsamples with dimensions of 50 μm and larger. By control of focus, the depth of the vaporization pits could be limited to about 10 μm. However, the laser beam eventually damaged the microscope optics, because of unavoidable internal reflections. It was also found that quantitative analysis produced coefficients of variation for analysis of 15 to 40%, which were apparently the result of variation in the laser pulse amplitude and duration. It is to be hoped that these difficulties can be overcome, because this technique could be used to advantage in many areas of investigation.

Microphotometers are available as accessories for microscopes and are used for the measurement of intensity and reflectivity, as well as color adsorption. Photometry has been used extensively to measure the reflectivity of ores, coal, and metals. The determination of reflectivity is very important in the description and identification of these materials, and numerous articles refer to this phase of microphotometry (50,243,260). Reflectivity, like many other properties, may vary somewhat in different specimens of the same material, and in different crystallographic orientations, but it seems to be less affected by methods of polishing than do properties that depend on the retardation of light.

Microphotometry has become very sophisticated in the last decade and measurement can be made on an area of the field as small as 0.5 μm. Microphotometers are also available to cover the ultraviolet region. Absorption analysis is conducted in many fields of research in the ultraviolet region and the UV microphotometer assists in these investigations.

With the introduction of automatic scanning stages for the microscope and computerized readout, it is possible to design microphotometers for fast and accurate analysis of microscopic images. Such instruments are capable of measuring and indicating numerically an optical density range of from 0 to 1.75, and with an accuracy of 0.01. Since scan and readout are fast (3200 measurements/min), a large amount of data can be accumulated quickly, accurately, and economically.

Since interference or polarizing microscopes convert phase differences into brightness differences, the microphotometer is also used successfully to read out this information. The microphotometer has been used to make densitometric measurements on photographic emulsions, and absorption measurements on microstructures. By measuring the absorption of certain dyestuffs applied to a cell, it has been possible to determine the cell's chemical content (14). The microphotometer has also been suggested as a microcolorimeter where only a drop or two of liquid is available for analysis (61). Microphotometry has also been suggested for concentration determinations in solid solutions as it was found that in certain systems (MnO-FeO, MgO-FeO, and MnS-FeS) there was a definite concentration dependence on the reflecting power (160). Microphotometry finds extensive use in metallography (162).

The index of refraction is a characteristic of any substance of definite composition and is important for identification. Furthermore, this property can be determined by microscopy with relative ease. The use of microscopy is also probably the most convenient way, within its limits of accuracy.

There are numerous ways that this determination may be made (61,148,157). The immersion method is based upon the fact that a material immersed in a liquid of the same refractive index becomes invisible. Since standard series of refractive index liquids are available, the method is relatively easy.

Viewing an object through a medium whose surface is normal to the line of vision will cause the image to appear to lie in a plane above that of the object, the amount of displacement being dependent on the thickness of the medium and its refractive index. This upward apparent displacement may be measured with the microscope fine focus and the depth of the cell can be standardized. Interferometry and phase contrast can also be used to determine the index of refraction (61,148,171).

B. Microradiography and Autoradiography. In contact microradiography a specimen is placed in direct contact with a recording material and x-ray exposure is made through the specimen into the recording material. After proper developing, the recording material image is examined with the microscope. Since this x-ray treatment normally produces no enlargement, the maximum resolution and magnification obtainable is dependent upon the granularity or structure of the recording material. Photographic film is the most commonly used recording medium. Ordinary films for photographic enlargement are designed for not more than 20 times enlargement and have a resolution range of from 60 to 150 lines/mm. There are, however, film materials available on special order from the principal film manufacturers that have a resolution of about 1000 lines/mm or even more, but even in this case, individual grains become visible at about 500×. Another problem is that most film emulsions are much too thick for the depth of field of the microscope at these magnifications.

There are other materials that can be used to record the x-ray image. Some ionic crystals develop a characteristic color after irradiation (266). Some plastics develop a change in solubility (177). Another possibility is photoreactive dye derivatives which become dyes when sufficiently irradiated (56,226).

These have the additional advantage that they are not affected by ordinary light, which simplifies preparation and handling. The use of a xerographic process has also been reported (17). Most of these methods still remain experimental but excellent microradiographs have been made and published, using photographic film, indicating the feasibility of the method.

In addition to the grain of the film emulsion, there is another factor that affects resolution. This involves the relations between source, sample, and recording medium distances. The source to specimen distance should be small to obtain the maximum intensity but not so small that the source penumbra limits the resolution. With a fixed source diameter and average specimen thickness of 10–100 μm, the minimum source distance is determined by the maximum allowable penumbra. The foregoing is mathematically expressed

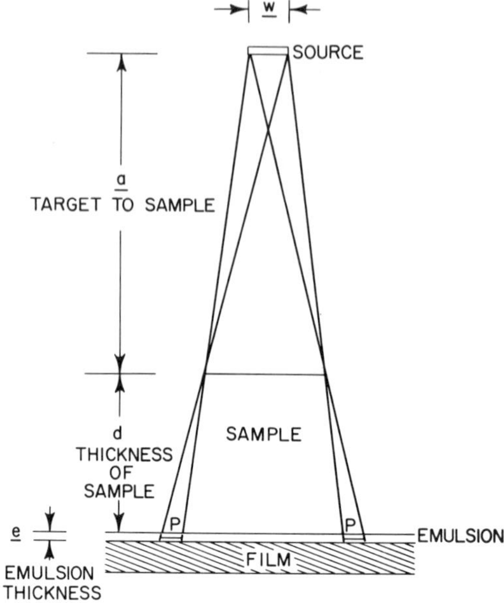

Fig. 16. The effect of the penumbra on image definition in microradiography.

as $P = s(d + e)/a$ where s = source focal spot width, d = thickness of sample, e = thickness of emulsion, and a is the distance from target to sample (Fig. 16). It can be seen that the penumbra (therefore, blurring) becomes larger as the emulsion is made thicker, or as the sample thickness increases. It should be noted that the distance between sample and emulsion is part of the distance d, so that contact between specimen and emulsion is also important. Conversely, it can be seen that blurring is reduced when the sample is thin or the emulsion is thin (or both) and when the distance between sample and emulsion is as small as possible. It would seem that increasing the target to sample distance or decreasing target width would also decrease blurring. However, this has the effect of increasing the exposure time and reducing the useful homogeneous image field registered by the emulsion.

The thickness of the emulsion, and the sample to emulsion distance, can be considerably reduced by painting a thin coating of melted emulsion, containing the proper silver salts, directly upon the specimen surface. This emulsion is subsequently removed, dried, and examined with microscopy (313).

A range of emulsions are obtainable as stripping films. These consist of an emulsion bonded to a gelatine base which is temporarily carried on a film support. The emulsion-gelatine layer is stripped off the film base before use, floated emulsion surface down on a water surface, and, after softening, is collected on the specimen, which is brought up underneath it. After drying, the emulsion is exposed and processed and finally removed for examination. Some of these films are only 2 μm thick when dried (313).

The necessity of preparing a thin sample is one of the principal difficulties of the method. With microtomed biological tissue, this is not a serious problem. However, metals, minerals, or impregnated coatings, of more interest to electrochemists, present a much more serious problem. Such specimens are required to be from 0.002 to 0.004 in. thick, and can present problems if uneven, stressed, cracked, or damaged during grinding and polishing.

The most obvious method of obtaining such thin films from ductile materials is by rolling, but this will show a highly distorted state. In many cases, it will be advantageous or necessary to reduce a specimen to the required thickness by grinding. The method is similar to that used to prepare thin sections of petrographic specimens (8,49,121,158,314,324).

A further factor that limits magnification is the simultaneous imaging in the emulsion of objects at different depths in the specimen. This confusion can be overcome by making stereo pairs in which two exposures are made with the specimen at different angles to the beam and then employing stereo viewing. Microradiography has been much used in petrographic mineralogical, and ceramic applications. Some of these have been discussed (8,9,146). Autoradiography is similar in contact radiography in that the emulsion is placed in contact with the specimen, but it differs in that no external x-ray source is used. The silver bromide grains in the emulsion are acted upon by radiation coming from radioisotopes within the sample. Autoradiography is of special interest to electrochemists because it offers many possibilities for electrochemical investigations, and has already been used in numerous applications. In addition, it has been applied to many metallurgical studies because qualitative and quantitative investigation of the distribution of elements are possible in studies of segregation, diffusion, oxidation, corrosion, and so forth. It has been used in the study of grain boundary diffusion (70,127,327), lattice self-diffusion (109,110), for investigating lead-acid battery reaction phenomena (36,96,182,303), to study the mechanism of leveling in bright nickel plating (85), for determining oxygen diffusion and locating oxide inclusions in metals (67), to determine the mechanism of the nickel-fluorine reaction (138), for following the mechanism of scaling in metals (42,131), for investigation of the oxidation of cobalt (53), to identify the diffusing species in uranium oxidation (261), and to detect nitrogen transport in solids (68).

Autoradiography has the advantage over contact microradiography in that very small concentrations of the radioactive element can be detected and any movement followed. In the majority of investigations, according to Ward (314), useful magnifications of about 20× have been used, although the capabilities are 20 times this figure or about 2.5 μm resolution, with careful technique.

The radiation from a naturally radioactive substance consists of α, β, and γ rays. The relative ionizing power, and hence, degree of photographic blackening is in the order α, 1; β, 100; γ, 10,000; whereas relative penetration is in the inverse ratio γ, 1; β, 100; α, 1000–10,000. When a photographic emulsion is placed in contact with the specimen, the particles effective in irradiating the emulsion will be those within the surface layer of the specimen (usually less than 50 μm), and these will have the greatest effect on the film. Fortunately, the β and γ radiations from deeper in the specimen, will cause relatively less darkening of the emulsion. Consequently, these radiations will cause relatively little blurring of the more pronounced image caused by the α particles near the surface. Ward discusses the various isotopes that can be used in autoradiography and methods for introducing them into the material to be studied (315).

It is not always necessary to introduce an isotope into the substance. When a substance is irradiated by neutrons in an atomic pile, or by ions from a cyclotron beam, many of the elements produce radioactive isotopes. Among the various isotopes, the half life and activation cross section will vary considerably, so that it is often possible to make use of especially radioactive elements for autoradiography without significant interference from lesser radioactive isotopes that may be present. It is also possible to introduce inactive tracers deliberately which can then be irradiated after insertion (67,131, 68).

The preparation of specimens for autoradiography may require melting, heat treatment, cutting, grinding, polishing, etc., which must be done with the greatest caution and with special handling if the radioactive isotope is already in the material. The great advantage of irradiation *in situ* is that all of these preparatory steps may be carried out without hazard or special precautions against radioactivity. Another advantage is that isotopes with very short half-lives may be used. Additional applications of microradiography are given in references (9,270,308).

C. Fiber Optics Microscopy. It would seem that electrochemists are somewhat backward in making use of fiber optics to observe phenomena at working electrodes, or in their porous interior. Although the medical profession is steadily advancing in this technique, it seems to have gone virtually unnoticed elsewhere. Kapany and co-workers appear to have made extensive use of both fiber optics microscopy and laser microsurgery (52,152). These workers have used fiber optics in endoscopy, oximetry, blood flow measurements, radiology, laser coagulation of tissues, and inspection of subcutaneous tissues. It has been also used for remote illumination, spectrophotometry, and fluorescence studies of tissue *in vivo*. These applications demanded high resolution imagery, appropriate spectral transmission characteristics, and ability to withstand and transport laser radiation through the fibers. Long, Brushenko, and Pontarelli (191) described a hypodermic fiber optics microscope making it possible to make observations through needles with inside diameters of 0.81 mm. Special features included a self-contained illuminating system within the same needle and a unique microscope eyepiece, which corrected for distortion caused by the necessary bevel at the needle point. A resolving power of about 10 μm was claimed. As far as human tissue is concerned, there is the limitation that living tissue presents little contrast or optical density difference by which to identify structures.

Innis (135) described an apparatus for producing high-speed motion pictures of the interior of the heart, showing the action of the endocardium, valves, and great vessels of living dogs. In this case, blood was displaced by a balloon at the tip of the instrument which was inflated and gently pressed against the object to be viewed. The illumination was also contained in the fiber sheath. A method of performing *in vivo* spectrophotometry of blood has been described by Polanyi (233) in which light was sent to the tip of the catheter and brought back by glass fibers. While these applications are strictly medical, they should suggest applications for electrochemical microscopy.

D. Micromanipulators and Micromanipulation. The object under the microscope is within reach of the microscopist and it is natural that there would be an urge to manipulate it for various purposes. For lower magnification and with binocular microscopes, a surprising amount of manipulation can be accomplished by hand, with fine needles. It is possible to sort a mixture of particles of sufficient size into its components at magnifications up to about 150×. Above this, it becomes more awkward both because small motions are greatly magnified and because the working space beneath the objective becomes smaller. For this reason, mechanical assist is required from instruments in which relatively large hand movements are converted into much smaller movements at the objective end of the instrument. These micromanipulators not only reduce hand movements but also act as extra hands to hold tools, and to manipulate them with great precision.

It is not possible to discuss here the many uses of micromanipulators, the varieties of instruments that have been designed for this purpose, or the ingenious tools for use with them. For this, the reader is referred to the book by El-Badry (89). There are, however, several areas in which micromanipulation would appear to be of particular interest to the electrochemist and these will be briefly discussed.

When observing materials under the microscope, it frequently happens that small amounts of unidentified material will appear. While various means exist to identify these materials *in situ*, it is frequently more desirable to remove them, or to sample them for identification by x-ray diffraction or spectrographic or chemical means. For mixtures, it may be possible to perform this operation by hand, using a small probe, but for inclusions in minerals, metals or other hard materials more elaborate means are required. One method of doing this is by drilling. Naturally, drills for this purpose must be very small, and the depth of penetration must be controlled. Means must also be provided for centering the drill precisely over the desired area. Several microdrilling machines have been described for use with the microscope (44,155,163,254,280). Drills may be from 25 μm to 0.12 mm in diameter. Larger sizes are, of course, available, but of less utility for this purpose because such large areas of the sample are seldom homogeneous. The smaller the microdrill, the more carefully it must be operated to avoid breakage. Accuracy of drill feed, speed of drill rotation, concentricity and precision of the drill rotation all become increasingly important with decreasing size.

A somewhat novel method of removing small areas is by the use of vibrating stylus oscillating at ultrasonic frequencies. It is claimed that inclusions as small as 10 μm can be removed with this machine (155). The throw of the stylus is about 25 μm each way from center.

All of these methods produce chips or dust in very small quantities and present almost as great problems in analysis as in sampling. Generally, these chips are collected on fibers and analyzed by long exposure in an x-ray powder camera, or by emission spectroscopic techniques.

Another area of interest is in the use of minute electrodes in conjunction with micromanipulators. This allows precise placement of the electrode coupled with the ability to make potential measurement in very small areas. Considering the advances made in micromanipulation, it is surprising that a greater effort has not been made to explore the various potential regions at the operating electrodes. The use of microelectrodes dates back to at least 1925 and, again, it was the medical profession that pioneered. Ettisch and Petérfi (95) and Taylor (296) described the preparation of such electrodes in the study of biological cells. Ettisch and Petérfi employed a relatively simple electrode consisting of a capillary, filled with 0.1 *N* KCl solution in agar, which was drawn out to a fine point. This capillary was connected, through a T tube to a 0.1 *N* calomel electrode. The details of preparation, including the task of preparing a continuous agar bridge in the fine capillary, are described. Taylor's electrodes were somewhat more ambitious, and details are given for preparing an electrode by fitting a platinum wire into a closely fitting quartz capillary, and then drawing out the combination to a fine tip, over a microflame. He thus produced a perfectly insulated electrode, with an exposed surface of less than 1.0 μm in diameter. He described a second type of electrode as nonpolarizable; this consisted of a finely drawn out capillary filled with 0.5% KCl solution in carefully dialyzed agar. This was connected to a larger bulb, also partially filled with agar jell of the same composition. On the surface of this agar jell was deposited a layer of moist, well-pulverized AgCl. Connection to the AgCl was made with a silver wire of convenient length. The bulb was sealed with deKhotinsky cement, which also firmly held the silver wire. He also described the construction of microelectromagnets. More sophisticated electrodes with the conducting film of silver, gold, platinum, or silver chloride being plated on the outer surface of the glass or quartz microneedles have been described by Sen (269). Although all of these microelectrodes were constructed for biological use, very little effort would be required to convert them for electrochemical studies. Mahin and Brewer

(195) used a microelectrode to explore the differences in potential between the surfaces of the variously oriented grains in alloys. The electrode was brought into contact with the grain under investigation. Mears and Brown (205) made similar studies but used the method of masking all but the desired part of the surface with wax.

Allied to micromanipulation is the determination of hardness by means of microindenters. Here, the movable stage acts as the micromanipulator and the desired area is brought under the cross hairs of the eyepiece. The fine focus of the microscope then acts as the vertical component of the micromanipulation to bring the indenter into contact with the specimen. Microhardness indenters are made by most of the makers of microscopes. Indenters are either the Knoop or the Vickers type. The diamond has been beveled in the Knoop indenter to give a pyramid with two opposed faces making an angle 130° to one another and the other two faces making an angle of 172° 30′. This results in a diamond-shaped indentation with the long diagonal about seven times that of the short diagonal and about 30 times the depth of the impression. The Vickers indenter has a quadratic base and the opposed faces make an angle of 136°. This indentation is square, and the diagonal is only seven times the depth of penetration. Both types of indenter can be used to measure the hardness of very small areas (down to approximately 1.0 μm diameter). Measurement is made by bringing the indenter into position over the desired area, and gently bringing the indenter into contact with the surface while applying a known load for a predetermined time. The load and hardness of the tested material determine the size of the indentation. After removing the indenter, the diagonals of the depression are measured by a micrometer eyepiece. The hardness is then computed from an appropriate formula. The method can also be used for identification but is subject to certain limitations. Metals or minerals in solid solution vary in hardness with composition, and most materials will vary somewhat depending upon the orientation of the face being measured. Thin crystals may differ from thick crystals of the same material. It would seem that, with a sufficient number of readings, a fairly accurate average value is possible. Efforts have been made to improve the accuracy by using interferometry (101), or by the scanning electron microscope (119), to measure the edges of the indentation, and by automating the application of the load.

Despite the difficulties mentioned, the microhardness determination has been applied with success to various investigations.

Gebhardt and Sighezza (111) for example used microhardness measurements as a means of following the increase of oxygen in tantalum. They found that the microhardness of tantalum increased almost linearly with increasing oxygen content. Microhardness traverse of a metallographic cross section could therefore be used to study the rate of diffusion of oxygen into the surface.

E. Heating Microscopes. The various changes that occur in materials subject to heating may all be observed under the microscope and, with the proper type of heating stage, these processes can also be accurately controlled and measured. The importance of microscopic examination of temperature effects is especially important in the case of heterogeneous material where different parts of the material will react to the heat at different rates or at different temperature. A heating stage equipped with accurate temperature control and measurement is an important accessory in almost any use of microscopy. Just how elaborate this must be, however, depends upon the use and the temperature range required. Quite simple equipment suffices for determining melting points of most organic materials and many inorganic materials. Much more elaborate stages are required for high temperature investigations of metals, ceramics, and many other materials.

For most purposes, the stage must be insulated in order to ensure uniform and gradientless heating, to protect the microscope, and for attainment of the desired temperature. The higher the temperature required the more difficult these requirements become. If too much insulation is used the response becomes very sluggish, but if too little is used control of temperature becomes poor. Since the object must be viewed and yet must be placed in a heating chamber, some sort of window must be provided. Two such windows are required for transmitted light. Under such conditions, gradients in heating are inevitable, and the window will also transmit heat to the lenses, particularly where high magnification is used, and the lenses must be very close to the window. Commercial heating stages are now available that have either solved these problems, or reduced them to reasonable compromises, by the development of long working distance objectives, etc. Special heating stages are required for different purposes, and many have provision for sudden quenching, introduction of reagents, prevention of evaporation, and for maintaining a vacuum or inert atmosphere. Usually, all these features are not to be found in a single instrument.

The first heating stages were designed solely for the purpose of melting point determination. Since the beginning of melting can be observed quite readily in individual crystals when sufficiently magnified, it is possible to provide very accurate melting point determinations in this way, particularly when the heating stage has been carefully calibrated. From melting point determination, the method was soon turned to fusion studies, and an extensive literature soon developed on this method of analysis. While very practical and easily applied, this method has been almost entirely used in the identification of organic substances. It has also been applied to studies of the kinetics of crystal growth, determining the stability of decomposable compounds, and recrystallization and grain growth studies. Most of these investigations have been used for organic materials and will not be discussed here. For those interested in this phase of temperature microscopy, details and references will be found in the book by McCrone (202).

The heating stages capable of operating up to about 2000°C are probably those of most interest to electrochemists. Microscopes equipped with this type of stage have been used for many types of investigation; studies of oxidation, grain boundary migration, recrystallization, phase changes, etc. A number of heating stages suitable for such studies have been described (24,55,58,61,80,115, 140,225,246,249,251).

Welch (317) has described what must be the simplest possible type of high temperature apparatus. This attachment cannot properly be called a heating stage, for it consists of a thermocouple of small diameter wire, bent into a loop at the junction so as to hold the specimen, which is in the form of a bead. The bead is observed through the microscope while the thermocouple is heated. The heating current is taken from an alternating supply and a vibrating reed driven from the same supply permits the current to flow only during one half of the cycle. During the other half cycle the current from the thermocouple is employed for measuring the temperature. The response in both heating and cooling is extremely rapid and the equipment has been used by that author to study melting points, glass devitrification, growth of single crystals, and the delineation of phase equilibria.

Mercer and Miller (208) used essentially the same apparatus in thermal analysis. By incorporating a cathode ray tube and photography to record the cooling curve, they were able to record cooling curves for melt droplets cooled at rates up to 20,000°C/sec. To obtain such rapid cooling a stream of water was impelled across the microfurnace. Another innovation worth noting was the use of two heating supplies that could be independently switched on, so that cooling could be observed over as small a range as desired. The simplicity of this equipment can be appreciated when compared with the much more elaborate equipment described in some of the foregoing references.

Low temperature stages have also been developed which are particularly useful in studies of substances that remain liquid at room temperature or in the observing of the effects of freezing on certain substances. Such stages are usually cooled by circulating, externally cooled liquids or by gases such as CO_2 or N_2. The insulating problems are similar to those of high temperature stages. An additional problem is the difficulty of preventing frost on viewing windows.

5. Automation in Microscopy

To reduce the labor, increase the efficiency, and make possible greater utilization of microscopy, a great deal of automation has been developed over the years. There are, for instance, automatic specimen polishers, automatic exposure controls, and many other devices that remove some of the labor and boredom of routine operations from microscopy. However, until very recently, all measurements or comparisons had to be made by the operator and because of the labor involved in obtaining it, quantitative data were limited.

It has been pointed out in the foregoing that many kinds of exact measurements can be made by microscopy. The problem is that the area measured represents such a small fraction of the total sample that the question arises as to whether the measurements are representative of the whole sample. The problem becomes greater as higher magnifications are used. This question can be resolved by making a sufficient number of measurements on various parts of the sample and on other samples from the same source. Naturally, such a process requires patience, time, and expense. So far, many types of measurements continue to require this type of procedure, but in the areas of counting and linear measurement the microscopist can now obtain relief.

The principle of stereometric analysis, the quantitative characterization of three-dimensional figures by linear measurements of planar surfaces, has been known for practically a century, but only in the last few years has it been widely acknowledged. Difficulty in obtaining recognition was based on the difficulty of gaining acceptance for the validity of two-dimensional determination of three-dimensional objects and also because of the great difficulty and labor involved in obtaining enough such measurements for them to be of statistical significance. Confidence in the method was gained after numerous publications had treated the method in detail (35, 37,78,90,97,104) and proven its merit.

It is possible to get data automatically in a number of ways from a polished cross section. In the simplest, the specimen is simply moved past a stationary detector capable of distinguishing between surface features. This detector may be the observer, a photometer, a microprobe, or an electron-optical instrument. The method is slow but permits long traverses, equipment required may be minimal or extensively electronic. Such a method is discussed by Dörfler (84).

A second possibility is to illuminate the object with a very small spot of light. The spot is reflected back into the microscope in a fine pencil of rays and makes a very accurate signal. The spot of light is moved at high speed across the specimen surface, as in the television raster, and the resulting microscope scan contains signals at every change of phase, as when moving into and out of particles. These impulses can be used for counting and other purposes.

A third method, introduced by Metals Research (103) simply scans the magnified image from the microscope and thus obtains its signals. Because it uses standard television and electronic equipment, this so-called Quantimet compares favorably with the flying spot microscopes. Other instruments have been produced in Germany which work on a similar principle (187,329) and also in the United States (25). With the popularity of using initials that is now in vogue, this system has already been referred to as QTM (Quantitative Television Microscopy).

In their present sophisticated forms, these instruments are capable of assessing the position, distribution, shape, quantity, and size of features in numerical terms which can then be fed into associated computers to give almost instantaneous analysis of volume fractions, particle size distributions, etc. Since the microscope stage is also equipped for automatic traverse in two directions, a large number of fields can be scanned and measured in very short times. Clark (65), speaking of the Micrometron, an automatic microscope with photoelectric detection and fully automatic stage traverse, compares its capabilities of thousands of single observations per second with the 50 to 100 observation events that the microscopist can study in a reasonable period. Speaking of the above instrument, Clark also reports that it can count 10,000 blood cells in 10–20 sec. Truly, this is automation.

Despite its many advantages, there are still sources of error that have not been eliminated. These are mainly caused by difficulties in specimen preparation, inadequate resolving and differentiation power, and insufficient contrast between phases, or a variation of contrast within a single phase.

III. Examples of Microscopy in Electrochemistry

1. Detection of Matter Transport

One of the most rewarding areas of microscopic investigation has been in the detection and measurement of material transport across an interface or through a film. When combined with the use of autoradiography and radioactive tracers, microscopy becomes a sensitive and useful tool for detecting and measuring diffusion and determining the diffusing species. One of the first applications was in the study of oxidation.

For example, prior to the work of Carter and Richardson (53) several investigators had reported that cobalt oxidized according to the parabolic law, but no detailed study had been made of the oxidation mechanism. In the referenced investigation, inert markers of radioactive Pt were applied as $(NH4)_2PtCl_6$ slurry in thin streaks across cobalt discs that were oxidized at 1200°C in pure oxygen. The discs were then mounted, sectioned, and autoradiographed to locate the markers. The radio-platinum markers were found at the metal-oxide interface, indicating that cation diffusion was the principal mechanism operating during oxidation. Having established this, radioactive cobalt was used to determine how the cobalt migrated into the oxide layer. The authors were able to determine that the cation was the diffusing species and to demonstrate that the concentration of vacancies varied in a complex manner across the growing oxide layer, but that it differed from the values that could be calculated from the Wagner (309) theory of oxidation.

A similar study was made of uranium oxidation. Schnizlein (261) and co-workers used two methods to determine the migrating species. In the first, the assumption was made that two oxide films growing toward each other would fuse into a continuous mass and that an inert marker would become buried in the oxide. On this basis, two uranium blocks were highly polished and the polished faces were wired together with Inconel wire. Before doing so, however, a platinum wire was placed across one end so that the two surfaces were separated by a wedge shaped space. The blocks were then oxidized. After mounting, microscope observation showed that the wedge shaped space had not been closed by oxide. In cross section, it was evident that the oxides had not filled the space and that the Inconel binding wires had not been buried by the oxide formed on the outside. In radioactive experiments, markers indicated that in this case oxygen, rather than uranium, diffused through the oxide layer.

Marker techniques are not necessarily confined to metals. Carter (54) for example, studied the mechanism of the solid-state reaction between magnesium oxide and aluminum oxide and between magnesium oxide and ferric oxide. MgO and Al_2O_3 can combine to form $MgAl_2O_4$ in a solid-gas reaction (213). Several strands of molybdenum were spaced across each side of an Al_2O_3 disc and held in place by Al_2O_3 rings. An excess of MgO and the above assembly were placed in a molybdenum boat but out of contact with one another. The boat was placed in a furnace at 1900°C for 48 hr, under H_2. The resulting product was then examined microscopically. Carter reasoned that if Al_2O_3 diffused through the $MgAl_2O_4$ product to the MgO interface, the formation of $MgAl_2O_4$ would bury the molybdenum wires, which would remain then at the original Al_2O_3 interface. On the other hand, if the MgO diffused through the $MgAl_2O_4$, the markers would be at the MgO-$MgAl_2O_4$ interface. If diffusion were in both directions, then the marker wires should appear in the $MgAl_2O_4$. In actuality, it appeared that the reaction was by the last mentioned mechanism and diffusion had indeed occurred in both directions.

In the case of reaction between MgO and Fe_2O_3, attempts to place markers were unsuccessful because of the tendency for markers to interact. To solve this difficulty, pores were used as markers. Discs of Fe_2O_3 were prepared by sintering. These were found to contain 17% pores. These Fe_2O_3 discs were placed between two slabs of dense MgO. This sandwich was then oxidized. The pores were found to terminate within the ferrite, indicating cation diffusion.

Slightly out of context but relevant to the problem is a paper by Burley and Dale (46) which relates to a technique for preserving fragile oxide-on-metal coatings for metallographic examination. To improve the support of fragile oxide coatings during impregnation, grinding, and polishing operations, these investigators used a chemically deposited silver coating, followed by electrolytic copper deposition, to strengthen and demarcate the oxide coating. This treatment not only strengthened the coating, but also helped to eliminate artifacts.

Nickel is a material of construction much used in plant operations where fluorine is present. Its resistance to fluorine attack is attributed to the formation of an adherent scale of nickel fluoride on the surface. The kinetics of this nickel-fluorine reaction would appear to be similar to those involved in oxidation, since it also involves metal-gas reaction at elevated temperature. The two processes were compared by Jarry and his co-workers (138). Oxidation and fluorination were first compared by the impinging scale method. Opposing polished faces of nickel coupons were placed parallel and about 0.1 mm apart. Such assemblies were then either oxidized or fluorinated. After reaction, the samples were mounted, sectioned normal to the scale between the parallel faces, and examined microscopically. The impinging fluoride scale was found to be separated by an interface, indicating that the migration of fluorine took place from the gas-solid interface, through the nickel fluoride film, and to the nickel-nickel fluoride interface, where reaction occurred. The above-described film appearance also indicated that metal diffusion was negligible.

In the case of the oxide film, it was found that the two oxides had grown together without sign of an interface and the layer was continuous from one metal surface to the other. In this case, the indication was that the metal had migrated through the metal oxide scale and then reacted at the metal oxide-gas interface to produce a continuous oxide film.

To check the accuracy of these assumptions, a somewhat similar investigation was performed using a radioactive tracer. Nickel-63, a low energy beta emitter, was plated upon nickel coupons, in a 2-μm thick layer. These coupons were then either fluorinated or oxidized. For the fluorination, the photomicrography showed a narrow band of activity at the position of the nickel fluoride-fluorine interface, indicating that the Ni^{63} had not migrated and that fluorine was the migrating species. In the case of oxidation, the opposite was found to be true, and the broad band of activity indicated extensive migration of the Ni^{63} to the nickel oxide-oxygen interface, confirming the correctness of the previous observations.

Condit and Holt (67) have discussed the feasibility of using the proton radioactivation of O^{18} to locate oxygen and tag its movements. They investigated the extent to which other elements might obscure the reaction and predicted that Be, C, N, F, Mg, Na, Al, Si, P, K, Mn, Co, As, Y, Nb, and those elements with atomic number above 50 would cause no difficulties. Some that would cause problems are Cr, Fe, Cu, Zr, Rb, and Pd, while serious difficulties would be presented by Li, B, Ca, Ti, Ni, Ge, Sr, Zn, Cd, Se, and Ag.

An essential point of using O^{18}, as pointed out by Condit and Holt, is that protons impinging on a solid surface should become ineffective within a 5-μm distance of the surface. Thus, activation would be confined to this surface layer and, as a consequence, the resolution of the autoradiographic film image should be high, as it would not be blurred by radiation from deeper in the surface. By making shallow angle bevels of a diffusion specimen, for example, and then irradiating it, the profile of the oxygen penetration can be established.

According to Pemsler (227) the oxidation of groups IVA and VA elements takes place with significant amounts of oxygen dissolving in the metal substrate, simultaneously with the formation of the oxide. This causes a metal zone, rich in oxygen, to form beneath the oxide layer. He measured the oxygen gradient beneath the oxide film on zirconium by a technique that involved the measurement of the rate of dissolution of oxide films, formed over sections made of the oxygen gradient in the metal. The oxide films used for measurement were formed by anodization.

The oxidation behavior of uranium in carbon dioxide becomes important when considering the consequence of leaks in the types of reactors using gas cooling. Stobbs (289) studied the weight gain/time curves as obtained on a thermal balance in relation to metallographic observations. Surface examination showed unequal interference colors as the oxide developed, with thickest oxide at grain boundaries, scratches, twin bands, and other surface discontinuities. As the oxide thickness increased, cracks appeared and the whole surface become a network of such cracks. Cross sections showed that these cracks did not extend to the metal-oxide interface. It was found that, while the oxide was initially protective, it finally obeyed a linear rate law, as a result of cracking.

Using metallographic techniques and weight gain determination, Stringer (290) made a study of the oxidation rate of tantalum, determining the effect of grain size, cold work, surface preparation, and specimen shape.

A general approach to the study of scaling reactions that dispensed with conventional inert markers was proposed by Holt and Himmel (131), using instead radioisotopes of the reacting materials. For the study of oxidation of metals, the already mentioned O^{18} was proposed, subsequently activated by proton bombardment. These authors point out some of the problems associated with the use of inert markers, including several examples where the use of such markers had led to erroneous conclusions. They advocated that both radioactive cations and anions should be used in the same investigation, since this would provide a rigid cross check of the consistency of the experiment. In addition to the nuclear reaction $O^{18}{}_{(p,n)}F^{18}$, other similar techniques might be employed. Amsel and Samuel (6), for example have utilized the nuclear reaction $O^{18}{}_{(p,\alpha)}N^{15}$ to study the mechanism of the anodic oxidation of aluminum, and Cox and Ray (71) and Ollerhead and co-workers (222) have used the $O^{17}{}_{(He^3,\alpha)}O^{16}$ reaction in a similar manner to explore the transport of oxygen in thin ZrO_2 films formed on Zr. These methods have shown great sensitivity. Holt and Himmel investigated the oxidation of iron and found that while a thin surface layer of the oxide was composed of Fe_3O_4 and Fe_2O_3, about 95% of the oxide consisted of wüstite (131). Autoradiographs obtained after proton activation showed that the O^{18} had not migrated from the surface of the metal. This offered proof that the growth of the wüstite layer is accompanied by outward transport of Fe ions. Since inert markers had given results not in agreement with this and other check methods, considerable doubt is cast upon the absolute reliability of inert marker methods. Using similar techniques, the authors determined that oxygen was the mobile species in thick ZrO_2 scales formed on Zr.

In a rebuttal to this paper, Brückman et al. (42) called attention to their own extensive work with radioactive isotopes and with sulfurization, and suggested that the readioactive isotope of the oxidizing element must be introduced in the second and not the first stage of reaction, as had been suggested by Holt and Himmel. The reader should by all means refer to this discussion and the reply by Holt and Himmel (132) if at all interested in the first paper by these authors.

Using thermogravimetric analysis, x-ray analysis, metallographic examination, and microprobe tests, Kofstad and Hed (164) investigated the oxidation of Co-25 w/o Cr. The scales formed at high oxygen pressure had a duplex structure with the outer layer consisting of CoO and the inner layer being the $CoCr_2O_4$ spinel. The mechanism of oxidation at high temperature and oxygen pressure involved cobalt ion diffusion through a CoO network in the inner layer.

There have also been a number of papers in which a very comprehensive oxidation study was made of certain oxidation processes, using only metallographic, x-ray, and microprobe techniques (72, 207,320). Since the data for these studies are contained in the photomicrographs accompanying the articles and, since no new techniques are involved, they are mentioned only as evidence that microscopy is used in such studies.

2. *Electroluminescence*

The phenomena of luminescence and electroluminescence have long been of interest and have formed the basis for considerable microscopic investigation. Early investigators were aware that not all phosphors became electroluminescent, and that there was a relation between crystal purity and the degree of electroluminescence. However, until microscopy was used, no very clear picture had emerged. Wayworth and Bitter (316) began their microscopic investigation by preparing a dilute dispersion of a phosphor in a Lucite sheet. It was concluded that the emission was essentially a two-stage process, that is, excitation occurred as the field was applied and the electroluminescence occurred as the field collapsed. This was to be confirmed by later workers.

Zahn and his co-workers (328) also arrived at about the same conclusions. For their microscopic examination, they constructed a small electroluminescent condenser by putting a small quantity of a suspension of phosphor particles (in castor oil) on a sheet of "conducting glass." In the ac field, the crystals appeared to "twinkle," and it seemed that light was being seen both from the crystal surface and from imperfections within the crystal. At these imperfections they reported that the light was as much as 100 times as bright as from the average surface. On the other hand, when the material was excited with light with 3650 Å wavelength, the crystals no longer "twinkled" but looked dull, with a fairly uniform light emission from the entire crystal.

Loebner and Freund (190) made similar observations but used a large single crystal in their experiments. While exciting this crystal with a sinusoidal 1000-Hz electric field they observed the appearance of small blue spots in the crystal. They described these spots as being only about a few micrometers in diameter and about 30 μm long.

The pulses appeared to occur at the 1000-Hz rate and had about a μsec duration.

Diener (82) suspected that impurities might be the cause of this phenomenon. He selected a ZnS single crystal that was not activated and which showed hardly any luminescence with uv radiation. After assuring himself that the crystal was transparent and free from surface defects, he used vacuum evaporation to deposit a very thin film of copper on the surface of the crystal. Microscopic examination now showed the surface to be covered with numerous dark spots that were described as being dendritic crystals. He found that the areas immediately below these specks now showed luminescence when irradiated with uv light. The use of ac fields now showed bluish-green lines, lying within the crystal. With higher magnification, these lines were found to consist of rows of individual dots, as can be seen in his photomicrographs. He assumed that copper had diffused into the crystal and was concentrated at interior imperfections.

Short and his co-workers (276) believed that this appearance of electroluminescence along parallel paths in the crystal might be associated with a one-dimensional disorder. They prepared needle-shaped ZnS crystals that gave the same parallel electroluminescent streaks as Diener had reported. Using a polarizing microscope, they then examined these crystals under crossed polars and found distinct bands of interference colors. By using x-ray examination they were able to determine that these bands were perpendicular to the *c* axis of the crystal and that they were the result of a one-dimensional stacking disorder. The electroluminescent streaks were observed to parallel the color bands.

Lehmann (183) made the next contribution to this study when he demonstrated that many powdered crystal phosphors which were normally nonelectroluminescent became electroluminescent when mechanically mixed with suitable powdered metals. A cell for microscopic observation was made up in which particles of phosphor were placed at one side of the field and particles of the metal were placed at the other. The central portion of the cell contained an intimate mixture of the two. All the particles were suspended in a castor oil dielectric. When excited by an ac field, only the central portion showed electroluminescence.

In a later paper (184) Lehmann attached phosphor particles to a thin plastic film on a microscope slide. The two ends of the slide each had a vacuum-evaporated film of aluminum which served as electrodes. The space between and above the particles was filled with a highly refractive liquid in order to better see into the ZnS particles. A cover glass was used as the upper surface of the cell. Since previous work had demonstrated that ZnS:Cu phosphors might contain CuS as a segregated or occluded phase, Lehmann made a careful search for such a phase. He was able to find what appeared to be black particles within the phosphor crystals and tentatively identified them as CuS occlusions (leaching the surface gave material identifiable as CuS). The linear dimensions of these black areas were found to be 5–10% of the particle diameter. He reported that the emission intensity in such areas was as much as 1000 times that found over the average particle surface.

Lehmann (185) also used a projection microscope to determine the particle size distribution of normal phosphors as obtained by the commonly employed preparation procedures. He measured 200–300 particle diameters for each preparation. From this work, he found that the particle size of electroluminescent zinc sulfide phosphors usually covered a broad range but the particle distribution curves always had the same shape. He determined that every phosphor having this "normal" particle size distribution curve showed a voltage (V) dependence for the electroluminescence brightness (L) which over many decades of L could be closely defined by $L = L \exp [(-V_0/V)^{0.5}]$. When the particle size distribution deviated from this normal form it also caused a deviation of the $L(V)$ dependence.

However, Kremheller (170) also made a microscopic study of various electroluminescent phosphors contained in liquid, plastic, and glass dielectric cells and arrived at other conclusions. After studying the brightness variations among and within many single electroluminescent particles, he decided that particle orientation, particle-to-particle distances, and particle-to-electrode contact all had an influence on the electroluminescence brightness.

Gillson and Darnell (113) have made a very detailed microscopic examination, studying the effect of field potential and frequency on numerous crystals. They found that the electroluminescence was usually confined to certain planes of the crystal, originating from tiny glowing lines and spots. The emission was definitely not from the surface but from the interior of the crystal. From these and many other observations, a model was proposed which required the existence of linear defects 100 μm in length, lying in particular directions which were related to the crystal structure. The model also required local variations from *p*- to *n*-type material, with consequent narrow *p-n* boundaries or junctions.

This paper was closely followed by one by Fischer (102), in which great pains were taken to match the embedding media with the electroluminescent particles. Several satisfactory media were found: (1) a red-orange mixture of sulfur and selenium; (2) a thallium bromide-thallium chloride mixture; and (3) an arsenic-chalcogen-halogen mixture. The third glass proved easiest to work with and was used in most of the work. While the examination of electroluminescent particles confirmed the observation of Lehmann that numerous black spots were present in the phosphors, the better resolution obtained by Fischer led him to the conclusion that these were not inclusions of CuS, as Lehmann had assumed, but were in actuality small voids in the crystal. It was decided that insulating luminescent ZnS particles contained conductive, copper sulfide decorated imperfection lines that, by means of geometrical field intensification at their ends, were capable of emitting both holes and electrons.

Gedney (112) made a microscopic examination of a commercial lamp under actual working conditions. The commercial lamp consisted of three parallel layers, the outer two being the electrodes and the middle layer consisting of phosphor particles held in a solid matrix. The cross section of the lamp was examined by reflected light, cross-polarized reflected light, and its own electroluminescence. The observations checked those made previously but served especially to show that many of the particles were inactive. It was also noted that the positioning of the particles with respect to the electrodes did not seem to have any bearing on the electroluminescent activity. This was at variance with the findings of Kremheller (170) that this factor was important.

The solid solutions of gallium arsenide-gallium phosphide, $GaAs_{1-x}P_x$ are of considerable practical importance. Throughout most of the composition range of this system, the band gap corresponds to energies associated with the visible spectral region. These solid solutions have potential, therefore, for the manufacture of such luminescent devices as electroluminescent diodes and injection lasers.

Rosenberg et al. (253) studied the strain patterns in $GaAs_{(1-x)}P_x$ alloy overgrowths on GaAs single crystal wafers. The ultimate purpose was to determine the effect of the strain field on the light emission characteristics of the heterojunction region, because of its disruption of the crystal regularity. A birefringence technique was used to study the strain patterns. On the basis of available evidence, it was deduced that segregation of the alloy species As and P during growth was responsible for the strains produced.

Tietjen and Amick (299) grew single crystals of $GaAs_{(1-x)}P_x$ epitaxially from the vapor phase by using gaseous arsine and phosphine as sources for arsenic and phosphorus. They used microscopy to study dislocation densities on the basis of etch pit densities. They determined that the density of dislocations in the epitaxially grown layers was determined by the density of dislocations present in the substrate.

Prener (239) used microscopy in a study of calcium fluor- and chlorapatite crystals. Fluorapatite is the major ore of phosphorus, and the mixed fluor-chlorapatites constitute the base for the major phosphor component of fluorescent lamps. Microscopic examination was made under crossed polaroids, which showed chlorapatite to be multiply twinned. When heated between 185–210°C, the crystals lost their birefringence and became optically isotropic in the *c* axis. Below the transition temperature, multiple twinning was found to recur, but this changed drastically with each heating cycle. The high temperature form was found to be hexagonal, while the room temperature form was monoclinic.

3. Cells for the Microscopic Examination of Electrochemical Processes

Electrochemists have been interested for many years in actually seeing what occurs during electrochemical reactions. This is particularly true in the cases of electrochemical deposition and electrochemical crystallization. Köhlschutter and co-workers (167) studied the dendritic growth of a number of metals with the microscope. The miscroscope was simply mounted horizontally to look through one of the flat sides of a cell. An anode was placed in the bottom of this cell and the exposed end of a platinum wire, inserted into an insulating glass sheath, was used as a cathode. This cathode was inserted into the cell from above and was mounted so that its end was directly in front of the microscope objective. The growth of dendritic crystals was thus observed as they moved out from the cathode filament. The arrangement permitted only low magnification (50–70 ×).

Erdey-Grüz and Volmer (93) studied the growth of similar electrodeposits with the microscope, but failed to give any experimental details of the microscope arrangement. Somewhat later, Kohlschütter and Torricelli (168) elaborated upon their original cell design, but were still able to obtain a magnification of only 80 ×. In this improved arrangement, the cathode consisted of a silver wire. Somewhat below its end in the cell was placed a ring-shaped silver anode. Deposition of metal upon the Ag was observed from above, through the surface of the electrolyte. Hoekstra (130) used a similar arrangement but employed a water immersion objective. With this arrangement he was able to obtain magnifications of 500 ×. Contrary to what might be thought, the objective was not immersed in the electrolyte. A cover glass was floated on the surface of the electrolyte and the objective was joined to this with a drop of water. A similar arrangement could be used with an oil immersion objective.

Bartlett and Stephenson (22) studied the anodic behavior of Fe in H_2SO_4. The arrangement was simple and the magnification was low. The microscope was placed horizontally, and observation was through the cell wall (a rectangular cell). Nevertheless, they were able to make out considerable detail in the anodic films. In a later paper (288), a motion picture camera was used with the same cell arrangement. They were also able to arrange a knife edge to create a schlieren cell. The schlieren knife edge was arranged so that the positive and negative concentration gradients could be shown with equal clarity. The knife edge was oriented so that the development of bright regions corresponded to an increase of refractive index toward the anode, while the development of dark regions corresponded to a decrease in refractive index toward the anode. Using this technique, they were able to take motion pictures that clearly showed the change in concentration around an electrode when current flowed.

Wranglen (322) used a lucite cell shaped like a microscope slide but considerably thicker. The lucite block terminated in copper blocks at each end. A shallow groove was cut longitudinally along the center line of the upper surface. An electrode of the desired material could be placed at each end of this groove, where it would contact the copper blocks, which were connected to the power supply. The crystals were allowed to grow horizontally in the groove between the electrodes. The electrolyte contained in the groove was protected by a cover glass. Grease was used to prevent the electrolyte from creeping. At the highest magnifications used, Wranglen immersed the objective directly into the electrolyte, first coating the metal parts with plastic. He found, however, that strong acids tended to etch the lenses and that silver deposited upon the metal parts of the objective regardless of the precautions taken. One observation especially worth noting was that, as the cell was made smaller, the crystals grown in the system were also smaller, so that nothing was gained by the higher magnification made possible by the smaller cell.

Graf and Morgenstern (118), in a study of the electrocrystallization of silver, embedded Pt and Ag wires of very small diameter (3–150 μm) in Araldite resin and then used these as cathodes, observing the squared off end of the electrode wire and the lateral growth of dendrites across the plastic from above the solution. Graf and Weser (117) improved on this method by the following construction. A finely drawn Pt wire was embedded vertically at the axis of an Araldite cylinder. Two Ag rings were placed concentric to the Pt wire in a horizontal position. The two rings were placed at different levels in the Araldite cylinder and the upper ring had a somewhat smaller diameter than the lower. A groove was turned into the Araldite cylinder upper surface with a diameter and side slope adjusted so that the outer edges of both rings were exposed at different heights in the groove. The upper ring served as reference electrode while the lower served as anode. The center of the disc, containing the platinum wire, was machined down in the area within the groove by a few tenths of a mm to form a plateau across which a shallow layer of the electrolyte would flow when the groove was filled (Fig. 17).

Froment (106) studied interference contrast at electrode surfaces by providing an upright cylindrical electrode which he then observed by immersing the lens directly into the electrolyte. In a following paper Epelboin and Froment (91) used the same technique: to study the changes taking place in the surface as observed by phase contrast; to obtain quantitative information on the surface contours, using differential interferometry; and to detect any layers that might form near the electrode, using polarized light. These authors made an extensive study of anodic dissolution and electrolytic deposition. Judging by the quality of their photomicrographs, their results probably justified the routine destruction of lenses that must have occurred.

Tint and Damiano (300) studied etch patterns produced on zinc, using a cinephotomicrographic method that involved the use of microscopy. This cell consisted of a rectangular Pyrex box, 10 × 4 ×

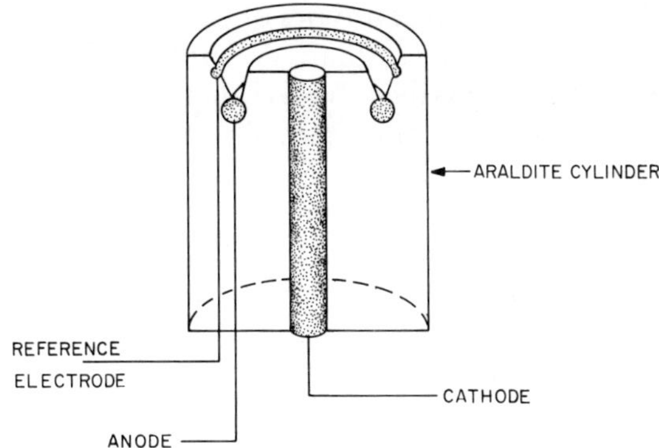

Fig. 17. Simplified cross-sectional view of the cell constructed by Graf and Morgenstern.

3 cm which had a wall thickness of 2 mm. In one face a 1.6-cm circular opening was cut and a thin window (0.15-mm-thick cover slip) was cemented in place with Dow-Corning A-400 silicone cement. The electrode to be examined was placed next to this window and viewed with a horizontally placed microscope. This microscope was coupled to the motion picture camera by an adaptor that contained a light-splitting prism. This made possible simultaneous viewing and photography.

Barton and Bockris (23) constructed a rather elaborate cell to observe the electrolytic growth of dendritic silver crystals. The cathodes and reference electrodes were prepared from spectroscopically pure Ag wire (0.2-mm diameter). This was placed in Pyrex tubing drawn down to the same diameter as the wire. After sealing the exposed tip into the glass, this tip was further reduced in diameter by electroetching, and a 0.1-mm-spherical electrode was formed by melting the exposed tip back to the glass. After suitable cleaning, the cathode, reference electrode, and anode were sealed in a small glass cell. This cell was thermostated in a larger glass vessel. Microscopic examination was made possible by using an auxiliary lens 8 cm from the cell wall, which formed a real image of the cathode before the objective of a horizontally mounted microscope. The system had sufficient definition to determine the velocities of dendritic growth and the general shape and size of dendrites. It was also possible to make cinemicroscopic measurements. This was the first cell that attempted to ensure against impurities in the system.

Simon (281) described several types of cells that could be used for electrodeposition, corrosion, or battery studies. One of these was quite similar to that described by Wranglen (322). Another was described that contained a miniature Pb-acid cell which could be examined while in charge-discharge operation. The principal contribution of this paper was in pointing out the superior results that could be obtained with an inverted metallurgical microscope and a cell with a bottom window. The use of such a cell and polarized vertical illumination provided by the microscope overcame many of the problems associated with the earlier cells, particularly the problems of viewing and illumination. These cells did not make provision for high purity environment, however.

Damjanovic et al. (74) took this problem into consideration. Their cell provided for magnification up to 600 × while maintaining a closed system. The material of construction was Teflon. A cylindrical Teflon body piece had a cylindrical hole placed vertically along its axis. A cylindrical-shaped copper single crystal fitted this cavity, flush with the upper surface of the body cylinder (Fig. 18). Contact to the copper crystal was by means of a screw threaded into the side of the Teflon body cylinder. Two addition holes, low on the sides of the body cylinder, and opposite to one another, were supplied as entrance and exit ports for the electrolyte. At the inner end of these bores, rectangular shafts connected with the upper surface of the Teflon body cylinder. These were at a distance of about 10 mm from the central bore. The electrolyte was forced through the inlet port and up its vertical shaft to the top surface, where it was channeled over the crystal surface to the other shaft, and then escaped through the outlet bore. A thin (0.15–0.2 mm) Teflon sheet covered the entire upper surface of the Teflon cylinder, except for a space passing between the rectangular shafts. On top of this was placed the counterelectrode, a thin copper sheet containing a very small hole (< 1-mm diameter) through which the copper crystal could be observed. A tightly fitting Teflon cap was placed over this assembly, its lid fitting closely to the Teflon body cylinder. This cap also contained a small central hole for viewing and provision for connection to the counterelectrode. Provision was also made for clamping the assembly tightly together. This assembly was intended for an upright microscope. There was no provision for a reference electrode.

Kruger (172) constructed an all glass cell to study the passivation film formed on iron single crystals. Provision was made to evacuate

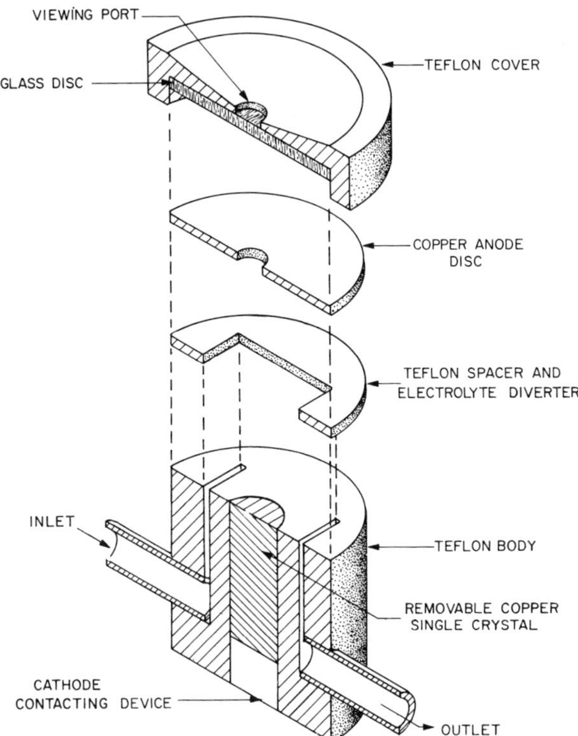

Fig. 18. Simplified cross-sectional view of the cell constructed by Damjanovic et al.

the system, and to introduce purified electrolyte. Polarized light and ellipsometry were used to measure the film thickness, and its optical properties, during the forming process. At the same time, the usual electrochemical measurements were being made. The optical measurements were made through the cylindrical walls of the cell, on the face of the iron single crystal. The crystal was suspended by a wire that also acted as an electrical connection. Manipulation of the crystal was by means of magnets operating on the outside of the cell. Great care was used to insure cleanliness. No greased joints were used, the specimen was sealed in for each run, and the crystal was annealed in hydrogen at 500°C before the system was evacuated.

Barak (19) developed a cell for the study of the $PbSO_4/PbO_2$ electrode. Unfortunately, this cell was not very well described. The electrode was contained in a cell made of Perspex with an optical window made as flat as possible. The counterelectrode was a flat plate at some distance behind the electrode under observation. A reference electrode made contact with the observed electrode, by means of a fine Luggin capillary. A mercury sulfate reference electrode was used. In investigations where gas measurements were involved, a different cell was used. In this case, the counterelectrode was in a side arm to prevent diffusion of gas from the counterelectrode to the reaction cell. This cell could be hermetically sealed.

Ogburn et al. (221) did not construct a microcell for examining dendritic growth, but made most of their observations after completion of the growth process. Their technique for *in situ* examination was nevertheless interesting. The cell consisted of a 100-ml beaker covered with a rubber stopper. The lead anode was introduced through a hole at one side of this stopper. A 4-mm-diameter glass tube, reaching to within a centimeter of the bottom of the beaker, was inserted at the other side of the stopper. The cathode was inserted into this tube, extending almost to the bottom. Dendritic growth was thus confined within the tube, and only one or two dendrites were able to grow under this condition. A telescope was mounted at the side of the beaker so that growth rate in the tube could be observed and measured.

O'Brien (218) described the construction of a cell, giving Fizeau type fringes, that was suitable for observing concentration changes at working electrodes. The cell construction consisted of two outside clamping plates, a center Teflon ring which supported and separated two glass optical flats, and two semicircles of thin copper, separated by a small gap at their major chord and contained within the space left between the glass plates. The two outer clamping plates were 89 mm square and 19 mm thick. They were suitably drilled and plugged so that the interconnecting holes allowed the circulation of a thermostated liquid. A cental aperture, 2×4 cm, was cut in each plate to allow passage of light through the cell. Four milled screws at the corners of the two plates allowed adjustment of the wedge angle between the glass flats. The flexibility of the Teflon ring allowed this movement, since the ring had already been cut to approximately the correct wedge angle.

R. Piontelli and co-workers (231) made a very extensive study of the nucleation and growth of gas bubbles in electrochemical cells. They used small electrodes arranged horizontally and vertically and also studied the formation of gas bubbles on wire electrodes. They used a cylindrical cell, thermostated in a glass outer chamber.

The high-speed motion picture camera used to record the action was equipped with a magnifying lens. During photography, the electrode was illuminated from the side and rear.

Powers (235) designed an elaborate cell for maintaining high purity in the system while at the same time permitting microscopic observation. This cell is too complex to be described fully in this article. Reference should be made to the original article to obtain fully dimensioned working drawings. The cell was built of Teflon with openings for viewing, thermocouple, counterelectrode, gas venting, and reference electrode. Provision was also made for a gas-driven lift pump to induce flow across the specimen surface. The author lists the following goals in the design of the cell and he appears to have accomplished them: (1) to be able to view the entire surface undergoing electrochemical reaction while maintaining good resolution, contrast, and avoiding vibration; (2) to satisfactorily measure the current-overvoltage relationships that take place; (3) to produce a cell that is easy to clean, quick to assemble, and easy to use; (4) to have control of the condition of purity; (5) to have temperature control; (6) to provide for stirring; and (7) to be reliable. Powers' cell was designed for an upright microscope.

A similar cell, designed for an inverted microscope, was described by Cammarota (51) and is perhaps somewhat simpler to build. With the possible exception of its all glass contruction, it should perform as reliably as the cell designed by Powers. Almost all of the essential parts are sealed into a large ground glass stopper which seals the reaction chamber. The following parts are fed through the stopper and extend to the bottom of the reaction vessel, directly above the viewing port: the connection to the electrode under observation, and the device which holds the electrode; a Luggin capillary; a gas inlet, for inert or reactive gas; and the counterelectrode. The electrode under observation is surrounded by the others mentioned. The glass stopper tightly seals the reaction vessel. A thermostated outer chamber surrounds the reaction vessel and a sealed window at the bottom of the cell permits observation, and illumination, by an inverted microscope. Cammarota used this cell to study the electrochemical surface phenomena of zinc, and to observe its reactivity in various etchants.

Tajima and Ogata (292) described a cell for the study of metal dendritic growth that was extremely simple. The cell was contained in a glass vial, 100 mm tall and 30 mm in diameter. The bottom was made optically flat. The cathode consisted of a rod 5.5 mm in diameter embedded in a cylinder of Araldite resin 18 mm in diameter and about 80 mm in length. This was suspended near the bottom of the cell, and the dendrites were observed as they grew out from the end surface of the rod.

Diggle et al. (83) prepared a cylindrical cell, double walled so that it could be thermostated, and having an optical window in the side wall. Spherical electrodes (0.5-mm radii) were used for anode, cathode, and reference electrodes. To make these electrodes, short lengths of 0.2-mm diameter wire were sealed into suitable bore glass tubing. The exposed end was then slowly melted in a torch flame until a sphere formed and contacted the glass tubing. The three electrodes were of the same length and arranged in a row in front of the optically flat window. At their upper end they passed through a stopper. Light for observation was directed through the side of the cell onto the electrode. Magnification up to about 800 × was possible.

Landolt and co-workers (179) built a cell with provision for high current density and high flow rate that allowed electrochemical measurements to be taken of the dissolution process under hydrodynamic conditions typical of those used for electrochemical machining (ECM). The cell had plane glass sides so that the flow of electrolyte and dissolution at the electrode could be observed. In a later paper (180), this same cell was used to measure the mechanism of hydrogen evolution under high rate electrolysis.

Jenkins (141) made up a simple microscopic cell for the study of electrochemical processes at the surface of a copper single crystal. The cylindrical copper body shell also served as a counterelectrode. The bottom of the reaction chamber consisted of the copper single crystal pressed into place against a "Vito" O-ring seal. The glass cover was also pressed into place by a similar O-ring seal. Solution was introduced by means of a small port in the side of the shell, normally closed by a Teflon screw. In some cases, this port was closed by a Teflon screw containing a copper wire sealed through its length to form a reference electrode. The system was observed through the viewing port at the top, and electrochemical reaction at the electrode surface could be thus observed.

4. *Electrodeposition and Electrocrystallization*

A great amount of microscopic study has been devoted to electrodeposition, whereas there has been comparatively little study of the electrocrystallization of isolated metal crystals. The electroplater is primarily interested in smooth, continuous deposits of metals and is concerned only with the avoidance of isolated, dendritic crystals. Since the advent of powder metallurgy, there has been increased interest in electrochemically produced powders, and the need for controlling the dendritic growth in certain electrochemical power sources has also generated interest in metal electrocrystallization. However, the primary interest has been in the theoretical aspects.

Early work with the microscope in this field (94,118,167,169) was largely confined to the observation of dendritic growth, without any serious effort to obtain quantitative data. Microscopy therefore played a most important part in these early observations but the results were of somewhat doubtful value because of this lack of quantitative data.

Wranglen (323) more or less summarized the important findings of the previous work, as well as his own, in the abstract of his paper. He found that there was a simple relation between dendritic growth and the space lattice. The planes resulting from the growth layers were found to be the closest packed lattice planes, and were: for Pb, Ag, and Cu, (111) and (100); for Cd, (001) and (100); for Sn, (101) and (100). Inhibition by various substances increased the thickness and number of growth layers but decreased the rate of growth.

Burton et al. (47) had treated the problem of crystal growth and had shown that growth could not be expected to occur at low supersaturation except in the presence of lattice imperfections, particularly screw dislocations. Examples of spiral growth observations have been given by Steinberg (287), who worked with titanium; by Kaishew et al. (151), working with silver; and by Seiter and Fischer (264,265), working with copper.

Economou et al. (88) studied the structure of copper electrodeposits using microscopy and transient dc. They observed the growth of pyramidal structures at low current densities and the formation of lines which they considered to be lines of dislocations.

Barton and Bockris (23) made a microscopic study of the electrochemical growth of dendrites from ionic solutions, and recorded the dendritic velocities and the dendritic morphology. Dendritic growth took place under potentiostatic conditions. After growth, the dendrites were collected, suitably mounted, and then polished and etched sections were examined for crystal twinning.

Faust and John (98) reported the finding of twin structures in silver dendrites. These dendrites were structures grown by electrodeposition from molten LiCl-KCl mixtures (325). This report added to the growing evidence that electrochemically grown dendrites were not necessarily single crystals.

The growth hillocks that form during the growth of copper single crystals by electrodeposition were of interest to Shanefield and Lighty (272). Considerable attention had been given to growth spirals in connection with dislocation theories of crystal growth (265), and it had also been suggested that these small steps could coalesce to form the growth hillocks seen on electrodeposition (196,228). Imperfections or impurities might be expected to impede the small steps and thus cause the bunching that leads to growth hillocks. Shanefield and Lighty tested this possibility by attempting to increase the purity of the plating bath by preelectrolysis, or by oxidation, as well as by deliberately adding impurities to determine their effect. They found that the concentration of growth hillocks was significantly decreased when the solution had been pretreated by oxidation or preelectrolysis. Since this had the opposite effect to that obtained by the deliberate addition of impurities, it was tentatively concluded that the growth hillocks were indeed caused by impurities.

Ogburn et al. (221) made studies of electrodeposited lead dendrites, mounting the crystals, and making microscopic examination. Contrary to previous reports they found no evidence of twinning, but did find a dendritic structure that was essentially that of two crystals having a twist relationship around a common [111] pole, which was normal to the flat surface of the dendrites. In cross section these bicrystals were found to have many small voids and channels in their interiors.

Shams El Din and Wranglen (271) studied the morphology of silver dendrites grown from a molten bath of either nitrate or chloride. The dendrites were found to grow in the [110] and [112] directions. Addition of $NaNO_3$ had very little effect on the structure of crystals growing in the nitrate melt. When the temperature was increased sufficiently, the growth direction of the dendritic stem changed from [110] to [112]. The faces of the crystals, however, were invariably (111). The growth layers on these faces were found to develop in a manner similar to that which occurred in aqueous electrolytic cells.

Damjanovic et al. (75) studied the propagation of macrosteps and the pyramidal type of growth that occurs on copper, using microscopy and Normarski interferometry *in situ*. They concluded that their results indicated the theory that trace impurities determine the kinetics and mechanism of electrochemical crystallization; that the origin of the microsteps is in the misorientations of the substrate; and that only a part of the electrodeposition takes place at macrosteps.

Tajima and Ogata (292) studied the dendrites of Ni, Co, and Fe deposited in hot aqueous solutions. Since Fe, Ni, and Co are metals with small exchange current density and high overvoltage during deposition, they usually produce compact and smooth electrodeposits, and only very small dendrites had been made prior to this work. Tajima and Ogata succeeded in preparing macrodendrites of all three, although this paper referred mostly to the results obtained with nickel. Their calculations indicated that the current density at the growth site of nickel amounted to about 2400 A/cm^2. This should be differentiated from the average cathode current density which was varied from 0.5 to 5.0 A/cm^2 during the different experiments.

Kilner and Plumtree (159) studied the formation of growth twins in electrodeposited silver. The importance of growth twins is that their boundary provides re-entrant grooves or nucleation sites. These sites become self-perpetuating when two or more twin planes are present in a dendrite, provided that the material crystallized in the body-centered cubic or face-centered cubic systems. Careful microscopic examination indicated that the growth twins formed out of a faulting sequence during deposition, rather than from epitaxial growth of the substrate.

Diggle et al. (83) studied the mechanism of the electrocrystallization that produced dendritic zinc crystals. They pointed out the relationship between electrodeposition from the melt and electrocrystallization from solution. The growth rate was measured as a function of overpotential, concentration, and temperature.

Jenkins (141) measured overpotential transients on copper single crystal faces at the various orientations of (100), (110), (111), and (321) while they were undergoing dissolution or deposition under constant current. Photomicrographs taken *in situ* and goniometric observations were used to determine rates of facet development and their orientations. He could find no evidence that defect structure played a significant role in crystal growth or dissolution.

Continuing their studies with macrodendrites, Tajima and

Ogata (293) gave an account of their further experiments with the macrodendrites of Ni and Co. Using the same type of *in situ* microscopic observation of dendritic growth employed previously, they obtained additional information concerning growth. When growth took place in a flowing stream of electrolyte, it was found that the dendrites grew in the same direction as the current rather than against it.

Mansfeld and Gilman (196) have studied the dissolution and deposition characteristics of a zinc single crystal. These studies were directed toward control of the deposition and corrosion of zinc as used in secondary electrochemical power sources. The use of Pb and Sn had previously been proposed as additives (34,223). They made their observations in a cell similar to that described by Powers (235). The samples were then removed and studied with the scanning electron microscope. It was concluded that the lead ions changed the deposition characteristics from dendrites to protrusions which consisted of many smaller crystals. The lead also transferred deposition to the more macroscopic physical defects of the crystal.

In a later paper (197) Mansfeld and Gilman tested the effect of tin and tetraethyl ammonium ions on the characteristics of the zinc deposit. The effect of Sn was found similar to that of Pb but the deposit formed was less compact. The effect of the tetraethylammonium ions was less than that of Sn or Pb. These ions were not able to completely suppress the dendritic growth of zinc, but produced dendrites that were much finer, with shorter stems and side branches.

Using *in situ* microscopy and scanning electron microscopy (SEM), a further study by these authors (198) was made of zinc in order to obtain greater details of the mechanism. It is to be noted that they never observed dendritic growth to start at the top of a growth pyramid, as would be the case in the mechanism proposed by Diggle et al. (83).

Liaw and Faust (188) reported on studies made of the electrodeposition of lead dendrites, after examining the process by *in situ* microscopy and x-ray techniques. It was found that both Ag and Cu could retard the start of dendritic growth but that only Ag could do this once the process had started.

The foregoing discussion in this section has been kept as much as possible confined to the growth or deposition of single crystals, thus justifying the term electrocrystallization. In actuality, however, the separation is artificial since all electrochemical crystallizations are in fact deposits. However, the foregoing has sufficed to show the type of studies that can be made of the actual crystallization processes, using *in situ* microscopy. In the actual deposition of dense deposits, electrodeposition, or plating, the interest is more confined to the characteristics of the total deposited layer than to its individual crystals. Here the use of microscopy has been very extensive, usually in the metallographic study of cross sections through film and base metal. The purposes of such investigations are to examine such characteristics as bonding between base and electrodeposit, to determine thickness and uniformity of the film, to detect possible diffusion between base and deposit, and to study the orientating influence of the base on the electrodeposit. Since the number of such studies is large and covers a variety of different deposited metals, the references are too extensive to be even partially covered here. Therefore, only a few examples are discussed. For those with greater interest it is suggested that the numerous journals devoted to electroplating should be consulted.

In the plating of chromium a study was undertaken to discover the factors that influence the initial deposition of chromium on nickel surfaces, and the subsequent growth pattern. For this purpose, Jones and Saiddington (147) constructed a microcell in which they could observe the actual initiation and development of deposition. The sulfate concentration, and the condition of the base metal, were the two factors that most influenced deposition.

Becker (27) made microscopic observation of the plating and deplating of silver. He studied the effect of cathode pretreatment in nitric acid or in sulfuric-chromate cleaning solution, the effect of flame heating, and of anodic treatment in nitric acid solutions.

The electrodeposition proceeded by the nucleation of the deposited metal at various sites on the substrate. The average current density corresponding to the total cathode area was obviously concentrated at a much higher value at the growth sites, while approaching zero over most of the surface.

Discrepancy between the coulombs of electricity used in plating out the silver at a platinum cathode and the coulombs obtained in its deplating was found by microscopic examination to be the result of loosening of the deposit by undercutting, with resultant loss of contact of a portion of the deposit during deplating.

Luner and Murray (194) were interested in the alloy ($FeSn_2$) formed when the electrodeposited tin is melted on a steel substrate (tinplate). These authors developed a method for stripping relatively large areas of the alloy from the tinplate, and then made optical and electrochemical measurement of the isolated alloy.

After detinning, the specimen was mounted in a Lucite block, with one alloy side bonded to the mount. Since alloy was on both sides, the exposed side was ground down to the steel. The specimen was then placed in a solution of 75% by volume of saturated copper sulfate and 25% by volume of concentrated nitric acid. When the specimen was immersed in this solution, copper deposited on the steel. Additional concentrated nitric acid was then added slowly enough to initiate dissolution of the steel but to prevent violent reaction. The completion of the reaction, the dissolution of all the steel, was indicated by the cessation of copper deposition. The specimen was then immediately removed and washed in distilled water. The alloy remained, still intact, bonded to the mount, and with a transparent backing of Lucite.

Microscopic examination showed highly porous areas in the layer, although analysis had indicated an average thickness of 1500 Å. The highly porous areas were found to correspond to substrate grains in the steel with a one-to-one correspondence. While this paper dealt more with the product of interaction between electrodeposit and substrate than with the electrodeposition, it is nevertheless an example of how microscopy can aid in investigation if sufficient ingenuity is applied to sample preparation.

The phenomenon of cracking, in connection with chromium plating, has been mentioned previously. Hardesty (123) studied the effect of surface preparation in the avoidance of this problem. The effect of the plated area, the effect of buffing of the substrate, grinding of the substrate, and electropolishing of the substrate were studied. The cracking was found to occur by polishing-induced residual stresses in a surface layer of the nickel substrate, which were subsequently transferred to the chromium electrodeposit. Electropolishing of the nickel to remove this stressed area eliminated this cracking. It was also found that chromium deposits on electropolished nickel were strongly epitaxial, a characteristic not noticeable when the highly stressed nickel was used.

The leveling that occurs in bright nickel plating was studied by the use of a radioactive addition agent. Beacom and Riley (26) studied the addition of sodium allyl sulfonate (SAS) using C^{14} and S^{35} labeled isotopes. The work demonstrated that more radioactivity was found in the electrodeposit next to peaks than in valleys, agreeing with the general theory that the thinner diffusion layer above peaks facilitated the diffusion of addition agents to those points. Bright nickel plating uses two or more addition agents, so Doty and Riley (85) repeated the above work with SAS in combination with *N*-allyl quinaldinium bromide (NAQ). C^{14} allyl or C^{14} quinaldine labeled *N*-allyl quinaldinium bromide samples were prepared and added to the bright nickel bath. Autoradiographs showed greater darkening next to peaks as previously. Since the darkening was the same with either marker, it was concluded that the whole NAQ ion is preferentially incorporated in the deposit at the peaks. The amount of NAQ that was incorporated was found to depend on the mass transfer processes, the diffusion layer thickness, and the catalytic state of the nickel deposit.

It is well known that, in the absence of additives, a bath having a relatively uniform macrocurrent distribution has a nonuniform microcurrent distribution. It was shown by Meyer (209) and Gardam (108) that uniform plating occurred in pores of the base metal when an acid electrolyte was used, even though the macrocurrent distribution was poor. Conversely, in a cyanide bath, the plating in microgrooves was poor, although the macrocurrent distribution was good. Somewhat arbitrary limits are set on micro and macro profiles. The linear dimension of a microprofile on a surface is about 50 μm, according to Cheh (62), whereas that of a macroprofile is above 1000 μm.

Cheh used two types of electrodes to study the current distribution during the electrodeposition of gold. One electrode was electroformed nickel with a wavy surface, having a wave amplitude of about 25 μm. The other electrode was polished copper with grooves 2.5 mm apart cut into the surface, with groove angles of 30°, 60°, and 90°, and groove depths of 50, 125, and 250 μm.

The plates were sectioned and polished and the distribution of gold determined from the thickness of the deposit in the various areas. It was found that the microcurrent distribution was relatively uniform on the undulating electroformed nickel. The microcurrent distribution was much less uniform on the copper, the deposit becoming less uniform at increasing V-angles and with increasing current densities. The deposit also became thinner with increasing groove depth. It was concluded that to improve microcurrent deposition it was necessary to have smooth surfaces and low current density, or else to raise the gold content of the bath.

Hardesty (122) became interested in the anion effect in copper deposition. He prepared sulfate, nitrate, chloride, and acetate solutions of copper and prepared electrodeposits from these solutions. The surfaces of the deposits were examined by microscopy and the facet angles were measured with a microgoniometer. Interference microscopy was also used to study the surface contours of those facets that were parallel to the surface. It was found that in solutions of nominally comparable compositions the copper deposits showed radically different surface morphology or growth habit, depending upon the ion that was present. The results were considered as evidence that the adsorption of the anion on the cathode was a factor in determining the growth habit.

Dhananjayan (81) used microscopy and x-ray analysis to study the mechanism of electrodeposition for manganese. The microscopy was confined to examination of cross sections through the deposit. He found that a thin layer of the amorphous manganese was first deposited from pure solutions, on which gamma manganese began to build up. In the presence of sulfur dioxide the amorphous layer became thicker, and eventually alpha manganese was obtained.

A microscopic study of the formation of cathodic films on iron during the electrolysis of chromium plating solutions was made by Saiddington and Hoey (258). The study was made *in situ* with various $CrO_3:SO_4$ ratios, in order to determine the type of film formed and the function of the sulfate. One school of thought maintains that pure chromic acid cannot be reduced to metal because a nonporous oxide film forms, which prevents all reduction processes, except that of the hydrogen ion. It is believed by this school that sulfate ion either penetrates and disrupts this film; causes its partial solution; or else modifies it so that reduction processes can take place. The second school does not believe that the same type of oxide film is originally formed in the presence of sulfate. In this case, the film is considered to be colloidal instead of amorphous. The above authors observed the formation of an amorphous film in both sulfate-containing and sulfate free plating baths. This film, from the sulfate-containing baths, disintegrated when plating began so that it could not be considered an intermediate step in the process of deposition. They also observed the formation of a second, transparent, viscous layer during chromium deposition in sulfate-containing solution that was not formed in sulfate-free electrolyte. They assumed that this film aided in the final reduction step of the chromate to metal, as had been proposed by the second school of thought.

One area where every effort is made to produce rough plating is in the preparation of high-surface-area aluminum capacitor electrodes. The surface area of these aluminum capacitors has been usually increased by etching, either chemical or electrochemical. Muhlhausser (211), however, has reported a means of electrodeposition of aluminum which yields large grains. Roethlein (250) used microscopy to study this process. This author attempted to find the parameters that would produce roughness suitable for high capacitance values. He observed that the largest capacitances were observed for aluminum electrodeposits obtained at high current densities. Initially a bright deposit would form to be replaced almost immediately by very fine, black powder. After a few minutes of continued electroplating the deposits would become metallic again. The highest capacities were from a combination of very fine, black aluminum particles overgrown by a porous dendritic deposit. Deposits grown at lower current densities to avoid this formation of black aluminum particles had much less capacity.

No discussion of electrodeposition and the use of microscopy in its investigation could be considered completed without mention of the extensive work at the Bureau of Standards on chromium (39), nickel (330), and copper (178).

5. Semiconductors

In many ways the microscopic study of semiconductors resembles that of metals and may be considered as direct metallography. In other ways the unique properties of semiconductors have initiated microscopic studies of a different nature.

Shaw and Busen (273) studied the defect found in epitaxial silicon layers, using optical microscopy and microprobe analysis. The most prominent defect, so-called hillocks, were generally believed to be due to impurities, such as α-SiC or SiO_2, on the surface of the substrate. Hillocks examined in cross section did not terminate at the surface of the substrate, as previously reported, but entered the substrate layer. Copper and chromium (Sirtl etch) decorations showed large concentrations of what appeared to be dislocations surrounding the hillocks. It was found with electron microscopy that the hillocks were polycrystalline. Microprobe analysis showed that high concentrations of impurities were present in both hillocks and pockets.

Silverman (277) has listed a number of etches suitable for delineating *p-n* junctions and impurity striations in GaSb, which are useful in microscopic examination. Pugh and Samuels (241) investigated the effect of etching abraded germanium surfaces. It had been established that abraded germanium surfaces contained numerous cracks. By using tapered sections, it was possible to show that the etch reagent penetrated quickly into the surface cracks. This caused widening of the cracks and rounding of their tips. It was concluded that the microscopic method was best for estimating the depth of damage.

Faust, Sagar, and John (99) have proposed molten metal etches suitable for the orientation of semiconductor devices by optical goniometry. Surfaces of semiconductor wafers were lapped to remove saw marks and surface damage. The surfaces were etched by placing them in contact with a metal foil or pellet and heating to the desired temperature, in a vacuum. The metal could also be coated on the surface with an ultrasonic soldering iron. Under controlled conditions, it was found that the dissolution and redeposition of semiconductors proceeded along definite crystallographic planes. These planes were found to be (111) for Ge and Si and both (111) and (100) for the III–IV groups of intermetallic compounds.

Meltzer (206) discussed methods of determining minute resistivity variations in germanium crystals by means of selective copper plating or micropotential scanning. In the selective copper plating method, the samples to be plated are covered on the back with solder and cast in an epoxy resin so that only the desired face is exposed. After etching, this face is plated as cathode in a copper sulfate solution. Both *n*- and *p*-type germanium, in from 0.03 to 40 ohm-cm resistivity ranges, were satisfactorily plated. The copper appears as striations representing resistivity variations. The micropotential probe consists of a two-point probe of minute dimensions, arranged in a jig, and drawn across the surface by a motor driven micromanipulator. A dc voltage is applied across the sample to be scanned. The sensitivity is limited by the spacing between the probes. Results of his work showed that the horizontal zone-leveled crystals had a much larger resistivity variation than did vertically grown crystals. It was also found that *p*-type crystals had lower resistivity variation than *n*-type.

Chu and Gavaler (63) investigated stacking faults in vapor-grown silicon on silicon substrates, making use of microscopy, chemical etching, and cross sectioning. They investigated the stacking faults on substrates at orientations of (111), (110), (100), and (211). The etch figures were found to be equilateral triangles on (111), isosceles triangles on (110), squares on (100), and isosceles on (211). The stacking faults in the majority of cases originated in the substrate-deposit interface.

Knopp (161) used microscopy to compare thermal and chemical etching of silicon, and to show the similarity of the two processes. Since thermal etching is due to the formation and subsequent evaporation of volatile SiO, thermal etching in sealed quartz tubes can be carefully controlled and made to yield clean surfaces.

Kyser and Millea (176) studied the etching of gallium arsenide with nitric acid, using x-ray analysis and microscopy. A study was made of the earlier reported formation of etch figures on GaAs after prolonged etching with dilute HNO_3. It was found from this investigation that the etch figures were in reality As_2O_3 crystals which formed preferentially on the (111) planes of Ga and As.

Dermatis et al. (79) studied the morphology, internal structure, and perfection of silicon ribbons pulled from a supercooled melt. To reveal growth history, ribbons were pulled from melts highly doped with boron and aluminum. The resultant microsegregation striations were examined by suitable etching of faces, potted and etched cross sections, and by fracturing. It was found that the basic growth sequence was similar to that of germanium.

Tausch and Lapierre (295) observed an overgrowth phenomenon while observing the epitaxial growth of GaAs in the windows of a SiO_2-masked GaAs substrate. Single crystals were observed to overgrow onto the amorphous SiO_2 mask. The overgrowth was a single crystal. It was determined that the mask of SiO_2 was still in good condition beneath the overgrowth.

Shaw (274) on the other hand, reported the selective epitaxial deposit of gallium arsenide in holes etched into gallium arsenide substrates, using the same SiO_2 masking films, and did not observe GaAs overgrowths.

The holes were etched into a (100) substrate. To examine the deposits, the slices were cleaved along the (110) planes to obtain cross sections. From these, it was possible to determine that side facets were formed at an interfacial angle of 55° to the upper surface, indicating slow growing (111) faces. These well defined side facets tended to prevent extensive overgrowth. However, epitaxial growth on accurately aligned (111) substrates tended to grow laterally over the SiO_2 mask, rather than having significant height above the mask.

Tucker (302) made a topographic study of silicon web crystals, using profilometer techniques, interference contrast, and interferometer methods. Silicon web crystals are grown by withdrawing parallel, spaced dendrites from a molten bath. Enumerating the advantages of such crystals, Tucker claimed virgin surfaces, free from contamination; substantial regions of extreme flatness (flat within several hundred angstroms over 12-mm widths); a usual roughness less than 100 Å, and that longitudinal surface deviations affected only small portions of the surface.

Primak et al. (240) used a peroxide etching technique to determine the depth of electrical conductivity change produced by ionic bombardment with H^+, Ne^+, and A^+ in the 100-keV range. Since weight loss determinations could not be made, it was necessary to know the etching rate of normal and ion-bombarded germanium. Etching experiments were made on mechanically polished flat plates so that interferometric observations could be made. One half of the surface was subjected to ion bombardment and the whole surface was etched. Positions of interference fringes indicated relative depths of etching.

Seltzer et al. (268) arranged an apparatus so that direct microscopic examination could be made of the movement of growth steps during the chemical vapor deposition of epitaxial layers of germanium on a germanium substrate. The design of the equipment is described elsewhere (267). Measurement of growth steps was made during growth. The topography was measured with interference microscopy and the hillock heights determined. Rate of step movement was determined as a function of gas composition, carrier gas flow rate, the pressure of the GeI_2, step height, and seed temperature.

Although the behavior of small-scale surface perturbations had been studied, the behavior of rather large surface perturbations during growth from the vapor phase might be different. Realizing this, Runyan et al. (256) made a study of larger scale surface perturbations by masking (111) orientated single crystals of silicon and etching holes approximately 125 μm in diameter and 25 μm deep. It was found that, at high concentrations of the silicon-bearing species in the gas stream, surface stability (the filling of holes and the leveling of hills) occurred.

An optical method and apparatus for determining haze and other defects in silicon surfaces, was described by Adley and Gorey (2). While not a microscopic method, it nevertheless made use of optical principles. The basic principle depended on the fact that when a collimated beam is normal to a surface, light reflected from off-normal surfaces represents surface defects. Tests showed that this method of inspection was more reliable and faster than previous microscopic inspection.

Gallium arsenide shows crystallographic polarity along [111] directions. This is because the outermost atom layer in each (111) face may consist of either Ga or As atoms, triply bonded to the lattice beneath. By convention, the Ga layer is designated as A(111) and the As layer is designated as B(111). It has been found that these two faces are physically and chemically different. Akasaki and Kobayasi (3) showed by etch pit orientations and light figures how these two types of (111) faces could be distinguished either on abraded or etched surfaces.

From this sampling of papers it can be seen that the use of microscopy in semiconductor work has been mostly applied to a study of surface defects in epitaxial layers, studies of growth kinetics, and similar relationships. Of similar interest is the development of the many highly specialized etches and stains that have been developed to make visible defects, *p*- and *n*-type structures, and segregation. Strictly speaking, these are of essential importance to microscopic techniques but unfortunately cannot be covered in the space of this review.

6. *Electrochemical Power Sources*

The microscopic work on various types of batteries falls into three main categories. First, there is the work that has been done in microcells, *in situ* studies of single crystal growth of substances like Zn, Ag, Pb, etc. Second, there is the work that has been done by impregnating actual samples of battery material, cross sectioning, polishing, and finally examining samples taken at various stages of operation, or after various pretreatments. Finally, there is the metallographic work that has been done in studying grid corrosion and grid alloy structure and in the determination of equilibrium conditions for various grid alloys.

An attempt to support battery-active material so that it could be sectioned and polished was first made by Buckle and Hanemann (45), who used carnuba wax to impregnate battery plates for microscopic examination. Simon and Jones (278) improved on this method with the use of vacuum impregnation of a catalyzed liquid resin into the pores of the active material structure. The Coil Seal epoxy formulation proposed in their original article was later replaced by a much less viscous polyester resin (116), which was able to give much better impregnation and was less reactive with the oxides of the active material. Simon and Jones used this method for observation of the changes in microstructure that took place in lead-acid battery plates during formation. They examined a series of samples removed from formation after various lengths of time, or formed under different conditions, so that the nature of the process and the effect of temperature, acid specific gravity, and current density could be studied. They observed that the lead dioxide formation appeared to nucleate on some basic sulfate crystal already present in the unformed active material and to then spread more slowly to surrounding areas (278).

Briggs and Wynne-Jones (40) studied the effects of aging on freshly prepared nickel hydroxide electrodes, using x-ray and microscope observation. On the aged $Ni(OH)_2$ electrode, oxidation began with the formation of a sharp boundary separating the oxidized and unoxidized part of the layer. The freshly prepared $Ni(OH)_2$ electrode showed a more complex reaction.

Stachurski et al. (286) designed and constructed a small cell wherein the growth of zinc deposits could be studied, especially in relation to their tendency to penetrate the cell separators. The electrodes were made of sheet zinc. All of the cathode except a small area in front of the anode was covered by an epoxy film. Provision was made to cover the cathode with separator material. The cell case was transparent and was made of plexiglass with polyethylene washers. A microscope cover glass was placed in one wall of the cell, by means of which microscopic observations could be made across the separator during the formation of dendrites. Although magnification was only $100\times$, a very good picture was obtained of the penetration process. It was observed that the growing dendrite actually grew through the separator pores rather than puncturing the separator mechanically.

Simon (282) made a microscopic study of the positive active material of the lead-acid battery in an effort to determine the factors involved in active material retention. A major problem was that, during prolonged periods of float potential (battery held at a fixed potential but below gassing), it was found that the positive active material softened and disintegrated, leading to major loss in capacity.

Simon (283) also studied the lead-acid battery *in situ,* using a small microcell. With this arrangement he studied the precipitation reaction that causes the formation of $PbSO_4$ upon discharge of a PbO_2 electrode. The theory had been advanced that the voltage drop and subsequent recovery that occurs at the beginning of discharge of the PbO_2 electrode is caused by an initial supersaturation. This prevents the formation of lead sulfate for an appreciable period so that equilibrium potentials cannot be obtained. Once this supersaturation is relieved by the nucleation of lead sulfate crystals, the potential would be expected to rise to its equilibrium value. With *in situ* microscopy it was possible to observe delayed nucleation; then, rapid, dendritic growth as nucleation was initiated; and finally, slow, regular growth of $PbSO_4$ crystals as the system once again returned to equilibrium.

Tuomi (304) made a rather detailed study of the forming process in the nickel positive electrode, using microscopy and x-ray diffraction examination. His investigations showed that the discharged state was a nickel hydroxide $Ni(OH)_2$, whereas the charged phase included a trivalent nickelate-beta phase and a tetravalent nickelate-alpha phase.

Tracey and Williams (301) studied the production and properties of porous nickel alkaline battery and fuel cell electrodes made from carbonyl nickel powders, using metallographic techniques to study the types of contact bridging that took place in loose sintering, slurry, roll compacting and pressing. Fundamental differences in battery and fuel cell electrodes dictated different methods of compacting for the two uses, since the battery required 70–90% porosity and not much strength, whereas fuel cell electrodes could be as low as 50% porous, but required greater strength.

Tudor et al. (303) applied autoradiographic techniques to the lead-acid battery in an effort to locate the source of capacity loss in floated Pb-Ca grid batteries. Radioactive S^{35} (as $H_2S^{35}O_4$) was added to the electrolyte of small cells, and the cells were then cycled until isotopic exchange was assured. Plates were removed from the cells in various conditions of charge or discharge. They were washed in isotope-free electrolyte to remove activity, blotted to remove excess liquid, and dried in vacuum at 60°C. After drying, the plates were vacuum impregnated with a low-viscosity epoxy resin. After the resin had hardened, the plates were sectioned and polished. Following this, autoradiographs were made of the sections. These authors found large areas of radioactive S^{35} surrounding the grids of the plates, indicating excessive sulfation in these areas. The presence of this excessively sulfated layer in batteries of low final capacity was taken as evidence of a barrier layer, probably low conductivity lead sulfate, which by its presence between grid and active material served as a barrier to current flow.

In a later paper (303), these same authors used radioactive P^{32} and S^{35} in further investigations of the lead-acid battery. Claims had been made that adding a small amount of phosphoric acid to the sulfuric acid battery electrolyte resulted in less sulfation and grid corrosion. It was also supposed to harden the positive active material and to reduce shedding. Autoradiographs showed that areas of the plate containing large concentrations of phosphate generally contained only small amounts of sulfate. It was found that the phosphate entered the positive plate during charge and left again during the subsequent discharge. It was concluded that phosphate additions were of advantage in maintaining good active material characteristics.

Krumbein and Russel (173) made metallographic cross sections of ion-exchange membrane fuel cells for microscopic examination. Such a fuel cell consists of powdered metal electrodes that are hot pressed on both sides of a wet ion-exchange membrane. The metal electrodes are a mixture of platinum black and a binder such as Teflon. The latter acts as a "wet proofing agent," to ensure that sufficient pore space will be present for gas reaction. The electrode is thus a complex porous structure, and its stability depends upon its water content. The problem of impregnating such a structure was difficult. It was accomplished by removal of the aqueous phase and replacing it with the embedding medium, using techniques similar to those employed in biology. Dehydration was accomplished with a graded series of alcohol-water solutions, followed by room-temperature vacuum impregnation. Grinding and polishing were regular metallographic procedures, except that very low pressures, slow speeds, and fine powders were used. It was noted from the study that the binder in aqueous emulsion formed uniform structures, that some cracking of the metal electrodes occurred, and that there were regions of poor adhesion between electrode and membrane. It was concluded that the method showed promise for this and similar systems where air drying would cause numerous changes in structure.

Bode and Euler (36) used radioactive S^{35} and Se^{75} to study the current distribution in the positive plate of the lead-acid battery. Since the formation of lead sulfate during discharge would result in a distribution of $PbSO_4$ throughout the plate and since this $PbSO_4$ would be radioactive, autoradiographs taken of the surface would indicate whether the $PbSO_4$ formation was uniform throughout the plate. Euler (96) has given a review of some of the uses made of radioisotopes in battery research.

Acton (1) described a method of impregnating plates and making very thin sections of lead dioxide deposits which could be examined by transmitted light. Several different lead dioxide growth habits were noted and polarized light was used in a partially successful effort to distinguish between α and βPbO_2.

Simon (284) used microscopy to investigate the cause of grid failure in the lead-acid battery. His investigation showed that grid corrosion occurred in two stages. The first stage gave a uniform layer of PbO_2, and segregated antimony disappeared from the corrosion product. After a certain thickness of deposit had built up, the corrosion attack began to be preferential. Faster corrosion occurred at eutectic areas between the lead primaries and, in particular, at grain boundaries.

Powers (235) used microscopy to observe the initial growth of protrusions in the electrodeposition of zinc. The unusually well-designed features of the cell he used for *in situ* observations have been mentioned in a previous section. The photomicrographs accompanying the article serve very well to illustrate the effectiveness of this cell in electrochemical investigations.

Pierson (229) used a somewhat unusual technique to study the crystallographic changes occurring in the formation of the positive plate of the lead-acid battery. The upright microscope was equipped with an oil immersion lens, a camera, and a special stage. The stage was made so that the sample could be held rigidly in place. The sample to be studied was impregnated with epoxy resin. After the resin had set, the sample was ground and polished so as to have two parallel faces. After polishing, leads were connected to the sample and counterelectrode. The sample was placed under the objective of the microscope, electrolyte was added to the cell, and the process of formation begun. At intervals the current was turned off, the electrolyte level was dropped below the surface of the specimen, the surface of the specimen was swabbed off and a drop of immersion oil was placed on the area under observation. The progress of the process was photographed, the immersion oil was swabbed off, and the electrolyte level was again raised to cover the surface of the specimen. Progress of the formation process could thus be easily followed.

Keralla and Lander (156) used microscopy and life cycle tests to determine the effect of surfactants on the zinc electrode of the silver oxide-zinc cells.

Pierson (230) made further studies on battery plate forming and curing. He showed by x-ray analysis that the material formed above 150°F contained considerable amounts of tetrabasic lead sulfate. The formation of this tetrabasic material was followed by microscopy. It was concluded that under certain conditions the presence of tetrabasic lead sulfate might be advantageous if the amount could be regulated.

Lange and Smith (182) combined autoradiography and microscopy to study the problems of the lead acid battery. They pointed out the advantages of using S^{35} in such studies, since it is a low energy beta emitter with an 87 day half-life. They maintained the structure and integrity of the active material by wax impregnation, using vacuum. The sections were then microtomed and successive sections were autoradiographed.

Euler (96) used radioactive sulfur-35 to study the negative plate. He found that details of negative plate sulfation were more diffuse and less strong than those found on the positive plate. The autoradiographs were taken from the surfaces of the plate.

Leibssle et al. (186) also used radioactive S^{35} in positive plate studies but furnished a few additional observations beyond those previously made. In plates discharged at low current densities, they found that the $PbSO_4$ concentration and therefore the current distribution was uniform. Plates discharged at high current densities, however, had a lead sulfate concentration at the surface much higher than in the interior.

Wales and Simon (310) describe a microscopic study of the morphological changes in the silver electrode during charge and discharge. They were able to impregnate the electrode with an epoxy resin, cross section and polish, and make metallographic type inspection. They reported that slow discharge resulted in the formation of large granular Ag particles.

Wales (311) has continued this microscopic study and has investigated the crystallographic effects of slow and fast discharge and of recharge upon the sintered silver electrode and of the effect that rate of discharge has upon future cycling (Fig. 19).

Diggle et al. (83) used *in situ* microscopy to study the growth rate of zinc dendrites from alkaline zincate solutions. It was observed that there first occurred a pyramidal growth, originating at separated growth centers, which proceeded under bulk diffusion control.

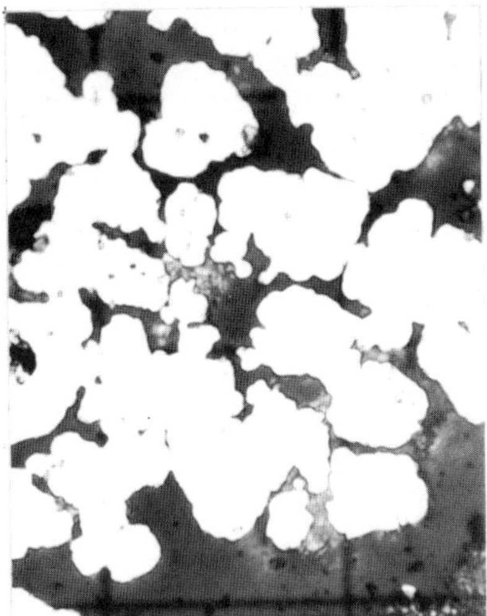

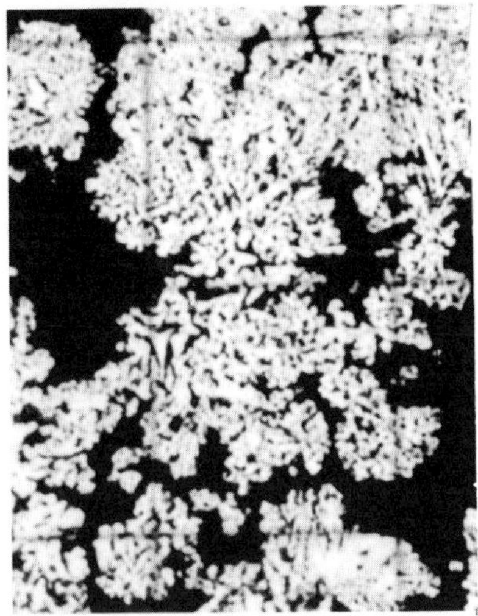

Fig. 19. Photomicrograph showing the effect of different discharge rates on the microstructure of the silver electrode. The appearance of the electrode cross section at the end of the second discharge, following charge at the 20-hr rate and discharge at the 20-hr rate (*a*), and after charge at 20-hr rate and discharge at 1-hr rate (*b*). Magnification 1600 ×. The inability of the large particles in (*a*) to become completely oxidized on charge seriously limits the discharge capacity. (Photomicrographs by courtesy of C. P. Wales).

Kukoz et al. (174) used stereometric metallography to determine the structural characteristics of porous sintered electrodes, such as those used in storage batteries. Specimens of plates were boiled in molten rosin for impregnation. Their surfaces were then leveled by grinding and finally polished for metallographic examination. They determined that the electrode pores and grains were uniformly distributed, so that it was practical to use stereometric procedures. Porosity, as determined by this method, was in very close agreement with that determined by other means. However, the specific surface as calculated from mercury poresimetric data was three times as high as that calculated from the method of random sections. Nevertheless, the difference was a constant factor so that a correction could be easily applied. This method, in the presence of quantitative television microscopy, might well be more applicable and easier than conventional methods of determining porosity and surface.

Mao et al. (199) made a metallographic study of lithium-lead alloys as grid material for lead-acid batteries. The investigations of this type make extensive use of metallography in the study of metal structure, phase relations, grain size, relationships between structure and corrosion, etc. In this sense they are metallurgical. In the ultimate application of the material they are electrochemical.

7. *Miscellaneous*

There are many other areas of electrochemistry where microscopy has been or could be used. Corrosion studies, for instance, have made frequent use of microscopy, usually in the metallographic sense. To do justice to the many papers that have appeared relating to corrosion studies and involving microscopy would require much more space than is available here. A few miscellaneous papers in this and other fields can be mentioned as examples, however.

O'Brien and co-workers (217,218,219,220) have made frequent use of interferometric determinations in their study of the processes of diffusion and the concentration changes at electrodes, usually of copper or zinc. According to O'Brien, refractive index may be related directly to concentration with an error no greater than 0.1% when dealing with simple systems where only one ion is changing. Interferometry can also supply information on processes such as the beginning and extent of convection, processes which indirectly affect the refractive index. Using special cells of their own design, they could make such measurements in only a few minutes, and with the use of motion picture photography they were able to record transient phenomena that occurred over very short intervals of time. A similar study, made by Tvarusko and Watkins (305), used interferometry to accurately measure the diffusion layer thickness on a vertical cathode in $CuSO_4$ solution, at various current densities and various H_2SO_4 concentrations.

Novak et al. (216) used ellipsometric methods to study the surface films formed on copper electrodes while they were undergoing electropolishing. A vertical copper electrode was observed simultaneously by current-potential scans and by *in situ* steady-state ellipsometric measurements. A novel technique was developed for displacing the visible, viscous film with glycerol to determine if a second film was sandwiched between it and the metal surface. They were able to show that two films were formed: (1) an adherent solid film about 40 Å thick next to the metal and (2) a removable viscous layer of highly concentrated electrolyte probably 2000–3500 Å thick.

The work of Froment and co-workers (92,107,106) should also be mentioned, not only because of its continuing interest but because their early studies pioneered in the anodic dissolution of metals, at least where microscopy was concerned. The technique of immersing the objective directly in the electrolyte, while somewhat unconventional, certainly achieved remarkable results and gave much insight into the process of anodic dissolution.

Bartlett and Stephenson (22) also did early work in combining microscopy and electrochemical observations. Using only low magnification, but combining it with motion picture photography, they were able to obtain a fair picture of the method of formation of anodic films on copper and iron in HCl and H_2SO_4, respectively.

Jenkins and co-workers (142,143,144) have made several investigations of the formation of films and dissolution of copper. These were not *in situ* studies, but the films and surfaces were examined by microscopy after the process was completed.

Kruger (172) also employed ellipsometry to measure the progress of film formation in a study of the passivation of iron single crystals in inorganic inhibitor solutions. His studies revealed that a film only 20–30 Å thick, next to the metal surface, was responsible for passivity. The number of sites of film breakdown were found to be highest for films located on $\{110\}$ iron surfaces, and these sites were not located at points of dislocations.

Powers and Breiter (236) and Powers (235,237) and Breiter (38) have made *in situ* microscopic investigation of anodic films on zinc, anodic dissolution of zinc, passivation of zinc, and electrodeposition of zinc. These studies have done much to reveal the mechanisms of such processes. Mansfeld and Gilman (196–198) have made similar *in situ* studies of the dissolution and deposition characteristics of zinc, using almost the same apparatus for their investigation.

Block (33,66) made a different type of study of anodized coatings on aluminum. Microscopy was used to study the manner in which such anodic coatings buckled. Buckling was evidenced by short vertical lines between the horizontal circumferential cracks caused by 2% extension of the substrate. Buckling was caused by reduction in circumference of the substrate as a result of linear extension.

High rate anodic dissolution of copper has been observed under the microscope by Landolt et al. (179,181) in connection with studies of high-speed electrolytic machining of metals. These studies also involved the construction of a cell suitable for high rate electrolyte flow and capable of sustained high current densities.

Powers and Jerabek (238) employed Normarski interferometry in a study of the electropolishing of zinc. Cammarota (51) employed *in situ* microscopy for investigation of the etching of zinc crystals, particularly as concerned the nucleation and growth of etch pits. Damiano et al. (73,300) developed a technique and instrumentation for a cinephotomicrographic study of etching phenomena on zinc single crystals, particularly the method of growth of etch pits.

Bakish et al. (18) made a microscopic study of the type of surfaces produced from the chemical etching of high purity aluminum, with major emphasis of the study being the obtainment of maximum increase of surface area.

A somewhat unusual application of microscopy was made by Biery (32) who determined the surface tension of three Pu-Co-Ce alloys and three Co-Ce alloys at their freezing point, from the curvature of their frozen menisci. Of the 28 menisci studied, 15 were found acceptable for such determination.

It can be seen from the foregoing discussion that, while extensive use of microscopy has been and is being made in electrochemical investigations, very limited use is being made of many of the applicable techniques of microscopy. The majority of investigations using microscopy that have been reviewed either deal with metallographic type studies or else with direct observation of processes as they occur. Neither type of microscopic procedure tends to become quantitative. In electrochemical studies very limited use has been made of polarization, interferometry, ellipsometry, stereometry, or other techniques that are capable of giving quantitative data.

It was noted repeatedly in the preparation of this review that, while authors frequently referred to their use of polarization, interference microscopy, or other quantitative techniques, they almost as frequently failed to give any data derived from such use.

With a few notable exceptions, it appears that most of the investigators making use of microscopy in their investigations are indeed the novices referred to at the beginning of this article and are using the research tools of optical microscopy in a very unsophisticated manner. This would appear to justify the somewhat elementary approach given here, and frequent citation of much more sophisticated texts that has been used in this article, for it was realized at the start that the area to be covered was too vast for other than very elementary coverage.

At the same time the author must hasten to add that the unsophisticated use of microscopy referred to above has in most cases been adequate to obtain the information most desired and has immeasurably added to our knowledge of electrochemical processes, and it therefore becomes debatable in many cases whether much more information of value to the solution of the concerned problems would have been obtained by a full array of more sophisticated microscopic techniques.

In any case, it is encouraging to see that the microscope techniques and instruments available in other fields of science are being progressively added to those available for the electrochemist. It is to be hoped that the period between inception of the method and adoption by the electrochemist will not be as long as has been the case in some other fields.

In conclusion, it must be evident that this article does not represent a complete coverage of the use of microscopy in electrochemistry. Some areas have been covered perhaps in more detail than necessary while other areas have received little more than mention. It is believed, however, that sufficient references have been included to aid considerably in suggesting areas of investigation with the microscope, in obtaining details of techniques and procedures, and in suggesting sources for more detailed information on theory.

IV. Prognostication

The foregoing has demonstrated the need for, and the use of, optical microscopy in electrochemical investigation. Undoubtedly, the use of optical microscopy will continue to increase in this field. Improvement in microscope accessories and in their ease of use can be expected to continue. Automation and the rapid development that has taken place in providing for quantitative methods of microscopic measurement will lead to more extensive use of microscopy for obtaining data that has in the past been considered too expensive and time-consuming to be practically used. Ellipsometry and interference methods may be expected to be used more frequently in electrochemistry as their use becomes better understood and their capabilities better appreciated.

Recent improvements in electron and electron scan microscopes offer interesting possibilities for seeing further into the microscopic and submicroscopic regions. In this type of investigation, the optical microscope will undoubtedly be required to aid in interpretation and identification. It has been shown that the optical microscope has many capabilities not shared by the other instruments and these will ensure a cooperation in the use of optical and electron instruments.

Progress in the theory and practice of holography offers fascinating glimpses of what may be very important future advances in both optical and electron microscopy. Foremost is the intriguing possibility that undistorted, magnified, three-dimensional images may be obtained by reconstructing the wave front obtained from the hologram. Not only will the hologram itself record many times the information of the present photomicrograph but its reconstructed wave front will give better and more easily understood images than are now available.

Perhaps even more important are several other potentials of holography that make it theoretically superior to conventional imaging techniques. First, the theoretically attainable resolution and field size are greater than are possible with conventional methods. Second, it is theoretically possible to free the image of aberrations by the use of a proper arrangement in the recording of the object and in the reconstruction of the wave front. Third, it is possible to obtain magnification by changing the wavelength between the recording of the object and the reconstruction of the wave front, thus dispensing with conventional means of magnification. Fourth, while conventional interferometry is limited to highly polished surfaces of simple shape, holography allows interferometry to be observed on surfaces of various shapes and condition as well as from various prospectives. Finally, from the standpoint of electrochemistry, it is of particular interest that imagery is theoretically possible through diffusing media such as thin films, or the turbulence produced in air or water.

While all of these potential applications are at present limited by many practical difficulties, it is expected that they will become more usable realities in the microscopy of the future.

Acknowledgments

The author wishes to express his appreciation to the Naval Research Laboratory for permission to publish this article and to Mr. C. P. Wales, of the same Laboratory, for his permission to use the photomicrographs of Fig. 19.

References

1. Acton, R. G., *Power Sources*, D. H. Collins, Ed., Pergamon, London, 1967, pp. 133–146.
2. Adley, J. M., and E. F. Gorey, *J. Electrochem. Soc.*, **117**, 971 (1970).
3. Akasaki, I., and H. Kobayasi, *J. Electrochem. Soc.*, **112**, 757 (1965).

4. Allen, R. M., *Photomicrography*, Van Nostrand, New York, 1958.
5. American Association for the Advancement of Science, "1970–71 Guide to Scientific Instruments," *Science*, **169A**, 3951A, Sept. 23, 1970.
6. Amsel, G., and D. Samuel, *J. Phys. Chem. Solids*, **23**, 1717 (1962).
7. Amstutz, G. C., *Am. Mineral.*, **45**, 1114 (1960).
8. Andrews, K. W., and W. Johnson, *The Encyclopedia of Microscopy*, G. L. Clark, Ed., Reinhold, New York, 1961, p. 587.
9. Andrews, K. W., and W. Johnson, *X-ray Microscopy and Microradiography*, Cosslett et al., Eds., Academic Press, New York, 1957, pp. 581–589.
10. *AB Metal Digest*, **15** (2) 22 (1969).
11. *AB Metal Digest*, **10** (1) 3 (1964); **11** (2) 2 (1965).
12. "Reflected Light Interference Microscope," *Tech. Brochure 860-6c*, E. Leitz, Rockleigh, N. J.
13. "Measurement of Surface Relief," *Tech. Inf. Bull.*, **10**, 1, April, 1970, E. Leitz, Rockleigh, N. J.
14. *Tech. Inf. Bull.*, **8**, (3) 1 (1968), E. Leitz, New York, 1970.
15. ASTM Special Technical Publication No. 285, "Methods of Metallographic Specimen Preparation," American Society for Testing Metals, Philadelphia, 1960.
16. ASTM Subcommittee 1 on Specimen Preparation, *Pract. Metallogr.*, **7**, 192, 268, 314 (1970).
17. Auld, J. H., and J. F. McNeill, *X-ray Microscopy and X-ray Microanalysis*, Cosslett et al., Eds., Elsevier, Amsterdam, 1960, p. 5.
18. Bakish, R., E. Z. Borders, and R. Kornhaas, *J. Electrochem. Soc.*, **109**, 791 (1962); *Electrochem. Tech.*, **3**, 116 (1965).
19. Barak, M., *Batteries*, D. H. Collins, Ed., Pergamon Press, London, 1963, pp. 61–72.
20. Barer, R., *Nature*, **169**, 366 (1952).
21. Barringer, A. R., *Bull. Inst. Mining Met.*, **63**, (1) 22–41 (1953).
22. Bartlett, J. H., and L. Stephenson, *J. Electrochem. Soc.*, **99**, 504 (1952).
23. Barton, J. L., and J. O'M. Bockris, *Proc. Roy. Soc.* (*Lond.*), **A268**, 485 (1962).
24. Baumann, I., *High Temperature Technology*, I. F. Campbell, Ed., Wiley, New York, 1956.
25. Bausch and Lomb, Rochester, N. Y., Manufacturer of the "Quantative Metallurgical System (QMS)."
26. Beacom, S. E., and B. J. Riley, *J. Electrochem. Soc.*, **106**, 309 (1959); **107**, 785 (1960); **108**, 758 (1961).
27. Becker, J. J., *J. Electrochem. Soc.*, **111**, 480 (1964).
28. Beckert, M., and B. Klemm, *Handbuch der metallographischen Aetzverfahren*, VEB Deutscher Verlag für Grundstoffindustie, Leipzig, 1966.
29. Behrndt, K. H., *AVS 9th Vacuum Symposium Transactions*, G. E. Bancroft, Ed., Macmillan, New York, 1962, p. 111.
30. Benedicks, C., and C. F. Metts, *Jernkontorets Ann.*, **118**, 4 (1934).
31. Bennett, A. H., H. Osterberg, H. Jupnik, and O. W. Richards, *Phase Microscopy*, Chapman and Hall, New York, 1951.
32. Biery, J. C., *J. Electrochem. Soc.*, **114**, 225 (1967).
33. Block, R. J., *J. Electrochem. Soc.*, **117**, 788 (1970).
34. Bockris, J. O'M., J. W. Diggle, and A. Damjanovic, First Quarterly Report to NASA, Contract NGR 39-010-002, 1-1-69 to 3-31-69.

35. Bockstiegel, G., *Z. Metallkd.*, **57**, 647 (1966); *Pract. Metallogr.*, **6**, 597 (1969).
36. Bode, H., and J. Euler, *Electrochim. Acta*, **11**, 1211, 1221, 1231 (1966).
37. Brandon, D. G., *Modern Techniques in Metallography*, Butterworth, London, 1966.
38. Breiter, M. W., *J. Electrochem. Soc.*, **117**, 738 (1970).
39. Brenner, A., P. Burkhead, and C. Jennings, *J. Res. Nat. Bur. Stand.*, **40**, 31 (1948).
40. Briggs, G. W. D., and W. F. K. Wynne-Jones, *Electrochim. Acta*, **7**, 241 (1962).
41. Brown, R., *Am. Ceram. Soc. Bull.*, **45**, 206 (1966).
42. Brückman, A., S. Mrowec, and T. Weber, *J. Electrochem. Soc.*, **117**, 955 (1970).
43. Bryan, F. R., and G. A. Nahstoll, *J. Opt. Soc. Am.*, **37**, 311 (1947); *Mater. Methods*, **25**, 90 (1947).
44. Bryan, F. R., and C. H. Neveu, *Met. Progr.*, **64**, 82 (1953).
45. Buckle, H., and H. Hanemann, *Z. Metallkd.*, **32**, 120 (1940).
46. Burley, N. A., and J. J. Dale, *Metallurgia*, **65**, 203 (1962).
47. Burton, W. K., N. Cabrera, and F. C. Frank, *Proc. Roy. Soc.* (*Lond.*), **A243**, 299 (1951).
48. Cameron, E. N., *Ore Microscopy*, Wiley, New York, 1961.
49. *Ibid.*, pp. 34–50.
50. *Ibid.*, pp. 86–140.
51. Cammarota, G. P., *Electrochim. Met.*, **2**, 29 (1967).
52. Capellaro, D. F., N. S. Kapany, and C. Long, *Nature*, **191**, 927 (1961).
53. Carter, R. E., and F. D. Richardson, *J. Met.*, **11**, 336 (1959).
54. Carter, R. E., *J. Am. Ceram. Soc.*, **44**, 116 (1961).
55. Cech, R. E., *Rev. Sci. Instr.*, **21**, 747 (1950).
56. Chalkley, L., *J. Opt. Soc. Am.*, **42**, 387 (1952).
57. Chalmers, B., and A. G. Quarrell, *The Physical Examination of Metals*, Edward Arnold, London, 1960, pp. 1–80.
58. Chalmers, B., R. King, and R. Shuttlework, *Proc. Roy. Soc.* (*Lond.*), **A193**, 465 (1948).
59. Chalmers, B., and A. G. Quarrell, *The Physical Examination of Metals*, Edward Arnold, London, 1960.
60. Chamot, E. M., and C. W. Mason, *Handbook of Chemical Microscopy*, Vol. 1, Wiley, New York, 1958, p. 91.
61. Chamot, E. M., and C. W. Mason, *Handbook of Chemical Microscopy*, Vol. I, Wiley, New York, 1958.
62. Cheh, A. Y., *J. Electrochem. Soc.*, **117**, 609 (1970).
63. Chu, T. L., and J. R. Galaver, *J. Electrochem. Soc.*, **110**, 388 (1963).
64. Clark, G. L., *The Encyclopedia of Microscopy*, Reinhold, New York, 1961.
65. *Ibid.*, p. 468.
66. Cochrane, N. J., and R. J. Block, *J. Electrochem. Soc.*, **117**, 225 (1970).
67. Condit, R. H., and J. B. Holt, *J. Electrochem. Soc.*, **111**, 1192 (1964).
68. Condit, R. H., J. B. Holt, and L. Himmel, *J. Electrochem. Soc.*, **114**, 1101 (1967).
69. Conn, G. K. T., and F. J. Bradshaw, *Polarized Light in Metallography*, Academic Press, New York, 1952.

70. Couling, S. R. L., and R. Smoluchowski, *J. Appl. Phys.*, **25**, 1538 (1954).
71. Cox, B., and C. Ray, *Electrochem. Tech.*, **4**, 121 (1966).
72. Dalvi, A. D., and W. W. Smeltzer, *J. Electrochem. Soc.*, **117**, 1431 (1970).
73. Damiano, V. V., and H. Herman, *J. Franklin Inst.*, **267**, 303 (1959).
74. Damjanovic, A., M. Paunovic, and J. O'M. Bockris, *Plating*, **50**, 735 (1963).
75. Damjanovic, A., M. Paunovic, and J. O'M. Bockris, *Electrochim. Acta*, **10**, 111 (1965).
76. Davies, H. G., and M. H. F. Wilkins, *Nature*, **169**, 541 (1952).
77. Delessé, F., *Compt. Rend.*, **25**, 544 (1847); *Ann. Mines*, **13**, 379 (1848).
78. Denhoff, R. T., and F. N. Rhines, *Quantitative Microscopy*, McGraw-Hill, New York, 1969; see also: *Microscope* **19**, 1–122 (1971); *Quantitative Stereology*, E. E. Underwood, Addison-Wesley, Reading, Mass. 1970.
79. Dermatis, S. N., J. W. Faust, and H. F. John, *J. Electrochem. Soc.*, **112**, 792 (1965).
80. Dewhirst, D. W., and M. J. Olney, *J. Iron Steel Inst.*, **169**, 221 (1951).
81. Dhananjayan, N., *J. Electrochem. Soc.*, **117**, 1006 (1970).
82. Diener, G., *Phillips Res. Rep.*, **10**, 194 (1955).
83. Diggle, J. W., A. R. Despic, and J. O'M. Bockris, *J. Electrochem. Soc.*, **116**, 1503 (1969).
84. Dörfler, G., *Pract. Metallogr.*, **6**, 144 (1969).
85. Doty, W. R., and B. J. Riley, *J. Electrochem. Soc.*, **114**, 50 (1967).
86. Dreeben, A., *J. Electrochem. Soc.*, **112**, 493 (1965).
87. Dudley, R. H., and T. H. Briggs, *Rev. Sci. Instr.*, **37**, 1041 (1966).
88. Economou, N. A., H. Fischer, and D. Trivich, *Electrochim. Acta*, **2**, 207 (1960).
89. El-Badry, H. M., *Micromanipulators and Micromanipulation*, Academic Press, New York, 1963.
90. Elias, H., and E. Weibel, *Quantitative Methoden in der Morphologie*, Springer-Verlag, Heidelberg, 1967.
91. Epelboin, I., and M. Froment, *Electrochim. Acta*, **6**, 59 (1962).
92. Epelboin, I., M. Froment, and G. Nomarski, *Rev. Met.*, **55**, 260 (1958).
93. Erdey-Grüz, T., and M. Volmer, *Z. Phys. Chem.*, **A157**, 165 (1931).
94. Erdey-Grüz, T., and R. F. Kardos, *Z. Phys. Chem.*, **A178**, 262 (1937).
95. Ettisch, G., and T. Petérfi, *Arch. Ges. Physiol. Pflüger's*, **208**, 454 (1925).
96. Euler, K. J., *Chem. Ing. Tech.*, **38**, 631 (1966); *Electrochim. Acta*, **13**, 2245 (1968).
97. Exner, H. E., and A. Fischmeister, *Pract. Metallogr.*, **3**, 18, 334 (1966); **6**, 639 (1969).
98. Faust, J. W., and H. F. John, *J. Electrochem. Soc.*, **108**, 109 (1961).
99. Faust, J. W., A. Sagar, and H. F. John, *J. Electrochem. Soc.*, **109**, 824 (1962).
100. Fedot'ev, N. P., and S. Ya Grilikhes, *Electropolishing, Anodizing and Electrolytic Pickling of Metals*, Robert Draper, Teddington, 1959.
101. Findeisen, B., and L. Illgen, *Pract. Metallogr.*, **5**, 186 (1968).
102. Fischer, A. G., *J. Electrochem. Soc.*, **109**, 1043 (1962).
103. Fischer, C. F., and M. Cole, *Microscope*, **16**, 81 (1968).

104. Fischmeister, H., *Pract. Metallogr.*, **2**, 251 (1965).
105. Freund, H., *Handbuch der Mikroskopie in der Technik, Band III, Mikroskopie der metallischen Werkstoffe, Teil 1: Die metallographische Proben-Präparation für die Mikroskopische Untersuchung*, Umschau Verlag, Frankfurt am Main, 1968, Ch. 2.
106. Froment, M., *Corros. Anticorros.* **7**, 46 (1959).
107. *Ibid.*, pp. 98, 134.
108. Gardam, G. E., *J. Electrodepositor's Tech. Soc.*, **22**, 155 (1947).
109. Gatos, H. C., and A. D. Kurtz, *Trans. Am. Inst. Mining Met. Engrs.*, **200**, 616 (1954).
110. Gatos, H. C., and A. Azzan, *Trans. Am. Inst. Mining Met. Eng.*, **194**, 407 (1952).
111. Gebhardt, E., and H. D. Seghezzi, *Z. Metallkd.*, **50**, 248 (1959).
112. Gedney, I. J., *J. Electrochem. Soc.*, **111**, 1093 (1964).
113. Gillson, J. L., and F. J. Darnell, *Phys. Rev.*, **125**, 149 (1962).
114. Glick, D., and R. C. Rosan, *Microchem. J.*, **10**, 393 (1966).
115. Graber, D. G., and W. C. McCrone, *J. Chem. Ed.*, **27**, 649 (1950).
116. Grace and Company, Marco Chemical Division, West Linden, New Jersey.
117. Graf, L., and W. Weser, *Electrochim. Acta*, **2**, 145 (1960).
118. Graf, L., and W. Morgenstern, *Z. Naturforsch.*, **10A**, 345 (1955).
119. Haglund, B. O., *Pract. Metallogr.*, **7**, 173 (1970).
120. Hallimond, A. F., *Mineral Mag.*, **95**, 271–276 (1956).
121. Hallimond, A. F., *The Polarizing Microscope*, Vickers Instruments, Malden, Mass., 1970, pp. 206–228.
122. Hardesty, D. W., *J. Electrochem. Soc.*, **117**, 168 (1970).
123. Hardesty, D. W., *J. Electrochem. Soc.*, **111**, 912 (1964).
124. Hartshorne, N. H., and A. Stuart, *Crystals and the Polarizing Microscope*, Edward Arnold, London, 1960.
125. Hartshorne, N. H., and A. Stuart, *Practical Optical Crystallography*, American Elsevier Pub. Co., New York, 1964.
126. Haselmann, H., *Z. Wiss. Mikrosk.*, **67**, 244 (1966).
127. Hayness, G. W., and R. Smoluchowski, *Acta Met.*, **3**, 130 (1955).
128. Herzog, A. H., *Semicond. Prod.*, **5**, 25 (1962).
129. Herzog, F., *Mikroskopische Untersuchung der Seide mit besonderer Berücksichtigung der Kunstseidenindustrie*, J. Springer, Berlin, 1924, p. 28.
130. Hoekstra, J., Dissertation (Amsterdam, 1932) referred to by G. Wranglen, *Trans. Roy. Inst. Technol. Stockholm*, **94**, 3 (1955).
131. Holt, J. B., and L. Himmel, *J. Electrochem. Soc.*, **116**, 1569 (1969).
132. Holt, J. B., and L. Himmel, *J. Electrochem. Soc.*, **117**, 957 (1970).
133. Illgen, L., H. Ringfeil, H. Hemschick, *Pract. Metallogr.*, **6**, 363, 421, 483, 539 (1969).
134. Image Analyzing Computers, Ltd., Cambridge, England. A division of Metals Research Instrument Corp., manufacturer of the "Quantimet 720," a Quantitative Television Microscope.
135. Innis, R. E., Symposium on Recent Developments in Research Methods and Instrumentation, Bethesda, Maryland, Oct. 4–7, 1965.
136. Inoue, S., *The Encyclopedia of Microscopy*, G. L. Clark, Ed., Reinhold, New York, 1961, pp. 481–485.

137. Jäniche, W., and J. Brauner, "Metallkundliche Untersuchungsverfahren," in *Hütte, Taschenbuch der Werkstoffkunde*, Verlag W. Ernst u. Sohn, Berlin, 1967, pp. 896–920.
138. Jarry, R. L., J. Fischer, and W. H. Gunther, *J. Electrochem. Soc.*, **110**, 346 (1963).
139. Jelley, E. E., *J. Roy. Microsc. Soc.*, **99**, 234 (1934); *Phot. J.*, **74**, 514 (1934); *J. Roy. Microsc. Soc.*, **56**, 101 (1936); *Anal. Chem.*, **13**, 196 (1941).
140. Jelley, E. E., *Physical Methods of Organic Chemistry*, Vol. 1, A. Weissberger, Ed., Interscience, New York, 1949, p. 847.
141. Jenkins, L. H., *J. Electrochem. Soc.*, **117**, 630 (1970).
142. Jenkins, L. H., and R. B. Durham, *J. Electrochem. Soc.*, **117**, 768 (1970).
143. Jenkins, L. H., *J. Electrochem. Soc.*, **113**, 75 (1966).
144. Jenkins, L. H., and U. Bertocci, *J. Electrochem. Soc.*, **112**, 517 (1965).
145. Johannsen, A., *Manual of Petrographic Methods*, McGraw-Hill, New York, 1918, p. 290.
146. Johnson, W., *The Encyclopedia of Microscopy*, G. L. Clark, Ed., Reinhold, New York, 1961, p. 574.
147. Jones, M. H., and J. Saiddington, *Tech. Proc. Am. Electroplat. Soc.*, **48**, 32 (1961).
148. Jonnard, R., *The Encyclopedia of Microscopy*, G. L. Clark, Ed., Reinhold, New York, 1961, pp. 485–524.
149. Joyce, B. A., *Solid State Electron.*, **5**, 102 (1962).
150. Just, J., and W. Altgeld, *Pract. Metallogr.*, **7**, 59 (1970).
151. Kaishew, R., B. Mutaftchiew, and D. Nenow, *Z. Phys. Chem.*, **205**, 341 (1955–1956).
152. Kapany, N. S., Proceedings of the New York Academy of Sciences Conference on the Laser, New York, May 4–5, 1964; N. S. Kapany, N. A. Peppers, H. C. Zweng, and M. Flocks, *Nature*, **199**, 146 (1963); N. S. Kapany and N. Silbertrust, *Nature*, **204**, 138 (1964).
153. Karyakin, A. V., M. V. Akhmanova, and V. A. Kaigorodov, *J. Anal. Chem.* (*USSR*), **20**, 133 (1965).
154. Kauczor, E., *Kleine Metallographie, Leitfaden für das Anfertigen von Makroschliffen*, Die Schweisstechnische Praxis, Band 5, Deutscher Verlag für Schweisstechnik, GmbH, Düsseldorf, 1969.
155. Kehl, L., H. Steinmetz, and W. J. McGonnagle, *Metall.*, **55**, 151 (1957).
156. Keralla, J. A., and J. J. Lander, *Electrochem. Tech.*, **6**, 202 (1968).
157. Kerr, P. F., *Optical Mineralogy*, McGraw-Hill, New York, 1959.
158. *Ibid.*, pp. 1–10.
159. Kilner, T., and A. Plumtree, *J. Electrochem. Soc.*, **116**, 810 (1969).
160. Knoop, H., *Z. Metallkd.*, **60**, 536 (1969).
161. Knopp, A. N., *J. Electrochem. Soc.*, **110**, 82 (1963).
162. Knosp, H., *Pract. Metallogr.*, **7**, 494 (1970).
163. Koch, W., H. Malissa, and D. Ditges. *Arch. Eisenhuettenw.*, **28**, 785 (1957).
164. Kofstad, P. K., and A. Z. Hed, *J. Electrochem. Soc.*, **116**, 1542 (1969).
165. Kohlhaas, E., and A. Fischer, *Pract. Metallogr.*, **6**, 339 (1969).
166. Köhler, A., *Z. Wiss. Mikrosk.*, **10**, 443 (1893).
167. Kohlschütter, V., and F. Übersax, *Z. Elektrochem.*, **30**, 72 (1924).
168. Kohlschütter, V., and A. Torricelli, *Z. Elektrochem.*, **38**, 218 (1932).
169. Kohlschütter, V., and A. Good, *Z. Elektrochem.*, **33**, 277 (1927).

170. Kremheller, A., *J. Electrochem. Soc.*, **107**, 8 (1960).
171. Krug, W., J. Rienitz, and G. Schulz, *Contributions to Interference Microscopy*, Hilger and Watts, London, 1964; see also S. Tolansky, *An Introduction to Interferometry*, Wiley, New York, 1962.
172. Kruger, J., *J. Electrochem. Soc.*, **110**, 654 (1963).
173. Krumbein, S. J., and R. R. Russell, *Electrochem. Tech.*, **4**, 471 (1966).
174. Kukoz, F. I., Z. I. Mityagina, M. F. Skalozubov, and V. D. Chebakov, *J. Appl. Chem.* (*USSR*), **42**, 762 (1969).
175. Kutzelnigg, A., *Testing Metallic Coatings*, Robert Draper, London, 1963, Ch. 3.
176. Kyser, D. F., and M. F. Millea, *J. Electrochem. Soc.*, **111**, 1102 (1964).
177. Ladd, W. A., W. M. Hess, and M. W. Ladd, *Science*, **123**, 3192 (1956).
178. Lamb, V. A., C. E. Johnson, and D. R. Valentine, *J. Electrochem. Soc.*, **117**, 291C (1970); see also V. A. Lamb and D. R. Valentine, *Plating*, **52**, 1289 (1965); **53**, 86 (1966).
179. Landolt, D., R. H. Muller, and C. W. Tobias, *J. Electrochem. Soc.*, **116**, 1384 (1969).
180. Landolt, D., R. Acosta, R. H. Muller, and C. W. Tobias, *J. Electrochem. Soc.*, **117**, 839 (1970).
181. Landolt, D., R. H. Muller, and C. W. Tobias, *J. Electrochem. Soc.*, **118**, 36 (1971).
182. Lange, W. H., and V. V. Smith, *Electrochem. Tech.*, **6**, 405 (1968).
183. Lehmann, W., *J. Electrochem. Soc.*, **104**, 45 (1957).
184. Lehmann, W., *J. Electrochem. Soc.*, **107**, 657 (1960).
185. Lehmann, W., *J. Electrochem. Soc.*, **107**, 20 (1960).
186. Leibssle, H., H. Reber, W. Herrmann, and E. Zehender, *Bosch. Tech. Ber.*, **2** (4) 159 (1968).
187. Leitz, E., GmbH, Wetzler, Germany, manufacturer of a quantitative television microscope, the "Classimat," first introduced as the "Integramat."
188. Liaw, H. M., and J. W. Faust, "Growth Kinetics on Morphology of Electrodeposited Lead Dendrites," Fall Meeting, the Electrochemical Society, Atlantic City, October 4–8, 1970.
189. Lichtenegger, P., A. Kulmburg, and R. Blöch, *Pract. Metallogr.*, **6**, 535 (1969).
190. Loebner, E. E., and H. Freund, *Phys. Rev.*, **98**, 1545 (1955).
191. Long, C., A. Brushenko, and D. Pontarelli, *Applied Opt.*, **3**, 1031 (1964).
192. Losinskij, M. J., *High-temperature Metallography*, Pergamon, London, 1961.
193. Lucas, F. F., *Am. Inst. Mining Met. Eng.*, Pamphlet no. 1654-E, February 1927; *Bell Telephone Laboratories*, Reprint B-241, March, 1927.
194. Luner, C., and M. V. Murray, *J. Electrochem. Soc.*, **110**, 176 (1963).
195. Mahin, E. G., and R. E. Brewer, *Ind. Eng. Chem.*, **12**, 1095 (1920).
196. Mansfeld, F., and S. Gilman, *J. Electrochem. Soc.*, **117**, 588 (1970).
197. Mansfeld, F., and S. Gilman, *J. Electrochem. Soc.*, **117**, 1154 (1970).
198. Mansfeld, F., and S. Gilman, *J. Electrochem. Soc.*, **117**, 1521 (1970).
199. Mao, G. W., T. L. Wilson, and J. G. Larson, *J. Electrochem. Soc.*, **117**, 1323 (1970).
200. Maréchal, J. R., and M. Doucet, *Rev. Met.*, **48**, 561 (1951).

201. Martin, G., C. E. Blythe, and H. Tongue, *Am. Ceram. Soc., Bull.*, **23**, 61 (1924).
202. McCrone, W. C., *Fusion Methods in Chemical Analysis*, Interscience, New York, 1957.
203. McDonald, B., and A. Goetzberger, *J. Electrochem. Soc.*, **109**, 141 (1962).
204. McLean, D., *J. Inst. Met.*, **81**, 133 (1952).
205. Mears, R. B., and R. H. Brown, *Ind. Eng. Chem.*, **33**, 1001 (1941).
206. Meltzer, G., *J. Electrochem. Soc.*, **109**, 947 (1962).
207. Menzies, I. A., and J. Lubkiewicz, *J. Electrochem. Soc.*, **117**, 1539 (1970).
208. Mercer, R. A., and R. P. Miller, *J. Sci. Instr.*, **40**, 352 (1963).
209. Meyer, W. R., *Mon. Rev. Am. Electroplat. Soc.*, **23**, 116 (1935).
210. Monsour, T. M., *Mater. Res. Stand.*, **3**, 29 (1963).
211. Muhlhausser, M., *Electrochem. Tech.*, **6**, 183 (1968).
212. Murray, W. M., B. Gettys, and S. E. Q. Ashley, *J. Opt. Soc. Am.*, **31**, 433 (1941).
213. Navias, L., "Preparation and Properties of Spinel made by Vapor Transport and Diffusion in the System $MgO\text{-}Al_2O_3$," Fall Meeting, American Ceramic Society, Detroit, Michigan, September 26, 1960.
214. Nelson, J., and R. M. Slepian, *Pract. Metallogr.*, **7**, 510 (1970).
215. Nelson, J. A., *Pract. Metall.*, **7**, 189 (1970).
216. Novak, M., A. K. Reddy, and H. Wroblowa, *J. Electrochem. Soc.*, **117**, 733 (1970).
217. O'Brien, R. N., and C. Rosenfield, *J. Phys. Chem.*, **67**, 643 (1963).
218. O'Brien, R. N., *Rev. Sci. Instr.*, **35**, 803 (1964).
219. O'Brien, R. N., *J. Electrochem. Soc.*, **111**, 1300 (1964).
220. O'Brien, R. N., D. Quon, and C. R. Darsi, *J. Electrochem. Soc.*, **117**, 888 (1970).
221. Ogburn, F., C. Bechtoldt, J. B. Morris, and A. de Koranyi, *J. Electrochem. Soc.*, **112**, 574 (1965).
222. Ollerhead, R. W., E. Almquist, and J. A. Kuehner, *J. Appl. Phys.*, **37**, 2440 (1966).
223. Oxley, J. E., and C. N. Fleischmann, Report No. 3, NASA Contract No. NAS-5-9591, March, 1966.
224. Padden, T. R., and F. M. Cain, *Met. Prog.*, **66**, 108 (1954).
225. Partridge, E. P., V. Hicks, and G. W. Smith, *J. Am. Chem. Soc.*, **63**, 454 (1941).
226. Pattee, H. H., *X-ray Microscopy and X-ray Microanalysis*, Cosslett, Engström, and Pattee, Eds., Elsevier, Amsterdam, 1960, p. 61.
227. Pemsler, J. P., *J. Electrochem. Soc.*, **111**, 381 (1964).
228. Pick, H. J., G. G. Storey, and T. B. Vaughan, *Electrochim. Acta*, **2**, 165 (1960).
229. Pierson, J. R., *Electrochem. Tech.*, **5**, 323 (1967).
230. Pierson, J. R., "Power Sources 2, 1968," D. H. Collins, Ed., Pergamon Press, Oxford, 1970.
231. Piontelli, R., A. Berbenni, B. Mazza, and P. Pedeferri, *Electrochim. Met.*, **1**, 279 (1966); **2**, 257, 377 (1967).
232. Pliskin, W. A., *IBM J. Res. Dev.*, **8**, 43 (1964); *J. Appl. Phys.*, **36**, 2011 (1965); *J. Phys.*, **25**, 17 (1964).

233. Polanyi, M. L., Symposium on Recent Developments in Research Methods and Instrumentation, Bethesda, Maryland, October 4–7, 1965.
234. Policard, A., *Compt. Rend.*, **194**, 201, 491, 1015 (1932); *Harvey Lect. Ser.*, **27**, 204 (1931/32).
235. Powers, R. W., *Electrochem. Tech.*, **5**, 429 (1967).
236. Powers, R. W., and M. W. Breiter, *J. Electrochem. Soc.*, **116**, 719 (1969).
237. Powers, R. W., *J. Electrochem. Soc.*, **116**, 1652 (1969).
238. Powers, R. W., and E. C. Jerabek, *J. Electrochem. Soc.*, **117**, 1099 (1970).
239. Prener, J. S., *J. Electrochem. Soc.*, **114**, 77 (1967).
240. Primak, W., R. Kampwirth, and Y. Dayal, *J. Electrochem. Soc.*, **114**, 88 (1967).
241. Pugh, E. N., and L. E. Samuels, *J. Electrochem. Soc.*, **109**, 409 (1962).
242. Radley, J. A., and J. Grant, *Fluorescence Analysis in Ultraviolet Light*, Chapman and Hall, London, 1954.
243. Ramdohr, P., *Die Erzmineralien und ihre Verwachsungen*, Akademie-Verlag, Berlin, 1960, pp. 270–1023.
244. Ramdohr, P., and G. Rehwald, *Handbuch der Mikroskopie in der Technik, Band I, Teil 2*, Umschau Verlag, Frankfurt am Main, 1954, pp. 189–257.
245. Rasberry, S. D., B. F. Scribner, and M. Margoshes, *Appl. Opt.*, **6**, 81, 87 (1967).
246. Reinacher, G., "Heizbare Objekttische für die Metallmikroskopie," *Handbuch der Mikroskopie in der Technik*, Band I, Teil 2, Umschau Verlag, Frankfurt am Main (1960), pp. 387–441.
247. Rexer, J., and M. Vogel, *Pract. Metallogr.*, **5**, 361 (1968).
248. Rexer, J., L. Gessner, and B. Jäntschke, *Pract. Metallogr.*, **7**, 473 (1970).
249. Roberts, H. S., and T. Stadnichenko, *J. Opt. Soc. Am.*, **10**, 605 (1925).
250. Roethlein, R. J., *J. Electrochem. Soc.*, **117**, 714 (1970).
251. Rogers, B. A., *Met. Alloys*, **2**, 9 (1931).
252. Rosenberg, R., *J. Electrochem. Soc.*, **112**, 459 (1965).
253. Rosenberg, R., M. Kozlowski, W. J. McAleer, and P. I. Pollak, *J. Electrochem. Soc.*, **112**, 459 (1965).
254. Runge, E. F., and F. R. Bryan, *Appl. Spectrosc.*, **13**, 74 (1959).
255. Runge, E. F., S. Bonfiglio, and F. R. Bryan, *Spectrochim. Acta*, **22**, 1678 (1966).
256. Runyan, W. R., E. G. Alexander, and S. E. Craig, *J. Electrochem. Soc.*, **114**, 1154 (1967).
257. Ryan, J. R., and J. L. Cunningham, *Met. Progr.*, **90**, 100 (1966).
258. Saiddington, J. C., and G. R. Hoey, *J. Electrochem. Soc.*, **117**, 1071 (1970).
259. Samuels, L. E., *Metallographic Polishing by Mechanical Methods*, Pitman, London, 1967.
260. Schneiderhöhn, H., *Erzmikroskopisches Praktikum*, E. Sweizerbart'sche Verlags-buchhandlung, Stuttgart, 1952.
261. Schnizlein, J. G., J. D. Woods, J. D. Bingle, and R. C. Vogel, *J. Electrochem. Soc.*, **107**, 783 (1960).
262. Schüller, H., and P. Schwaab, *Pract. Metallogr.*, **7**, 679 (1970).
263. Schwartz, B., and N. Schwartz, *Measurement Techniques for Thin Films*, The Electrochemical Society, New York, 1967.
264. Seiter, H., and H. Fischer, *Z. Elektrochem.*, **63**, 249 (1959).
265. Seiter, H., H. Fischer, and L. Albert, *Electrochim. Acta*, **2**, 97 (1960).

266. Seitz, F., *Rev. Mod. Phys.*, **18**, 384 (1946).
267. Seltzer, M. S., N. Albon, B. Paris, and P. C. Himes, *Rev. Sci. Instr.*, **36**, 1423 (1965).
268. Seltzer, M. S., N. Albon, B. Paris, and R. C. Himes, *J. Electrochem. Soc.*, **114**, 102 (1967).
269. Sen, B., *Proc. Soc. Exp. Biol. Med.*, **27**, 311 (1930).
270. Sharpe, R. S., *X-ray Microscopy and Microradiography*, Cosslett et al., Eds., Academic Press, New York, 1957, pp. 591–602.
271. Shams El Din, A. M., and G. Wranglen, *Electrochim. Acta*, **7**, 79 (1962).
272. Shanefield, D., and P. E. Lighty, *J. Electrochem. Soc.*, **110**, 973 (1963).
273. Shaw, E. R., and K. M. Busen, *J. Electrochem. Soc.*, **107**, 872 (1960).
274. Shaw, D. W., *J. Electrochem. Soc.*, **113**, 904 (1966).
275. Shillaber, C. P., *Photomicrography in Theory and Practice*, Wiley, New York, 1944.
276. Short, M. A., E. G. Steward, and T. B. Tomlinson, *Nature*, **177**, 240 (1956).
277. Silverman, S. J., *J. Electrochem. Soc.*, **109**, 166 (1962).
278. Simon, A. C., and E. L. Jones, *J. Electrochem. Soc.*, **102**, 279 (1955).
279. Simon, A. C., unpublished work.
280. Simon, A. C., and D. A. Gildner, *Rev. Sci. Instr.*, **29**, 1125 (1958).
281. Simon, A. C., *Electrochem. Tech.*, **1**, 82 (1963).
282. Simon, A. C., *Batteries 2*, D. Collins, Ed., Pergamon, London, 1965, p. 63.
283. Simon, A. C., *Electrochem. Tech.*, **3**, 305 (1965).
284. Simon, A. C., *J. Electrochem. Soc.*, **114**, 1 (1967).
285. Smith, F. H., and S. Iverson, *Quart. J. Microsc. Sci.*, **98**, 151 (1957).
286. Stachurski, Z. O. J., G. A. Dalin, and F. Solomon, "Investigation and Improvement of Zinc Electrodes for Electrochemical Cells," Nat. Aero. Space Adm., Contract No. NAS 5-3873, Quarterly Report No. 1, July to September, 1964.
287. Steinberg, M. A., *Nature*, **170**, 1119 (1952).
288. Stephenson, L., and J. H. Bartlett, *J. Electrochem. Soc.*, **101**, 571 (1954).
289. Stobbs, J. J., *J. Electrochem. Soc.*, **112**, 916 (1965).
290. Stringer, J., *J. Electrochem. Soc.*, **114**, 428 (1967).
291. Sullivan, M. V., and W. R. Bracht, *J. Electrochem. Soc.*, **114**, 295 (1967).
292. Tajima, S., and M. Ogata, *Electrochim. Acta*, **13**, 1845 (1968).
293. Tajima, S., and M. Ogata, *Electrochim. Acta*, **15**, 61 (1970).
294. Taoki, T., T. Tameto, K. Ogasa, E. Furubayashi, S. Takeuchi, J. Imamura, and H. Hayakawa, *Pract. Metallogr.*, **5**, 22 (1968).
295. Tausch, F. W., and A. G. Lapierre, *J. Electrochem. Soc.*, **112**, 704 (1965).
296. Taylor, C. V., *Soc. Exp. Bio. Med. Proc.*, **23**, 147 (1925).
297. Tegart, W. G. McG., *The Electrolytic and Chemical Polishing of Metals*, Pergamon, New York, 1959.
298. *Thomas Register of American Manufacturers. Vol. IV, Products and Services*, Thomas Publishing Co., New York, 1970.
299. Teitjen, J. J., and J. A. Amick, *J. Electrochem. Soc.*, **113**, 724 (1960).
300. Tint, G. S., and V. V. Damiano, *Rev. Sci. Instr.*, **32**, 325 (1961).
301. Tracey, V. A., and N. J. Williams, *Electrochem. Tech.*, **3**, 17 (1965).
302. Tucker, T. N., *Electrochem. Tech.*, **4**, 546 (1966).

303. Tudor, S., A. Weisstuch, and S. H. Davang, *Electrochem. Tech.*, **3**, 90 (1965); **4**, 406 (1966).
304. Tuomi, D., *J. Electrochem. Soc.*, **112**, 1 (1965).
305. Tvarusko, A., and L. S. Watkins, *J. Electrochem. Soc.*, **118**, 248 (1971).
306. Unitron Instrument Co., Newton Highlands, Mass., manufacturers of a microgoniometer.
307. Vanark, R. J., *Metallography*, **1**, 439 (1969).
308. Votava, E., A. Berghezan, and R. H. Gillette, *X-ray Microscopy and Microradiography*, Cosslett et al., Eds., Academic Press, New York, 1957, pp. 606–616.
309. Wagner, C., Addendum to article by R. E. Carter and F. K. Richardson, *J. Met.*, **11**, 336 (1955).
310. Wales, C. P., and A. C. Simon, *J. Electrochem. Soc.*, **115**, 1228 (1968).
311. Wales, C. P., *J. Electrochem. Soc.*, **116**, 729 (1969); **118**, 7 (1971).
312. Walker, D. E. Y., *Pract. Metallogr.*, **7**, 445 (1970).
313. Ward, R. G., *The Physical Examination of Metals*, Chalmers and Quarrell, Eds., Edward Arnold, London, 1960, pp. 846, 850.
314. *Ibid.*, pp. 834, 838.
315. *Ibid.*, pp. 825–850.
316. Wayworth, J. F., and F. Bitter, *Phys. Rev.*, **95**, 941 (1954).
317. Welch, J. H., *J. Sci. Instr.*, **31**, 458 (1954).
318. Wilska, A., *The Encyclopedia of Microscopy*, G. L. Clark, Ed., Reinhold, New York, 1961, pp. 476–480.
319. Winterbottom, A. B., *The Physical Examination of Metals*, Chalmers and Quarrell, Eds., Edward Arnold, London, pp. 1–77.
320. Wlodek, S. T., "Oxidation of Cobalt-base Alloys," Fall Meeting, The Electrochemical Society, Atlantic City, October 4–8, 1970.
321. Woodbury, J. C., *AB Met. Dig.*, **11**, (3) 2 (1965); **12**, (1) 3 (1966); **13**, (1) 3 (1967), Buehler, Ltd., Evanston, Illinois.
322. Wranglen, G., *Trans. Roy. Inst. Tech. (Stockholm)*, **94**, 3 (1955).
323. Wranglen, G., *Electrochim. Acta*, **2**, 130 (1960).
324. Wright, R. E., *The Encyclopedia of Microscopy*, G. L. Clark, Ed., Reinhold, New York, 1961, pp. 473–476.
325. Yang, L., C. Chien, and R. G. Hudson, *J. Electrochem. Soc.*, **106**, 632 (1959).
326. Yeh, T. H., *J. Electrochem. Soc.*, **110**, 1018 (1963).
327. Yukawa, S., and M. J. Sinott, *Trans. Am. Inst. Min. Met. Eng.*, **203**, 996 (1955).
328. Zahn, P., G. Diemer, and H. A. Klasens, *Philips Res. Rep.*, **9**, 81 (1954).
329. Zeiss-Siemens, Oberkochen, West Germany, manufacturers of the "Microscan," a quantitative television microscope.
330. Zetner, V., A. Brenner, and C. W. Jennings, *Plating*, **39**, 865 (1952).

SUBJECT INDEX

ADVANCES IN ELECTROCHEMISTRY AND ELECTROCHEMICAL ENGINEERING

Cumulative Index, Volumes 1–9